Martin Oliver Steinhauser
Computer Simulation in Physics and Engineering

Martin Oliver Steinhauser

Computer Simulation in Physics and Engineering

De Gruyter

Physics and Astronomy Classification 2010: 82.20.Wt, 89.20.Ff, 83.10.Rs.

ISBN: 978-3-11-025590-4
e-ISBN: 978-3-11-025606-2

Library of Congress Cataloging-in-Publication Data

A CIP catalog record for this book has been applied for at the Library of Congress.

Bibliographic information published by the Deutsche Nationalbibliothek

The Deutsche Nationalbibliothek lists this publication in the Deutsche Nationalbibliografie; detailed bibliographic data are available in the internet at http://dnb.dnb.de.

© 2013 Walter de Gruyter GmbH, Berlin/Boston

Typesetting: Da-TeX Gerd Blumenstein, Leipzig, www.da-tex.de
Printing and binding: Hubert & Co. GmbH & Co. KG, Göttingen
∞ Printed on acid-free paper
Printed in Germany

www.degruyter.com

Preface

This monograph is intended as a course textbook for students majoring in physics, engineering, or computer science in the final year of their undergraduate studies, as well as for graduate students – either those engaged in their first research projects involving scientific computation, or those who want to really understand the "black box" of simulation programs for use on their desktop computers. Since this book discusses both the engineering issues in program design and the mathematical and physical foundations of scientific computing and because it is supplemented with many end-of-chapter exercises, it should be well-suited for self-study. The author hopes to convince the readers that a behind-the-scenes look at computer simulations is well worth the effort and will prove a rewarding and enjoyable experience (compared to e.g. just learning to click the right buttons of a GUI of a commercial simulation package). It is by all means an absolute prerequisite for being able to write their own simulation programs and to adapt existing programs to their specific needs. Thus, the focus of this book is to put the reader in a position where she is able to design and write her own scientific programs in a high-level language. What this book does *not* provide is a discussion of how to use one of the existing commercial or academic simulation programs or packages.

The book aims to provide an overview of the basic principles underlying modern computer simulations as applied to physics and engineering, represented here by the unit of three disciplines: physics, mathematics, and computer science. As no book can really discuss *all* methods currently in use without itself becoming of encyclopedic size, I decided to emphasize the foundations and typical applications of one of the most important classes of simulation techniques, namely *mesh-free, particle based methods*, in particular the Molecular Dynamics (MD) and Monte Carlo (MC) methods.

This textbook differs from many other books on the subject in that it utilizes a practical learning-by-doing approach to computational science. It presents the field by providing not only many algorithms in pseudo code notation, but also many ready-to-use code examples that show how to actually implement them in a high-level language. I will discuss not only the mathematics and physics on which both the outlined general principles and the presented algorithms are based, but will also show how to design and write your own scientific computing programs. I will also discuss many important, real-life, practical issues pertaining to code implementation, which are very often neglected in textbooks on scientific computing. Becoming an expert in scientific computing requires above all practice and experience and this book discusses some good practices, learned from professional code development and reveals a lot of important tricks of the trade in scientific coding, that are usually scattered throughout the original literature. It is important to understand that the concepts laid down in an algorithm cannot be grasped by just looking at the algorithm alone. It is necessary to do some work with actual source code.

By studying the examples and solving the exercises the reader will be able to fully understand and appreciate the type of problems that must be solved when writing scientific programs. Along the way, the reader will be exposed to many different physical problems, ranging from the classical harmonic oscillator to the features of a macromolecule. We will use some of the most popular algorithms found in numerical mathematics and computer science to treat these problems numerically. Finally, we will implement these algorithms in stand-alone programs using C/C++ as a high-level programming language. C and C++ are regarded for most of the book as essentially the same language and the reasons for picking C/C++ as a vehicle for teaching scientific computing are illustrated in Chapter 2.

A note is in place here as to the organization of the book and the prerequisites expected from the reader. Besides curiosity, the reader should have some elementary background in calculus and classical mechanics. No prior programming knowledge is required, since the necessary skills in this area are developed on a learning-by-doing basis, as the reader proceeds through the text. Chapter 1 provides a thorough introduction to the field of scientific computing, also from a historical perspective. Chapter 2 is a key chapter, because it provides a concise introduction to the high-level programming language C and to some additional features of C++, which do not pertain to object-oriented code design. We focus on those elements of the language that are really necessary to be able to write your own scientific code. Chapter 3 is an introduction to statistical physics, which in turn forms the basis of the MD and MC methods. In Chapter 4 we discuss the origin of molecular forces and how to model them. Chapter 5 introduces the basics of the MD method and provides many source code and case studies, which should be well suited for self-study. Chapter 6 does the same with the MC method. Finally, in Chapter 7 we introduce some advanced topics and suggest a few projects before finally discussing typical applications of computer simulations in macromolecular physics.

Much of the material included in this volume emerged from lecture notes to a course on "Molecular Dynamics with Applications in Soft Matter" which I taught in the Faculty of Mathematics and Physics at the University of Freiburg (Albert-Ludwigs-Universität) in Germany, in the winter term 2011/2012, to a group of international students. The course was designed completely from scratch for beginning graduate students, majoring in either physics or engineering, who already had some basic knowledge of statistical mechanics and thermodynamics. The course was taught in two lectures per week, each lasting two hours, and involved two additional hours of recitation class per week for discussion of homework assignments and questions (with a total of 10 ECTS credits to earn). The homework assignments included theoretical (pencil and paper) exercises to further develop theoretical understanding, but also programming exercises on a learning-by-doing basis. Many of these exercises have been included in this book.

All of the algorithms and source codes presented in this book have been carefully selected and, additionally, the source code of more than two dozen of the program listings is available on the Walter de Gruyter webpage http://www.degruyter.com[1] Code listings which can be downloaded from the book's website are marked in the listing caption with a bold-face asterisk (*) as indicated in this example:

Listing 0. This is a sample caption of a code listing. (*)

```
/* This is a comment */
Here comes the source code
...
```

Furthermore, several different icons are used throughout the book to indicate boxes that summarize or expand on important issues.

I hope that this book will be useful in many ways: as a first introduction to the subject for students wishing to enter this field of research, as a companion text for beginning to intermediate-level courses at universities, and as a reference for those who use computer simulations in their everyday work. While attending international scientific conferences in the past two years, I was on several occasions able to discuss relevant parts of the book manuscript with researchers around the world (too many to be mentioned by name!) who work in the field of computational physics and engineering. I hope that their suggestions for improvements have indeed enhanced the overall readability of this text. As the only exception, I want to mention Professor John Borg of Marquette University in Milwaukee, Wisconsin, USA, who spent his sabbatical year in my research group at EMI in 2011 and 2012, and who read through parts of the manuscript at a close to final stage, but still pointed out possibilities for lots of improvements. Thank you John!

The PhD, graduate, and undergraduate students of the University of Freiburg who braved my initial course on computational physics and the MD method deserve my sincere thanks. This book is mainly based on my lecture notes to this graduate course. Finally, working together with Christoph von Friedeburg of the Walter de Gruyter publishing house has been a very pleasant experience, as every issue concerning the manuscript was taken care of very smoothly and professionally. It was not at all an easy task for me to finish the book manuscript within the allotted time frame set by the publisher, whilst having to handle the tremendous work load of teaching at university, leading the Fraunhofer research group "Shock Waves in Soft Biological Matter", and taking care of all the many other project related tasks that are associated with a position at a Fraunhofer research institute.

---

[1] http://www.degruyter.com/staticfiles/pdfs/9783110255904AccompanyingPrograms.zip

Last but not the least I thank my lovely family Katrin, Pia and Sven (not to forget our cats Charly and Micky and our Mongolian gerbils) for their enjoyable mix of distraction, curiosity, support and encouragement while working on this book project day and night.

Gundelfingen im Breisgau, June 2012   Martin Steinhauser

Contents

Preface		v
1	**Introduction to computer simulation**	1
	1.1 Physics and computational physics	1
	1.2 Choice of programming language	5
	1.3 Outfitting your PC for scientific computing	13
	1.4 History of computing in a nutshell	17
	1.5 Number representation: bits and bytes in computer memory	22
	1.5.1 Addition and subtraction of dual integer numbers	24
	1.5.2 Basic data types	29
	1.6 The role of algorithms in scientific computing	41
	1.6.1 Efficient and inefficient calculations	43
	1.6.2 Asymptotic analysis of algorithms	51
	1.6.3 Merge sort and divide-and-conquer	56
	1.7 Theory, modeling and computer simulation	59
	1.7.1 What is a theory?	59
	1.7.2 What is a model?	67
	1.7.3 Model systems: particles or fields?	72
	1.7.4 The linear chain as a model system	74
	1.7.5 From modeling to computer simulation	78
	1.8 Exercises	81
	1.8.1 Addition of bit patterns of 1 byte duals	81
	1.8.2 Subtracting dual numbers using two's complement	81
	1.8.3 Comparison of running times	81
	1.8.4 Asymptotic notation	82
	1.9 Chapter literature	83
2	**Scientific Computing in C**	84
	2.1 Introduction	84
	2.1.1 Basics of a UNIX/Linux programming environment	87
	2.2 First steps in C	99
	2.2.1 Variables in C	101
	2.2.2 Global variables	103

2.2.3	Operators in C	104
2.2.4	Control structures	108
2.2.5	Scientific "Hello world!"	111
2.2.6	Streams – input/output functionality	116
2.2.7	The preprocessor and symbolic constants	119
2.2.8	The function *scanf()*	122
2.3	Programming examples of rounding errors and loss of precision	125
2.3.1	Algorithms for calculating e^{-x}	130
2.3.2	Algorithm for summing $1/n$	133
2.4	Details on C-Arrays	137
2.4.1	Direct initialization of certain array elements (C99)	141
2.4.2	Arrays with variable length (C99)	141
2.4.3	Arrays as function parameters	142
2.4.4	Pointers	144
2.4.5	Pointers as function parameters	152
2.4.6	Pointers to functions as function parameters	154
2.4.7	Strings	159
2.5	Structures and their representation in computer memory	161
2.5.1	Blending structs and arrays	163
2.6	Numerical differentiation and integration	165
2.6.1	Numerical differentiation	166
2.6.2	Case study: the second derivative of e^x	169
2.6.3	Numerical integration	176
2.7	Remarks on programming and software engineering	181
2.7.1	Good software development practices	181
2.7.2	Reduction of complexity	184
2.7.3	Designing a program	188
2.7.4	Readability of a program	189
2.7.5	Focus your attention by using conventions	190
2.8	Ways to improve your programs	191
2.9	Exercises	193
2.9.1	Questions	193
2.9.2	Errors in programs	194
2.9.3	*printf()*-statement	197
2.9.4	Assignments	198
2.9.5	Loops	199
2.9.6	Recurrence	199
2.9.7	Macros	200
2.9.8	Strings	200
2.9.9	Structs	201

		2.10 Projects	203
		2.10.1 Decimal and binary representation	203
		2.10.2 Nearest machine number	203
		2.10.3 Calculating e^{-x}	203
		2.10.4 Loss of precision	204
		2.10.5 Summing series	204
		2.10.6 Recurrence in orthogonal functions	205
		2.10.7 The Towers of Hanoi	205
		2.10.8 Spherical harmonics and Legendre polynomials	207
		2.10.9 Memory diagram of a battle	208
		2.10.10 Computing derivatives numerically	208
	2.11 Chapter literature		210
3	**Fundamentals of statistical physics**		**211**
	3.1 Introduction and basic ideas		212
		3.1.1 The macrostate	216
		3.1.2 The microstate	218
		3.1.3 Information conservation in statistical physics	219
		3.1.4 Equations of motion in classical mechanics	225
		3.1.5 Statistical physics in phase space	229
	3.2 Elementary statistics		235
		3.2.1 Random Walk	236
		3.2.2 Discrete and continuous probability distributions	241
		3.2.3 Reduced probability distributions	242
		3.2.4 Important distributions in physics and engineering	244
	3.3 Equilibrium distribution		249
		3.3.1 The most probable distribution	251
		3.3.2 A statistical definition of temperature	253
		3.3.3 The Boltzmann distribution and the partition function	255
	3.4 The canonical ensemble		258
	3.5 Exercises		261
		3.5.1 Trajectories of the one-dimensional harmonic oscillator in phase space	261
		3.5.2 Important integrals of statistical physics	261
		3.5.3 Probability, example from playing cards	261
		3.5.4 Rolling dice	262
		3.5.5 Problems, using the Poisson density	262
		3.5.6 Particle inside a sphere	262

4 Inter- and intramolecular potentials 264

4.1 Introduction 265
4.2 The quantum mechanical origin of particle interactions 266
4.3 The energy hypersurface and classical approximations 270
4.4 Non-bonded interactions 271
4.5 Pair potentials 274
    4.5.1 Repulsive Interactions 275
    4.5.2 Electric multipoles and multipole expansion 280
    4.5.3 Charge-dipole interaction 280
    4.5.4 Dipole-dipole interaction 283
    4.5.5 Dipole-dipole interaction and temperature 284
    4.5.6 Induction energy 285
    4.5.7 Dispersion energy 287
    4.5.8 Further remarks on pair potentials 288
4.6 Bonded interactions 291
4.7 Chapter literature 292

5 Molecular Dynamics simulations 294

5.1 Introduction 295
    5.1.1 Historical notes on MD 299
    5.1.2 Limitations of MD 303
5.2 Numerical integration of differential equations 309
    5.2.1 Ordinary differential equations 309
    5.2.2 Finite Difference methods 310
    5.2.3 Improvements to Euler's algorithm 316
    5.2.4 Predictor-corrector methods 317
    5.2.5 Runge–Kutta methods 318
5.3 Integrating Newton's equation of motion: the Verlet algorithm 320
5.4 The basic MD algorithm 323
5.5 Basic MD: planetary motion 327
    5.5.1 Preprocessor statements and basic definitions 327
    5.5.2 Organization of the data 328
    5.5.3 Function that computes the energy 328
    5.5.4 The Verlet velocity algorithm 329
    5.5.5 The force calculation 329
    5.5.6 The initialization and output functions 332
    5.5.7 The *main()*-function 332
5.6 Planetary motion: suggested project 335

	5.7 Periodic boundary conditions	337
	5.8 Minimum image convention	338
	5.9 Lyapunov instability	338
	5.10 Case study: static and dynamic properties of a microcanonical LJ fluid	341
	5.10.1 Microcanonical LJ fluid: suggested projects	346
	5.11 Chapter literature	356
6	**Monte Carlo simulations**	**357**
	6.1 Introduction to MC simulation	357
	6.1.1 Historical remarks	360
	6.2 Simple random numbers	362
	6.2.1 The linear congruential method	365
	6.2.2 Monte Carlo integration – simple sampling	369
	6.3 Case study: MC simulation of harddisks	378
	6.3.1 Trial moves	378
	6.3.2 Case study: MC simulation of harddisks – suggested exercises	380
	6.4 The Metropolis Monte Carlo method	381
	6.5 The Ising model	383
	6.5.1 Case study: Monte Carlo simulation of the 2D Ising magnet	386
	6.6 Case Study: NVT MC of dumbbell molecules in 2D	392
	6.7 Exercises	394
	6.7.1 The GSL library	394
	6.7.2 Calculating π	395
	6.7.3 Simple and importance sampling with random walks in 1D	399
	6.7.4 Simple sampling and importance sampling with random walks in 1D	399
	6.8 Chapter literature	407
7	**Advanced topics, and applications in soft matter**	**408**
	7.1 Partial differential equations	408
	7.1.1 Elliptic PDEs	410
	7.1.2 Parabolic PDEs	411
	7.1.3 Hyperbolic PDEs	411
	7.2 The finite element method (FEM)	412
	7.3 Coarse-grained MD for mesoscopic polymer and biomolecular simulations	414
	7.3.1 Why coarse-grained simulations?	414
	7.4 Scaling properties of polymers	415

	7.5 Ideal polymer chains	416
	7.6 Single-chain conformations	420
	7.7 The ideal (Gaussian) chain model	421
	7.8 Scaling of flexible and semiflexible polymer chains	422
	7.9 Constant temperature MD	429
	7.10 Velocity scaling using the Behrendsen thermostat	430
	7.11 Dissipative particle dynamics thermostat	431
	7.12 Case study: NVT Metropolis MC simulation of a LJ fluid	434
	7.13 Exercise	443
	7.13.1 Dumbbell molecules in 3D	443

A The software development life cycle   444

B Installation guide to Cygwin   445

C Introduction to the UNIX/Linux programming environment   448

C.1 Directory structure 448

C.2 Users, rights and privileges 450

C.3 Some basic commands 453

C.4 Processes 455

    C.4.1 Ending processes 455

    C.4.2 Processes priorities and resources 455

C.5 The Bash 457

C.6 Tips and tricks 458

C.7 Useful programs 458

    C.7.1 Remote connection: ssh 459

    C.7.2 Gnuplot 459

    C.7.3 Text editors: vi, EMACS and others 459

D Sample program listings   470

D.1 Sample code for file handling 470

E Reserved keywords in C   473

F Functions of the standard library *<string.h>*   474

G Elementary combinatorial problems   475

G.1 How many differently ordered sequences of N objects are possible? . 475

G.2 In how many ways can N objects be divided into two piles, with n and m objects, respectively? 475

Contents xv

    G.3 In how many ways can N objects be arranged in $r+1$ piles with n_j objects in pile number j with $j \in [0, 1, \ldots, r]$? 476

    G.4 Stirling's approximation of large numbers 476

H Some useful constants   477

I Installing the GNU Scientific Library, GSL   478

J Standard header files of the *ANSI-C* library   479

K The central limit theorem   480

Bibliography   481

Acronyms   505

Index   506

Authors   509

The source code of more than two dozen of the program listings is available on the Walter de Gruyter webpage:

`http://www.degruyter.com/staticfiles/pdfs/`
`9783110255904AccompanyingPrograms.zip`

List of Algorithms

0	Sample caption of a code listing	vii
1	Accuracy of *float* and *double*.	37
2	Euclidean algorithm (pseudocode).	43
3	Insertion sort algorithm (pseudocode).	47
4	Merge sort applied to an Array A[1,...,N] (pseudocode).	56
5	Sample code for lexical analysis.	94
6	The simplest C *main()* function	100
7	Simple example of a *for*-loop.	110
8	Scientific "Hello World!" C program.	112
9	Scientific C++ "Hello World!".	116
10	Scientific C++ "Hello World!" without namespace.	117
11	Scientific C++ "Hello World!" with exception handling.	118
12	Use of user defined macros.	121
13	Use of predefined macros.	123
14	Example for the use of the C *scanf()* function.	123
15	Plotting of an address.	125
16	Improved calculation of e^{-x}.	132
17	Calculation of the exponential function e^{-x}.	135
18	Calculation of the standard deviation.	136
19	The use of the `sizeof` operator.	138
20	An array of variable size (only *C99*).	142
21	An array as function parameter.	143
22	Simple illustration of the use of pointers in C.	146
23	The use of pointers as function parameters.	153
24	Passing of function names as arguments to other functions via pointers.	156
25	Initialization of strings.	159
26	Reading strings.	161
27	*main()*-function for calculating the second derivative of $\exp(x)$.	171
28	The *Initialize()*-function.	172
29	The function *SecondDerivative()*.	172
30	The function *InputOutput()*.	173
31	The header file *Initialization.h* with the prototype definition of *Initialize()*.	174
32	The C++ version of the *main()*-function to compute $\exp(x)''$.	175

33	The C++ version of the *Output()*-function, called *Output2()*.	176
34	Trapezoidal rule. (pseudocode).	177
35	Trapezoidal rule detailed. (pseudocode).	179
36	Function that implements the trapezoidal rule.	179
37	Function that implements the rectangle rule.	180
38	Simpson's rule. (pseudocode)	181
39	Implementation of the C *strcpy()* function.	182
40	Comparison of two integer values.	196
41	Structures.	197
42	Printing integers.	201
43	Macros multiplication.	201
44	Use of macros.	202
45	Memory diagram of a battle.	209
46	Part 1 – Generation of random numbers according to the Gaussian distribution.	247
46	Part 2 – The main file.	248
47	Predictor-corrector algorithm.	317
48	Basic MD algorithm (pseudocode).	324
49	Part 1 – Planetary motion code. Basic definitions.	328
49	Part 2 – Planetary motion code. Data organization.	328
49	Part 3 – Planetary motion code. Energy calculation.	329
49	Part 4 – Planetary motion code. Integration.	330
49	Part 5 – Planetary motion code. The $O(N^2)$ force calculation.	331
49	Part 6 – Planetary motion code. Initialization and output.	333
49	Part 7 – Planetary motion code. The *main()*-function.	334
50	Function *WritePDBOutput()*.	335
51	Code fragment for calculating LJ forces.	343
52	Code fragment for implementing periodic boundary conditions.	344
53	Code fragment for calculating LJ forces.	344
54	Calculation of the radial density distribution function $\rho(r)$.	353
55	Usage of *rand()*.	362
56	Part 1 – Pseudorandom numbers limited to a certain interval.	364
56	Part 2 – The header file *Random.h*.	365
57	Floating point pseudorandom numbers. Header file *"Random.h"*.	366
58	Floating point pseudorandom numbers. *main()*-function.	367
59	Park and Miller pseudorandom number generator.	374
60	Linear congruential pseudorandom number generator.	375
61	Brute force MC integration of π.	376
62	The library function ran0().	377
63	Part 1 – *main()*-program for calculating the 2D Ising model, using NVT MC.	388

List of Algorithms

63	Part 2 – *main()*-program for calculating the 2D Ising model, using NVT MC, continued	389
63	Part 3 – *main()*-program for calculating the 2D Ising model, using NVT MC, continued	390
64	Function *EnergyChange()* for the Ising model.	390
65	Function *SampleSystem()* for the Ising model.	390
66	Function *Init()* for the Ising model.	391
67	File *gslDemo.c* which uses the GSL library.	396
68	Part 1 – Calculating π with the circle method.	397
68	Part 2 – Calculating π with the circle method.	398
69	Part 1 – Random walk in one dimension.	401
69	Part 2 – Random walk in one dimension.	402
70	Part 1 – Random walk in two dimensions.	403
70	Part 2 – Random walk in two dimensions.	404
70	Part 3 – Random walk in two dimensions.	405
70	Part 4 – Random walk in two dimensions.	406
71	Part 1 – *main()*-program for calculating the phase diagram of an LJ system.	437
71	Part 2 – *main()*-program for calculating the phase diagram of an LJ system continued	438
71	Part 3 – *main()*-program for calculating the phase diagram of an LJ system continued	439
71	Part 4 – *main()*-program for calculating the phase diagram of an LJ system continued	440
72	Function *TotalEnergy()* called from *main()* for calculating the phase diagram of an LJ system.	441
73	Function *Init()* called by *main()* for calculating the phase diagram of an LJ system.	442
74	Part 1 – Sample Code for File Handling.	470
74	Part 2 – Sample Code for File Handling continued	471
74	Part 3 – Sample Code for File Handling continued	472

Chapter 1

Introduction to computer simulation

"I think there is a world market for maybe five computers."

Statement attributed to Thomas Watson, IBM chairman, 1943

Summary

This chapter aims at providing a basic introduction to the field of computer simulation. I will first discuss various aspects of computer languages, suitable for high-performance computing, that find applications in this book. I will switch from the history of computing to discussing how numbers are actually represented in computer memory and how numerical round off errors arise. Next, I will introduce the definition of an algorithm and show how to analyze algorithms, with use of the important idea of asymptotic notation. Finally, I will discuss modeling processes in physics and engineering and look at the two basic modeling paradigms, namely the concepts of particles and fields.

Learning targets

✓ Learning about the diversity of different computer languages.

✓ Learning how to prepare your PC to be used for scientific computing.

✓ Learning the history of computing.

✓ Understanding how to analyze algorithms using asymptotic notation.

✓ Understanding the process of model building in physics

1.1 Physics and computational physics

Solving a problem in any field of physics or engineering very often amounts to merely solving ordinary or partial differential equations. This is the case in, for example, classical mechanics, electrodynamics, quantum mechanics, fluid dynamics, and so on. In all of these areas of Assembly language has some of the physics, we usually have some fundamental equations at hand, that were originally found by generalization from corresponding experiments and which (after they had been established) could be introduced to the theory in an axiomatic way.

Our insight in a physical system, combined with numerical mathematics, gives us the rules for setting up an *algorithm*, or a set of rules for solving a particular problem. Our understanding of a physical system is obviously determined by the laws of nature, i.e. the fundamental laws, the initial conditions, the boundary conditions and additional external constraints[1] which determine the ultimate behavior of the system.

After spelling out the physics of the system, for example, in the form of a set of coupled partial differential equations, we need efficient methods for setting up the final algorithm for solving the problem numerically. This algorithm is in turn coded[2] into a computer program and executed on the available computing facilities.

> Computer simulations not only link analytic theory and experiment in order to test theories, but can also be used as an exploratory tool in "computer experiments" under conditions which would be unfeasible, too expensive, or too dangerous for real experiments in the laboratory. Finally, computer simulations can conduct "thought experiments", which are commonly used in physics to check fundamental ideas and to illustrate their logical consequences.

Nowadays, computer simulations are an integral part of contemporary basic and applied scientific research. In engineering and physics, computational, theoretical and experimental explorations are equally important to our daily research and studies of physical systems. Moreover, the ability to perform computations is part of the essential repertoire of engineers and research scientists today. Several new fields within computational science have emerged and strengthened their positions in the last years, such as computational mathematics, mechanics, biology, chemistry and physics, materials science and bioinformatics. These fields underpin the ever increasing importance of computer simulations as a tool to gain novel insights into physical and biological systems, in particular in those cases where no analytical solutions can be found or experiments are too complicated, too dangerous or simply impossible to carry out.

Computational science as a field of research unites theory, experiment and computation. In order to master this field, research scientists have to expand both their

---

[1] For example, an external load such as pressure or shear forces.
[2] Generally speaking, the term "coded" merely refers to the actual process of typing in computer code (software), with use of a text editor, whereas the term "programming" is generally used in a broader sense, which also refers to the process of planning and designing software. A more sophisticated technical term for the latter is "software engineering", which also refers to the whole *software development life cycle*, see Appendix A.

understanding and the repertoire of tools common to physics, engineering, mathematics and computer science, see Figure 1.1.

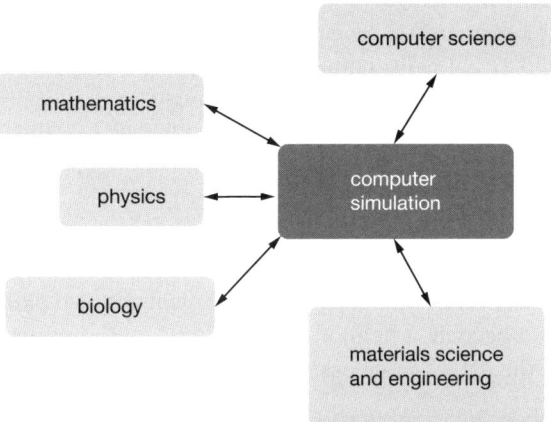

Figure 1.1. Computer simulation and its fields of application. Computer simulation usually involves knowledge about a variety of scientific areas, among them mathematics, computer science, physics, biology and engineering.

In a typical situation, several such tools are combined. For example, suppose one is working on a project which demands extensive visualizations of the results, say, the accumulation of stresses and pressures in the material during a simulation of a car crash. To obtain these results, i.e., to solve a physics problem, we need a fairly fast program, because here computational speed matters. In such a case, one does not want to simply use some black-box commercial program, but would almost certainly write a high-performance simulation in a compiled language[3] such as C or C++[4]. However, for the visualization of results, we would probably use one of the many free open source tools available on the Internet. For appropriately preparing data we may find *interpreted* scripting languages such as JAVA, PYTHON, R, S or MATLAB quite useful. Used correctly, these tools, ranging from scripting languages to compiled high-performance languages and free visualization programs, considerably enhance our capability to solve complicated problems.

---

[3] In Section 2 I will discuss in detail the compilation process of a computer program, i.e. its translation into machine language.

[4] In Section 1.2 I discuss various choices of programming language and my preference for C/C++.

Think!

> **The limits of computer simulation.** So-called "brute-force" simulations will never be able to bridge the gap between scales of length relevant for a description of natural phenomena on a microscopic (nanometers and picoseconds) and macroscopic (biological cells, humans, planets) level. Hence, in computer simulations, we need different, "coarse-grained" levels of description and we need input from experiments at many different spatial levels, in order to test and validate the models that lie at the bottom of computer simulations.

Scientific visualization, i.e. the evaluation and visualization of large data sets, which can emerge from numerical simulations or experiments, is also part of computational science. It has even become a field of research in its own right in high performance computing, where the data volume easily exceeds many terabytes, see Figure 1.2.

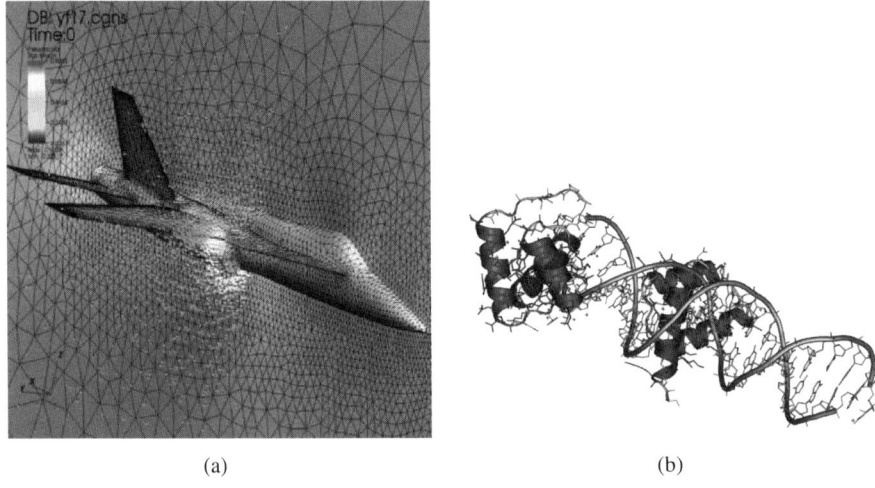

(a)   (b)

Figure 1.2. Illustration of the visualization of very large data sets. (a) Plot of a CFD General Notation System (CGNS) [1] dataset representing a YF-17 jet aircraft. The dataset consists of an unstructured grid. The free tool *VisIt* created the image using a pseudocolor plot of the dataset's Mach variable, a mesh plot of the grid, and a vector plot of a slice through the velocity field. Taken from [5]. The CGNS data format has recently been substituted by the HDF5 [7] format, which has parallel input/output (I/O) and data compression capabilities and has rapidly grown to become a world-wide format standard for storing scientific data. In (b) a PDB file of the DNA double helix is displayed with the sugar-phosphate backbone and the different bases including the secondary protein structure as *ribbon diagram* [353, 354]. The visualization is done using the free *Visual Molecular Dynamics*, *VMD* tool [9]. The Protein Databank Format is a standard data format used for the description of three-dimensional structural data of large biological molecules, such as proteins and nucleic acids.

This text aims to teach you how to write *your own* high-performance computer programs in a high-level language and to show several examples of the questions that can be tackled with these programs. What we will *not* cover is a discussion of Graphical User Interfaces (GUIs) of typical commercial products such as ANSYS[5], COMSOL[6], LS-DYNA[7], PATRAN[8], ACCELRYS[9], ABAQUS[10] or GAUSSIAN[11], to name just a few. Most of these commercial products were developed decades ago using FORTRAN code and were then visually enhanced by adding a modern GUI[12]. When using such tools, one usually has no access to the source code and usually the only way to change anything in the code – if at all – is provided by so-called user subroutines, which have to be written (often in FORTRAN) and then linked to the main program binary[13]. The commercial market for scientific molecular dynamics (MD) code is much less developed, probably because applying such code still requires considerable expert knowledge on the side of the user. As a consequence, the barrier hindering use of such code in commercial industry projects is relatively high. The general problem with commercial code is (due to hidden, unavailable source code) that they usually cannot be used for real scientific work. Thus, in this book we are not interested in discussing the numerous random interfaces provided in GUIs or how to write user subroutines in commercial packages, but rather intend to discuss the basic physical and algorithmic principles that are at the bottom of any such simulation package. This knowledge will remain valid and valuable completely independent from any software version or arbitrary changes to any particular GUI. Our overall goal is to encourage you to write your own simulation programs by teaching you the mathematical, physical and computational foundations. Our objective is always to develop a fundamental understanding and the purpose of computing is to provide further insight, *not* mere "number crunching"!

1.2 Choice of programming language

In scientific computing, the choice of a programming language is sometimes highly contentious. Since the advent of the first high-level language FORTRAN (FORmula TRANslator) in 1954[14], literally hundreds of high-level languages have been devel-

---

[5] http://www.ansys.com
[6] http://www.comsol.com
[7] http://www.ls-dyna.com
[8] http://www.mscsoftware.com
[9] http://www.accelrys.com
[10] http://www.3ds.com
[11] http://www.gaussian.com
[12] This is not to diminish the practical importance of many such tools, which often are an engineer's workhorse when a fast solution to a problem is needed.
[13] Often, the linking can only be done by using an expensive commercial compiler. For example, with ANSYS products one has to use the costly Intel FORTRAN compiler.
[14] Before FORTRAN, all programming was done directly in assembly, i.e. machine code.

oped, some of them for highly specialized purposes. Only a few of them have survived the test of time. Higher level computer languages were developed to allow the programmer to focus on the problem[15] at hand in a high-level notation and not on the translation of ideas into machine code. The latter is done by a translation program called a *compiler*. Several of the best known languages are listed in Table 1.1. They can also be assigned to different generations as shown in Table 1.2.

Think!

> What is a computer ?

A computer is a machine that processes bits. A bit is an individual unit of computer storage which can take on two values, 1 or 0. We use computers to process information, but all information is represented as bits. Collections of bits can represent characters, numbers, or any other information (compare our discussion of computer memory in Section 1.5). Humans interpret these bits as pieces of information, whilst computers merely manipulate the bits.

Each type of computer has a collection of instructions that it can execute. These instructions are stored in memory and fetched, interpreted and executed during the execution of a program. This sequence of bytes is called a "machine program". It would be quite painful to use machine language for programming on this most elementary level. You would have to enter the correct bytes for each single instruction of your program and know the addresses of all data used in your program. The very first computers were programmed in machine language (by literally plugging cables and wires) but people soon figured out ways to make things easier. People started to use symbolic names to represent particular instructions, addresses of instructions and data. This led to the development of symbolic assembly languages in the very early history of computing, in the 1950s. Assembly quickly replaced machine language, thus eliminating a lot of tedious work. Machine languages are considered "first-generation" languages, whereas assembly languages are considered "second generation". Many programs from that era continued to be written in assembly even after the invention of the first "third-level" languages, FORTRAN and COBOL (see Table 1.1), in the late 1950s. In particular, operating systems were typically nearly 100% assembly until the creation in 1969 of C[16] as the primary language for the UNIX[17] operating system.

---

[15] This is why high-level languages are also called *problem oriented* languages. They allow for coding algorithms from different areas in *one* computer language, no matter which computer system one uses, i.e. high-level languages help the programmer to abstract away from the computer hardware.

[16] C was also the primary language in which the much more recent free Linux operating system was written, by a Finnish student, Linus Torvalds, who created a new operating system kernel, which he published for the first time on 25 August 1991.

[17] UNIX is a computer operating system originally developed in 1969 by a group of AT&T employees at Bell Labs, including Ken Thompson, Dennis Ritchie, Brian Kernighan, Douglas McIlroy, and Joe Ossanna. During the late 1970s and early 1980s, the influence of UNIX in academic circles led to its large-scale adoption (particularly of the BSD variant, originating from the University of California,

Section 1.2 Choice of programming language

Table 1.1. List of some of the best known computer languages.

language	year published	typical area of application
assembly	1950s	low level language
FORTRAN[a]	1957	science-oriented
LISP[b]	1958	artificial intelligence
COBOL[c]	1959	data processing / accounting
ALGOL[d]	1960	science-oriented
BASIC[e]	1963	private use, many dialects
PL1[f]	1960s	IBM language, developed as a "universal" language
APL[g]	1960s	IBM language
PASCAL[h]	1968	teaching
PROLOG	1972	artificial intelligence
MODULA[i]	1975	teaching
ADA[j]	1975	security applications
C	1969	UNIX operating system, compilers, developed as "super-assembly"
C++[k]	1980	general applications
JAVA	1995	Internet, distributed networks
C#[l]	2001	Internet, distributed networks

[a] *FOR*mula *TRAN*slator. This language was developed in 1954 and introduced to the public in 1957.
[b] *LIS*t *P*rogramming. Invented by John McCarthy in 1958 and published in 1960 [289]. There are many dialects of LISP. Linked lists are one of Lisp languages' major data structures, and Lisp source code is itself made up of lists.
[c] *CO*mmon *B*usiness-*O*riented *L*anguage.
[d] *ALG*ebraic *O*riented *L*anguage. ALGOL is the ancestor of all procedural languages. It implements elementary loops and the concept of blocks.
[e] *B*eginner's *A*ll-Purpose *S*ymbolic *I*nstruction *C*ode.
[f] *P*rogramming *L*anguage *1*.
[g] *A P*rogramming *L*anguage.
[h] Developed by Niklaus Wirth in honor of the French mathematician Blaise Pascal.
[i] Descendant of the PASCAL language.
[j] Used by the US military today.
[k] Includes features to aid object-oriented program design.
[l] C# was developed by the Microsoft corporation and published in 2001. It contains many elements from JAVA and C++.

Berkeley) by commercial start-ups, the most notable of which are Solaris, HP-UX and AIX. Today, in addition to certified UNIX systems such as those already mentioned, UNIX-like operating systems such as Linux and BSD are commonly encountered. The term "traditional UNIX" may be used to describe a UNIX or an operating system that has the characteristics of either Version 7 UNIX or UNIX System V. Today the term UNIX is used to describe any operating system that conforms to UNIX standards, meaning that the core operating system operates the same as the original UNIX operating system.

Table 1.2. Important computer languages assigned to different generations.

generation	paradigm	example	characteristics
1st	machine	machine code	binary commands
2nd	assembly	assembly code	symbolic commands
3rd	procedural	FORTRAN COBOL	hardware independent
3rd+	functional, object oriented	PASCAL, C, C# C++, Java, C	structured and object oriented
4th	application oriented	SQL, Oracle, ABAP	transactions oriented
5th	declarative	Visual Basic, PROLOG	target oriented

Table 1.3. Assembly language and machine code (displayed as hexadecimal number), compare also Figure 1.9 in Section 1.5. It is obvious from these few examples that understanding the code of a machine language program is virtually impossible. As there is a one-to-one correspondence between the instructions in assembly and machine language, it is fairly straightforward to translate instructions from assembly into machine language, using a compiler. However, when the first microprocessors were introduced in the early 1970s, due to the lack of compilers, some programming was, in fact, still done in machine language.

assembly	operation	machine language	C/C++ code
nop	no operation	90	–
inc result	increment	FF060A00	result++;
mov variable, 45	copy	C7060C002D00	variable = 45;
and mask, 120	logical and	80260E0080	mask1 = mask1 & 120;
add marks, 10	integer addition	83060F000A	marks = marks + 10;

We may appreciate the readability of a high-level programming language vs. assembly and machine code by looking at the equivalent machine language instructions of a simple example. In Table 1.3 I show both assembly and machine code of an Intel Pentium CPU for a number of very simple instructions. Figure 1.3 finally presents a comprehensive view of the different possible levels of programming and the degree of abstraction they provide.

Assembly language has some of the worst features known in computing:

- assembly produces absolutely non-portable code[18].

---

[18] That is because every CPU has its own assembly language.

Section 1.2 Choice of programming language 9

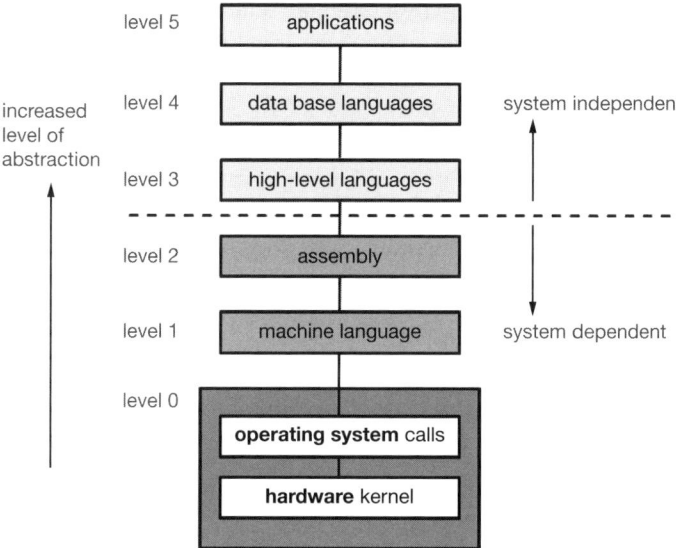

Figure 1.3. Different levels of abstraction in programming, at which users can interact with computers. The lowest abstraction level is formed by the built-in commands in the hardware kernel and the operating system. On top of that there is machine language and assembly which allow a user to program the machine with elementary machine commands and which require detailed knowledge of the complete address space of the specific machine. High-level languages allow the programmer to abstract from the machine hardware and to code commands in languages with a syntax akin to mathematical statements. Data base languages simplify the commands of high-level languages by providing very powerful commands for retrieving, processing and manipulating data stored in special databases. Finally, on the mere application level, the programmer is not only completely oblivious to the computer hardware but also in most cases completely unaware of any programming language. Manipulation of data works in a completely intuitive way by making use of graphical interfaces.

- assembly leads to bad reliability and maintainability of programs. In modern languages like JAVA the programmer is protected from many possible problems, such as pointer errors. Pointers do exist in JAVA, but the programmer can be totally oblivious to them. In assembly language however, *every* variable access is a pointer access. Also, assembly language is a sequence of individual symbolic machine instructions, whose syntax bears no resemblance to the problem being solved, whereas high-level language syntax resembles mathematical syntax.

- assembly is much slower to write than high-level languages.

- assembly language instructions are cryptic.

In a time when the latest fads in programming tend to be object-oriented high-level languages of at least the third generation, the trend is clearly to learn how to write portable programs, with high reliability, in record time. However, today, studying assembly languages still may teach you how a CPU actually works. This helps when programming in high-level languages. For example, understanding how the compiler implements the features of a high-level language can aid in programing efficiency. More importantly, understanding the translation from high-level language to machine language is fundamental in understanding why bugs behave the way they do.

> What is the best high-level language to use in scientific computing?

In essence, one could think of at least three different possible methods to perform numerical computations of the kind that we are interested in, in this book.

First, we could use a mathematical software package, such as MATHEMATICA[19], MAPLE[20], MATLAB[21] or R[22]. The main advantage of these ready-made packages is that they facilitate very rapid coding of numerical problems by providing special data types for complex numbers or matrices and matrix manipulations. The main disadvantage, besides the cost, is that they produce executable code that is *interpreted*, rather than *compiled*. Compiled code is translated directly from a high-level language into machine code instructions which, by definition, are platform dependent, whereas interpreted code is translated from a high-level language into a set of meta-code instructions which are platform independent. Each meta-code instruction is then translated into a fixed set of machine code instructions specific to the particular hardware platform on which the code is run. In general, interpreted code is nowhere near as efficient as compiled code in terms of computer resource utilization, i.e. interpreted code runs much slower than equivalent compiled code. Hence, although MATHEMATICA, MAPLE, MATLAB and R are useful environments for performing relatively *small* calculations, they are not at all suitable for full-blown research projects, since the code they produce generally runs far too slow.

Second, we could write our own programs in a high-level language, but only use calls to pre-compiled routines in freely available subroutine libraries, such as LAPACK[23], LINPACK[24], ODEPACK[25] or GMP[26], the *GNU Multiple Precision* arithmetic library, to perform all of the real numerical work. This is an approach used by the majority of research physicists.

---

[19] See http://www.wolfram.com/mathematica
[20] See http://www.maplesoft.com
[21] See http://www.mathworks.com
[22] See http://www.r-project.org
[23] See http://www.netlib.org/lapack
[24] See http://www.netlib.orglinpack
[25] See http://www.netlib.org/odepack
[26] See http://www.gmplib.org

Third, we could write our own programs completely from scratch in a high-level language. This is the approach used in this course. I have often opted *not* to use pre-written subroutine libraries, simply because this approach helps the reader to think for herself about scientific programming and numerical techniques. The reader should however realize that, in most cases, pre-written library routines offer solutions to numerical problems which are very hard to improve upon.

Although interpreted languages like PYTHON, which by now has replaced its older cousin PERL[27], can be used to solve computational problems, computational speed and the capability to write efficient code are topics which still do matter. Many other languages, such as ALGOL, PASCAL, MODULA, OBERON, or HASKEL can be dismissed as ephemeral computer science fads. Others, such as COBOL, LISP, or ADA, are too specialized[28] to adapt them for scientific use. The only serious remaining options are FORTRAN and C, for which compilers are available on all supercomputer architectures:

FORTRAN was the first high-level programming language to be developed. It was introduced in 1957. Before the advent of FORTRAN, all programming was done in machine code (only using zeroes and ones) and then in assembly, which used the hexadecimal number system and symbolic commands to summarize the machine code instructions. FORTRAN was specifically designed for scientific computing. Indeed, *all* computing in the early days, in the 1940s and 1950s, was scientific in nature; physicists and mathematicians were the original computer scientists. FORTRAN's main advantage is that it is very straightforward to program, and that it interfaces well with many commonly available, pre-written subroutine libraries, which are often compiled FORTRAN code. FORTRAN's main disadvantages are all associated with its antiquity. For instance, FORTRAN's control statements are very rudimentary and its I/O facilities are truly paleolithic. FORTRAN 77 and in particular FORTRAN 90 were major extensions to FORTRAN, which did away with some of the many objectionable features and introduced modern concepts, such as dynamic memory allocation.

C was developed at the Bell Telephone Laboratories, between 1969 and 1973, from the older language B, in the development of the UNIX operating system. C is one of the most widely used languages of all time. There are few computer architectures for which no C compiler exists. C has good control statements and excellent I/O capabilities. C's main advantage may be that it was not specifically written to be a scientific language, so in its original version there was, e.g., no built-in support for complex numbers[29]. C incorporates very low-level features such as pointers and bit

---

[27] PERL was originally developed by Larry Wall in 1987 as a general-purpose procedural UNIX scripting language to make report processing easier, which is the reason why PERL was very popular for UNIX administration during the 1980s and 90s.

[28] See Section 1.4 for a discussion of many different computer languages in a historical context.

[29] Built-in support for complex numbers has been included in the *C99* standard.

operations, which allow a good programmer to get very close to the physical hardware. The low-level features of C sometimes make scientific programming a little more complicated than necessary, which facilitates programming errors. On the other hand, these features allow scientific programmers to write extremely efficient code. Since, generally speaking, efficiency is the most important concern of scientific computing, the low-level features of C are, on balance, advantageous.

C++ is a major extension of C whose main focus is to facilitate programming and to manage very large software packages by introducing object-oriented features. Object-orientation is a completely different approach to programming than the traditional approach of procedural languages. C++ is well suited for very large software projects involving millions of lines of code and hundreds or thousands of programmers who are all working on different segments of the same code. For scientific computation, object orientation usually represents a large, and somewhat unnecessary overhead with respect to the straightforward, single or few-user programming tasks considered in this monograph. This also represents the more typical situation in science and engineering. I mention, however, that C++ includes several non-object-oriented extensions to C (for example, I/O handling) which are extremely useful. Actually, many MD programs freely available on the world wide web, e.g. LAMMPS,[30] are written in C++ without really taking advantage of a truly object-oriented code design. Rather, they are simple C, but make use of the methods, classes and libraries available in C++.

Some remarks on C and C++:

- There are many languages newer than C or C++ but both are still very popular, because programmers stick with good things they know well.

- When one executes a program written in C/C++ it's not C/C++ anymore, but is all 1s and 0s (machine code).

- Pure C, in contrast to C++, is a *procedural-oriented, verb-oriented*[31] or *imperative* language. Thus, the language C follows the *imperative* or *procedural paradigm*.

- With C++, the first thing you usually see, is the data, there is a focus on data organization, not on functions. Hence, C++ follows the *object-oriented paradigm*.

To summarize, of the above languages, we can rule out C++, because object orientation really is an unnecessary complication when taking into account the simplicity of the data structures usually involved in Finite Element Method (FEM)[32], Smooth Particle Hydrodynamics (SPH)[33] or MD codes used in scientific computing. The remaining options would be FORTRAN and C. Many old books on molecular dy-

---

[30] See http://lammps.sandia.gov
[31] That is because what you usually see first, is a function name describing what the function does (usually a verb).
[32] Finite Element Method.
[33] Smooth Particle Hydrodynamics.

namics [24, 348], Monte Carlo [170] and finite element simulations that include code snippets or even complete programming samples use FORTRAN: they face a very uncertain future, since FORTRAN as a programming language, taught in engineering or science undergraduate programs in Germany and the USA[34], has practically disappeared. So, how attractive can a book that uses a dead language for teaching be?

> In this book, we will use C as a high-level language to implement algorithms and write programs. Occasionally we will also use C++, without making particular use of its specific object-oriented features. In this book, C and C++ will be considered to be the same language. [35]

I have chosen to use C (and C++ only in a few examples, where C++ provides easier to understand code) for the programing examples in this book, simply because I find all the archaic features[36] and awkward control structures of FORTRAN too embarrassing to teach this computer language to students in the 21st century. I think it is probably safe to say, that hardly anyone knowledgeable in computer science and software engineering who intended to start a *new* scientific software project would even think for a second about employing FORTRAN as a computer language. To provide further support for this statement, I provide in Table 1.4 a May-2012 snapshot of ongoing open source software projects listed in http://www.freecode.com. This conveys the present relative importance of various languages.

1.3 Outfitting your PC for scientific computing

To really understand computer simulation in physics and engineering, one has to work with the code. This means reading it, modifying it, and sometimes writing it from scratch. It also means compiling code, running it, and analyzing the results, often using programs that also have to be written by you, for example, to convert the format

---

[34] FORTRAN still survives in Physics and Chemistry departments because of the large amounts of legacy code. It is still used particularly among "technical" programmers, who learned their programming skills in the 1970s and 1980s.

[35] Of course we know that this is not really true. What is meant by this statement is, that sometimes we will use C++ just to make use of some its additional features that are more convenient to use than in C, for example, input/output handling or program libraries. However, we will not make use of object-oriented programming design for reasons outlined in this section. In fact, this is what a lot of scientific C++ programmers actually do. They program in C and just make use of the objects available to them through the C++ syntax, but don't really do object-oriented design of scientific code, because in most cases this is simply overdone and unnecessary.

[36] For example, FORTRAN 77 completely lacks any modular software design and dynamic memory allocation. This is the reason why, for example, in an LS-DYNA simulation run, the program used to restart several times until the initially allocated memory for all fields and variables is large enough for the program to run, which is simply ridiculous.

Table 1.4. Snapshot of ongoing open source projects listed in `http://www.freecode.com`. Among the listed languages there are some quite exotic ones and I have highlighted the 5 most often used languages in bold.

language	number of projects	percentage
Ada	83	0.199
ABAP	3	0.007
ASP	77	0.185
Assembly	65	0.156
Awk	77	0.185
BASIC	40	0.960
C	**10 046**	**24.09**
C#	413	0.990
C++	**5 687**	**13.64**
Cold Fusion	22	0.053
Common Lisp	69	0.165
D	18	0.043
Delphi	75	0.180
Eiffel	31	0.074
Emacs-Lisp	33	0.079
Euphoria	2	0.005
Forth	29	0.070
FORTRAN 77	1	0.002
FORTRAN	29	0.070
Haskell	88	0.211
JAVA	**6 441**	**15.45**
JavaScript	1 635	3.921
Lisp	114	0.273
Logo	7	0.017
MATLAB	7	0.017
MAPLE	1	0.002
ML	31	0.074
MODULA-2	2	0.005
Object-PASCAL	7	0.017
Objective C	353	0.850
Octave	6	0.002
Owl	5	0.012
PASCAL	72	0.173
PERL	**4 309**	**10.333**
PHP	**5 393**	**12.933**
PROGRESS	2	0.005
PROLOG	21	0.050

Table 1.4. Snapshot of ongoing open source projects – continued from previous page.

language	number of projects	percentage
Pico	1	0.002
PLC	5	0.012
PLM	1	0.002
psql	1	0.002
PYTHON	4 176	10.014
QML	5	0.012
R	6	0.014
RDQL	2	0.005
Rexx	10	0.024
Ruby	644	1.544
Scheme	76	0.182
Scilab	4	0.010
sed	6	0.014
SGML	23	0.055
shell script	54	0.129
Smalltalk	24	0.056
SQL	741	1.780
Tcl	506	1.213
Visual Basic	42	0.101
YACC	31	0.074
Zope	48	0.115
Total Projects	**41 700**	

of generated data for use with freely available plotting and visualization tools such as *xmgrace* [4], *gnuplot* [6], *vmd* [9], *ovito* [10] or *visit* [5]. You will profit most from the program examples in this book when you have a computer system available where you can implement, compile and run the programs yourself, for example, a PC or notebook with either Linux or UNIX installed on it.

There is probably no estimate for such a number, but nearly 100% of serious research in scientific computing is conducted and tested under UNIX/Linux operating systems[37]. If need be, you can also work with a Windows operating system and use a standard commercial tool such as Visual Studio. For Windows systems however, I recommend installing the freely available Cygwin[38] instead, which provides you with a Linux-like (POSIX) environment on top of Windows.

---

[37] This is definitely true for supercomputing applications which are virtually *only* based on UNIX with a strong shift to Linux in the last decade. Linux is a completely free operating system that has essentially the same power, reliability, multi-user/multi-tasking features and file organization as a proprietary commercial UNIX system.

[38] http://www.cygwin.org

If you have a running Linux or UNIX system, you are likely ready to edit, compile and run scientific C/C++ code, so you need not do anything more. Here, I provide, in a nutshell, what you can do to set-up your hardware (PC or laptop) for running simulation code provided in this book, only using *freely available* options.

Install a free Linux distribution according to your taste. Currently, most popular are probably Ubuntu[39], SuSE Linux[40] or Fedora[41]. You will find installation instructions for all of these distributions on the corresponding websites. Hence, we will not go into any details here, except for noting that, sometimes, considerable experience with the Linux/UNIX environment is necessary to really get things running. On the other hand, in recent years, many distributions have become very user-friendly and provide installation packages that take care of (almost) all administrative tasks necessary to obtain a running native Linux system. The different Linux distributions are all based on the same Linux kernel (probably using different versions of it), and their major differences are usually the packaged software and user comfort when performing administrative tasks.

You may also consider installing a Linux operating system as an additional *virtual machine* on your Windows PC, using the freely available Vmware Player[42] and thus have a native Linux system running in a virtual machine on top of Windows. For good performance, this requires a PC or laptop well-equipped in terms of hardware.

Install Cygwin on Windows which is freely available. This provides a Linux-like environment on top of Windows. It is relatively small, easy to manage, and can be simply uninstalled later, if you wish, like any other Windows application. It will give you almost all the functionality of a full Linux PC right in windows, and it is certainly adequate for implementing the examples in this book. Cygwin is a port of POSIX[43] compatible software to the Windows system. Cygwin has been designed as a dynamic library which translates from the POSIX-API to the Windows API. Besides a number of developer tools, there is a variety of software packages available which can be run under windows, e.g. a full X-Windows server, which supports graphic applications from the UNIX/Linux world. Since I feel that some of the readers might opt for installing Cygwin, because it requires minimal effort for a spectacular amount of functionality, I provide a very brief description of the download and install procedure in Appendix B on Page 445.

---

[39] http://www.ubuntu.com
[40] http://software.opensuse.org
[41] http://fedoraproject.org/en/get-fedora
[42] http://downloads.vmware.com/d
[43] POSIX is a UNIX standard for commands and parameters.

1.4 History of computing in a nutshell

The quotation of Thomas Watson's 1943 statement at the very beginning of this chapter is usually taken as a typical example of the complete failure, even of the leading pioneers of the first era of electronic computers in the 1940s, to foresee the impact computers would have on society and on the way life is organized in the future. Sure enough, computers by today have thoroughly changed the way science and engineering is done. One may excuse Watson's statement in light of the fact that one of the very first calculators built in the United States, the Mark I weighed 5t and was 15m long, see Figure 1.4.

Surprisingly, the history of computing began about 3000 years ago with the Chinese abacus, a wooden frame with beads sliding on parallel wires, each wire representing one digit. In the following, we list some further milestones in computing from 1500 until 1930:

- A. Riese (1492–1559): Introduction of the decimal system. Adam Riese publishes a mathematics book in which he describes the Indian decimal system which then held sway in Europe. This facilitated the use of automated calculations.

- W. Schickard (1592–1635): First calculator. In 1623 Wilhelm Schickard constructed a machine for his friend Johannes Kepler which was able to perform addition, subtraction, multiplication and division. However, this machine did not attract general interest.

- C. Babbage (1792–1871): Analytic engine. In 1838, Charles Babbage developed the principle of his "analytic engine", which could perform all sorts of calculations. In essence, he had developed the principle of a programmable computer but, due to the limited technical possibilities of his era, the machine was never fully built. Babbage is generally considered the "father of computers" because his draft of a calculator already contained all basic elements of the *von Neumann architecture*[44] utilized in modern computers, see Figure 1.5.

---

[44] In 1945, John von Neumann wrote an incomplete 101-page document entitled "The First Draft of a Report on the EDVAC" (commonly shortened to "First Draft"). This document was distributed on June 30, 1945 by Herman Goldstine, security officer on the classified ENIAC project. The draft contains the first published description of the logical design of a computer, using the stored-program concept, which controversially has come to be known as the "von Neumann architecture". The treatment of the preliminary report as a publication (in the legal sense) was the source of bitter acrimony between factions of the EDVAC design team for two reasons. First, publication amounted to a public disclosure that prevented the EDVAC from being patented; second, some on the EDVAC design team contended that the stored-program concept had evolved out of meetings at the University of Pennsylvania's Moore School of Electrical Engineering predating von Neumann's activity as a consultant there, and that much of the work represented in the First Draft was no more than a translation of the discussed concepts into the language of formal logic in which von Neumann was fluent. Hence, failure of von Neumann and Goldstine to list others as authors on the First Draft led credit to be attributed to von Neumann alone.

Figure 1.4. MARK I computer built in 1944 by Howard H. Aiken in cooperation with IBM and Harvard University. Photo courtesy of IBM.

- H. Hollerith (1860–1929): Inventor of the punch card. Hermann Hollerith invented and constructed a machine for the evaluation of punch cards, used for a population census in the year 1890. The punch cards used for this purpose can be seen as ancestors of the punch cards of the late 1960s and 70s, that were used as a program storage medium on IBM mainframe computers, see Figure 1.6. In 1896 Hollerith founded the *Tabulating Machine Company* (TMC). Later TMC was merged with the *Computing Scale Corporation* and the *International Time Recording Company* to form the *Computing Tabulating Company* (CTR). In 1924 finally, CTR was renamed to *International Business Machines Corporation* (IBM).

The origins of modern computer systems, i.e. systems that are not based on purely mechanical devices, lie in the development of the *first functional programmable computer* Z3 by the mechanical engineer Konrad Zuse in 1941 in Germany. This computer was based on an *electromechanical switch* – the relay – which at that time was known from telephony. This switch was used for performing calculations and is crucial for the speed of a computer. The Z3 used 600 relays for the CPU and roughly 2000 relays for 64 memory cells with 22 bits each. The computer had a clock speed[45] of 5 Hz, weighed about 500kg and could perform 20 additions per second and one multiplication in roughly 3 seconds. In the 1940s Zuse also developed the first ever universal

---

[45] The number of pulses per second.

Section 1.4 History of computing in a nutshell

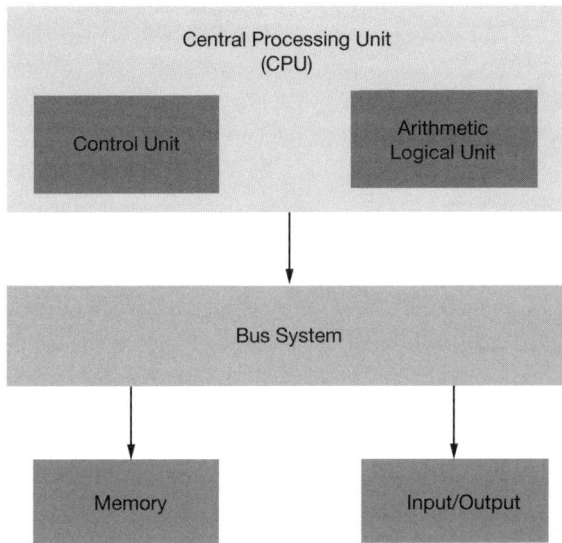

Figure 1.5. Basic computer architecture. Almost all modern computers today are designed using the *von Neumann architecture* of 1954. In the von Neumann architecture, the computer is divided into a Central Processing Unit (CPU), and memory, connected to the CPU, with a bus system that transfers data in both directions. The CPU contains all the computational power of the system and is usually split into the control unit, which manages and schedules all machine instructions, and the arithmetic logical unit (ALU) which performs the actual arithmetic operations $(+, -, :, \times)$. All arithmetic in the ALU of a computer are reduced to addition operations stored in a special *accumulating register*. Von Neumann's innovation was to use memory to store both the program instructions and the program's data. Before that time, "programming" meant, that cables were put together in the hardware. The "program" was not part of memory. The instructions that constitute a program are laid out in consecutive words in memory, ready to be executed in order. The CPU runs in a "fetch-execute" cycle where it retrieves and executes program instructions from memory. The CPU executes the current instruction, then fetches and executes the next instruction, and so on.

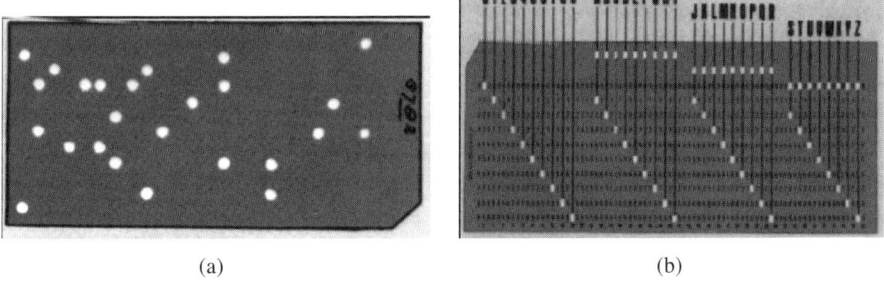

Figure 1.6. Computer Punch Cards dating from roughly 120 years ago and from the 1960s. (a) Hollerith's punch card from the 11th US population census of 1890. (b) Computer punch card of IBM from the 1960s. Photos courtesy of IBM.

Figure 1.7. Programming panels and cables of the ENIAC on the left. To reprogram the ENIAC one had to rearrange the patch cords, which can be seen on the left, and the settings of 3 000 switches at the function tables, which can be seen on the right. The very first problem run on the ENIAC concerned the feasibility of the hydrogen bomb. After processing half a million punch cards for six weeks, it needed only 20 seconds for the computations. This very first ENIAC program remains classified even today. US Army Photo.

programming language which he called "Plankalkül". This language could, however, not be implemented on the Z3.

More well known than Zuse's Z3 is the American ENIAC (*E*lectronic *N*umerical *I*ntegrator *A*nd *C*omputer), which was developed during war time at the Los Alamos Laboratories (founded in 1943) and which started to operate in 1946. It was the *first real electronic* calculator not based on mechanical devices (relays). The ENIAC was used for military calculations of ballistic curves and for hydrodynamics calculations of shock waves in the context of thermonuclear reactions. The ENIAC weighed 30 tons, used 17, 486 vacuum tubes as switches (thus constituting the "*first generation*" of electronic computers) and computed 1, 000 times faster than its electro-mechanical competitors. It was programmed by plugging cables and wires and setting switches, using a huge plugged that was distributed over the entire machine, see Figure 1.7. Hence, being a programmer in that era mostly meant being a technician. ENIAC's basic clock speed was 100, 000 cycles per second. For comparison, a typical PC today employs clock speeds of at least 2, 000, 000, 000 cycles per second.

Section 1.4 History of computing in a nutshell

(a)

(b)

Figure 1.8. First generation computers from the early 1950s. (a) The MANIAC computer, developed by John von Neumann and completed in 1952, was responsible for the calculations of "Mike", the first hydrogen-bomb. Von Neumann also helped Ulam and Metropolis develop new means of computing on such machines, including the Monte Carlo method, which has found widespread application and is today considered among the top 10 algorithms of the 20th century [100]. (b) The first commercial (non-military) computer – UNIVAC – built in the United States. UNIVAC was designed by J. Presper Eckert and John Mauchly, the designers of the ENIAC. The machine was 7.62m by 15.24m in length, contained 5,600 tubes, 18,000 crystal diodes, and 300 relays. It utilized serial circuitry, 2.25MHz bit rate, and had an internal storage capacity of 1,000 words, or 12,000 characters. The UNIVAC was used for general purpose computing with large amounts of input and output and also was the first computer to use buffer memory and to come equipped with a magnetic tape unit. It had a reported processing speed of 0.525 milliseconds for arithmetic functions, 2.15 milliseconds for multiplication, and 3.9 milliseconds for division.

Another famous first generation (vacuum tube) computer was the MANIAC[46] (*M*athematical *A*nalyzer *N*umerical *I*ntegrator *A*nd *C*omputer), which was developed by John von Neumann and built under the direction of Nicholas Metropolis who went on to use the MANIAC for the very first Monte-Carlo simulation (see Chapter 6) in 1952 [298, 30, 29, 297], although the method itself had already been published [300] and presented on a conference in 1950 [418]. On this computer, the expression "bug" was coined for anything that goes wrong in a computer program. The first computer bug was actually a moth causing a shunt to fault. Also in this era, in 1951 the first *non-military* computer, the UNIVAC (*UNI*versal *A*utomatic *C*omputer) was built, see Figure 1.8.

---

[46] Metropolis chose the name "MANIAC" in the hope of stopping the rash of silly acronyms for machine names [299]. The MANIAC was succeeded by MANIAC II in 1957 and by a third version MANIAC III in 1964 which was built at the Institute for Computer Research at the University of Chicago.

In the 1950s, the *second generation* of computers emerged, due to the invention of the transistor (a semiconductor), which could be used as a new computer switch for performing calculations. The first transistor computer ever built was the 48-bit *CDC 1604* constructed by Seymour Cray using 60,000 transistors. The transistor switch was 100 times faster than vacuum tubes.

At the beginning of the 1960s the *third generation* of electronic computers emerged, with the invention of yet another new computer switch, the integrated circuit (chip)[47]. A chip combines transistors and resistors on *one single* component. A major advantage is that chips can be mass-produced. With mass production of chips the computer market in the 1960s was basically split into computers used *commercially* for data processing and "fast" computers used for *science and research*. The first high-level languages were developed (COBOL for business applications and FORTRAN for science applications)[48] for both areas of application. Since then, the fastest computers of any era are called "supercomputers" and there has been a website keeping track of the latest developments in the field of supercomputing[49] since the mid 1990s.

The *fourth generation* of computers pertains to miniaturization of microchip components[50]. This development in miniaturization finally led to the use of microchips in personal computers (PCs) in the early 1980s.

In 1981 the "International Conference on 5th Generation Computer Systems" [308] tried to establish new ideas for the future, next generation of computers but could not really establish a clear border between 4th and 5th generation computers. The easiest way to delimit 5th generation computers is based on computer architecture and data processing by assigning all *parallel computers* with many CPUs to the 5th generation[51].

1.5 Number representation: bits and bytes in computer memory

In this section we will take a glimpse into the inner workings of a computer. The goal of this exercise is to show you the basics of how the computer represents numbers in memory, which will help you better understand how language features (not only in pure C or C++) actually work. Also – in the long run – this knowledge will make you a much better scientific programmer.

Almost all elementary counting systems are based on counting with fingers, which is why the decimal system is used for counting purposes almost everywhere. The

---

[47] The integrated circuit was invented by Jack Kilby in 1958 at Texas Instruments during summer, when he was alone in the labs, because – as a starter – he could not take a summer vacation in his first year.
[48] Compare also Tables 1.1 and 1.2.
[49] http://www.top500.org
[50] VLSI (Very Large Scale Integration) and Ultra Large Scale Integration (ULSI).
[51] Hence, with most computers manufactured today having at least two cores and multithreading built in, virtually all of them belong to the 5th generation of computers.

Section 1.5 Number representation: bits and bytes in computer memory

decimal system is a position system with basis ten, so for each digit you need ten figures, which is difficult to achieve with an electronic computer. Number systems have been constructed differently in the past. A number system with basis B is a number system in which a number x is expressed as a power series of B:

$$n = \sum_{i=0}^{N-1} b_i B^i, \tag{1.1}$$

where B is the basis of the number system ($B \in \mathbb{N}$, $B \geq 2$), b is the number of available figures ($b_i \in \mathbb{N}$, $0 \leq b_i < B$) and N is the number of digits. Depending on the value of B you get, e.g. the *dual* (or *binary*) *system* ($B = 2$), the *octal system* ($B = 8$), the *decimal system* ($B = 10$) or the *hexadecimal system* ($B = 16$).

The smallest unit of memory in a computer is the "bit". A bit can be in one of two states – on vs. off, or alternately, 1 vs. 0. This is the reason for the use of the dual system in computers. Technically any object that can switch between two distinct states can remember one bit of information. Almost all computers use little transistor circuits called "flip-flops" to store bits. The flip-flop circuit has the property that it can be set to be in one of two states, and will stay in that state and can be read until it is reset. Most computers do not work with bits individually, but instead group eight bits together to form a "byte", which at some point in computing history was commonly agreed upon as the basic unit of digital information. At the beginning of computing history, the size of the byte was hardware dependent and no definitive standards existed that dictated the size.

Each byte maintains one eight-bit pattern. A group of N bits can be arranged in 2^N different patterns. So a byte can hold $2^8 = 256$ different patterns. The memory system as a whole is organized as a large array of bytes, called its address space. Every byte has its own "address" which is like its index in the array. Strictly speaking, a program can interpret a bit pattern any way it chooses. By far the most common interpretation is to consider the bit pattern to represent a number written in base 2. In this case, the 256 patterns a byte can hold map to the numbers $0, \ldots, 255$. The CPU can retrieve or set the value of any byte in memory and it identifies each byte by its address.

The byte is sometimes defined as the "smallest addressable unit" of memory. Most computers also support reading and writing of larger units of memory, for example, 2 byte "half-words", see Figure 1.9, (in computer languages often called a "short" word) and 4 byte "words", often known as "*int*" or "*float*" words. This is the reason why the *hexadecimal system* (with the numbers {1, 2, 3, 4, 5, 6, 7, 8, 9, A, B, C, D, E, F}) also is of great importance for computer science. In fact, in many computer languages there are prefixes for representing hex numbers[52].

Both half-words and words span consecutive bytes in memory. By convention the address of any multiple-byte entity is the address of its lowest byte – its "base-

---

[52] For example in C/C++: "0x", in PASCAL: "$", or in BASIC: "&".

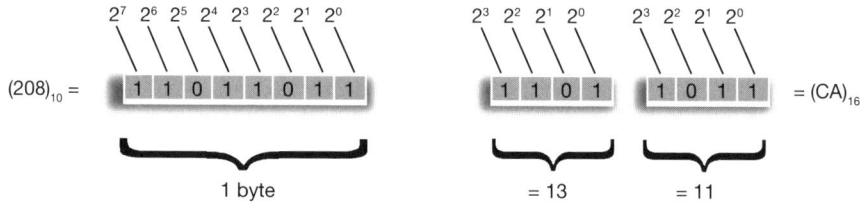

Figure 1.9. Representation of a decimal number $(208)_{10}$ as a dual $(11011011)_2$ and hexadecimal $(CA)_{16}$ number.

address". So the 4-byte word at address 600 is composed of bytes 600, 601, 602, and 603. Most computers restrict half-word and word accesses to be "aligned", i.e. a half-word must start at an even address and a word must start at an address that is a multiple of 4.

Most programming languages shield the programmer from the detail of bytes and addresses. Instead, programming languages provide the abstractions of *variable* and *type* for the programmer to manipulate. In the simplest scheme, a variable is implemented in the computer as a collection of bytes in memory. The type of the variable determines the number of bytes required.

> Why do high-level-programming languages provide different data types?

> On the machine level, there are no data types but only bits grouped in bytes. The concept of "data type" is an abstraction level provided by high-level languages to help make programming easier and more intuitive for human beings. Different data types arise by *interpreting* the *same* bit patterns in different ways for different purposes.

1.5.1 Addition and subtraction of dual integer numbers

The rules for adding dual numbers are the following:

$$0 + 0 = 0 \qquad (1.2a)$$
$$0 + 1 = 1 \qquad (1.2b)$$
$$1 + 0 = 1 \qquad (1.2c)$$

$$1 + 1 = 0 \quad \text{carryover 1} \tag{1.2d}$$
$$1 + 1 + 1 \,(\text{carryover}) = 1 \quad \text{carryover 1} \tag{1.2e}$$

Example 1.1 (Addition in the dual system). Adding the decimal numbers $(45)_{10}$ and $(54)_{10}$ as dual numbers according to the rules laid out in equation (1.2) above, yields:

$$\begin{aligned} (45)_{10} &= (00101101)_2 \\ + (54)_{10} &= (00110110)_2 \\ \hline = (99)_{10} &= (01100011)_2 \end{aligned}$$

The carried over 1's at each step of this addition are (00111100).

Negative decimal numbers are commonly represented by their absolute value preceded by a minus sign. One way to realize this with binary numbers for use in machines is to use the *first bit* as *sign bit*, where 0 means "positive" and 1 means "negative".

Example 1.2 (Subtraction of two 2 byte dual numbers.). The arithmetic logic unit of a von Neumann computer can only execute additions and not subtractions, so the way to represent subtractions is to map them to the addition of a negative number. In this example we add the two numbers $(+7)_{10}$ and $(-7)_{10}$ in the binary system which, of course, ought to add up to zero. However, when we adhere to the idea of representing negative numbers by just using a sign bit, and apply the rules for addition laid out in equation (1.2), we have:

$$\begin{aligned} (+7)_{10} &= (00000000 \mid 00000111)_2 \\ +(-7)_{10} &= (10000000 \mid 00000111)_2 \\ \hline = (0)_{10} &\neq (10000000 \mid 00001110)_2 = -(14)_{10} \end{aligned}$$

Thus, simply adding up bit patterns this way wrecks simple addition rules in that the result of adding a negative number $-a$ to the same positive number $+a$ does not result in zero. This way of representing negative numbers is therefore *not* realized in computers, because it wrecks simple rules for addition.

Think!

Negative numbers, as represented inside of a computer have to be *engineered*, such that addition follows simple rules. The question now is which bit pattern – interpreted as negative number – we have to add to the bit pattern of e.g. $(+7)_{10}$ to obtain zero?

$$(00000000 \mid 00000111)_2 = \quad (+7)_{10}$$
$$+? \quad +(-7)_{10}$$
$$= (00000000 \mid 00000000)_2 = \quad (0)_{10}$$

It is a little easier to first figure out what you have to do to get all 1's as a result of an addition of two duals, i.e.:

$$(00000000 \mid 00000111)_2 = +(7)_{10}$$
$$+ (11111111 \mid 11111000)_2$$
$$= (11111111 \mid 11111111)_2$$

Hence, to get all 1's as a result of addition we have to *reverse the bit pattern*. This automatically also reverses the first bit, i.e. it changes sign, if we take the first bit as sign bit. Consequently, we can interpret the above bit pattern $(11111111 \mid 11111000)_2$ as $(-7)_{10}$ and the resulting bit pattern $(11111111 \mid 11111000)_2$ as $(-0)_{10}$. Technically, this interpretation of a negative number is called "*one's complement*". One's complement thus leads to a *symmetric bit representation* of numbers with two representations of zero, as shown in equation (1.3) for 4 bit numbers.

$$(0000)_2 = (+0)_{10} \quad \mid \quad (1111)_2 = (-0)_{10} \quad (1.3a)$$
$$(0001)_2 = (+1)_{10} \quad \mid \quad (1110)_2 = (-1)_{10} \quad (1.3b)$$
$$(0010)_2 = (+2)_{10} \quad \mid \quad (1101)_2 = (-2)_{10} \quad (1.3c)$$
$$(0011)_2 = (+3)_{10} \quad \mid \quad (1100)_2 = (-3)_{10} \quad (1.3d)$$
$$(0100)_2 = (+4)_{10} \quad \mid \quad (1011)_2 = (-4)_{10} \quad (1.3e)$$
$$(0101)_2 = (+5)_{10} \quad \mid \quad (1010)_2 = (-5)_{10} \quad (1.3f)$$
$$(0110)_2 = (+6)_{10} \quad \mid \quad (1001)_2 = (-6)_{10} \quad (1.3g)$$
$$(0111)_2 = (+7)_{10} \quad \mid \quad (1000)_2 = (-7)_{10} \quad (1.3h)$$

Section 1.5 Number representation: bits and bytes in computer memory

> What number do you have to add to a dual number that is all 1's, e.g. to $(11111111 \mid 11111111)_2$ to get zero as a result?

> You simply have to add $(1)_{10} = (00000000 \mid 0000001)_2$.

All carried-over 1's in the addition will automatically generate all 0's. Thus,

$$(11111111 \mid 11111111)_2$$
$$+(00000000 \mid 00000001)_2$$
$$=(00000000 \mid 00000000)_2$$

The carried-over 1 that exceeds the size of the binary number is thrown away. Consequently, to represent a negative number, one first inverts the bit pattern and then adds one to the result. This way of representing negative numbers is called *"two's complement"* and this is how subtraction of dual integers is realized in computers. Consequently, in two's complement a binary number with all 1's represents $(-1)_{10}$ and there is only *one* representation of zero, namely $(00000000)_2$. equation (1.4) displays all possible bit combinations for a 4 bit dual number in two's complement.

$$(0000)_2 = (0)_{10} \mid \tag{1.4a}$$
$$(0001)_2 = (1)_{10} \mid (1000)_2 = (-1)_{10} \tag{1.4b}$$
$$(0010)_2 = (2)_{10} \mid (1001)_2 = (-2)_{10} \tag{1.4c}$$
$$(0011)_2 = (3)_{10} \mid (1010)_2 = (-3)_{10} \tag{1.4d}$$
$$(0100)_2 = (4)_{10} \mid (1011)_2 = (-4)_{10} \tag{1.4e}$$
$$(0101)_2 = (5)_{10} \mid (1100)_2 = (-5)_{10} \tag{1.4f}$$
$$(0110)_2 = (6)_{10} \mid (1101)_2 = (-6)_{10} \tag{1.4g}$$
$$(0111)_2 - (7)_{10} \mid (1110)_2 = (-7)_{10} \tag{1.4h}$$
$$\mid (1111)_2 = (-8)_{10} \tag{1.4i}$$

> Negative ("signed") integer numbers in computers are represented in *two's complement*, which is built according to the following rules:
>
> 1. If the first bit is 1, then this number is negative.
>
> 2. First, each bit is inverted and then 1 is added to this bit pattern.
>
> The smallest negative number that can be represented using the two's complement, in a number system with basis B and s digits, is $= -B^{s-1}$. The largest one is $B^{s-1} - 1$: thus, with $B = 2$ the intervals for different dual numbers that can be represented with s bits are:
>
> half-byte $s = 4$: $[-2^3 = -8, \quad 2^3 - 1 = 7]$
> byte $s = 8$: $[-2^7 = -128, \quad 2^7 - 1 = 127]$
> 2 bytes $s = 16$: $[-2^{15} = -32\,768, \quad 2^{15} - 1 = 32\,767]$
> 4 bytes $s = 32$: $[-2^{31} = -2\,147\,483\,648, \quad 2^{31} - 1 = 2\,147\,483\,647]$

Hence, with 4 bits (a half-byte) one can represent the numbers of the codomain $[-8, \ldots, 0, \ldots 7]$ and the pattern with which the bit combinations are distributed among the negative numbers can be represented by a representation of bits on a circle. In Figure 1.10 we display a 4 bit number where the first bit denotes sign.

The advantage of two's complement is that there is only *one* representation of zero and that a machine does not have to be able to subtract numbers. Instead, subtraction, multiplication and division can all be reduced to addition of positive and negative numbers in the ALU. Any carry-over of the largest bit of the sum has no influence on the correctness of the result of the addition, except if the result overflows the representable range of numbers, as illustrated in the following example:

Example 1.3. Bit overflow in a 4 bit number. With 4 bits one can represent the number range $[-8, \ldots, +7]$. What happens, if we add $(-7)_{10} + (-5)_{10}$?

Inverting the bit pattern of $(+7)_{10} = (0111)_2$ yields $(1000)_2$. Adding $(0001)_2$ to the inverted bit pattern finally yields $(1001)_2 = (-7)_2$, compare Figure 1.10. Analogously, $(-5)_{10} = (1011)_2$ and hence, we have

$$\begin{aligned}(-7)_{10} &= (1001)_2 \\ +(-5)_{10} &= (1011)_2 \\ \hline = (-12)_{10} &\neq (0100)_2 = (+4)_{10}\end{aligned}$$

because the overflow bit is thrown away. Thus, the result of this addition overflows the representable range of numbers.

Section 1.5 Number representation: bits and bytes in computer memory   29

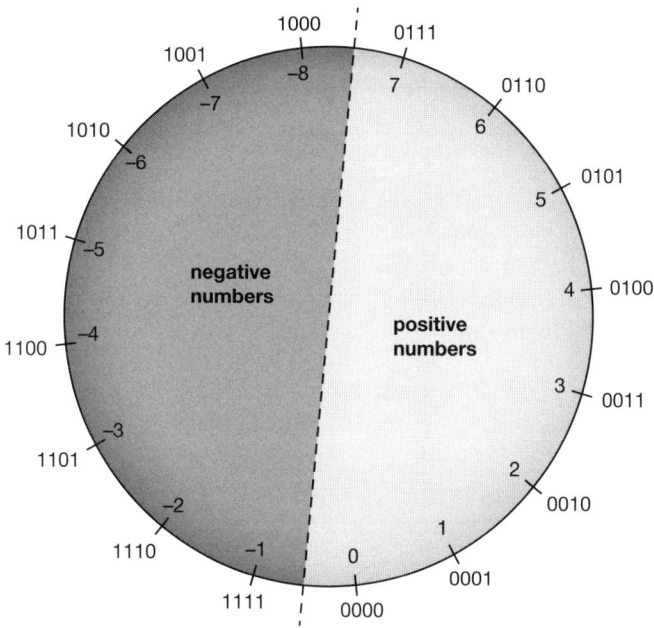

Figure 1.10. Distribution of bits for a 4 bit figure where the first bit denotes the sign. Note that the representation of numbers with the two's complement which is realized in computers is *asymmetric* because zero is represented by the bite pattern $(0000)_2$ alone. Representing numbers in one's complement is symmetric, but then zero is represented by *two* bit patterns, in the case of 4 bits by $(0000)_2$ and $(1111)_2$.

> In C/C++ there is no checking of number overflow. So, if an addition (or a different arithmetic operation) exceeds the representable range of numbers, the result will be incorrect.

1.5.2 Basic data types

In the following we discuss some basic data types and their typical sizes on most computer systems. The typical size of standard data types on most systems are listed in Table 1.5.

Characters

A character in C/C++ takes up one byte in memory. The ASCII code[53] defines 128 characters and a mapping of those characters to the numbers $0, \ldots, 127$. For example,

---

[53] *American Standard Code for Information Exchange* published for the first time in 1963.

Table 1.5. Typical ranges for C datataypes on 64 bit systems. The two most important data types, **int** and **double** are emphasized in bold.

Data type	Size [Bytes]	Range
char, signed char	1	$-128 \ldots 127$
unsigned char	1	$0 \ldots 255$
short, signed short	2	$-32\,768 \ldots 32\,767$
unsigned short	2	$0 \ldots 65\,535$
int, signed int	4	$-2\,147\,483\,648 \ldots 2\,147\,483\,647$
unsigned, unsigned int	4	$0 \ldots 4\,294\,967\,295$
long, signed long	4	$-2\,147\,483\,648 \ldots 2\,147\,483\,647$
unsigned long	4	$0 \ldots 4\,294\,967\,295$
float	4	$1.2 \times 10^{-38} \ldots 3.4 \times 10^{38}$
double	8	$1.2 \times 10^{-308} \ldots 3.4 \times 10^{308}$
long double	12	$3.4 \times 10^{-4932} \ldots 1.1 \times 10^{4932}$

the letter "A" is assigned the number 65 in the ASCII table. Expressed in binary, that is $2^6 + 2^0 = (64 + 1)$, so the byte that represents the letter "A" is:

$$(65)_{10} = (01000001)_2.$$

Since they only span the range $(0, \ldots, 127)$, all standard ASCII characters have a zero in the uppermost bit, called the "most significant" bit. Some computers use an extended character set, which adds characters like "é" and "ö" using the previously unused numbers in the range $(128, \ldots, 255)$. Other systems use the 8th bit to store parity information, so that, e.g., a modem can notice if a byte has been corrupted.

Short integer

A short integer in C/C++ takes up 2 bytes or 16 bits. 16 bits provide $2^{16} = 65\,536$ patterns. This number is known as "64k", where "1k" of something is $2^{10} = 1\,024$. For nonnegative, "unsigned" numbers these patterns map to the numbers $(0, \ldots, 65\,535)$. Consider, for example, the 2-byte short representing the value 65. It has the same binary bit pattern as the "A" above in the lowermost (or "least significant") byte and zeros in the most significant byte.

However, if a short occupies the 2 bytes at addresses 650 and 651, is the most significant byte at the lower or at the higher numbered address? Unfortunately, this is not standardized. Systems that are *big-endian* (PowerPC, Sparc, most RISC chips) store the most significant byte at the lower address, so 65 as a short would look like in Figure 1.11.

Section 1.5 Number representation: bits and bytes in computer memory   31

> **Character spelling in C.** There are several character encoding systems in use, but the two most widespread are the *ASCII* code and *UNICODE*. The *ASCII* code was introduced in 1963 and originally only used 7 bits of each byte. Thus, it encoded 128 characters, corresponding to the standard Latin alphabet and a few special characters. Later, the 8th bit of the byte was also used and since then the *ASCII* code contains 256 characters, among them also unprintable control characters.
>
> In 1991 the *UNICODE* consortium[a] was founded, to enable people to exchange information with computers in any language in the world. Initially, *UNICODE* was restricted to two bytes[b] (65 536 characters) but *UNICODE* Version 3.0, from September 1999, already listed 49 194 characters. In order to compile a complete collection of all written characters, from all present and past cultures, one had to change to a four-byte scheme. So far, about 10% of characters available with 32 bit schemes have been listed.
>
> [a] www.unicode.de
> [b] The two-byte scheme is called *Basic Multilingual Plane* (BMP).

Figure 1.11. Example of a big-endian system. The MSB is stored at the lower address. Thus, on such a system, the displayed byte pattern, interpreted as a number indeed represents the (decimal) integer 65.

A so-called *little-endian*[54] (Intel x86, Pentium) system arranges the bytes in the opposite order, which looks like in Figure 1.12. This means that, when exchanging data through files or over a network between different endian machines, because of the lack of a standard, there is often a substantial amount of "byte swapping" required to rearrange the data. To obtain *negative* numbers, there is a slightly different system, which interprets the patterns as the numbers $(-32\,768, \ldots, 32\,767)$, with one bit reserved for storing sign information. The "sign bit" is usually the most significant bit of the most significant byte.

---

[54] Both the big and little endian names originate from Gulliver's Travels. In the novel, there were two political factions, the "Big Endians" and the "Little Endians." The difference between the two was how they broke their eggs. The big endians choose to do so on the large end of the egg, while the little endians on the small end.

Figure 1.12. Example of a little-endian system, where the byte order is exchanged. Now, the MSB is stored at the upper address and the byte pattern – again interpreted as a number – represents the decimal number $2^{14} + 2^8 = 16\,640$.

> The *most significant bit* (MSB) is the bit in a multiple-bit binary number with the largest value. This is usually the leftmost bit, or the first bit in a transmitted sequence. For example, in the 4-bit binary number $(1000)_2$, the MSB is 1, and in the binary number $(0111)_2$, the MSB is 0. The most significant byte is the byte in a multiple-byte word with the largest value. As with bits, the most significant byte is normally the leftmost byte, or the first byte in a transmitted sequence.

> What happens when a *char* variable is assigned to a *short*?

Let's do a simple pseudocode example to see what happens in memory when defining and assigning variables. Let's do the following declarations and assignments:

```
char ch = 'A';
short s = ch;
print s;
```

What will be the output when you print the variable s? In Figure 1.13, we see that the bit pattern of the number 65 (interpreted as the character 'A') is simply copied into the short.

> What happens when the value of a *short* is assigned to a *char*-variable?

Next, we want to assign a *short* to a *char*-variable, as in the following pseudocode:

```
short s = 67;
char ch = s;

print ch;
```

Section 1.5 Number representation: bits and bytes in computer memory 33

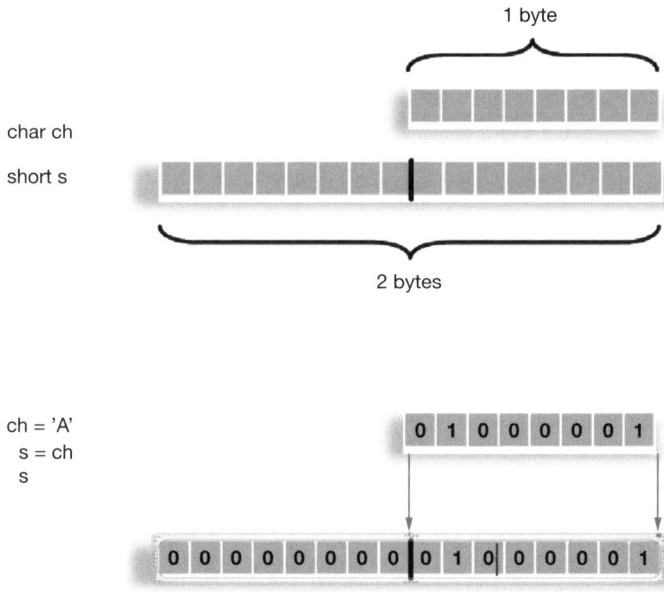

Figure 1.13. Bit pattern representation of *short* and *char*. When a character (1 byte) is assigned to a *short*, the bit pattern of the character 'A' which is 65 in ASCII-code, is simply copied into the two byte pattern of the *short*. The first byte of the two-byte pattern is simply filled with zeros.

> When assigning values of variables of different type, the corresponding bit pattern gets copied byte by byte.

In Figure 1.14 we provide further examples of copying bit patterns of variables by making assignments between different data types.

Long integer

A long integer in C/C++ takes up 4 bytes or 32 bits of memory on most systems. 32 bits provide $2^{32} = 4,294,967,296$ different patterns. Most programmers just remember this number as "about 4 billion". The signed representation can deal with numbers in the approximate range ± 2 billions. 4 bytes is the contemporary recommended default size for an integer, also denoted as a "word"[55]. The representation of

---

[55] So the *word length* on such a system is 4 bytes.

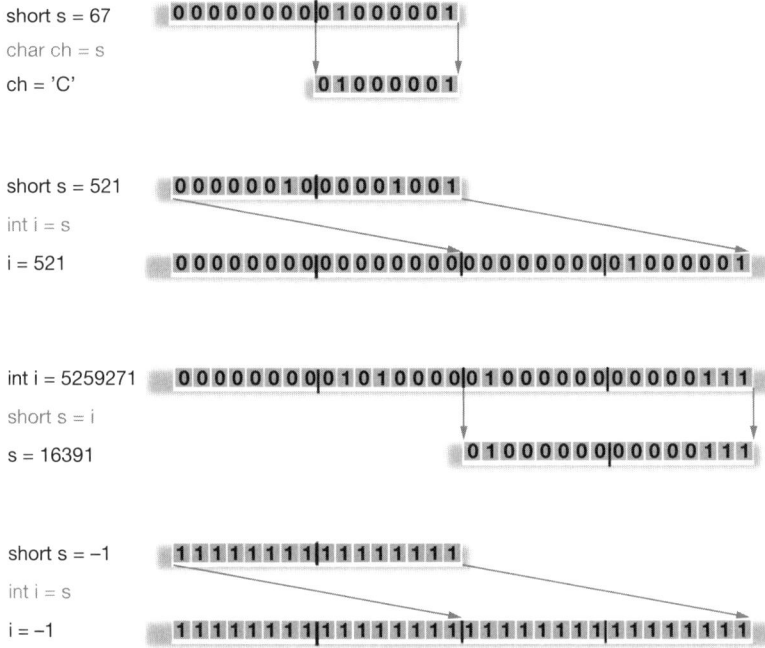

Figure 1.14. Bit pattern representation of *short*, *char* and *int*. From top to bottom: The bit pattern for the number 67 is simply copied into the memory space of variable ch, which is interpreted as *char*. Thus, when printing ch the result is the character 'C'. The *short* bit pattern of $2^{10} + 2^3 + 2^0 = 521$ gets copied bitwise into the first two bytes of i, which is interpreted as a four-byte *int*. Thus, the first two bytes are filled with zeros and printing i yields the same numerical value. When assigning the integer i= $2^{22} + 2^{20} + 2^{14} + 2^2 + 2^1 + 2^0 = 5\,259\,271$ to the *short* s, the first two bytes get lost in the assignment. Hence, when printing s, the result is the number $2^{14} + 7 = 16\,391$ which is not the same as the content of i. Finally, assigning the *short* s=-1 to an integer int i yields the same result, namely i=-1. The reason for this is that filling the first two bytes with zeros would lead to a change of sign (1 is the positive sign bit, 0 is the negative sign bit). Hence, the 1 is replicated in the bit pattern to conserve the sign. This is called *sign extension*.

a long is just like that of a short. On a big-endian machine, the four bytes are arranged in order of most significant to least and vice versa for a little-endian machine.

Floating point numbers

Almost all computers use the standard IEEE[56] representation of floating point numbers, which is a system much more complex than the scheme for integers, see Figure 1.15.

---

[56] *I*nstitute of *E*lectrical and *E*lectronic *E*ngineers (IEEE), pronounced "Eye-triple-E" (http://www.ieee.org).

Section 1.5 Number representation: bits and bytes in computer memory

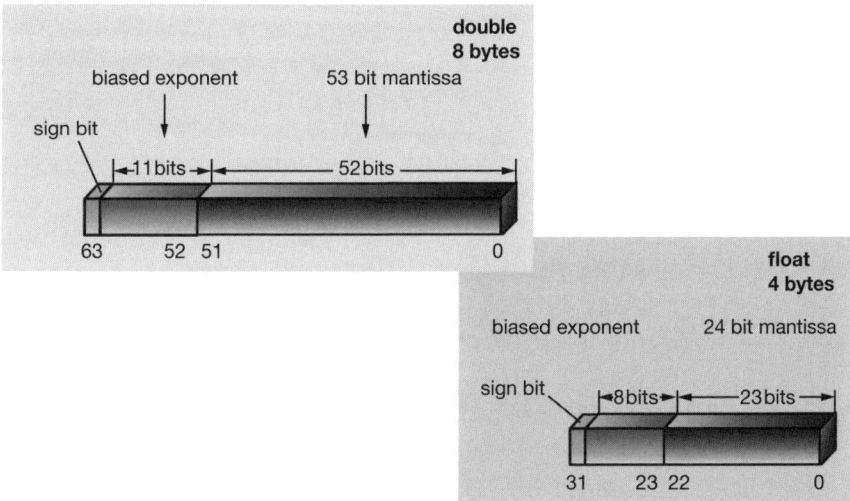

Figure 1.15. IEEE floating point number format for the JAVA and C/C++ data types *float* (4 bytes) and *double* (8 bytes).

Figure 1.16. Bit pattern of the floating point number 65.0.

The important thing to note here, is that the bit pattern for the floating point number 1.0 is not the same as the pattern for the integer 1. For example, 65 expressed as a floating point value 65.0 has the bit pattern displayed in Figure 1.16. Interpreted as a big-endian long, this pattern would be 1,079,001,088, which is not at all the same as the integer 65.

IEEE floats are represented in scientific notation: a 4-byte *float* uses 23 bits for the mantissa, 8 bits for the exponent, and 1 bit for the sign. It can represent values as large as 3×10^{38} and as small as 1×10^{-38} (both positive and negative). Clearly, there are many more floating point numbers in that range than the number of distinct patterns that can be represented with a 4-byte *float* (which is $\approx$ 4 billion), so floats are necessarily approximate values. A floating point value is usually only accurate up to a precision of about 6 decimal digits, and any digits after that must be considered suspicious. Using 8 bytes doubles the range up to around 10^{308} with numbers having a reliable precision of 15 digits.

> The IEEE bit pattern of Figure 1.16 tries to represent a floating point number as
>
> $$(\underbrace{-1)^S}_{\in[-1,+1]} \times 1.\underbrace{xxxxx...}_{\in[0,0.\bar{9}]} \times 2^{(\text{EXP}-\text{bias})}. \qquad (1.5)$$
>
> Thus, each floating point number is represented as a power series with base 2.
>
> - The bias is 127 for *float* and 1023 for *double*.
>
> - The exponent EXP is in the range $[-126, \ldots, 127]$ for *float* and $[-1022, \ldots, 1023]$ for *double*.
>
> - The sign bit S is either 0 or 1

> What happens, when the number 2.1 is represented as a *double* in the computer?

Let us work out two complete examples:

Example 1.4. Bit representation of 7.0. The *float* 7.0, written as a series of powers of 2, i.e. written as a binary number, is:

$$(7.0)_{10} = 1 \cdot 2^2 + 1 \cdot 2^1 + 1 \cdot 2^0 + 0 \cdot 2^{-1} = (111.0)_2 \cdot 2^0.$$

In the last notation $(111.0)_2 \cdot 2^0$, the comma separates positive and negative powers of 2 in the representation of a floating point number in the binary system. In normalized form (with shifted comma) this can be written as:

$$(7.0)_{10} = (1.11)_2 \cdot 2^2.$$

Comparing this expression with equation (1.5) in the info box above shows that EXP= $127 + 2 = 129$. Written in binary, this is $(129)_{10} = (10000001)_2$. Thus, the bit representation of 7.0 is:

$$(7.0)_{10} = (0 \mid \underbrace{10000001}_{\text{exponent}} \mid \underbrace{11000000000000000000000}_{\text{mantissa}})_2.$$

Example 1.5. Bit representation of 17.625 as *float*. 17.625 written as a series of powers of 2, i.e. as a dual number is:

$$(17.625)_{10} = 1 \cdot 2^4 + 0 \cdot 2^3 + 0 \cdot 2^2 + 0 \cdot 2^1 + 1 \cdot 2^0 + \underbrace{1 \cdot 2^{-1}}_{=\frac{1}{2}=0.5} + 0 \cdot 2^{-2} + \underbrace{1 \cdot 2^{-3}}_{=\frac{1}{8}=0.125}$$

$$= (10001.101)_2 \cdot 2^0.$$

Section 1.5 Number representation: bits and bytes in computer memory

In normalized form this is:

$$(17.625)_{10} = (1.0001101)_2 \cdot 2^4.$$

Comparing this expression again with equation (1.5) above shows that EXP= $127 + 4 = 131$. Written in binary, this is $(131)_{10} = (10000011)_2$. Thus, the bit representation of 17.625 is:

$$(2.1)_{10} = (0 \mid 10000011 \mid 0001101000000000000000)_2.$$

> Representing 2.1 as a *double* number in the computer leads to the bit pattern
>
> $$(2.1)_{10} = (0 \mid 10000000 \mid 0000110011001100110011001100110)_2.$$

Hence, the decimal number 2.1 cannot be exactly represented with a bit pattern. In Listing 1 we provide a very simple C-program, which allows us to verify the limited accuracy of the representation of the number 2.1 as a bit pattern, interpreted as IEEE *float* or *double*[57]. When you compile and execute this program listing, the output will

Listing 1. Accuracy of *float* and *double*. (*)

```
1  /* Check accuracy of bit representation of 2.1 */
2  #include <stdio.h>
3
4  int main(void)
5  {
6     float  f;
7     double d;
8
9     f = d = 2.1;
10    printf("Float %.40f \nDouble %.40f\n",f,d);
11
12    return 0;
13 }
```

be:

Float 2.0999999046325683593750000000000000000000
Double 2.1000000000000000888178419700125232338905

We find that the representation of a floating point number as *float* is not sufficient for the purpose of computer simulations. This is the reason, why one actually *only* uses doubles to represent floating point numbers in scientific applications. Floats are

---
[57] For an introduction in basic C syntax to help explain this program listing, see Chapter 2.

practically never used due to their limited accuracy, which would soon lead to considerable rounding errors when, for example, integrating the Equations of Motion (EOM) using millions of time steps in a typical MD simulation. In computer simulations, it is important to be aware of such possible sources of errors. We will return to this point when we present a few explicit source code examples in the next section on Page 38.

> To design a good algorithm, one needs to have a basic understanding of propagation of inaccuracies and errors involved in calculations. There is no magic recipe for dealing with underflow, overflow, accumulation of errors, and loss of precision; only a careful analysis of the functions involved can save one from serious problems.

Floating point operations usually are considerably slower than the corresponding integer operations. Some processors have a special hardware Floating Point Unit (Floating Point Unit (FPU)), which substantially speeds up floating point operations. With separate integer and floating point unit processing units, integer and floating point computations can often be processed in parallel. This has greatly clouded the old "integer is faster" rule; on some computers, a mix of 2 integer and 2 floating point operations may be faster than 4 integer operations. So, if you are really concerned about optimizing a piece of code, then you will need to run some tests. Most of the time, you should just write the code in whatever way you like best and let the compiler deal with the optimization issues.

Real numbers and numerical precision

As we have seen, numerical precision is an important aspect of computer simulation. In this section, we look at an example of the finite precision of representations of numbers, when solving calculus equations numerically. Since the typical computer communicates with us in the decimal system, but works in the dual system internally, the computer must execute conversion procedures, which hopefully only involve small rounding errors.

Computers also are not capable of operating with real numbers expressed with more than a fixed number of digits. The set of possible values is only a subset of the mathematical integers or real numbers. The word length that we reserve for a given number places a restriction on the precision with which a given number is represented. This in turn means that, for example, floating point numbers are always rounded to a machine dependent precision, typically with 6–15 leading digits to the right of the decimal point. Furthermore, each such set of values has a processor-dependent smallest negative and a largest positive value.

Section 1.5 Number representation: bits and bytes in computer memory

> **Why do we care at all about rounding and machine precision?**

The best way to see this, is to consider a simple example. In the following, we assume that we can represent a floating point number with a precision of only 5 digits to the right of the decimal point. This is an arbitrary choice, but mimics the way numbers are represented in a computer.

Example 1.6 (Numerical precision). Suppose we wish to evaluate the function

$$f(x) = \frac{1 - \cos(x)}{\sin(x)}, \tag{1.6}$$

for small values of x. If we multiply the denominator and numerator with $1 + \cos(x)$ we obtain the equivalent expression

$$f(x) = \frac{\sin(x)}{1 + \cos(x)}. \tag{1.7}$$

Let us now choose $x = 0.007$ (in radians). This choice of x means that our choice of precision results in

$$\sin(0.007) \approx 0.69999 \times 10^{-2},$$

and

$$\cos(0.007) \approx 0.99998.$$

The first expression for $f(x)$ in equation (1.6) results in

$$f(x) = \frac{1 - 0.99998}{0.69999 \times 10^2} = \frac{0.2 \times 10^{-4}}{0.69999 \times 10^{-2}} = 0.28572 \times 10^{-2}, \tag{1.8}$$

while the second expression in equation (1.7) results in

$$f(x) = \frac{0.69999 \times 10^{-2}}{1 + 0.99998} = \frac{0.69999 \times 10^{-2}}{1.99998} = 0.35000 \times 10^{-2}, \tag{1.9}$$

which also happens to be the "exact" result, but, obviously $0.35000 \times 10^{-2} \neq 0.28572 \times 10^{-2}$. So, what happened here?

In the first expression, after the subtraction, we have only *one* relevant digit in the numerator, due to our choice of precision. This leads to a loss of precision and an incorrect result due to *cancellation of two nearly equal numbers* in equation (1.8). If we had chosen a precision of *six* leading digits instead of five, both expressions would have had the same answer. If we were to evaluate $x \sim \pi$ (for which $\cos(x)$ will be close to -1), then the second expression of $f(x)$, i.e. equation (1.9), can lead to potential loss of precision, due to cancellations of nearly equal numbers.

This simple example demonstrates the loss of numerical precision due to rounding errors, where the number of leading digits is lost in a subtraction of two nearly equal numbers. The lesson to be learned from this example is that computers not only just add 0's and 1's, organized in groups of bytes, but that we cannot blindly trust the calculated numbers, due to their being represented in memory with finite precision. Thus, we also cannot blindly compute a function. We will always need to carefully analyze our employed algorithms in search of potential pitfalls. There is no magic recipe: the only guideline is an understanding of the fact that a machine cannot represent *all* numbers correctly. We will look at a few explicit code examples in Section 2.3 of Chapter 2, after having introduced the basic syntax of C/C++.

Arrays

The size of an array is the number of elements times the size of each element. So if, for example, one declares an array of 10 integers with the C command line

```
int array[10];
```

then the array size will be 4×10 bytes (4 bytes for each integer multiplied by the number of components). The elements in the array are laid out consecutively, starting with the first element and working from low memory to high. Given the base address of the array, the compiler can generate code and figure out the address of any element in the array. For more details on C/C++ arrays, see Section 2.4.

Pointers

A pointer is simply an address in computer memory. The size of the pointer depends on the range of addresses on the machine. Currently, almost all machines use 4 bytes to store an address, creating a 4GB addressable range. There is actually very little distinction between a pointer and a 4 byte unsigned integer. They both just store integers – the difference is in whether the number is *interpreted* as a number or as an address.

Machine code or assembly instructions

Machine instructions themselves are also encoded using bit patterns, most often using the same 4-byte native word size. The different bits in the instruction encoding indicate things such as the type of instruction (load, store, multiply, etc.) and the registers involved. We will not occupy ourselves further with this most elementary level of programming. For further reading on machine code or assembly programming for certain processors see e.g. [229, 113, 140, 291].

1.6 The role of algorithms in scientific computing

Before there were computers, there were algorithms[58]. Now that there are computers, there are even more algorithms; they are at the heart of computing. Generally speaking, an algorithm is any well-defined computational procedure that takes some value, or a set of values, as *input* and produces some value, or a set of values, as *output*. Thus, an algorithm is a sequence of computational steps to transform the input into the output. We can also view an algorithm as a tool for solving a well-specified computational problem. Stating the problem specifies in general terms the desired input/output relation, see Figure 1.17.

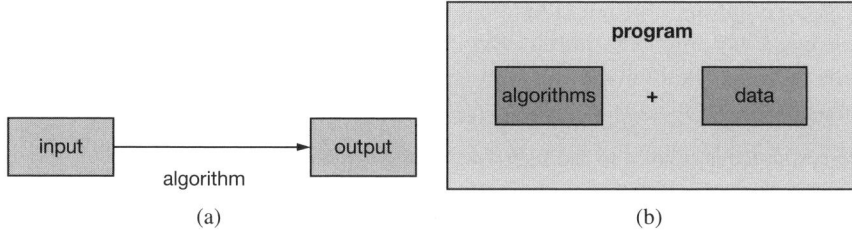

Figure 1.17. Relation of input data, algorithm and output data. (a) An algorithm transforms input into output. (b) A computer program consists of algorithms that solve problems, and data on which these algorithms operate.

The nice thing about algorithms is that they have to be formulated in such a way, that a person (or computer) executing the consecutive steps of an algorithm doesn't need to have an intellectual understanding of the individual steps, as long as the algorithm describes a specific computational procedure for achieving that input/output relation. For example, we might need to sort a sequence of numbers into nondecreasing order. In practice, this problem frequently arises in computer simulations and provides fertile ground for introducing many standard design techniques and tools for analysis. Here is how one formally defines the sorting problem:

Input: A sequence of n numbers $(a_1, a_2, \ldots, a_n)$.

Output: A permutation (reordering) $\langle a_1', a_2', \ldots, a_n' \rangle$ of the input sequence such that $(a_1' \leq a_2' \leq \cdots \leq a_n')$.

For example, given the input sequence $(31, 41, 59, 26, 41, 58)$, a sorting algorithm returns the sequence $(26, 31, 41, 41, 58, 59)$ as output. Because many programs use sorting as an intermediate step, it is a fundamental operation in computer science[59].

---

[58] According to Knuth [250], the word "algorithm" is derived from the name "al-Khwarizmial", a ninth-century Persian mathematician.
[59] As a result, we have a large number of good sorting algorithms at our disposal.

Which algorithm is best for a given application depends on – among other factors – the number of items to be sorted, the extent to which the items are already sorted, possible restrictions on the values of each item, the computer architecture, and the kind of storage devices to be used.

An algorithm is said to be *correct* if, for every input instance, it halts with the correct output. We then say that a correct algorithm solves the given computational problem. An incorrect algorithm may not halt at all on some inputs, or it might halt with an incorrect answer. An algorithm can be specified, for example, in English, in a mixture of ordinary language and elements of computer languages (often called *pseudocode*), or as a computer program, i.e. explicit source code that may be implemented, compiled and executed. The only requirement is that the specification must provide a precise description of the computational procedure to be followed.

The *efficiency* of an algorithm is also called its *performance*. Analyzing and knowing the performance of a specific algorithm is of great importance for computer simulations because:

- Performance is related to the user-friendliness of a program[60].

- Performance measures the line between the feasible and unfeasible[61].

- Performance is often at the edge of scientific innovation, when it comes to simulating things that have never been done before.

- The performance of algorithms provides a common language for analyzing and talking about program behavior[62].

---

[60] For example, a program that causes less waiting time until a result is achieved is more user-friendly than a program with excessive waiting time.

[61] For example, if a program uses too much memory or if it is not fast enough to really achieve a solution of a problem within an acceptable period of time.

[62] In particular the so-called "Big-O" notation which provides an upper bound (worst case) for the efficiency of an algorithm, see Page 55.

Section 1.6 The role of algorithms in scientific computing

> **Definition of an algorithm:** "An algorithm is a *finite, deterministic, effective* and *efficient* step-by-step instruction for solving a problem, during which some input data is transformed into some output data."
>
> - "Finite" here means that the algorithm comes to an end after some finite time and a finite number of steps.
>
> - "Deterministic" means that each *next* step in the algorithm is defined unambiguously.
>
> - "Effective" means that each individual step is unambiguously defined.
>
> - "Efficient" means that the algorithm uses few resources in terms of memory, data, and computing time.

One might think that algorithms were only invented along with the rise of the era of electronic computing in the 20th century. This is, however, not true at all. As an example we provide in Program Listing 2 a pseudocode[63] of the Euclidean algorithm for determining the greatest common divisor of two integers a and b.

Listing 2. Euclidean algorithm (pseudocode).

```
1 Input:   Read in integer numbers a and b
2 while (b != 0)
3    substitute the pair of numbers (a,b) by (b, a mod b).
4    //Remark: (a mod b) is the rest of the division of a by b
5 Output:  The number a, which is the greatest common divisor
```

> Possessing basic knowledge of algorithms is a characteristic that distinguishes the truly skilled scientific programmer from the novice.

1.6.1 Efficient and inefficient calculations

To classify the algorithms' efficiency we consider in Table 1.6 five different algorithms A_1, A_2, A_3, A_4, A_5 with corresponding running times $N, N^2, N^3, 2^N, N!$, where N

---

[63] In pseudocode we use indentation without brackets, which normally is a bad idea. The idea of pseudocode however, is to put real emphasis on the basic, *essential* steps in the algorithm, neglecting for example proper declaration and initialization of variables. Thus, to be succinct, we do not use brackets for code indentation in pseudocode.

Table 1.6. Overview of typical running times of algorithms in computer simulations. Depicted are the number of elementary steps (ES) and the corresponding elapsed real time for the different algorithms, under the assumption that one ES takes 10^{-9} seconds.

algorithm	runtime	$N = 10$	$N = 20$	$N = 50$	$N = 100$
A_1	N	10 ES 10^{-8} s	10 ES 2×10^{-8} s	10 ES 5×10^{-8} s	10 ES 10^{-7} s
A_2	N^2	100 ES 10^{-7} s	400 ES 4×10^{-7} s	2 500 ES 2.5×10^{-6} s	10 000 ES 10^{-5} s
A_3	N^3	1 000 ES 10^{-6} s	8 000 ES 8×10^{-6} s	10^5 ES 10^{-4} s	10^6 ES 0.001 s
A_4	2^N	1 024 ES 10^{-6} s	10^5 ES 0.001 s	10^{15} ES 13 d	10^{30} ES $\sim 10^{13}$ a
A_5	$N!$	$\sim 10^6$ ES 0.003 s	$\sim 10^{18}$ ES 77 a	$\sim 10^{64}$ ES 10^{48} a	10^{158} ES $\sim 10^{141}$ a

is the size of the system under consideration, e.g. the number of particles, the number of finite elements, or the number of discrete locations where a partial differential equation is solved in a simulation program. These running times are typical for different applications in computational materials science. We assume that on a real-life computer, one elementary step in the calculation takes 10^{-9} seconds.

It is obvious from Table 1.6 that *exponential* running times (algorithms A_4 and A_5) are generally not acceptable for all practical purposes. For these algorithms, even with very small system sizes N, one reaches running times larger than the estimated age of the universe (10^{10} years). Algorithm A_5 – which is $\propto N!$ – could be a solution of the Traveling Salesman Problem[64] [265, 205, 33]. A running time of 2^N as in A_4 is typical for problems where the solution space of the problem consists of a subset of a given set of N objects; There are 2^N possible subsets of this basis set. The "efficient" algorithms A_1, A_2, A_3 with runtimes of at most N^3 are those most commonly used in engineering and physics applications.

Usually, in MD simulations (see Chapter 5 for a thorough discussion) one assumes the interactions between particles to be pairwise additive, in order to reduce the complexity of the calculation to a running time at least proportional to N^2, albeit still with a large prefactor. Hence, the interaction of particles in an MD simulation depends only on the current position of *two* particles. Sometimes, however, three-body interactions

---

[64] Given a set of cities, and known distances between each pair of cities, the Traveling Salesman Problem is the problem of finding a tour that visits each city exactly once, and minimizes the total distance traveled. A more formal definition of the Traveling Salesman Problem speaks of "vertices" and "edges" rather than "cities" and "roads": "Given an undirected graph and a cost for each edge of that graph, find a Hamiltonian circuit of minimal cost."

Table 1.7. Speedup of the runtime of various typical algorithms, assuming a speedup factor of 10 and 100. The efficiency of polynomial algorithms is shifted by a factor, wheres exponential algorithms only improve by an additive constant.

algorithm	runtime	efficiency	speedup factor 10	speedup factor 100
A_1	N	N_1	$10 \times N_1$	$100 \times N_1$
A_2	N^2	N_2	$\sqrt{10} \times N_2 = 3.16 \times N_2$	$10 \times N_2$
A_3	N^3	N_3	$\sqrt[3]{10} \times N_3 = 2.15 \times N_3$	$4.64 \times N_3$
A_4	2^N	N_4	$N_4 + 3.3$	$N_4 + 6.6$
A_5	$N!$	N_5	$\approx N_5 + 1$	$\approx N_5 + 2$

have to be included, e.g. when considering bending and torsion potentials in chain molecules (compare Section 4.6 on Page 291). These potentials depend on the position of at least three different particles. Solving the Schrödinger equation in ab initio simulations also leads to at least a N^3-dependency of the running time. This is the main reason why ab initio methods are restricted to very small system sizes (usually not more than a few hundred atoms can be considered, even on the largest computers). Solving the classical Newtonian equations of motion with a "brute-force" strategy leads to a N^2-efficiency $\left(\frac{N \times (N-1)}{2}\right)$ of the algorithm that calculates the interactions of particles. This is also generally true in finite element codes, where special care has to be taken if elements start to penetrate each other. This case is usually treated by so-called *contact-algorithms* which employ a simple spring model between penetrating elements. The spring forces try to separate the penetrating elements and the core of the contact algorithm is a lookup-table of element knots, which is used to decide whether two elements penetrate each other or not. This algorithm, in its plain form, has an efficiency of N^2. Because N^2 efficiency of an algorithm still restricts the system size to very small systems of a few thousand elements or particles, one uses several methods to speed up the efficiency of algorithms in computer simulations. Usually, this is done by using sorted search tables, which can be processed linearly (and thus reaching an efficiency $\propto N \log N$, which is typical of "divide-and-conquer" algorithms). A discussion of several of the most important speedup techniques commonly used in MD simulation programs is provided in Chapter 5.

Algorithms A_1, A_2 and A_3 in Table 1.6 are called *efficient*, because they have a polynomial runtime, whereas the exponential algorithms A_4 and A_5 are called *inefficient*. One way to see why this is the case, is the following consideration, displayed in Table 1.7. Assuming that – due to a technology jump – available computer systems will be 10 or 100 times faster than today, the efficiency of algorithms A_1, A_2 and A_3 will be shifted by a *factor*, whereas for the exponential algorithms A_4 and A_5 the efficiency will be shifted only by an *additive constant*.

Generally speaking, an algorithm is said to be *efficient* if its runtime – which depends on some input N – has a *polynomial upper bound*. For example, the runtime

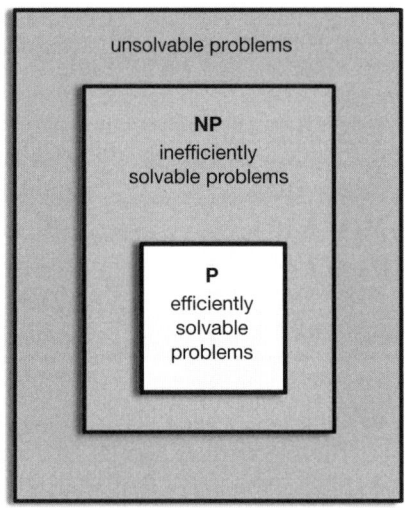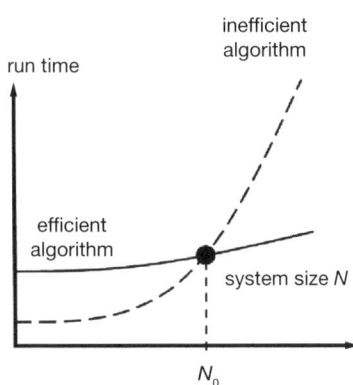

Figure 1.18. Efficiently and inefficiently solvable problems. All *polynomial time solvable* problems pertain to class *P*, whereas the set *NP* is the set of all problems which are solved by a *nondeterministic polynomial algorithm*. The problem here lies in the term "nondeterministic" which – as a mathematical concept – cannot be simply translated into a computer program. Thus, in order write a program to solve an *NP* problem, one has to somehow map the nondeterministic branching of the solution into a sequential program. This is the reason why, so far, only inefficient algorithms of at least exponential running time are known for solving *NP* problems.

function $2N^3 (\log_2 N)^6 + 3\sqrt{N}$ has a polynomial upper bound N^4 for large N. In "Big-O" notation [65], this is expressed as $O(N^k)$ with k the degree of the polynomial. Algorithms A_4 and A_5 on the other hand have no polynomial upper limit, because an exponential function always grows larger than a polynomial for large N. Thus, they are called *inefficient*. The class of problems that can be solved with efficient algorithms – i.e. algorithms that are polynomially bounded – are denoted with the letter *P*, cf. Figure 1.18. As the set of polynomials is closed under addition, multiplication and composition, *P* is a very robust class of problems. Combining several polynomial algorithms results into an algorithm which again exhibits a polynomial running time.

---

[65] In essence, "Big-O" means "on the order".

Section 1.6 The role of algorithms in scientific computing 47

> According to the above definition of an "efficient" algorithm, the algorithm with running time, e.g. $1\,000 \times N^{1000}$ falls into the class P, whereas an algorithm with running time 1.1^N is exponential and thus inefficient. However, in this case, the exponential algorithm only exhibits running times longer than the efficient one, for system sizes $N > N_0 \sim 123,000$ (see Figure 1.18).

Efficiency analysis of the insertion sort algorithm

The crucial steps of the insertion sort algorithm, for sorting an array $A[1,\ldots,N]$ of N numbers, can be written in pseudocode as in Program Listing 3.

Listing 3. Insertion sort algorithm (pseudocode).

```
1  Input:   An array A[1,...,N] of N numbers
2  for (j=2 to N)
3    key = A[j]
4    //Remark: Insert A[j] into the sorted sequence A[1,...,j-1]
5    i = j - 1
6    while ( (i > 0) and (A[i] > key) )
7      A[i+1] = A[i]
8      i = i - 1
9    A[i+1] = key
10 Output:  The sorted array A[1,...,N]
```

> What does this algorithm actually do?

Let's figure out what this algorithm does:

1. It takes an array A.

2. It takes some element j of the array, i.e. it pulls out a value which is called the "key".

3. It copies the values until a place to insert this key is found and then inserts it[66].

4. Now, we have the array sorted from beginning (element 1) to j^{th} element, see Figure 1.19, and then we take $j + 1$ as a key.

---

[66] This is the reason for the name "*insertion* sort".

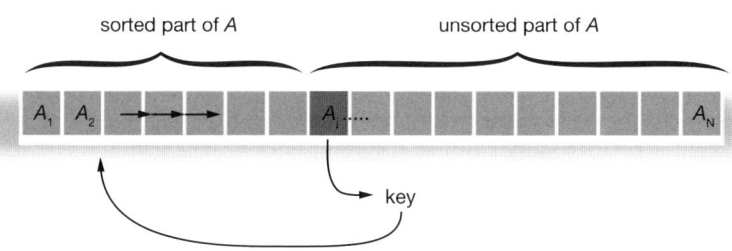

Figure 1.19. The invariant of insertion sort. In all cases, the part $A[1, \ldots, j-1]$ of the array is sorted. The goal of the loop in the algorithm of Listing 3 is to add one element to the length of the part of the array that is sorted.

Let's assume the original array was $A = \{8, 2, 5, 9, 3, 6\}$. Then, the algorithm of Program Listing 3 performs the following steps:

$$\begin{aligned} A &= \{8, \mathbf{2}, 5, 9, 3, 6\} \\ &= \{2, 8, \mathbf{5}, 9, 3, 6\} \\ &= \{2, 5, 8, \mathbf{9}, 3, 6\} \\ &= \{2, 5, 8, 9, \mathbf{3}, 6\} \\ &= \{2, 3, 5, 8, 9, \mathbf{6}\} \\ &= \{2, 3, 5, 6, 8, 9\} \quad \text{done} \end{aligned}$$

> **?** What is the worst case for insertion sort?

Obviously, the running time of insertion sort depends on the input itself (whether it is already sorted) and on the input size (sorting more elements takes more time). "Running time" of an algorithm means the number of elementary operations (or "steps") executed. It is convenient to define the notion of step so that it is as machine independent as possible, since we want to talk about software and not hardware. Let us adopt the following view: For each line i of the pseudocode in Listing 3, a machine dependent constant amount of time c_i is needed. For each $j = 2, 3, \ldots, N$, where N is the length of the array $A[]$, we let t_j denote the number of times the *while* loop test is executed in *line 6*, for that value of j. When a *for* or a *while* loop exits in the usual way, it is executed once more than the loop body. As for comments like in line 4, we assume that they are not executable statements, so they take no time.

Table 1.8. Cost analysis of insertion sort according to Listing 3.

line in Listing 3	cost	times
1	c_1	N
2	c_2	N
3	c_3	$N-1$
4	0	$N-1$
5	c_5	$N-1$
6	c_6	$\sum_{j=2}^{N} t_j$
7	c_7	$\sum_{j=2}^{N} (t_j - 1)$
8	c_8	$\sum_{j=2}^{N} (t_j - 1)$
9	c_9	$N-1$
10	c_{10}	N

Average case of insertion sort Let's start with analyzing each line of pseudocode in Listing 3 to determine its time "cost". The result of this analysis is displayed in Table 1.8. To compute the total running time $T(N)$ we sum the products of the cost and times columns and yield the average case, i.e. the *expected time over all inputs of size N*. With expected time, we mean the weighted average of the probability distribution for a certain input to occur. Of course, one does, in general, not know the weight of every input, so one has to make an assumption about the statistical distribution of inputs. Usually, one will simply assume that all inputs are equally likely (compare our discussion of a priori probabilities in Chapter 3). From the detailed line by line cost analysis in Table 1.8 we obtain for $T(N)$:

$$T(N) = c_1 N + c_2 N + c_3(N-1) + c_5(N-1) + c_6 \sum_{j=2}^{N} t_j \\ + c_7 \sum_{j=2}^{N} (t_j - 1) + c_8 \sum_{j=2}^{N} (t_j - 1) + c_9(N-1) + c_{10} N. \quad (1.10)$$

The average case of the running time of an algorithm is often as bad as the worst case. Assume that we randomly choose N numbers and apply an insertion sort. How long does it take to determine where in subarray $A[1, \ldots, j-1]$ to insert the key $A[j]$?

On average, half the elements in $A[1, \ldots, j-1]$ are less than $A[j]$, and half the elements are greater. So, on average, we will have to check half of the subarray $A[1, \ldots, j-1]$, so t_j in this case is about $j/2$. The resulting average case running time turns out to be a quadratic function in N, the input size, just as in the worst case, as we will see in the next paragraph. It turns out that, even for inputs of given size N, an algorithm's running time may depend on *which* input of this size is given. For example, the *best case* occurs, when the array is already sorted.

> Why is the "best case" scenario for an algorithm bogus?

In the best case, for each $j = 2, 3, \ldots, N$, we find that $A[i] \leq$ key in *line 6* when i has its initial value of $j - 1$. Hence, $t_j = 1$ for $j = 2, 3, \ldots, N$, and the best-case running time is:

$$T(N) = c_1 N + c_2 N + c_3(N-1) + c_5(N-1) + c_6 \sum_{j=2}^{N} t_j +$$
$$+ c_7 \sum_{j=2}^{N} (t_j - 1) + c_8 \sum_{j=2}^{N} (t_j - 1) + c_9(N-1) + c_{10} N \propto N. \quad (1.11)$$

This running time can generally be expressed as $aN + b$, for some constants a and b that depend on the statement costs c_i. Thus, it is a *linear function* of N.

> The *best case scenario* of an algorithm is bogus, because you can cheat. The purpose of a running time analysis is to provide a guarantee as to the time needed to run this algorithm. However, a best case analysis only looks at *one* case that works well, but nothing is said about *most of the other cases*!

Worst case of insertion sort The worst case scenario of a computational procedure is usually the one which is most important because it constitutes a guarantee to the user in terms of running time $T(N)$, where $T(N)$ in this case is the *maximum time on any input size N*. Knowing this time provides a guarantee that the algorithm will *never* take any longer. For a reverse sorted array, we must compare each element $A[j]$ with each element in the entire sorted subarray $A[1, \ldots, j-1]$, and hence $t_j = j$ for $j = 2, 3, \ldots, N$.

> The worst case for an insertion sort is when the array $A[]$ is inverse sorted.

By using

$$\sum_{j=2}^{N} j = \frac{N(N+1)}{2} - 1,$$

and

$$\sum_{j=2}^{N}(j-1) = \frac{N(N-1)}{2},$$

we see that in the worst case, the insertion sort has a running time of

$$T(N) = c_1 N + c_2 N + c_3(N-1) + c_5(N-1) + c_6\left[\frac{N(N+1)}{2} - 1\right]$$
$$+ c_7\left(\frac{N(N-1)}{2}\right) + c_8\left(\frac{N(N-1)}{2}\right) + c_9(N-1) + c_1 N$$
$$= \left(\frac{c_6}{2} + \frac{c_7}{2} + \frac{c_8}{2}\right) N^2 \qquad (1.12)$$
$$+ \left(c_1 + c_2 + c_3 + c_4 + \frac{c_6}{2} - \frac{c_7}{2} - \frac{c_8}{2} + c_9 + c_{10}\right) N$$
$$- (c_3 + c_5 + c_6 + c_9) \propto N^2.$$

This worst case running time can be expressed as $aN^2 + bN + c$ for constants a, b, and c, which again depend on the constant statement costs c_i. Hence, this is a *quadratic function* of N.

> **?**
> Why are we mostly interested in the worst case behavior of algorithms?

> **✓**
> The worst case usually is the most interesting because:
> - The worst case running time gives an upper bound on the running time for *any* input. It gives a guarantee that the algorithm will never take any longer.
> - For some algorithms, the worst case occurs fairly often.

1.6.2 Asymptotic analysis of algorithms

In the analysis of insertion sort we used some simplifying abstractions. First, we ignored the actual cost of each statement, using the constants c_i to represent these costs. Then, we observed that even these constants give us more detail than we really need. We expressed the worst case running time as $aN^2 + bN + c$ for constants a, b, and c that depend on the statement costs c_i. The statement costs c_i are machine dependent. So, when comparing two algorithms, one would always have to use the same machine, so that the costs c_i stay the same for all algorithms. But when you run one of the algorithms on a faster machine, with $c_i' < c_i$ then it will run faster. So, the question is how to compare the *absolute speed* of algorithms on different machines.

> The big idea for comparing the absolute efficiency (or running time $T(N)$) of algorithms, no matter on which machine they are run, is the *asymptotic analysis of algorithms*. This idea includes the following steps:
> 1. Ignore machine dependent constants.
> 2. Drop lower order terms and look at the *growth* of $T(N)$ as $N \to \infty$ (and do not consider $T(N)$ itself).

We now introduce one more simplifying abstraction: as we are really interested in the *rate of growth* of the running time, we consider only the leading term of a formula (e.g. aN^2 in the worst case example above), since the lower-order terms are relatively insignificant for large values of N. We also ignore the leading term's constant coefficient, since in determining computational efficiency for large inputs constant factors are less significant than the rate of growth. For insertion sort, when we ignore the lower-order terms and the leading term's constant coefficient, we are in the worst case left with the factor N^2 from the leading term.

The notation predominantly used in computer science is the so-called Θ-notation. In the natural sciences the "Big-O" notation is more common. So, for example, one would say that insertion sort has a worst case running time of $\Theta(N^2)$ – pronounced "theta of N-squared") – or $O(N^2)$. Both notations have a precise mathematical definition provided in equations (1.13) and (1.16) below.

Usually, one algorithm is considered to be more efficient than another if its worst case running time has a lower order of growth. Due to constant factors and lower order terms, an algorithm whose running time has a higher order of growth might take less time *for small inputs* than an algorithm whose running time has a lower order of growth. But for large enough inputs, an $O(N^2)$ algorithm, for example, will run more quickly in the worst case than a $O(N^3)$ algorithm (compare our example on Page 47).

The notations used to describe the asymptotic running time of an algorithm are defined in terms of functions whose domains are the set of natural numbers $\mathbb{N} = \{0, 1, 2, \ldots\}$. Such notations are convenient for describing the worst case running-time function $T(N)$, which is usually defined only for integer input sizes. This section defines the basic asymptotic notations for the running time of algorithms. Even when we use asymptotic notation to apply to the running time of an algorithm, we need to understand which running time we mean. Most often, we are *only* interested in the worst case.

Θ-notation

In Section 1.6.1 we found that the worst case running time of the insertion sort is $T(N) \propto N^2$. This can also be written as $T(N) = \Theta(N^2)$. Let us define what this notation means.

Section 1.6 The role of algorithms in scientific computing   53

> **Definition of Θ-notation.** For a given function $g(N)$, we denote by $\Theta(g(N))$ the set of functions
>
> $$\Theta(g(N)) = \{f(N) : \text{there exist positive constants } c_1, c_2, N_0$$
> $$\text{such that } 0 \leq c_1 g(N) \leq f(N) \leq c_2 g(N) \quad (1.13)$$
> $$\text{for all} \quad N \geq N_0\}.$$

A function $f(N)$ belongs to the set $\Theta(g(N))$ if there exist positive constants c_1 and c_2 such that the function is limited within $c_1\, g(N)$ and $c_2\, g(N)$, for sufficiently large N. Because $\Theta(g(N))$ is a set, one could also write $f(N) \in \Theta(g(N))$ to indicate that $f(N)$ is a member of $\Theta(g(n))$.

Figure 1.20 provides a picture of functions $f(N)$ and $g(N)$, where $f(N) = \Theta(g(N))$. For all values of N at and to the right of N_0, the value of $f(N)$ lies at or above $c_1\, g(N)$ and at or below $c_2\, g(N)$. In other words, for all $N \geq N_0$, the function $f(N)$ is equal to $g(N)$ to within a constant factor. We say that $g(N)$ is an *asymptotically tight bound* function for $f(N)$. The definition of $\Theta(g(N))$ requires that every member $f(N) \in \Theta(g(n))$ be *asymptotically nonnegative*, that is, that $f(N)$ be nonnegative whenever N is sufficiently large[67]. Consequently, the function $g(N)$ itself must be asymptotically nonnegative, or else the set $\Theta(g(N))$ is empty. Thus, one can generally assume that every function used within Θ-notation is asymptotically nonnegative. This assumption also holds for the other asymptotic notations defined in this section.

On Page 52 we introduced an informal notion of Θ-notation that amounts to throwing away lower-order terms and ignoring the leading coefficient of the highest-order term. Let us briefly justify this intuition by using the formal definition, to show that $\frac{1}{3}N^2 - 4N = \Theta(N^2)$. To do so, we must determine positive constants c_1, c_2, and N_0 such that

$$c_1 N^2 \leq \frac{1}{3} N^2 - 4N \leq c_2 N^2 \quad (1.14)$$

for all $N \geq N_0$. Dividing equation (1.14) by N^2 yields

$$c_1 \leq \underbrace{\frac{1}{3} - \frac{4}{N}}_{-4 \leq N(c_2 - \frac{1}{3})} \leq c_2. \quad (1.15)$$

We can now make the right-hand inequality hold for any value of $N \geq 1$ by choosing any constant $c_2 \geq 1/3$. Likewise, we can make the left-hand inequality hold for any value of $N \geq 12$ by choosing any constant $c_1 \leq 1/4$. Thus, by choosing $c_1 = 1/4$,

---

[67] An *asymptotically positive* function is one that is positive for all sufficiently large N.

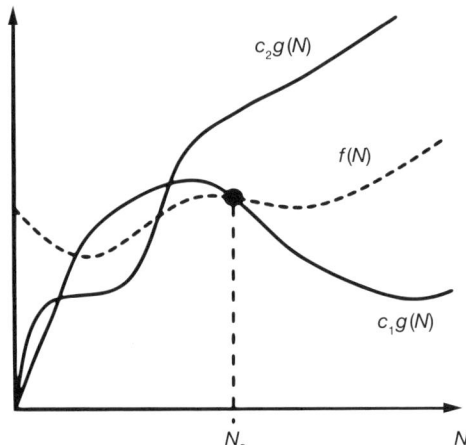

Figure 1.20. Graphical representation of Θ-notation. This notation bounds a function (representing the runtime of some algorithm) to within constant factors. One writes $f(N) = \Theta(g(N))$ if there exists positive constants c_1, c_2, N_0 such that at and to the right of N_0, the value of $f(N)$ always lies between $c_1 g(N)$ and $c_2 g(N)$. The value N_0 is the minimum possible value. Any value greater than N_0 also works.

$c_2 = 1/3$, and $N_0 = 12$, we have verified that $\frac{1}{3}N^2 - 4N = \Theta(N^2)$. Certainly, other choices for the constants exist, but the important thing is that *some* choice exists. Note that these constants explicitly depend on the function $\frac{1}{3}N^2 - 4N$. A different function belonging to $\Theta(N^2)$ would normally require different constants. We can also use the formal definition to verify that, e.g. $6N^2 \neq \Theta(N^2)$. Suppose, for the purpose of proof by contradiction that c_2 and N_0 exist such that $6N^2 \leq c_2 N^2$ for all $N \geq N_0$. But then, dividing by N^2 yields $N \leq c_2/6$, which cannot possibly hold for arbitrarily large N, since c_2 is constant. Intuitively, in determining asymptotically tight bounds, the lower-order terms of an asymptotically positive function can be ignored because they are insignificant for large N. When N is large, even a tiny fraction of the highest-order term suffices to dominate the lower-order terms. Thus, setting c_1 to a value that is slightly smaller than the coefficient of the highest-order term and setting c_2 to a value that is slightly larger permits the inequalities in the definition of the Θ-notation to be satisfied. The coefficient of the highest-order term can likewise be ignored, since it only changes c_1 and c_2 by a constant factor equal to the coefficient.

O-notation

The O-notation is used when a function describing an algorithm's runtime behavior only has an *asymptotic upper bound*.

Section 1.6 The role of algorithms in scientific computing

> **Definition of O-notation.** For a given function $g(N)$, we denote by $O(g(N))$ the set of functions
>
> $$O(g(N)) = \{f(N) : \text{there exist positive constants } c \text{ and } N_0$$
> $$\text{such that } 0 \leq f(N) \leq cg(N) \qquad (1.16)$$
> $$\text{for all} \quad N \geq N_0\}.$$

O-notation is used to provide an *asymptotic upper bound* on a function, to within a constant factor. For all values N at and to the right of N_0, the value of the function $f(N)$ is on or below $cg(N)$. Note that $f(N) = \Theta(g(N))$ implies $f(N) = O(g(N))$, since Θ-notation is a stronger notation than O-notation. Written set-theoretically, we have $\Theta(g(N)) \subseteq O(g(n))$. In the literature (in particular in physics literature), we sometimes find O-notation informally describing asymptotically tight bounds, that is, what we have defined using Θ-notation. However, in the literature on algorithms, it is standard procedure to distinguish asymptotic *upper* bounds from asymptotically *tight* bounds.

Using O-notation, one can often describe the running time of an algorithm merely by inspecting the algorithm's overall structure. For example, the doubly nested loop structure of the insertion sort algorithm from Listing 3 immediately yields an $O(N^2)$ upper bound on the worst case running time: the cost of each iteration of the inner loop is bounded from above by $O(1)$ (constant), the indices i and j are both at most N, and the inner loop is executed at most once for each of the N^2 pairs of values for i and j. Since O-notation describes an upper bound, when we use it to bound the worst case running time of an algorithm, we have a bound on the running time of the algorithm for every input. Thus, the $O(N^2)$ bound on the worst case running time of the insertion sort also applies to its running time for every possible input. The $\Theta(N^2)$ bound on the worst case running time of the insertion sort, however, does *not* imply a $\Theta(N^2)$ bound on the running time of the insertion sort for every input. For example, we saw in Section 1.6.1 that when the input is already sorted, the insertion sort runs in $\Theta(N)$ time.

Ω-notation

Just as O-notation provides an asymptotic upper bound on a function, Ω-notation provides an *asymptotic lower bound*.

> **Definition of Ω-notation.** For a given function $g(N)$, we denote by $\Omega(g(N))$ (pronounced "big-Omega of g of N" or sometimes just "Omega of g of N") the set of functions
>
> $$\Omega(g(N)) = \{f(N) : \text{there exist positive constants } c \text{ and } N_0 \tag{1.17}$$
> $$\text{such that } 0 \leq cg(N) \leq f(N)$$
> $$\text{for all} \quad N \geq N_0\}.$$

When we say that the running time of an algorithm is $\Omega(g(N))$ we mean that no matter what particular input of size N is chosen for each value of N, the running time on that input is *at least* a constant times $g(N)$, for sufficiently large N. Equivalently, we are giving a lower bound on the best case running time of an algorithm. For example, the best case running time of the insertion sort is $\Omega(N)$, which implies that the running time of the insertion sort is $\Omega(N)$. The running time therefore belongs to both $\Omega(N)$, and $O(N^2)$, since it falls anywhere between a linear function and a quadratic function of N. Moreover, these bounds are asymptotically as tight as possible: for instance, the running time of the insertion sort is *not* $\Omega(N^2)$, since there exists an input for which the insertion sort runs in $\Theta(N)$ time (e.g., when the input is already sorted).

With this, we leave the theoretical considerations of the running time behavior of algorithms. For more details on this topic, the interested reader is referred to the literature suggestions provided in Section 1.9 on Page 83.

1.6.3 Merge sort and divide-and-conquer

In section 1.6.1, we analyzed the efficiency of the insertion sort algorithm and found that it is $\Theta(N^2)$. When testing the insertion sort (try it!) it turns out that it is modestly fast for small input N, but not fast at all for large N. In this section, we want to introduce another algorithm that is faster than insertion sort called the *merge sort*. The merge sort operates on an array $A[1, \ldots, N]$ in the following way (see Listing 4).

Listing 4. Merge sort applied to an Array A[1,...,N] (pseudocode).

```
1. if n = 0 , done.
2. Recursively sort A[1,...,N/2] and A[N/2+1,...,N].
3. Merge the two sorted lists.
```

If there is only one element, then it is already sorted. In step 2 in Listing 4 one splits the array A in two halves and sorts these halves *separately* and *recursively*, for example, by using the following recursive function *MergeSort()*:[68]

---

[68] For an introduction in C syntax, see the next chapter.

```
void MergeSort(int a[], int left, int right){
   if (left < right){
   int middle = (left + right)/2;
MergeSort(a, left, middle);
MergeSort(a, middle + 1, right);
MergeSort(a, left, middle, right);
 }
}
```

The time T needed for this sort is obviously $2 \times T(N/2)$, i.e. double the time that is needed to sort each half of the array $A[]$.

Step 3 of the algorithm in Listing 4 puts the two separately sorted lists together by comparing in *each step* only two elements of both lists and selecting the smaller as input into the final list one at a time. That is, at each step there is a constant number of operations. For example, assume the two sorted arrays contain as elements $A_1 = [19, 12, 6, 3]$ and $A_2 = [11, 10, 8, 1]$. In the first step one compares 3 with 1 and puts the smaller of the two numbers, i.e. 1, in the final list $A[]$. In the next step one compares 3 with 8 and outputs 3 into $A[]$. The next comparison is between 8 and 6, the output being 6, then 10 and 12 and so on. The final list is then $A = [1, 3, 6, 8, 10, 11, 12, 19]$ and the key point here is, that each comparison is $\Theta(1)$, i.e. each step takes a constant amount of time, *independent of the list size N*. Hence, the total sorting time is $\Theta(N)$ (linear time) for a list with N elements. A function *Merge()* that performs this merging of two sorted lists, could be:

```
void MergeSort(int z[], int l, int m, int r){
   int i, j, k;
   for (i = m + 1; i > 1; i--)
     helper[i-1] = z[i-1];
 for (j = m; j < r; j++)
     helper[r + m - j] = z[j + 1];
 for (k = 1; k <= r; k++)
     z[k] = helper[i] < hilf[j]) ? helper[i++] : helper[j--];
}
```

Hence, considering the time T for each step in the above algorithm, we can write down a recurrence for merge sort: The first step is done in constant time, which is written $T_1 = \Theta(1)$. For the second step we have: $T_2 = 2\Theta(N/2)$, and the third step is $\Theta(N)$. So, we finally have:

$$T(N) = \begin{cases} \Theta(1) & \text{if } N = 1, \\ 2T(N/2) + \Theta(N) & \text{if } N > 1. \end{cases} \quad (1.18)$$

Usually, one omits the case $N = 1$ in equation (1.18) because it has no impact on the asymptotic solution of the recurrence, and just writes $T(N) = 2T(N/2) + cN$

with some positive constant c. We now use a graphical method – a *recursion tree* – to resolve this recurrence, i.e. we try to understand its asymptotic running time behavior by visual inspection of the recurrence. At the beginning, we have $T(N) = cN$ which is split into two parts which have $T(N/2)$ and $T(N/2)$. We then work on each of these two leaves and further split them into two halves, for which we have 4 times $T(N/4)$, and so on, until we end up at 1 element, which is $T(1)$ (see Figure 1.21). As one can see from drawing the recurrence, the height of the recurrence tree is given by $N \log_2 N$, because the problem is split into *two* halves at each recurrence step. The number of leaves is N and the overall running time, or efficiency, of this algorithm is $N \log N$. This is a typical running time for algorithms that are based on consecutively splitting a problem into smaller problems until the trivial case is reached. Due to the particular strategy applied, which in step 2 splits the array into two halves ("Divide") and then – after sorting the parts separately – merges them together, such algorithms are called *divide and conquer*.

> **Divide-and-Conquer** is a class of algorithms with a typical running time of $\Theta(N \log N)$.

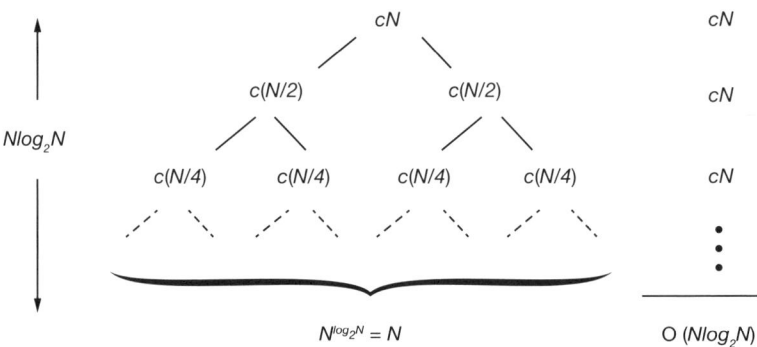

Figure 1.21. Constructing a recursion tree for the recurrence of Merge sort in equation (1.18).

1.7 Theory, modeling and computer simulation

Think!

> "Science is facts; just as houses are made of stones, so is science made of facts; but a pile of stones is not a house and a collection of facts is not necessarily science."
>
> *Henri Poincaré*

The main purpose of *fundamental* physical theories is to *explain observations* (and not so much to deliver predictions). Observations are required to test a physical theory – after all, it might prove wrong. Yet, if the only reason that we trust a theory is based on the fact that it always was right so far, we haven't learned very much, because then the theory is just a duplicate of nature. In order to *explain* observations, physical theories introduce *abstractions* and *simplifications* – generally called *principles* – that reduce the complexity of the real system. Such principles usually can be formulated and understood without the symbolic language of mathematics. Among the most important fundamental principles in classical physics are Albert Einstein's *principles of special and general relativity*, which are complemented by Heisenberg's *uncertainty principle* in quantum theory, the *principle of entanglements of quantum states* and *statistical physics*. These general principles form the essential framework of physical theories and are generally assumed to be applicable to and valid for *any* physical system[69].

One crucial step in obtaining a conclusive description of a system's behavior in physical theories, is the *isolation* of the considered system from its environment. This procedure is called *preparation of a system*. For example, we will see in Chapter 3 that, in thermodynamics, an *isolated system* describes a situation where the system can exchange neither energy nor particles, thus keeping the total energy of the system constant. In such idealized situations, where the system's parameters can be controlled, one can hope to find the basic laws underlying the dynamics of the system.

1.7.1 What is a theory?

Before expanding on the various general principles mentioned above, let us introduce several common primitive concepts in physical theories. The subject of a theory is usually called a *physical system*, which takes the place of what we ordinarily call a "thing" but avoids the idea of being spatially "well packaged." A system may or may not consist of parts; a partless system is called *single*. The *state* of a physical system

---

[69] In Chapter 3 we will introduce yet another important fundamental principle which is constantly used in statistical physics, namely the *principle of conservation of information*, which is normally so deeply and implicitly assumed that it is rarely mentioned.

provides an abstract summary of all its characteristics at a specific time, without committing to any specific way of characterization. Definite characterizations are made by *dynamical variables*. There can be many variables that describe the same state. The state and the dynamical variables are equally important in physical theories. A system can assume different states at various times. All possible states of the system are encompassed in its *state space*, or *phase space*, as it is often called in physics[70]. The state spaces of all systems of the same kind have the same mathematical structure, usually representable by mathematical group theory. The states or dynamical variables of a system evolve according to certain *equations of motion*. The changing states are represented by a curve in the state space. For example, the state space of a single classical particle is a six-dimensional differentiable manifold, spanned by the three axes of space coordinates $\{x, y, z\}$ and the three axes of components of momenta $\{p_x, p_y, p_z\}$, when using a Cartesian coordinate system.

> A *theory* is a logical system of general principles of reduced complexity, compared to the "real" object studied, which helps to understand observations.

Particles, classical physics and the N-body problem

One very important theoretical principle that we haven't mentioned yet is the notion that everything in the world is made of small, indivisible particles called *atoms*[71]. This assumption about the inner structure of matter has by itself been extremely successful in explaining many observed phenomena, e.g. the properties of gases and fluids and the abstract notions of "energy" and "heat" in statistical mechanics. The doctrine of the atom is ascribed to the Greek philosophers Leucippus (5th century BC) and his pupil Democritus of Abdera (460–370 BC), but only led to real progress in terms of a unified understanding and interpretation of many observations of the behavior of solid bodies through the works of Galileo Galilei and Sir Isaac Newton. Galilei formulated the law of inertia in his major work "Discorsi" published for the first time in Florence, in 1632 [139], which was re-formulated roughly 55 years later as "Lex I" by Newton in his "Principia".

The importance of the law of inertia lies in the fact that it constitutes a preferred system for the description of laws of nature, namely the so-called *inertial systems*, which is an equivalence class of systems which are either at rest relative to each other or in relative motion at constant speed. Newton, with his law of motion[72] and the

---

[70] Compare to this our expanded discussion of a priori probabilities and phase spaces used in thermodynamics in Chapter 3.

[71] From the Greek word "ατομο" meaning "indivisible".

[72] Newton's three laws are Lex I (law of inertia), Lex II (law of motion $\vec{F} = m\vec{a} = \dot{\vec{p}}$) and Lex III (principle of actio=reactio).

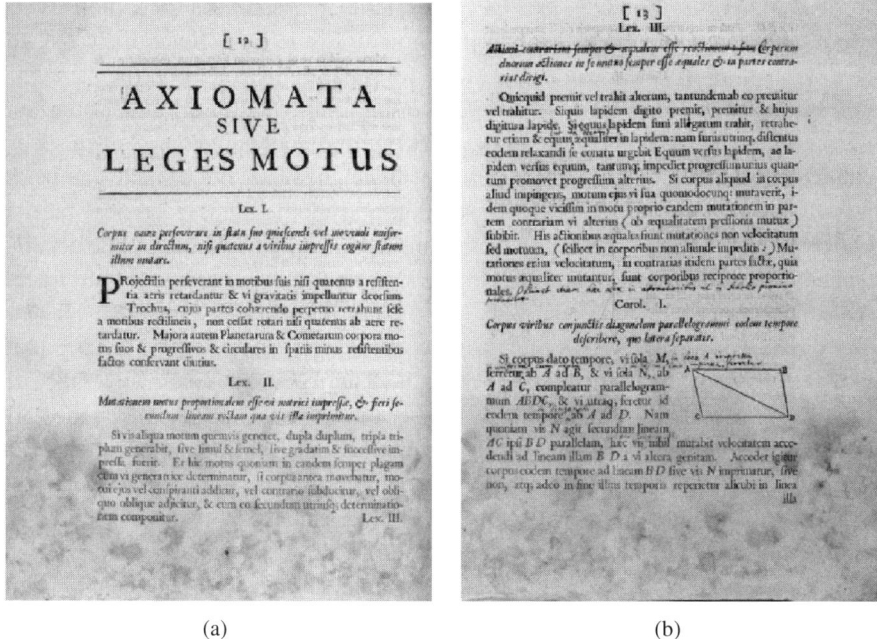

Figure 1.22. Scanned copy of the pages displaying Newton's laws in the original first issue of Isaac Newton's personal copy of the Principia from 1687. (a) Laws I and II. (b) Law III with annotations by Isaac Newton. Photocopy provided by the "digital library" of the University of Cambridge (http://cudl.lib.cam.ac.uk/collections/newton).

universal force law of gravitation between pairs of particles i and j

$$\vec{F} = Gm_i m_j \frac{(\vec{r}_j - \vec{r}_i)}{|\vec{r}_j - \vec{r}_i|^3}, \tag{1.19}$$

could introduce dynamical variables (position $\vec{r}$, velocity $\vec{v}$ and acceleration $\vec{a}$), with which the motion of particles – culminating in the concept of a "mass point" (a point with definite dynamic variables $(\vec{r}, \vec{v}, \vec{a})$ and mass m but no extension in space) – could be described. The N-body problem of celestial mechanics formulated in the second-order system of equations in (1.19) is an initial-value problem for ordinary differential equations and was the subject of many investigations by the best minds of the 18th and 19th centuries. For $N = 2$, the problem was solved by Johann Bernoulli in 1710 [440]. For $N > 3$ it was Poincaré who proved in 1896 that no general solution exists [333].

Newton was the first one with the idea to use the general principles of a theory and combine them with mathematics. The invention of calculus allowed him to describe the motion of particles (and more complex solid bodies, so-called *point systems*) by the formulation of *differential equations* with initial and boundary conditions, which was probably *the* major breakthrough in the natural sciences. Since then and until today, the method of science consists of the formulation of general principles which are then used to derive equations for the description of the systems under consideration, usually in the form of (ordinary or partial) differential equations. When specific dynamical variables are chosen as described above, the state of a particle can be definitely described, say, in terms of its position and momentum. Classical particles are governed by Newton's equation of motion (EOM)[73], which are completely time-reversible and deterministic, and the temporal variation of a particle's state traces a curve in state space.

The theory of special relativity rests on two postulates: the principle of special relativity and the constancy of the speed of light. The theory of general relativity also has two basic postulates: the principle of general relativity and the equivalence principle. The constancy of the speed of light is a feature of electromagnetism. It asserts that the speed of light is the same in all coordinate systems[74] and is independent of the motion of the light source. Of the four fundamental postulates, it alone is not controversial; the other three have spawned a great deal of analysis, connected with *symmetry*, which is another very important principle, often used in theories, to *impose restrictions on the possible mathematical formulations* (or realizations) of the various structures deduced from the basic principles.

In mathematics, basic principles that are not subject to further questioning are called "axioms". At the beginning of the 20th century the mathematician David Hilbert in Göttingen pursued the idea of deducing every theorem in mathematics by logical steps from the postulates of a given axiomatic system [207, 208], e.g. the Zermelo axioms [449] of the natural numbers. Later, he also tried to apply this conviction to the fundamental equations of theoretical physics [209] and, as a matter of fact, one hallmark of a "theory" is that the statements of the theory can be ascribed to a basic set of principles, or axioms, which are then used to formulate laws for dynamical variables, using the symbolic language of mathematics and differential equations. In 1931 however, Hilbert's program received a sharp blow when the Austrian logician Kurt Gödel published his incompleteness theorem [173]. Gödel proved that *any sufficiently rich, sound, and recursively axiomatizable theory is necessarily incomplete* and showed this for the example of the axiomatic system of the natural numbers.

---

[73] Newton's EOM are ordinary differential equations of second order, i.e. they need two initial constants to uniquely specify a solution.

[74] Coordinate systems are arbitrarily chosen frameworks that we use to denote the position and time (called "space-time") coordinates of "events", the dynamics of which is described in a theory. The coordinates themselves have no physical meaning – they are just numbers denoting points in space-time and when changing from one coordinate description to another, one has to adhere to certain "transformation rules".

Quantum physics, entanglement and wave-particle duality

The structure of quantum mechanics has more elements. More specifically, its dynamical variables, called *observables*, have a more complicated conceptual structure – the Hilbert space [427, 178] – which is also the space in which the fundamental properties of finite elements[75] are described. In addition to describing the state, an observable also provides the possible outcomes of measurements, a role it shares with classical dynamical variables.

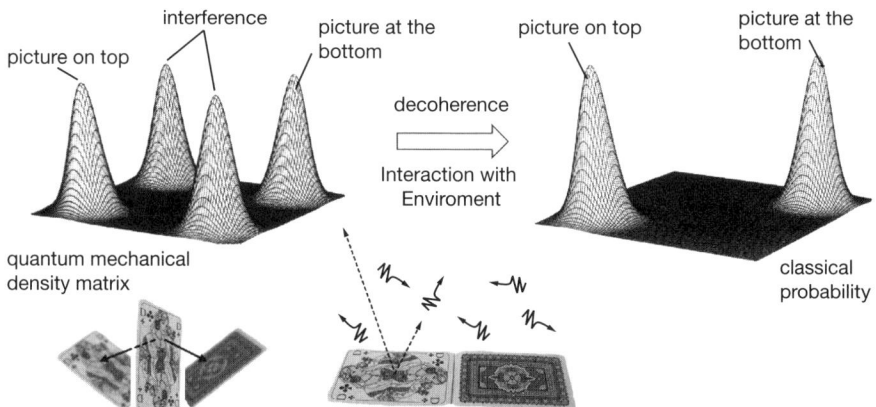

Figure 1.23. Illustration of quantum mechanical entanglement. At the beginning of the experiment, the falling card in a quantum mechanical description should be in an entangled state, which means that it is at the same time both falling to the left and falling to the right, until a measurement, i.e. an observation is made. According to the decoherence theory of quantum mechanics, even the weakest interaction of the system (in this case the falling playing card) with the environment – say with a single photon or a single molecule of the surrounding air – changes the quantum mechanical density matrix in such a way that only the *classical* probabilities remain. In this case, these are the (a priori) probabilities for the picture of the card to be on top or at the bottom. This decoherence of all truly macroscopic systems which cannot really be isolated enough from their environment is made responsible for the fact that in the classical macroscopic world, entangled states are actually never observed.

The development of quantum theory in the beginning of the twentieth century obliged scientists (and philosophers) to radically change the concepts they used to describe the world [224][76]. For example, to take into account the particle-wave duality of matter, quantum mechanics had to renounce the idea of a classical particle

---

[75] "Finite elements" is a certain form of spatially discretizing a system in space, endowed with particular properties the theory of which can be most easily described in Hilbert space.

[76] A few decades later, this conceptual revolution enabled a series of technological revolutions, for example, semiconductors, the transistor, and the laser which are at the root of today's information-based society.

trajectory. This renunciation is probably best stated in Werner Heisenberg's *uncertainty principle* published in 1925 [204] and the idea of mutually exclusive pairs of observables, say A and B, which cannot be measured both at the same time with arbitrary exactness[77], sets an absolute lower limit to what we can know (by measurements) about quantum systems. The uncertainty relation can be expressed as:

$$\Delta A \, \Delta B \geq |\langle [A, B] \rangle|, \tag{1.20}$$

which expresses the fact that for two mutually exclusive Hermitian operators A and B (e.g. position and velocity)[78], there does not exist an eigenvector in Hilbert space for which the statistical spread $\Delta A = \sqrt{\langle (A - \langle A \rangle) \rangle^2}$ and ΔB both vanish. The uncertainty principle is also closely connected to the representation of the state function of a quantum mechanical system by a function $\Psi(\vec{x}, t)$, as a Fourier integral of many waves and with the so-called *particle-wave duality* of quantum objects [35, 419, 192], which refers to the fact that the measurement process itself determines whether a quantum object shows its particle or wave properties. The first comprehensive paradigm of quantum theory was centered around the Heisenberg and Schrödinger [370, 371] formalisms of 1925 and 1926. The latter was a wave equation for matter, completing a beautiful duality: like light, matter can behave as either a particle or a wave. The wave-particle duality was originally L. de Broglie's 1924 proposition in his audacious PhD thesis [123][79], and remains completely incomprehensible to the classical way of thinking.

After an impressive accumulation of successes in the explanation of chemical, electrical and thermal properties of matter at the microscopic level, and eventual technological successes based on quantum theory, one might think that by 1960 all the interesting conceptual questions about quantum theory and quantum mechanics of materials had been raised and answered. However, in his now famous paper of 1964 [50], John Bell drew physicists' attention to the extraordinary features of *entanglement*. The idea of entanglement [68, 373] (and its immediate destruction in interacting objects) was crucial to explain the existence of a "classical world", where no superposition (entangled) states of macroscopic objects are observed, which clashes with the claim that quantum theory is *in principle* valid also for *classical*, i.e. *macroscopic* systems.

What one has to understand here, is that classical objects are not just *enlargements* of small quantum objects, but simply *many* quantum objects, which *interact* with each other. It is this interaction (even with a single photon) that immediately destroys the

---

[77] This is actually a crucial point in the so-called "Copenhagen interpretation" of quantum theory put forward by Niels Bohr and his collaborator Werner Heisenberg in the 1930s [66].

[78] Mathematically, mutual exclusiveness means that the commutator $[A, B] = AB - BA$ of the operators A and B, $[A, B] \neq 0$, i.e. that $AB \neq BA$.

[79] Due to the extraordinary nature of de Broglie's proposition, the PhD committee asked Albert Einstein in Berlin for advice and when his answer was positive, the committee accepted de Broglie's work as PhD.

quantum entanglement of the constituents of macroscopic objects. What remains are only those parts of the wave functions that correspond to the classical probabilities for finding the system in a certain state. See the example with two "quantum cards" in Figure 1.23. Quantum theory describes a pair of entangled objects as a *single global quantum system*, impossible to think of as two individual objects, even if the two components are far apart. John Bell demonstrated that there is no way to understand entanglement within the framework of the usual ideas of a physical, *local* reality in space-time that obeys laws of causality. This result was opposite to the expectations of Einstein, who had initially pointed out, in yet another famous publication, with his collaborators Podolsky and Rosen [145] in 1935, the strong correlations in the framework of ideas of a local physical reality. The most remarkable feature of Bell's work was undoubtedly the possibility it offered to determine *experimentally* whether or not Einstein's ideas would hold. Experimental tests of *Bell's inequalities* in the 1970s, 80s and 90s [169, 172, 101, 38, 37, 97] gave an unambiguous answer, namely that Einstein was not right and the idea of *local realism* has to be abandoned [192].

Statistics constitutes another important very general principle of theories, built into the mathematical structure of quantum mechanics, in an object called "density operator", the quantum mechanical analog of a classical distribution function. Statistics determine the mathematical structure of statistical physics – a theory that deals with the description of classical N-particle systems – which in turn is the theoretical foundation of two important simulation techniques, MD and Monte Carlo, which we will study in Chapters 5 and 6.

Symmetries in theories

One common meaning of "symmetry" is balance and proportion. This meaning originates from the Greeks and is closely associated with the notion of beauty. This is not the sense used in physics, where we are concerned with symmetry as in "bilateral" or "radial" symmetry. Symmetry pertains to a system as whole. When we say that something has bilateral symmetry, we refer not to its left or right side but to the whole figure. The features characterized by symmetries are meaningful only within the context of the whole, and they are recognizable without reference to external factors.

Technically, the symmetry of an object is defined in terms of the transformations that transform the object back into itself, or that leave the object unchanged or invariant. Usually, the larger the group of transformations, the more symmetric the object is. Thus a circle is more symmetric than a triangle. There are infinitely many ways to transform a circle into a position indistinguishable from the original: rotating about the center through any angle or flipping about any axis. Only six transformations can bring an equilateral triangle back into itself: rotations through 120, 240, and 360 degrees about the center, and rotations through 180 degrees about the three bisectors. If no transformation other than the identity can bring a figure back into itself, then the figure is asymmetric.

The set of symmetry transformations forms a group. A group is an algebraic structure; algebraic structures are defined by abstract operations. Abstractly, a group is a set of rules with a single rule of composition. A simple example of a group is the set of integers under the rule of addition. Under a symmetry transformation, the initial and final configurations of the system are identical. The concept of symmetry contains a concept for difference, another for identity, and a third one relating the two. The difference is marked by some kind of labels or names, usually called coordinates in physics. The identity is achieved by the invariance of the system under transformations. An equivalence class of coordinates, a group of transformations, and a system with certain invariant features form the triad of the concept of symmetry.

> **The importance of symmetries.** Because symmetries reveal the overall structures of physical systems, they provide a powerful tool for extracting information about a system without knowledge of the details of its dynamics. They come in handy, because equations of motions are often difficult or impossible to solve[a]. Symmetry considerations cannot replace solutions, but if exact solutions cannot be found they enable us to classify possible solutions, exclude certain classes as forbidden, and find selection rules for various transitions. Symmetry is even more useful when we want to find some fundamental laws. We know that if the unknown law has certain symmetries, its solution would exhibit certain patterns. The solutions of the unknown laws correspond to experimental data. Thus, one starts by gathering data, and organizes it into a large pattern whose symmetries provide a handle to help guess at the fundamental law[b].
>
> ---
> [a] In fact, Newton's equations are not solvable analytically for $N \geq 3$ [333, 365, 304].
> [b] The development of the quark model for the strong interaction is a good example of the synoptic power of symmetries, pointing the way to the fundamental laws [176, 177].

Transformations in physics are not arbitrary. They are generated by dynamical variables, through which symmetries are associated with conservation laws. For example, linear momentum is the generator of translations. The invariance of a system under transformations implies the conservation of the associated generator. If a system is invariant under temporal translations, its total energy is conserved; if it is invariant under spatial translations, its linear momentum is conserved; if it is invariant under rotations, its angular momentum is conserved. Conservation laws are already great generalizations sweeping across many branches of physics, but they are only one aspect of symmetries. The ascendancy of symmetries signifies intensive abstraction and generalization in physics.

Einstein stated the idea of symmetry in his famous principles of special and general relativity. These principles state that the mathematical form of the laws of nature are the same (they are invariant) in a certain class of coordinate systems. Each principle actually specifies an equivalence class of coordinate systems, which constrains the content of the physical theory. The constraint can be severe, as in special relativity, or it can be unobtrusive, as in general relativity. Whatever the specific constraints, the general idea of the principles is that certain physical quantities are invariant under certain groups of coordinate transformations. Stated this way, the relativity principles are *symmetry principles*. The revolutionary concepts they encapsulate, symmetry and invariance, quickly go beyond spatio-temporal characteristics and find application in many areas. The symmetry structure of physical theories unites many broad principles, of which conservation laws are one example and the coordinate-free expression of the laws another. The three most important conservation laws of classical physics – conservation of energy, momentum and angular momentum – are fundamentally based on symmetries of space-time, mathematically proved in Emmy Noether's theorem [317]. The coordinate-free, or coordinate-invariant[80] formulation of physical laws gave rise to the use of tensor analysis and tensor equations in physics, which – before Einstein's theory and the formulation of the *principle of general covariance* were only used to describe point-dependent stresses and tensions in elasticity theory.

1.7.2 What is a model?

Modeling dynamic processes and solving the models by using algorithms are the two essential parts that make up the field of "Computer Simulation" or "Scientific Computing". When modeling the complexity of nature one often has to make do with a simplification in an attempt to understand the original problem by solving a simpler problem of reduced complexity, but with well-defined (and thus, controllable) boundary conditions. Typical simplifications are *linearization* and *discretization* of equations. As one prominent example of the difficulty of solving model systems, we may mention Albert Einstein's general relativistic field equations, which are *nonlinear* and *coupled* partial differential equations, to which we are still trying to find solutions, almost 100 years after they were published for the first time in 1915 [144].

Scientific abstraction consists in replacing the part of the real world under investigation by a model [395, 361]. This process of designing models by abstraction and complexity reduction can be regarded as the most general and original principle of modeling. It describes the classical scientific method of formulating a simplified imitation of a real situation with preservation of its essential features. In other words, a model describes a part of a real system by using a *similar but simpler* structure. Abstract models can thus also be regarded as the basic starting point of a theory. However, we have to stress here that there exists no such thing as a unified exact method of deriving models, see Figure 1.24. This applies particularly to materials science, where one deals with a large variety of length and time scales, and mechanisms.

---

[80] The more general term, relating to tensor equations is "covariant".

> How do "good" models distinguish themselves from "bad" models?

Models can take many forms, including analytical, stochastic, declarative, functional, constraint, spatial or multimodel. A multimodel is a model containing multiple integrated models, each of which represents a level of granularity for the physical system.

The words "modeling" and "simulation" are often distinguished by somewhat arbitrary arguments or simply used as synonyms. This lack of clarity reflects the fact that theoretical concepts and technical facilities encountered in computational materials science develop faster than semantics. A less ambiguous definition of both concepts might be helpful, in order to elaborate on a common language in this field. In current scientific understanding the word "modeling" often has two quite different meanings, namely, *model formulation* and *numerical* modeling. The latter term is frequently used as a synonym for numerical simulation[81].

> **A model**, in essence, is an idealized representation or example that provides a benchmark for thinking about a real-life phenomenon. A *mathematical model* serves exactly the same purpose: it is the description of an experimentally delineated phenomenon by means of mathematics, with a view to capturing the salient aspects of the phenomenon at hand. A model is both more, and more ambitious than a mere mathematical description of observations in functional form – it aims at more generality than the description of a particular occurrence, but is still restricted to a modest range of similar situations, where first principles are not applicable or unavailable.

> In contrast to bad models, good models satisfy such criteria as being coherent, efficient and consistent with as small a number of free parameters as possible[a].
>
> ---
>
> [a] This idea to remove, or avoid from the beginning, redundant elements in models is called *Occam's razor* after the 14th century monk William of Occam who often used the principle of unnecessary plurality of medieval philosophy in his writings, such that his name eventually became connected to it.

Although a large number of articles and monographs exist on modeling and simulation, only a few authors have reflected on the nature and fundamental concepts of modeling. Interesting contributions in that context were made by Rosenblueth and

---

[81] *simulare* (Latin): fake, duplicate, mimic, imitate.

Section 1.7 Theory, modeling and computer simulation 69

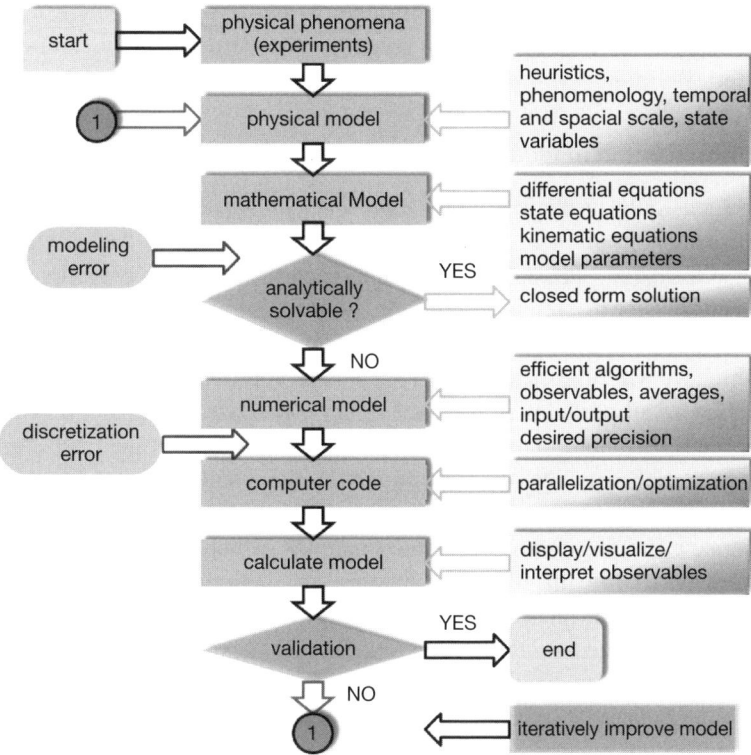

Figure 1.24. Modeling and simulation scheme. The process of modeling and simulation displayed as flow chart. At the beginning there are observations of the physical phenomena in experiments. The purpose of theories is to explain these observations within the framework of general principles. Usually these principles are too general to actually calculate anything, so one develops a simplified version of the theory, which is called a model. The model is most often formulated in the symbolic language of mathematics in the form of differential equations. The approximation (or the error) at this stage of the process enters by way of the implicit assumptions that were put into the physical/mathematical model. If the equations can be solved analytically, one is done. In all other cases one can try to find efficient algorithms that solve the equations numerically with a certain precision, indicated by the discretization error that occurs due to fact that decimal numbers can be represented in the binary number system with only finite precision. The algorithm is coded in a programming language and executed on a computer, which then calculates an approximate solution of the dynamic equations of the underlying model, by the step-by-step procedure specified in the algorithm. The numerical results are finally compared with either analytical solutions (limiting cases) or with experimental data. This allows for iteratively improving the model parameters, in order to ultimately obtain a numerical result close to the observed behavior of the experimental system.

Wiener [361], who contemplated the philosophical side of modeling, by Koonin [256], Doucet [138], and by Padulo [323]. A recent and modern treatment can be found in Steinhauser [395].

Spatial discretization in models

Here, we focus only on the *two most important categories of models* which can be identified, namely *continuum* and *particle models* according to the considered *spatial scale* in a system. Possibly the best approach to model the evolution of microstructures consists of discretely solving the equations of motion of all atoms, or particles in the segment of the material of interest. This is the major approach of MD simulations which we will discuss in Chapter 5. Such an approach provides the coordinates and velocities of all particles at any instant, i.e. it predicts the temporal evolution of the entire micro-structure. The more detailed the description of the inter-atomic particle forces in such simulations (see Chapter 4), the smaller the need for additional empirical model ingredients. In contrast, models beyond the nanometer scale have to incorporate averaging continuum approximations, which are often much more phenomenological in nature than the more ab initio assumptions made at the atomic scale. While atomistic particle methods thus indeed prevail in micro-structure simulations at the nanoscale, they are usually not tractable at the meso- and macroscale where at least 10^{23} atoms are involved[82]. Even when using simple radially symmetric interatomic pair potentials, ab initio atomistic simulations are at present confined to not much more than a few hundred atoms. Thus, for modeling microstructural phenomena beyond the nanoscopic scale, one must drop the idea of predicting the motions of single atoms and switch to the coarse-grained models that we will discuss in Chapter 7 in the context of soft matter applications.

Typical examples of particle models are all classical MD and MC methods. In contrast to classical MD [21, 179, 423, 424, 346, 345, 120, 121], modern atomistic particle approaches have increasingly gained speed through the use of more realistic potential functions and substantially increased computer power [161, 160, 406].

First-principles ab intio[83] models even aim at approximate solutions of the Schrödinger equation for a (very) limited number of atoms. Variants of the ab initio approach are provided by MD in conjunction with tight-binding or local density functional theory [214, 255, 87, 44]. The numerical performance of virtually all first-principles methods is still far too slow and too much restricted to overly simple systems for computation of any microstructural evolution of real material properties, which is why these methods are usually confined solely to academic toy models in computational chemistry and physics.

---

[82] This enormous number is due to the size of Avogadro's constant, see Appendix H.
[83] Also, so-called *first-principles* models based on e.g. local density functional theory must incorporate some simplifying assumptions and principal approximations. For instance, all of these methods use the adiabatic Born-Oppenheimer approximation [71].

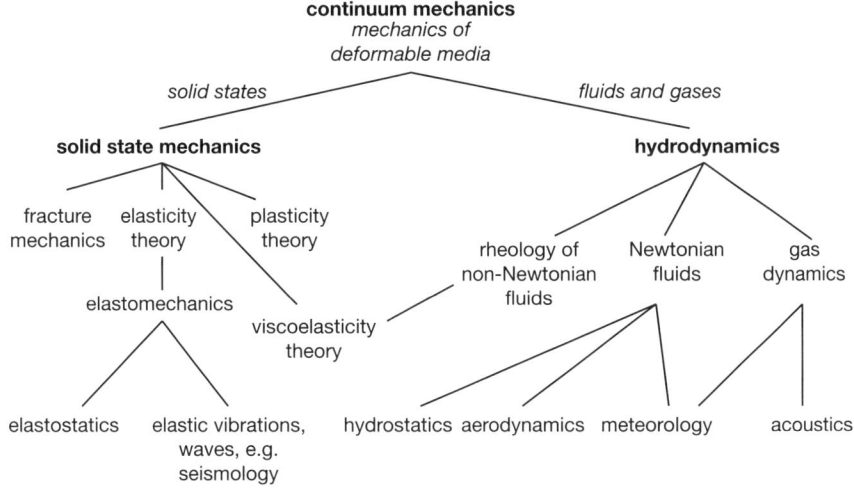

Figure 1.25. Various subfields of continuum theory. The major distinction for applications in continuum mechanics (or the *mechanics of deformable media*) is whether the theory is applied to solid states (solid state mechanics) or to fluids and gases (hydrodynamics).

Another choice is formed by *continuum models*, the detailed discussion of which is somewhat beyond the scope of this book, see Figure 1.25. We will only touch briefly on the finite element method in Chapter 3 as the most important example for numerically solving boundary value problems of partial differential equations. Since real microstructures tend to be highly complex, it can be a non-trivial and at the same time crucial task to extract those state variables out of the many observables that properly characterize microstructural phenomena at a continuum scale. From the physical picture thus obtained, one has to derive a phenomenological *constitutive description*, including the main physical mechanisms, that allows one to characterize the system behavior at a level beyond the atomic scale. The phenomenological picture must then also be translated into a mathematical model, in order to make it accessible to computer simulation. Typical examples of continuum models are classical finite element approaches [110, 453, 275, 347], polycrystal models [409, 65], self-consistent models [149, 258], certain deterministic cellular automata [426], dislocation dynamics [131], and phase field models [91]. In recent years, various refinements have been introduced particularly to enrich large-scale continuum models with microstructural ingredients. Typical examples of such advanced methods are anisotropic finite element models, which consider crystal plasticity, crystallographic texture, and the topology and morphology of the microstructure [49, 442, 443, 122, 397, 232, 398] as displayed in Figure 1.26.

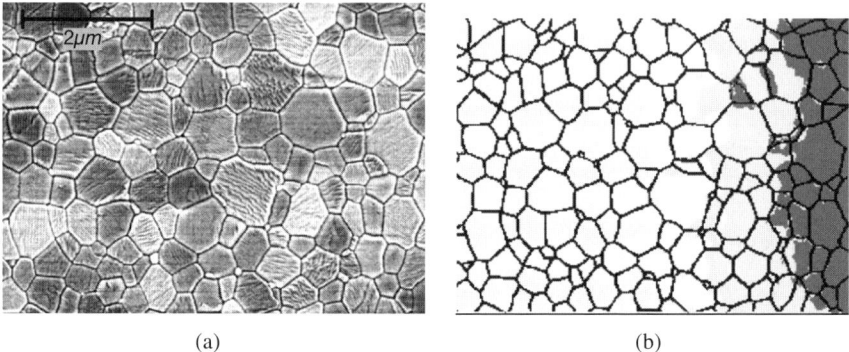

Figure 1.26. Example of the inclusion of microstructural details in FEM simulations. (a) SEM micrograph of a Al$_2$O$_3$ ceramic specimen displaying granular microstructural features. (b) Impact load simulation of a small ceramic specimen exhibiting the microstructural features of grains and grain boundaries obtained by digitization of the SEM micrograph. Displayed is the pressure level of a shock wave traversing the sample from left to right. Micrograph and simulation from unpublished data by M. O. Steinhauser, Fraunhofer EMI.

1.7.3 Model systems: particles or fields?

"Field" has at least two connotations in the physics literature. A field is a *continuous* dynamic system, or a system with *infinite degrees of freedom*. A field is also a dynamic variable characterizing such a system or an aspect of the system. Fields are continuous but not amorphous.

The description of field properties is *local*, concentrating on a point entity and its infinitesimal displacement. Physical effects propagate continuously from one point to another and with finite velocity. The world of fields is full, in contrast to the mechanistic Newtonian world, in which particles are separated by empty space, across which forces act instantaneously at a distance[84].

To get some idea of the notion of a field, consider an illustration found in many textbooks. Imagine the oscillation of N identical beads attached at various positions to an inelastic thread, whose ends are fixed. For simplicity, the bead motion is assumed to be restricted to one dimension and the thread to be weightless. We index the beads by integers $i = 1, 2, \ldots, N$ as in Figure 1.27 (a). The dynamic variable $\psi_i(t)$ characterizes the temporal variation of the displacement of the ith bead from its equilibrium position. The dynamic system is described by N coupled equations of motion for $\psi_i(t)$, taking into account the tension of the thread and the masses of the beads. Now imagine that the number of beads N increases but their individual mass m decreases, so that the product mN remains constant. As N increases without bound, the beads

---

[84] The mechanistic view of the world goes back to Issac Newton with the publication of his *Philosophiae naturalis principia mathematica* [313].

Section 1.7 Theory, modeling and computer simulation

are squashed together and in the limit we obtain an inelastic string with a continuous and uniform mass distribution, as in Figure 1.27 (b). Instead of using integers i that parametrize the beads, we use real numbers x, $0 \leq x \leq L$ as indices, for a string of length L. A particular value for x designates a specific point on the string, just as a value for i designates a specific bead. Instead of $\psi_i(t)$, the dynamics of the string are characterized by $\psi(x,t)$, called the *displacement field* of the string.

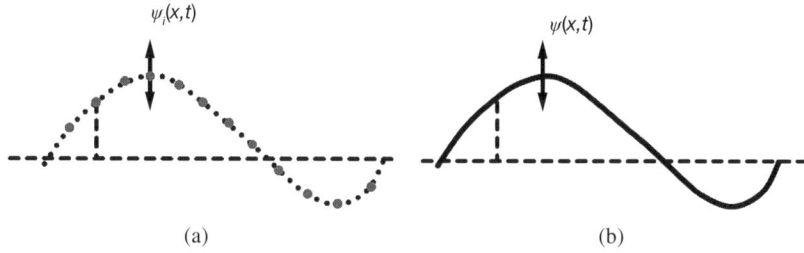

Figure 1.27. Notion of a field. (a) Beads on a vibrating thread are designated by discrete indices i. (b) Points on a vibrating string are designated by continuous indices x. The displacement of the string at any time t, $\psi(x,t)$ is called the displacement field of the string.

Parameter x does not denote position in space but position on the string; x is simply a continuous index having the same status as the discrete i. We can call x a spatial index but not an index of space points. The continuity of the parameter x should not mislead us into thinking that its values are not distinct. The real line continuum is rigorously a point set satisfying the condition of completeness. The N beads on the thread in Figure 1.27 (a) form a system with N degrees of freedom. Similarly, the string in Figure 1.27 (b) is a dynamic system with infinite degrees of freedom (as $N \to \infty$), characterized by the dynamic variable $\psi(x,t)$. In the case of the string, the parameter x is a number, so we have a one-dimensional field. The same meaning of dynamic variables with continuous indices applies in cases with more dimensions, where the spatial parameter becomes an ordered set of numbers. Consider a vibrating membrane such as the surface of a drum. Physicists describe the displacement of the membrane by a field $\psi(\vec{x},t)$ with $\vec{x} = (x_1, x_2)$. Again, x_1 and x_2 are parameters indexing points on the drum, not points in space. In the examples of the string and the drum, the dynamic variables ψ represent spatial displacements of matter, and are explicitly called displacement fields. Fields generally are not spatial displacements. To arrive at the general concept of fields, we abstract from the spatial characteristic of ψ.

> A field $\psi(\vec{x}, t)$ is a dynamic variable for a continuous system whose points are indexed by the parameters t and $\vec{x}$. A fundamental field of physics is a freestanding and undecomposable object by itself. It cannot be taken apart materially.

The spatial meaning of a field is exhausted by the spatial parameter x, which has the same status as the temporal parameter t. With $\vec{x}$ fixed, ψ varies, but generally does not vary spatially. The field variable $\vec{x}$ need not be a scalar, as it is in the previous example. It can be a vector, tensor, or spinor[85]. The parameter $\vec{x}$ allows us to distinguish points, but they are points-in-the-field and are meaningless when detached from the field.

1.7.4 The linear chain as a model system

In contrast to the theory and models of mass points and rigid bodies (point systems), the *theory of elasticity* investigates the mutual movements of the constituent particles of a body. If the mutual movements of particles are confined to a few lattice spacings, one can ignore the discreteness of the body and make use of a continuum model. The advantage of continuum models is that the number of relevant equations is considerably reduced. We illustrate this with the model of a linear chain, cf. Figure 1.28.

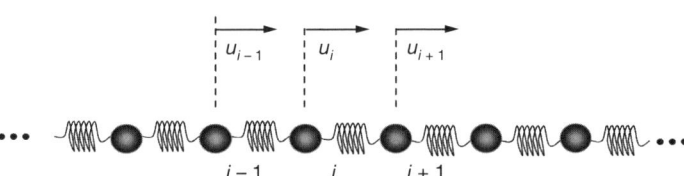

Figure 1.28. Model of a linear chain. The linear chain was introduced as a simple model for a solid state by Albert Einstein in 1907 (Einstein model [143]).

The linear chain consists of N mass points m connected via springs with spring constant λ. The chain is translation invariant in one dimension. In a particle treatment, one obtains the following equation of motion:

$$m\ddot{u}_i + \lambda(2u_i - u_{i+1} - u_{i-1}) = 0. \tag{1.21}$$

These are N coupled ordinary linear differential equations. Assuming that the movement of neighboring mass points is almost identical, one can set

$$u_i = u(x_i, t) = u(x, t), \tag{1.22}$$

---

[85] These three different geometrical quantities are generally defined by their transformation behavior under coordinate transformations.

Section 1.7 Theory, modeling and computer simulation

with t as time parameter. It follows that

$$2u_i - u_{i+1} - u_{i-1} = 2u(x) + u(x + \Delta x) - u(x + \Delta x) = -\frac{\partial^2 u}{\partial x^2}(\Delta x)^2, \quad (1.23)$$

with Δx being the mutual particle distance. In the limit of $\Delta x \to 0$, and at the same time keeping the product $\Delta x \cdot \lambda = f$ constant, one obtains with $c^2 = f/\rho = \lambda(\Delta x)^2/m$ the wave equation

$$\frac{\partial^2 u}{\partial x^2} - \frac{1}{c^2}\frac{\partial^2 u}{\partial t^2} = 0. \quad (1.24)$$

Thus, in a continuum description, instead of having to solve $3N$ coupled equations, there is only *one* (albeit partial) differential equation.

Exercise 1.7 (Linear Chain). Solve the differential equation (1.21).
The total force on the chain is given by

$$F_s = \sum_n f_n(u_{s+1} - u_s). \quad (1.25)$$

A possible ansatz for the solution is

$$u_{s+1} = u \exp(\pm i(qa - \omega t)) \quad (1.26)$$

with wave vector q, distance a of neighboring atoms on the lattice and angular frequency ω. Inserting this ansatz into (1.21) yields

$$\omega^2 M = f(2 - \exp(iqa) + \exp(-iqa)) = 2f(\cos(qa) - 1) = 4f \sin^2\frac{qa}{2}. \quad (1.27)$$

Thus, as a result, one obtains for the angular frequency of the wave that propagates through the crystal, the *dispersion relation*

$$\omega = 2\sqrt{\frac{f}{M}} \left|\sin\frac{qa}{2}\right|. \quad (1.28)$$

Looking at the elongations of neighboring particles

$$\frac{u_{s+1}}{u_s} = \frac{u \exp(i(s+1)qa)}{u \exp(isqa)} = \exp(iqa) \quad (1.29)$$

one realizes, that the useful range of the exponential function with phase qa is restricted to the interval $(-\pi = qa = \pi)$. A larger difference in phase than $\pm\pi$ makes no sense physically, as all behavior can always be reduced to this interval by consecutive subtraction of 2π from the phase. Thus, physically useful values of the wave vector q are restricted to the *Brillouin-zone*:

$$-\frac{\pi}{a} \leq q \leq \frac{\pi}{a}. \quad (1.30)$$

From equation (1.28) it follows that, in the limit of $q \ll 1/a$, i.e. for wave lengths ω that are large compared to a lattice constant a, the angular frequency is directly proportional to q, i.e.,

$$\omega = \sqrt{\frac{fa^2}{M}} q. \qquad (1.31)$$

In this case, the crystal lattice behaves like a continuum and deviations from continuum behavior are stronger, the closer the wave vectors are to the edges $\omega \pm \pi/a$ of the Brioullin-zone. The group velocity v_g of the wave packet, which transports energy through the lattice, is given by

$$v_s = \frac{d\omega}{dq} = \sqrt{\frac{fa^2}{M}} \cos\frac{qa}{2}. \qquad (1.32)$$

In the case that $q \ll 17a$, v is independent of the wave number q. The expression of v_g then corresponds to the propagation of a longitudinal sound wave. Assuming for the velocity of sound $v_s = 4 \times 10^3$ ms^{-1} and a lattice constant $a = 2 \times 10^{-10}$ m, one obtains for the phonon frequency

$$\omega \approx v_s q = v_s \frac{\pi}{a} = 2\pi \times 10^3 \text{ Hz}. \qquad (1.33)$$

To experimentally probe the properties of lattice oscillations of solids and fluids, different elastic scattering techniques are used, e.g. X-rays, photons or neutrons. Different scattering techniques only differ by the amount of energy that is carried by the scattering particles.

The continuum hypothesis

Continuum theory and continuum mechanics are based on the assumption of the existence of a continuum, in engineering textbooks sometimes called the "Continuum Hypothesis". Mathematically, the notion of a continuum is provided in real analysis by the fundamental limiting process used in a Cauchy series. This expresses the idea that the real numbers are "arbitrarily dense", that is, there is "no space left" to fit another number between two real numbers. Another way of expressing this idea in the language of sets is to say that the real numbers are an uncountable infinite set.

The existence of mathematical objects does not necessarily mean that these objects also exist (or have any meaning) in the "real" physical world of observation. Physics is an experimental science, and all theorems and ideas ultimately have to pass the experimental test. Using the real numbers as an example we want to expand on this idea.

One of the most fundamental notions used in mathematics is the axiom of the existence of real numbers[86], denoted with the letter $\mathbb{R}$. Real numbers are called "real"

---

[86] It is attributed to Leopold Kronecker (1823–1891) who once said: "The natural numbers are God-given. The rest is the work of man" [206].

Section 1.7 Theory, modeling and computer simulation

because they seem to provide the magnitudes needed for the measurement of distance, angle, time, or numerous other physical quantities. Hence, are real numbers "real"?

Real numbers refer to a *mathematical idealization*, a free invention of the human mind, rather than to any actual, objective, physical reality. For example, the system of real numbers has the property, that between any of them, no matter how close they are, there is a third[87]. It is highly questionable whether physical distances or time intervals actually have this property. If one continues to divide up the physical distance between two points, one eventually reaches scales so small that the whole concept of distance, in the intuitive primitive sense, ceases to have any meaning. In this context it is interesting to mention that, as early as 1899, Max Planck realized that the fundamental physical constants G (Newton's gravitational constant), c (velocity of light), and $\hbar = \frac{h}{2\pi}$ (later called Planck's quantum of action) could be combined in such a way as to build a length, a time and a mass [330]. These fundamental values are given by:

$$\text{(Planck length)} \quad l_P = \sqrt{\frac{\hbar G}{c^3}} \quad \approx 1.62 \times 10^{-35} \text{ m}, \quad (1.34a)$$

$$\text{(Planck time)} \quad t_P = \frac{l_P}{c} = \sqrt{\frac{\hbar G}{c^5}} \quad \approx 5.40 \times 10^{-44} \text{ s}, \quad (1.34b)$$

$$\text{(Planck mass)} \quad m_P = \frac{\hbar}{l_P c} = \sqrt{\frac{\hbar G}{c}} \quad \approx 2.17 \times 10^{-8} \text{ kg}. \quad (1.34c)$$

It is anticipated that on this "quantum gravity scale" or "Planck scale", due to Heisenberg's uncertainty principle, the ordinary notion of distance and also the notion of a continuum becomes meaningless. At this scale, one will start to notice the quantum nature of time, which is connected with energy via an uncertainty relation $\Delta E \, \Delta t \geq \hbar$. Obviously, distances smaller than the order of magnitude of Planck's constant do not make any physical sense anymore, because the uncertainty principle as a fundamental physical law of nature prevents any useful measurements of length at this scale. But in order to really mirror the real numbers one would have to go far beyond that scale, e.g. to distances 10^{300}th l_P or $10^{10^{2000}}$th l_P. Thus, are the real numbers to be considered "real"?

The example shows that theoretical constructs or models are only "real" to the extent that they lead to falsifiable consequences. In physics, the real numbers have been chosen, because they are known to agree with physical continuum concepts used for e.g. distance and time but not because they are known to agree with physical concepts over *all* scales. However, real numbers are highly applicable to the description of a continuum in the natural sciences and they have been applied to the smallest subatomic scales (quarks) $\sim 10^{-19}$ m, and to large scale structures in the universe $\sim 10^{25}$ m, thus extending the applicability range to the order of $\sim 10^{44}$. Assuming

---

[87] This property is a consequence of *Dedekind's Schnittaxiom*, see e.g. [106] or the original sources by R. Dedekind [126] (1892) and [127] (1893).

that reals are a useful concept down to the Planck scale, their range of applicability and thus their "reality" is extended to at least $\sim 10^{60}$.

In mathematics, the *continuum hypothesis* was proposed by Georg Cantor [85], after he showed that the real numbers cannot be put into one-to-one correspondence with the natural numbers. Cantor suggested that the cardinality of real numbers is the next level of infinity beyond the natural numbers. He used the Hebrew letter "aleph" to name different levels of infinity: $\aleph_0$ is the number (or cardinality) of the natural numbers or any countable infinite set, and the next levels of infinity are $\aleph_1, \aleph_2, \aleph_3, \ldots$. Since the reals form the quintessential notion of continuum, Cantor named the cardinality of the reals "c", for continuum. The continuum hypotheses is the assumption that there is no set with a cardinal number in between the number of natural numbers and the number of reals (the continuum), or simply stated: $\aleph_1 = c$.

1.7.5 From modeling to computer simulation

To simulate something physical, one first needs a mathematical model. Once a model has been developed, the next task is to execute the model on a computer – that is, to create a computer program which steps through time whilst updating the state and event variables in your mathematical model. There are many ways to "step through time." You can, for instance, leap through time using event scheduling or you can employ small time increments using time slicing. You can also execute the program on a massively parallel computer. This is called parallel and distributed simulation. For many large-scale models, this is the only feasible way of getting answers in a reasonable amount of time.

Figure 1.29 displays the principal design of most commercial simulation programs. Usually, before running a simulation, one prepares the system (variables, size of the system, number of particles, geometry, boundary conditions, etc.) with the help of a preprocessing tool that stores this information in external files. These files are then read by the main simulation program at runtime. After the simulation run, one analyzes the results, often using a special tool for data postprocessing (visualization, extraction of information, plots of functions extracted from the data). This splitting of tasks has the advantage that the main program (the solver) can be run for example on a supercomputing cluster whereas the pre- and postprocesing might be done on a local PC or laptop, provided that the amount of data can still be handled without a supercomputing system.

In order to demonstrate the increasing importance of computer simulations for research and development, we display in Figures 1.30 and 1.31 snapshots of database searches in the SCOPUS[88] database, with several keywords, as indicated in the figures for the time span 1960–2011. As one easily recognizes, the number of publications in the field of "computer simulation" has risen exponentially since the 1960s, although

---

[88] Scopus is the world's largest abstract and citation database of peer-reviewed literature in science and engineering.

Section 1.7 Theory, modeling and computer simulation 79

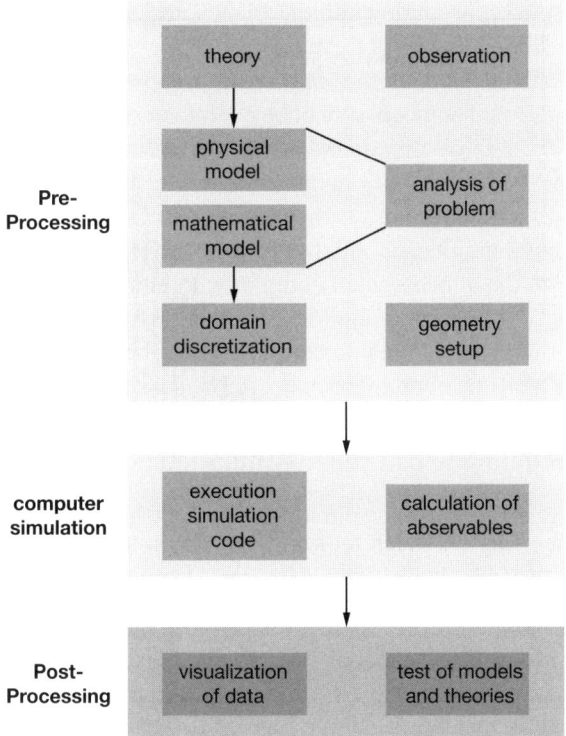

Figure 1.29. Principal design of a computer simulation. In virtually all cases the computer simulation process is split into three parts. Pre- and postprocessing, and the execution of the simulation itself. Preprocessing is done in order to set up the system geometrically, for example to construct a finite element mesh or to set up the initial particle configuration of an MD simulation, along with appropriate boundary and initial value conditions. Normally, these things are stored in external files, which are read by the computer program at runtime. This has the additional advantage that preparing the system and its analysis in the postprocessing stage are separated from each other and from the main simulation code (in commercial programs often just called "solver"). Very often, the pre- and postprocessing tools are implemented in separate main programs, isolated from the main solver that actually solves the equations of the model system.

there has been a considerable drop in the number of publications in the year 2006, leading the growth to stagnate at roughly 60 000 publications per year with "computer simulation" in either the title, abstract, or keywords. As for publications containing "finite element simulation" or "molecular dynamics simulation", there is an exponential increase in the number of publications up to the present day.

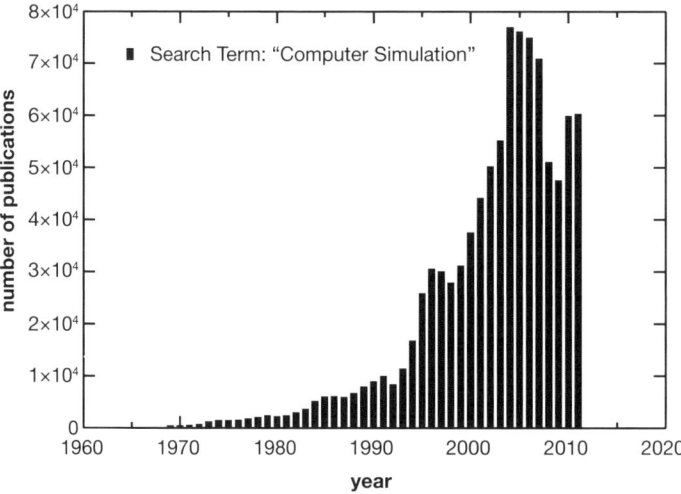

Figure 1.30. Scopus database search for keywords "Computer Simulation" in the title, abstract or keywords of publications from 1960–2011. The snapshot was generated in May 2012.

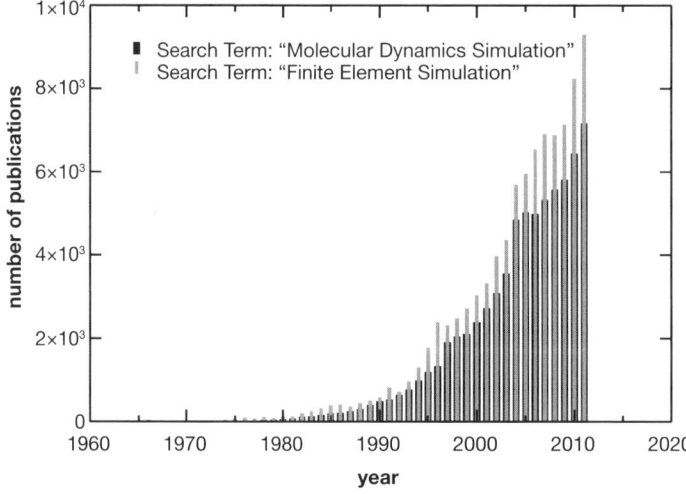

Figure 1.31. Scopus database search for keywords "Finite Element Simulation" and "Molecular Dynamics Simulation" in the title, abstract or keywords of publications from 1960–2011. The snapshot was generated in May 2012.

1.8 Exercises

1.8.1 Addition of bit patterns of 1 byte duals

Add the following decimal numbers, by first expressing them as dual numbers with 8 digits and then performing the addition.

$$(123)_{10} + (204)_{10} =?$$
$$(15)_{10} + (31)_{10} =?$$
$$(105)_{10} + (21)_{10} =?$$

1.8.2 Subtracting dual numbers using two's complement

Subtract the following decimal numbers, by first expressing them as dual numbers in two's complement with 8 digits and then performing the subtraction.

$$(57)_{10} - (122)_{10}$$
$$(43)_{10} - (11)_{10}$$
$$(17)_{10} - (109)_{10}$$

1.8.3 Comparison of running times

For each function $f(N)$ and time t in the following table, determine the largest size of a problem that can be solved in time t, assuming that the algorithm to solve the problem takes $f(N)$ microseconds.

	1 second	1 minute	1 hour	1 day	1 month	1 year	1 century
$\log N$							
$\sqrt{N}$							
N							
$N \log N$							
N^2							
N^3							
2^N							
$n!$							

1.8.4 Asymptotic notation

Prove the following equation:

$$f(N) = aN^2 + bN + c = \Theta(N),$$

where a, b, c are constants and $a > 0$.

For each of the following C-functions, determine its computational complexity, expressed in Big-O notation:

a)

```
int OddFunction (int n){
int i, j;
int sum = 0;
  for (i = 0; i < n; i++){
    for (j = 0; j < i; j++){
    sum += i * j;
    }
  }
return (sum);
}
```

b)

```
int VeryOddFunction (int n)
{
int i,j;
int sum = 0;
  for (i = 0; i < 10; i++){
    for (j = 0; j < i; j++){
    sum += i * j;
   }
  }
return (sum);
}
```

c)

```
int ExtremelyOddFunction (int n){
   if (n <= 1) return (1);
return (ExtremelyOddFunction( n / 2 ) + 1);
}
```

1.9 Chapter literature

There are many excellent texts on the general topic of algorithms, including those by Aho, Hopcroft, and Ullman [17, 18], Baase and Van Gelder [39], Brassard and Bratley [75], Chabert [90], Dasgupta, Papadimitriou, and Vazirani [116]; Goodrich and Tamassia [187], Hofri [213], Horowitz, Sahni, and Rajasekaran [220], Johnsonbaugh and Schaefer [228], Kingston [245], Kleinberg and Tardos [249], Kozen, [257], Levitin [271], Manber [281], Mehlhorn [293, 294, 295], Purdom and Brown [343], Reingold, Nievergelt, and Deo [352], Sedgewick [376], Sedgewick and Flajolet [377], Skiena [388] and Wilf [439].

In 1968, Donald Knuth published the first of a series of three volumes with the general title "The Art of Computer Programming [250, 251, 253]. The first volume displays the modern study of computer algorithms, with a focus on the analysis of running time. Knuth traces the origins of the Θ-notation to a number-theory text by P. Bachmann in 1892 [41]. The full series by Knuth still is a tremendously helpful reference for all kinds of problems related to algorithms. Sorting algorithms in particular are treated in an encyclopedic manner. The Ω and Θ-notations were advocated by Knuth in [253] to connect the popular, but technically sloppy, practice in the literature of using O-notation for both upper and lower bounds. Many people continue to use the O-notation where the Θ-notation is more technically precise. Further discussion of the history and development of asymptotic notations appears in the work of Brassard and Bratley [75]. Not all authors define the asymptotic notations in the same way, although the various definitions agree in most common situations.

The early history of proving programs' correctness is described by Gries [191]. The textbook by Mitchell [302] describes more recent progress. Some of the more practical aspects of algorithm design are discussed by Bentley [55, 56] and Gonnet [185]. Overviews of the algorithms used in computational biology can be found in textbooks by Gusfield [196], Pevzner [328], Setubal and Meidanis [378], and Waterman [434].

There are several excellent books and research papers that cover the debate on the interpretation of quantum theory [51, 351, 52, 350, 53, 303]. Information on quantum computing, quantum computers and quantum information can be found in [276, 20, 73, 314].

Chapter 2

Scientific Computing in C

"To put it quite bluntly; as long as there were no machines, programming was no problem at all; when we had a few weak computers, programming became a mild problem; and now we have gigantic computers, programming has become an equally gigantic problem."

E. Dijkstra, Turing award lecture, 1972

Summary

> This chapter aims at introducing the essential features of the programming language C for those who are new to it and have never programmed before. We start with very simple issues but will then rapidly progress and provide many exercises for the less experienced reader throughout and at the end of this chapter. We will also stress problems like overflow, underflow, rounding errors, and eventually loss of precision due to the finite amount of numbers a computer can represent. The program listings we discuss are tailored to these aims. Finally, we address some important issues of software design and programming.

Learning targets

✓ Learning the basics of the programming language C.

✓ Learning how numbers and datatypes are represented in computer memory.

✓ Learning about software design and -engineering

2.1 Introduction

The language C is a flexible, extremely powerful, high-level programming language, which was initially designed for writing operating systems and system applications such as compilers. In fact, all UNIX operating systems, as well as most UNIX applications (text editors, window managers, etc.) are written in C. But C is also an excellent vehicle for scientific programming because of its many low-level features

Section 2.1 Introduction   85

that allow for writing highly optimized, machine oriented code, which is particularly advantageous in high-performance supercomputing.

C was developed by two computer scientists at the Bell Laboratories, Dennis Ritchie and Ken Thompson in 1969, because they had had enough of the restrictions of its predecessor language, B(CLP)[1].

> At first, the language C was called NB (for *New* B), but eventually was renamed C after the next letter in the alphabet. As C++ was originally only an extension to C, they used the increment operator "++", which increases the value of a variable by 1. A stable version of yet another new programming language D [23, 2] has been around since 2008 after 8 years of development and is also a fully compatible extension to C. Up until now however, the language D hasn't been particularly noticed by the computational physics community.

C was originally developed to ease writing compilers and operating systems. However, it didn't take long until software companies, working for themselves, started to expand or remove parts of the C language at will, which soon resulted in many different C dialects, which made things worse for programmers. To do away with this problem, in 1978 Brian W. Kernighan and Dennis Ritchie (one of the developers of C) published a book, "The C Programming Language", thus initiating a first *C standard*, which became known as the *K&R standard*. In 1983, the American National Standard Institute (ANSI) created a commission, whose sole purpose was to standardize the language C. This was finally done in 1989 and this standard was also adopted by the International organization of Standardization (ISO) – the *C89*, or *ANSI-C standard* was born, which also includes the various C libraries.

At a later revision of the C standard, new header files were added. For example, in 1995 the header files <iso646.h>, <wchar.h>, <wctype.h> were added. This addition was called *Normative Amendment 1*. Finally, in 1999, the header files <complex.h>, <fenv.h>, <inttypes.h>, <stdbool.h>, <stdint.h>, <tgmath.h> were added. This revision is known as the *C99 standard* which is still the up-to-date standard (in the year 2012) and has been implemented in most current C compilers. Since 2007 the standardization commission is working on a new revision of the C standard which is currently called *C1x* which means that it will be released anytime before 2020. In this future standard, besides unicode support and functions for scanning memory and buffer overflows, thread programming, which is mandatory in light of the widespread use of multicore systems today, will also be included.

---

[1] This is short for *Basic Combined Programming Language*.

What do you need to write, compile and execute programs in C?

To write and compile a simple C source code program and to obtain an executable, one has, in essence, two options;
The first of these options consists of the following steps:

1. Type the source code using any ASCII[a] text editor and store it.

2. Translate the source code, using a compiler, and obtain an object file (with the ending ".o").

3. Bind the object file with a *linker* and thus generate an executable file. The linker also includes all functions needed from the standard C-libraries in the final executable. Both the compiler and linker have to be called from the command line.

The second option is using an all-inclusive solution to programming, i.e. an *Integrated Development Interface (IDE)*[b]. An IDE includes every tool that is needed for writing source code in *one* single window application.

[a] See also the info box in Section 1.4 on Page 31
[b] One option here is to use a commercial product such as Visual C++; However, we recommend using the freely available tool *Eclipse* which can be downloaded from http://www.eclipse.org. Eclipse can be readily run on a native Linux system or on Windows, where it integrates either into the POSIX environment provided by Cygwin or can be run with MinGW (*Mini*malist *GNU for Windows.*) MinGW (http://www.mingw.org/ provides a complete, open source programming tool set which is suitable for the development of native MS-Windows applications.

There is probably no number to back up this claim, but almost all serious scientific software development has been and is still done in a UNIX[2] environment, with a strong drift to the free Linux operating system in recent years. Linux has basically the same shell command syntax and file system organization, so the reader is strongly recommended to become acquainted with the UNIX/Linux programming environment.

---

[2] UNIX used to be *the* standard multi-user/multi-tasking operating system during the computer era of the 1980s and 1990s when workstations, instead of PCs, were the workhorses of scientific computation at universities and research institutes. Today, Linux has probably become more popular than UNIX, because it is completely free and offers the same, if not more functionality than previous commercial UNIX systems.

2.1.1 Basics of a UNIX/Linux programming environment

In writing C/C++ programs to run under UNIX or Linux, there are several concepts and tools that turn out to be quite useful. The most obvious difference, if you are coming from a Windows or Macintosh programming background, is that the tools are separate entities, not components of a tightly coupled environment like Metroworks CodeWarrior or Microsoft Visual C++. Instead, commands to the operating system are issued in a text console, a *command line interpreter* called the "shell". There are many different shells (ksh, csh, tclsh, etc.) with slight differences in syntax, but the most popular is probably still the "Bash", the "Bourne again shell". Figure 2.1 shows a snapshot of the Bash command line on a Linux system. In Appendix C.5 on Page 457, we provide a short introduction to the Bash and its special characters. In Figure 2.1, the command "which $SHELL" has been issued, which shows the full path of shell commands, in this case the path /bin/bash of the Bash program that provides the text console. If you want to find out about the available options for a particular Linux command, e.g. ls, you can type either "man ls" or "info ls" which will display the manual or info page of the respective command.

The most important tools in the UNIX/Linux domain are the *text editor*, the *compiler*, the *linker*, the *make utility*, and the *debugger*. There are a variety of choices as to "which compiler" or "which editor" to use, but the choice is usually one of personal preference. The choice of editor, however, is sometimes almost a religious issue. EMACS – or the more convenient version of it, XEMACS – integrates well with the other tools, has a nice graphical interface, and is almost an operating system in itself, so we encourage its use. Appendix C provides a summary of the most important basic UNIX and EMACS commands to get you started.

For further introductions to EMACS and the Unix/Linux operating system programming environment we refer to standard UNIX [40, 94, 327] and Linux [174, 389, 241] literature and to the many freely available online tutorials in the Internet (just perform a search for the appropriate keywords). We particularly recommend the classic "UNIX System V.4" by Jürgen Gulbins [195] and the two excellent O'Reilly books [338, 355].

The Compilation process

It is useful to review what happens during the building of an executable program. There are actually two phases in the process, *compilation* and *linking*. The individual source files must first be compiled into object modules, see Figure 2.2. These object modules contain a system dependent, relocatable representation of the program, as described in the source file. The individual object modules are then linked together to produce a single executable file (in machine code, i.e. containing only 0's and 1's), which the system loader can use when the program is actually invoked. The compilation of source code is done with the command "gcc", the GNU C-compiler. You can also compile C code using the command "g++". Both commands – gcc and

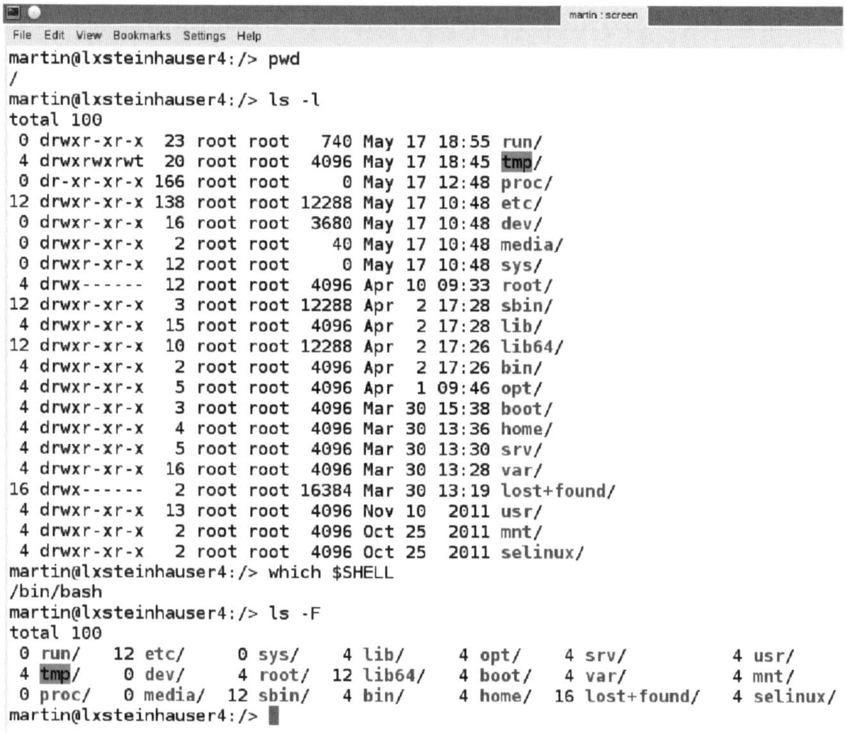

```
martin@lxsteinhauser4:/> pwd
/
martin@lxsteinhauser4:/> ls -l
total 100
 0 drwxr-xr-x  23 root root   740 May 17 18:55 run/
 4 drwxrwxrwt  20 root root  4096 May 17 18:45 tmp/
 0 dr-xr-xr-x 166 root root     0 May 17 12:48 proc/
12 drwxr-xr-x 138 root root 12288 May 17 10:48 etc/
 0 drwxr-xr-x  16 root root  3680 May 17 10:48 dev/
 0 drwxr-xr-x   2 root root    40 May 17 10:48 media/
 0 drwxr-xr-x  12 root root     0 May 17 10:48 sys/
 4 drwx------  12 root root  4096 Apr 10 09:33 root/
12 drwxr-xr-x   3 root root 12288 Apr  2 17:28 sbin/
 4 drwxr-xr-x  15 root root  4096 Apr  2 17:28 lib/
12 drwxr-xr-x  10 root root 12288 Apr  2 17:26 lib64/
 4 drwxr-xr-x   2 root root  4096 Apr  2 17:26 bin/
 4 drwxr-xr-x   5 root root  4096 Apr  1 09:46 opt/
 4 drwxr-xr-x   3 root root  4096 Mar 30 15:38 boot/
 4 drwxr-xr-x   4 root root  4096 Mar 30 13:36 home/
 4 drwxr-xr-x   5 root root  4096 Mar 30 13:30 srv/
 4 drwxr-xr-x  16 root root  4096 Mar 30 13:28 var/
16 drwx------   2 root root 16384 Mar 30 13:19 lost+found/
 4 drwxr-xr-x  13 root root  4096 Nov 10  2011 usr/
 4 drwxr-xr-x   2 root root  4096 Oct 25  2011 mnt/
 4 drwxr-xr-x   2 root root  4096 Oct 25  2011 selinux/
martin@lxsteinhauser4:/> which $SHELL
/bin/bash
martin@lxsteinhauser4:/> ls -F
total 100
 0 run/   12 etc/     0 sys/     4 lib/    4 opt/   4 srv/            4 usr/
 4 tmp/    0 dev/     4 root/   12 lib64/  4 boot/  4 var/            4 mnt/
 0 proc/   0 media/  12 sbin/    4 bin/    4 home/ 16 lost+found/     4 selinux/
martin@lxsteinhauser4:/>
```

Figure 2.1. The Bourne again shell (Bash) command line under Linux. Several commands have been issued in this screenshot. "pwd" prints the current working directory in the directory tree, in this case the lowest (root) directory, indicated with a single slash /. ls lists the directory content. If this command is used with the option -l (for "long"), "ls -l" lists an extended directory content, displaying the block size as an integer, in multiples of kilobytes, file permissions, the owner of the file, and the group to which the owner belongs (here: both are root), the size of the file in bytes, the last access date and the name of the file. By default, directories are displayed with a slash at the end. The command "which $SHELL" lists the path of the currently used shell (stored in the predefined shell variable "SHELL"), which can be printed by prefixing a $ sign to the variable and then issuing the command "echo $SHELL".

g++ do very similar things and they are really the same program. Both decide whether to compile a program as C or as C++ based on the filename extension (.c versus .c++ or .cpp). Both are capable of linking against the C++ standard library, but only g++ does this by default. So if you have a program written in C++ that doesn't happen to need to link against the standard library, gcc will do the right thing. Even though it is called a compiler, gcc and g++ are used as both compiler and linker. The general form for invoking g++ or gcc is:

```
g++ <option flags> <file list>
```

Section 2.1 Introduction 89

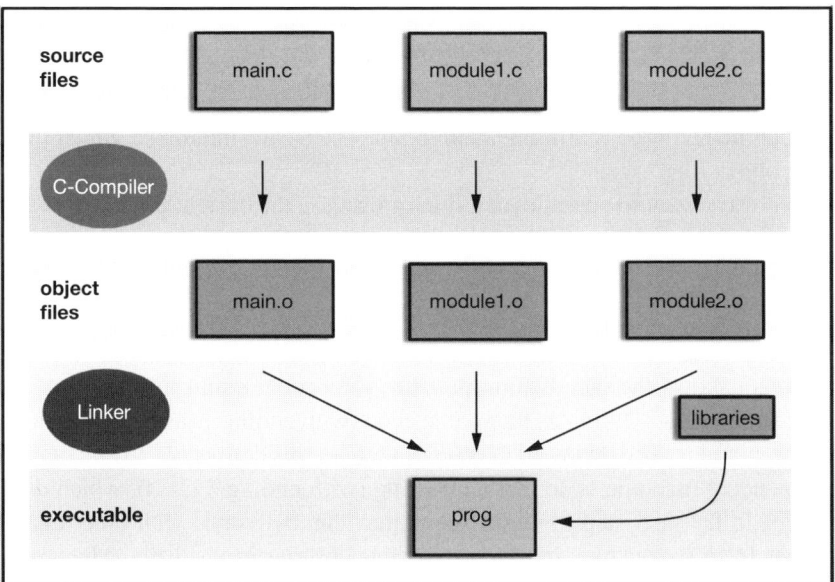

Figure 2.2. The compilation process in C. Assuming that the source code has been modularized by splitting it into three different modules, *main.c*, containing the *main()*-function (explained in the next section), *module1.c* and *module2.c*, invoking the C-compiler generates three different object (*.o) files which are then linked together, possibly with other needed libraries, to one executable file named "prog". If no name for the executable is provided upon calling the compiler, the executable name defaults to "a.out".

where <option flags> is a list of command flags that control how the compiler works, and <file list> is a list of files, source or object, that g++ is being directed to process. It is not, however, commonly invoked directly from the command line, that is what *makefiles*[3] are for. If g++ is unable to process the files correctly it will print error messages on standard error. Some of these error messages, will however be caused by g++ trying to recover from a previous error so it is best to try to tackle the errors in order. The two phases, compiling and linking, are often combined with the gcc command, but it is quite useful to separate them when using the *make* utility. For example, the command:

```
gcc -o prog main.c module1.c module2.c
```

which produces an executable file named "prog", can actually be broken down into the four steps of:

```
gcc -c main.c
gcc -c module1.c
```

---
[3] See Page 96

```
gcc -c module2.c
gcc -o prog main.o module1.o module2.o
```

The first three command lines, using the option "-c", just generate the object (*.o) files and the command in the fourth line, with option "-o", calls the linker, which links the different object files together.

If you always want to compile *all* C-files present in the current directory you could abbreviate the above commands within one command line using a *regular expression* as:

```
gcc -c *.c; gcc -o prog *.o
```

Figure 2.3 shows the compilation process when programming in JAVA, which is different from C/C++. In JAVA, the source code (with ending *.java) is compiled with the JAVA compiler. The output of the compilation in this case is *not* an object file with CPU-dependent machine code, but a class file (with ending *.class), which contains *byte code*. In the next step the linker combines this byte code with other class files from other JAVA source files and adds respective libraries in a new file, which is called a *JAR*[4] archive. The JAR archive contains the complete byte-code of the program. This byte code is not machine-specific, but a collection of instructions for a so-called *virtual machine*. The virtual machine, which has to be installed on every system on which the code is supposed to be run, then interprets the byte code of the JAR file of the program *at runtime* and translates it into machine-specific instructions (which are again machine-dependent). Thus, the advantage of this compilation procedure is that one has to compile a JAVA program on some operating system only once. It can then readily be run on any other operating system, provided a virtual machine is installed on that system, which interprets the byte code instructions according to the CPU on which the code is actually run[5]. The disadvantage is that the code is not compiled into CPU-specific machine language *before* the program is run, but is *interpreted at runtime*. This makes such code run much slower and thus rules out interpreted languages for genuine scientific high-performance computing.

Command line options of gcc/g++ Like almost all UNIX programs, gcc/g++ have myriad options that control almost every aspect of its actions. However, most of these options deal with system dependent features and do not concern us here. The most useful option flags for us are: -c, -o, -g, -Wall, -I, -L, and -l.

-c requests that gcc compile the specific source file directly into an object file without going through the linking stage. This is important for compiling only those files that have changed rather than the whole project.

---

[4] JAR stands for *Java AR*chive.
[5] This is the reason why JAVA is primarily cut out and used for distributed systems applications on the Internet. Basically the same byte code can be run on any system that has a virtual machine installed, no matter if it is UNIX, Linux, Windows, Macintosh or some other operating system.

Section 2.1 Introduction

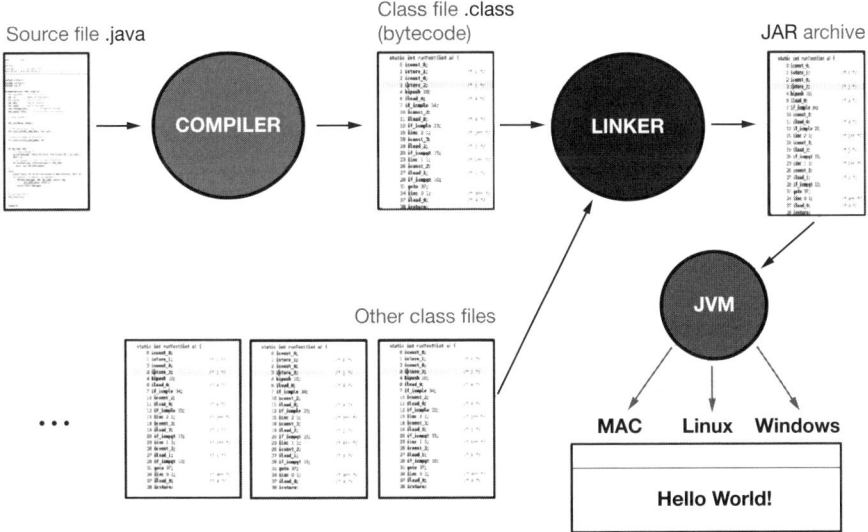

Figure 2.3. The compilation process in JAVA. The JAVA virtual machine *interprets* the byte code of the JAR archive, which was generated by the compiler and linker, *at runtime*.

-o file specifies that you want the compiler's output to be named "file". If this option is not specified, the default is to create a file "a.out", if you are linking object files into an executable. If you are compiling a source file (with suffix .c for files written in C) into an object file, the default name for the object file is simply the original name with the ".c" replaced with ".o". This option is generally only used for creating an application with a specific name (during linking), rather than for making the names of object files differ from the default source-filename.

-g directs the compiler to produce debugging information. I recommend that during testing one should always compile ones sources with this option, since I encourage the reader to gain proficiency using a debugger (I recommend gdb). Note that the debugging information is generated for gdb, and could possibly cause problems with dbx, because there is typically more information stored for gdb, the GNU Project debugger, which dbx will have difficulty processing[6].

-Wall produces warnings about a lot of syntactically correct but dubious constructs. Think of this option as being a way to do a simple form of style checking. Again, I highly recommend that the reader compiles her code with this option. Most of the time the constructs that are flagged are actually incorrect usages, but

---

[6] Additionally, on some systems, e.g., some MIPS based machines, the stored information cannot encode full symbol information, and some debugger features may be unavailable.

occasionally there are instances where they are what you really want. Instead of simply ignoring these warnings, there are simple workarounds for almost all of the warnings, if you insist on doing things this way. The following, sort of contrived snippet is a commonly used construct in C to set and test a variable in as few lines as possible:

```
int flag;
if (flag = IsPrime(13)) {
...
}
```

The compiler will produce a warning about a possibly unintended assignment. This is because it is more common to have a Boolean test in the if clause, using the equality operator "==", rather than to take advantage of the return value of the assignment operator. This code snippet should be better written as

```
int flag;
if ((flag = IsPrime(13)) != 0) {
...
}
```

so that the test for the 0 value is made explicit. The generated code will be the same, and it will make both programmer and compiler happy at the same time. Alternately, one can enclose the entire test in another set of parentheses to indicate one's intentions.

-Idir adds the directory "dir" to the list of directories searched for include files. There are a variety of standard directories that will be searched by the compiler for standard library and system header files by default, but since we do not have root access, we cannot just add our files to these locations. There is no space between the option flag and the directory name.

-llib searches the library named "lib" for unresolved names during linking. The actual name of the file will be "liblib.a", and must be found in either the default locations for libraries or in a directory added with the -L flag. The position of the -l flag in the option list is important because the linker will not go back to previously examined libraries to look for unresolved names. For example, if you are using a library that requires the math library it must appear before the math library on the command line otherwise a link error will be reported. Again, there is no space between the option flag and the library file name.

-Ldir adds the directory "dir" to the list of directories searched for library files specified by the -lflag. Here too, there is no space between the option flag and the library directory name.

Section 2.1 Introduction

Compiling in Emacs The EMACS editor (see Section 2.1.1) provides support for the compile process. To compile your code from within EMACS, type M-x compile. You will be prompted for a compile command. The EMACS buffer will split at this point, and compile errors will be brought up in the newly created buffer. In order to go to the line where a compile error occurred, place the cursor on the line which contains the error message and hit^c^c. This will jump the cursor to the line in your code where the error occurred.

The different phases of a compiler Generally speaking, the translation of source code by a compiler is separated into different phases, which are displayed in Figure 2.4 for the simple example of the following C source code:

```
newValue = old + x * 10;
```

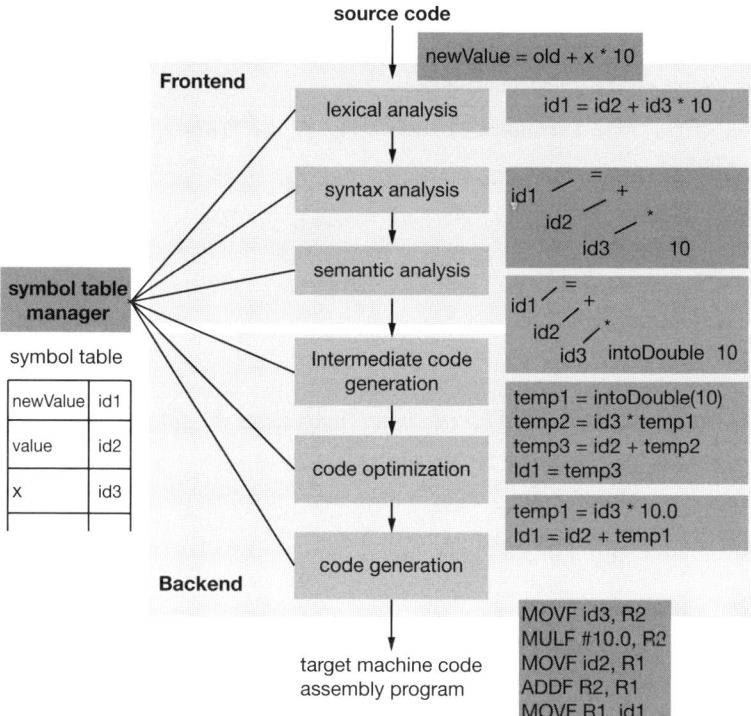

Figure 2.4. The typical phases of a compiler.

We first look at the phase during which the source program is analyzed (also called "frontend"). When analyzing source code, the compiler checks whether the sequence of characters in the program text follows the rules of the corresponding programming language. This analysis process is itself divided into three parts:

1. **Lexical analysis,** also called *linear analysis* or *scanning*.

   Here, the given program text is split into a sequence of single words by removing all *whitespace* (blanks, tabs and newlines) between these words. Also, all commentaries are removed (which are included only for human readers of the source code). Next, the stream of words of the source code is read from left to right and assigned to symbols, so-called *tokens*, which are stored in a symbol table. Each token corresponds to a keyword of the respective language and denotes which *category of words*[7] the respective word belongs to. For example, the source code in Listing 5 would be split into the following sequence of words:

   ```
   main ( ) { int i , sum = 0 ; for ...
   ```

   with the token sequence *id, bo, bc, BO, int, id, comma, id, equal, integer, semicolon, for* ....

Listing 5. Sample code for lexical analysis.

```
1  /* Sum of all even numbers between 1 and 100 */
2  main(void)
3  {
4     int i, sum = 0;
5     for (i=1; i < 100; i = i + 2)
6        sum = sum +1;
7  }
```

---

2. **Syntax analysis,** also called *hierarchical analysis* or *parsing*.

   Here, the compiler checks whether the sequence of tokens abides by the grammar of the corresponding language, i.e. it checks if the token sequence does not violate the syntax (for example by typing `sum = sum +*1;`.

3. **Semantic analysis.**

   Here, the compiler checks whether an operand has first to be converted before it can be linked with the second operand of an expression. For example, converting an integer into a floating point value, and then linking it with another floating point value.

After these three steps of analysis follows the *synthesis*, also called "backend", which constructs the desired program from the *parse tree*. Usually, this step is also subdivided into three different steps:

---

[7] for example, *integer, for,* bracket open (*bo*), bracket close (*bc*).

Section 2.1 Introduction

1. **Intermediate code generation.**

    Here, the source code is translated into an "intermediate" language. In Figure 2.4 a so-called *three-address-code* is used, which resembles assembly (machine language). This code is a sequence of commands, where each command has at most three operands and one operator (*,+,-,...), besides the equal sign, for example,

    ```
    temp1 = intoDouble(10)   /* transform int to double */
    temp2 = id3 * temp1
    temp3 = id2 * temp2
    id1   = temp3
    ```

    Variables `temp1`, `temp2`, and `temp3` are temporary names for memory addresses (registers) which can be quickly accessed by the CPU. The advantage of intermediate code is, that it can be re-used in case one wants to write compilers for several different languages.

2. **Code optimizer.**

    The generated intermediate code is improved by removing redundant commands. In our example, one can reduce the above four lines of code to just two:

    ```
    temp1 = id3 * 10.0
    id1   = id2 + temp1
    ```

3. **Code generator.**

    Finally, the code generator translates the intermediate code into machine code of the respective CPU, as shown in Figure 2.4 with pseudo assembly code:

    ```
    MOVF id3, R2;    load contents of register id3 into register R2
    MULF #10.0, R2;  multiply content of register R2 with 10.0
    MOVF id2, R1;    load content of id2 into register R1
    ADDF R2, R1;     add the content of R1 to the content of R2
    MOVF R1, id1;    store the content of R1 in register id1
    ```

    Besides these major tasks, a compiler has the following two additional tasks:

- **Management of the symbol table.**

    The symbol table is a central data structure, that lists all names, data, types and attributes of all identifiers in the source code.

- **Exception handling.**

    All possible errors have associated error codes and messages. After most errors, the compiler may continue to find more errors.

Editing files

The first obstacle a person is confronted with when programming is the editing environment. The standard editing tool available in a UNIX/Linux environment, if you don't want to use an IDE, is the EMACS or XEMACS editor mentioned above, which has pretty much all the features you will possible need in an editor, and then some. The only problem is finding out how to invoke them as needed. For those readers new to EMACS, we provide some basic EMACS commands to get you started in Appendix C.

The make utility

Typing the entire command line to compile a program turns out to be a somewhat complicated and tedious affair. What the make utility does is to allow the programmer to write out a specification of all the modules that go into creating an application, and how these modules need to be assembled to create the program. The make facility manages the execution of the necessary build commands (compiling, linking, loading etc.). In doing so, it also recognizes that only those files which have been changed need be rebuilt. Thus, a properly constructed makefile can save a great deal of compilation time. Invoking the make program is really simple, just type "make" at the shell prompt. This command will cause make to look in the current directory for a file called either "Makefile" or "makefile" for the build instructions. If there is a problem building one of the targets along the way, error messages will appear on standard error or the EMACS "compilation" buffer if you invoked make from within EMACS.

A makefile consists of a series of make variable definitions and dependency rules. A variable in a makefile is a name defined to represent some string of text. This works much like macro replacement in the C compiler's preprocessor. Variables are most often used to represent a list of directories to search, options for the compiler, and names of programs to run. A variable is "declared" when it is set to some value. For example, the line: CC = gcc will create a variable named CC, and set its value to gcc. The name of the variable is case sensitive and traditionally make variable names are in all caps. While it is possible to define your own variables, there are some that are considered "standard", and using them along with the default rules makes writing a makefile much easier. These important variables are: CC, CFLAGS, and LDFLAGS, which are discussed in the following:

1. CC is the name of the C compiler, this will default to cc or gcc in most versions of make.

2. CFLAGS is a list of options to pass to the C compiler for all source files. This is commonly used to set the include path to include non-standard directories or build debugging versions with the -I and -g compiler flags. For C++ programs, this is sometimes called CPPFLAGS.

Section 2.1 Introduction 97

3. LDFLAGS is a list of options to pass to the linker. This is most commonly used to set the library search path to non-standard directories and to include application specific library files, the -I and -L compiler flags. Referencing the value of a variable is done by having a $ followed by the name of the variable within parentheses or curly braces. For example:

```
CFLAGS = -g -I/usr/home/book/teaching
$(CC) $(CFLAGS) -c program.c
```

The first line sets the value of the variable CFLAGS to turn on debugging information (-g) and add the directory /usr/home/book/teaching/ to the include file search path. The second line uses the value of the variable CC as the name of the compiler and passes it the compiler options (CFLAGS), set in the previous line. If you use a variable that has not been set previously in the makefile, make will use the empty definition, an empty string. The second major component of makefiles are dependency/build rules. A rule tells how to make a target based on changes to a list of certain files. The ordering of the rules in the makefile does not make any difference, except that the first rule is considered to be the default rule. The default rule is the rule that will be invoked when make is called without arguments, which is the usual way. If, however, you know exactly which rule you want to invoke you can name it directly with an argument to make. For example, if your makefile had a rule for "clean", the command line "make clean" would invoke the actions listed after the clean label. A rule generally consists of two lines: a dependency list and a command list.

Here is an example:

```
program.o : program.c program.h anotherHeader.h
<tab>$(CC) $(CFLAGS) -c program.c
```

The first line says that the object file program.o must be rebuilt whenever any of program.c, program.h, or anotherHeader.h are changed. The target program.o is said to depend on these three files. Basically, an object file depends on its source file and any non system files that it includes. The second line lists the commands that must be taken in order to rebuild program.o, invoking the C compiler with whatever compiler options have been set previously. These lines must be indented with a <tab> character – just using spaces will not work! Because of this, you have to make sure not to be in an editor mode that may substitute space characters for an actual tab. In particular, the "indented-text-mode" always "tabs" to the same indent level as the previous line, using spaces if the previous line was indented less than the standard full 8 spaces. This is also a problem when using copy/paste from some terminal programs. To check whether you have a tab character on that line, move to the beginning of that line and try to move "right" (^f). If the cursor skips 8 spaces to the right, you have a tab. If it moves space by space, then you need to delete the spaces and retype a tab character. For "standard" compilations, the second line can be omitted, and make will use the default build rule for the source file, based on its extension, which is .c for C files. The default build rule that make uses for C files looks like this:

```
$(CC) $(CFLAGS) -c <source-file>
```

Here is a "complete" sample makefile that can be used as a template:

```
CC = gcc
CFLAGS = -g -I /usr/home/book/teaching/
LDFLAGS = -L /usr/home/book/teaching/lib -lm

PROG = nameOfExecutable
HDRS = program.h another.h defs.h
SRCS = main.c program.c another.c
OBJS = main.o program.o another.o

$(PROG) : $(OBJS)
$(CC) -o $(PROG) $(LDFLAGS) $(OBJS)

clean:
rm -f core $(PROG) $(OBJS)

main.o: program.h another.h defs.h
program.o: program.h
another.o : another.h defs.h
```

This makefile includes one extra target, in addition to the one building the executable, namely clean. The clean target is used to remove all object files and the executable so that you can start the build process from scratch[8]. You will need to do this if you move to a system with a different architecture from where your object libraries were compiled originally, because source code is compiled differently on different types of machines.

Debugging source code with the GNU debugger (gbd)

There is a variety of different techniques for finding bugs, i.e. "anomalies", but a good debugger can make the job a lot easier and faster. I recommend the GNU debugger, since it works nicely with the recommended gcc compiler. While gdb does not have a flashy graphical interface like, e.g. ddd (Data Display Debugger), it is a powerful tool that provides the knowledgeable programmer with all information she could possibly want and then some. There is on-line help for gdb which can be seen by using the

---

[8] The command "make clean" also removes any "core" files that might be lying around – not that there should be any.

help command from within gdb. If you want more information, try xinfo if you are logged onto the console of a machine with an X display, or use the info-browser mode from within XEMACS. A debugger is invaluable to a programmer because it eases the process of discovering and repairing bugs at run-time. In most non-trivial programs of any significant size, it is not possible to determine all bugs in a program at compile-time, because of oversights and misconceptions about the problem that the application is designed to solve. The way debuggers allow you to find bugs is by allowing you to run your program on a line-by-line basis, pausing the program at specified times or conditions, and allowing you to examine variables, registers, the runtime stack and other facets of the program's state, whilst execution is paused. Sometimes these bugs result in program crashes ("core dumps", "register dumps", etc.) that bring your program to a halt with a message like "Segmentation Fault" or the like. If your program has a severe bug that causes a program crash, the debugger will "catch" the signal, sent by the processor, which indicates the error it found, and allow you to further examine the program. This information can be quite valuable when trying to reason about what caused your program to die. After all, all segmentation faults sort of look the same. In Appendix C.7.3 on Page 464 I will point out some of high points of gdb, without coming anywhere close to describing all of the features of gdb, of course.

2.2 First steps in C

We will now dive into programming with the language C and learn how to write and compile structured programs. Those readers who have previous programming experience may safely skip this section, which is mainly introductory. For beginners in programming, it is recommended to type the listings in this section themselves and only use the provided source code from the book's website if they really cannot get a program to compile and run.

In the language C, all commands are merged within command blocks. The largest possible framework for this merging of commands into blocks are *functions*. Of all functions, the function has a particular importance, as this function is always called first when a C program starts[9]. Reaching the end of *main()* is equivalent to the program finishing.

Each function in C may return a value and, likewise, each function may be passed several values as arguments in a parameter list. The provided arguments to a function are written in parentheses, i.e. with "()" behind the function name as the type of the returned value is provided in front of the function name, as exemplified in our first C code in Listing 6.

Program Listing 6 is an example of the simplest possible executable program in C which does absolutely nothing. Let's analyze this program a bit. Any C program starts

---

[9] The brackets "()" with *main()* indicate that this is actually a function (and e.g. not a variable called "*main()*".

Listing 6. The simplest C *main()* function.

```
1 /* Comments in C begin like this and end with */
2 void main(void) /* The simplest main function */
3 {
4 // This is a C++ comment
5 }
```

with the *main()*-function and all other functions are called from within the *main()*-function. Thus, in a C program, the function *main()* must always be present. Without the *main()*-function, the linker cannot generate an executable program. The C keyword *void* in line 1, meaning "empty" in the parameter list of *main()*, stands for an undetermined datatype, meaning here that *no* parameter is passed to *main()*. A function in C can have a return value (e.g. an integer value – *int* – or a floating point value – *float* or *double* – but also derived datatypes[10]), the type of which is written in front of the function name.

In Listing 6 the return value of *main()* is also given by the keyword *void*, meaning that *main()* does not return a value. If you forget to provide a return value for *main()* (i.e. if you leave out the keyword *void* in front of *main()* (i.e. by simply typing "main(void)", or "main()"), the compiler will most likely issue a warning because *main()* is always expected to return a value, the default being *int*. After the function call in line 1 follows a commentary, which is removed from the source code by the preprocessor just before it is compiled. In line 1 you also see that commentaries in C always start with "/*" and end with "*/". Here, an important rule applies:

Commentaries in C may not be nested. This means that a construct like /* Commentary 1 with /* commentary 2 */ included */ is not valid.

In line 3 there is another commentary, written in C++ style, using "//", which is a valid comment in *C99* but not in *ANSI-C*. If your compiler does not show a compile error with Listing 6, it probably uses the *C99* standard by default. You can enforce the *ANSI* standard by using the option "-ansi" in the gcc command[11] and the *C99* standard by using the option "-std=c99"[12]. The simple example shown above also shows that command blocks (in this case an empty command block which does nothing) are delimited by curly brackets.

---

[10] Derived datatypes are datatype definitions, structures, or abstract datatypes such as lists, stacks or queues.
[11] An important Linux command to see all options of a command is "man *command*", i.e. in this case "man gcc" will show you a detailed description of the command gcc including its options.
[12] The option "-std=c99" in C++ is equivalent to "-std=c++98".

Section 2.2 First steps in C

> Curly brackets summarize individual commands in a command block.

Assuming that the simplest C program of Listing 6 has been stored under the name "test.c", it can be compiled with the command line

`gcc -c -Wall test.c`

As a result of the compilation, an executable file named by default "a.out" will be generated. If you want to change the default binary name to e.g. "MyBinaryName" you can change the compile command accordingly to

`gcc -o MyBinaryName -Wall test.c`

The compiler option "-Wall" switches on all compiler warnings, so in this case – depending on your compiler version – you should actually get a warning like

`warning: return type of ''main'' is not ''int'' [-Wmain]`

This is because by default the *main()*-function is expected to *always* return a value (different from *void*). Should this be the case on your system, you can for once ignore this warning here. Generally speaking, a warning during compilation will not prevent the compiler from producing a working binary, whilst a compile *error* actually stops the compilation process. No binary is generated.

> In general, one should always try to make all compiler warnings disappear by appropriate changes to your program, as these warnings can hint to problems in your program design and lead to serious problems later on.

2.2.1 Variables in C

The choice of names in C, be it for a function, for variables, or for constants, is subject to a special rule:

> Function or variable names in C must start with a character and may further only contain letters, characters or underscores.

With respect to the length of names in C, the following rule applies:

> The length of a name in C can be arbitrarily long, but only the first 32 characters are used to discriminate between different names.

All names in C are *case-sensitive*. In addition, variable names in C are not allowed to clash with *C keywords* that have a special role, such as *int*, *double*, *if*, *return*, *void*, etc. A list of reserved C-keywords can be found in Appendix E. The following names are all examples of valid and *different* variable names in C:

```
SpeedOfLight
speedOfLight
variable
Variable
sound_velocity
soundVelocity
```

Integer constants in C are denoted in the regular fashion, by strings of Arabic numbers, e.g.:

```
0    12    34567    89012345
```

Floating point constants can be written in either regular or scientific notation, e.g.:

```
0.01    70.567    3e+5    .5678e-15
```

> What exactly is a "variable"?

> A variable is nothing other than an address in computer memory.

When you assign a value to a certain variable, you actually store a value at a certain address in the address space of your computer's memory and access it later, when needed. For easy access to the contents of memory one needs a unique identifier, i.e. the name of the variable. The compiler later translates this name into an address in memory. Each variable uses some space in memory, depending on its datatype and the type of system (16-bit, 32-bit, 64-bit)[13] you are using. The size a certain datatype – for example an integer – uses in memory is not explicitly specified in any C standard[14], so this may vary on different systems. The typical sizes of standard datatypes on most systems are listed in Table 1.5 on Page 30.

> Individual commands in C and C++ are finished with a semicolon.

---

[13] Today (in the year 2012) 64-bit systems are standard.
[14] The C standard only provides "recommendations".

Section 2.2 First steps in C

A *declaration* associates a group of variables with a specific datatype, compare Table 2.2 on Page 119. All variables in a C program must be declared before they can appear in executable statements. A declaration consists of a datatype followed by one or more variable names, *ending in a semicolon*. A semicolon concludes a command. For example:

```
int     a, b, c;
double abb, epsilon, force, t;
```

In the above code snippet, a, b and c are declared to be integer variables, whereas abb, epsilon, force and t are declared to be floating point variables, in this case doubles[15].

A type declaration can also be used to assign initial values to variables. Some examples of how to do this are given below:

```
int     a = 3, b = c = 7;
double abb = epsilon \
            =.0003,   \
 force = t = 3.2e+3;
```

Note that there is no real restriction on the length of a type declaration; it can be even split over several lines by using the slash "\", as in the example above. However, all declaration statements in a program (or program segment) must occur *prior* to the first executable statement.

2.2.2 Global variables

A C program is basically a collection of functions and variables. Furthermore, the variables used by these functions have *local scope*: i.e. a variable defined in one function is not recognized in another. The main method of transferring data from one function to another is via the argument lists in function calls. Arguments can be passed in one of two different ways. When an argument is *passed by value*, then the value of a local variable (or expression) in the calling routine is copied to a local variable in the function which is called. When an argument is *passed by reference*, a local variable in the calling routine shares the same memory location as a local variable in the function which is called. Hence, a change in one variable is automatically reflected in the other. However, there is a third method of transferring information from one function to another: It is possible to define variables which have *global scope*. Such variables are recognized by all functions making up the program, and have the same value in all of these functions. The C compiler recognizes a variable as global, as opposed to local, when its declaration is located outside the scope of any of

---

[15] Note that variables of the most basic floating point datatype *float* are generally not stored in a computer with sufficient precision to be of much use in scientific programming, see our discussion of number representation in Section 1.5.

the functions making up the program. Of course, a global variable can only be used in an executable statement after it has been declared. Hence, the natural place to put global variable declaration statements is before any function definitions: i.e., right at the beginning of the program. Global variable declarations can be used to initialize such variables in the usual manner. However, the initial values must be expressed as constants, rather than expressions. Furthermore, the initial values are only assigned *once*, at the beginning of the program. Generally speaking, the use of global variables is an indication of poor programming style, because it contradicts the principle of encapsulation of data, modularization of code, and because it impairs maintainability.

2.2.3 Operators in C

As we have seen, general expressions in C are formed by joining together constants and variables via various operators. Operators in C fall into five main classes: arithmetic operators, unary operators, relational and logical operators, assignment operators, and the conditional operator. Let us now examine each of these classes in detail.

There are four main *arithmetic operators* in C. These are:

- addition +

- subtraction −

- multiplication ∗

- division /

There is no built-in exponentiation operator in C. Instead, there is a library function *pow()* – defined in the library file *<math.h>* which carries out this operation. It is poor programming practice to mix types in arithmetic expressions. In other words, the two operands operated on by the addition, subtraction, multiplication, or division operators should both be either of type *int* or type *double*. The value of an expression can be converted to a different datatype by prefixing the name of the desired datatype, enclosed in parentheses. This type of construction is known as a *cast*. Thus, to convert an integer variable j into a floating point variable with the same value, one would write

(double) j

Finally, to avoid mixing datatypes when dividing a floating point variable f by an integer variable i, one would write

f / (double) i

Of course, the result of this operation would be of type *double*. The operators within C are grouped hierarchically according to their precedence (i.e., their order of evaluation). Amongst the arithmetic operators, ∗ and / have precedence over + and −. In other words, when evaluating expressions, C performs multiplication and division operations with higher priority than addition and subtraction operations. Of course,

Section 2.2 First steps in C

the rules of precedence can always be bypassed by judicious use of parentheses. Thus, the expression

```
a - b / c + d
```

is equivalent to the unambiguous expression

```
a - (b / c) + d
```

since division takes precedence over addition and subtraction.

The distinguishing feature of *unary operators* is that they only act on single operands. The most common unary operator is the *unary minus*, which occurs when a numerical constant, variable, or expression is preceded by a minus sign. Note that the unary minus is distinctly different from the arithmetic operator (−), which denotes subtraction, since the latter operator acts on two separate operands. The two other common unary operators are the *increment operator*, ++, and the *decrement operator*, −−. The increment operator causes its operand to be increased by 1, whereas the decrement operator causes its operand to be decreased by 1. For example, -i is equivalent to i = i - 1. A cast is also considered to be a unary operator. Note that unary operators have precedence over arithmetic operators. Hence, - x + y is equivalent to the unambiguous expression (-x) + y, since the unary minus operator has precedence over the addition operator. Note that there is a subtle distinction between the expressions a++ and ++a. In the former case, the value of the variable a is returned before it is incremented. In the latter case, the value of a is returned after incrementation. Thus, b = a++; is equivalent to

```
b = a;
a = a + 1;
```

whereas

```
b = ++a;
```

is equivalent to

```
a = a + 1;
b = a;
```

There is a similar distinction between the expressions a- and -a.

There are four *relational operators* in C. These are:

- less than <

- less than or equal to <=

- greater than >

- greater than or equal to >=

The precedence of these operators is lower than that of arithmetic operators. Closely associated with the relational operators are the two equality operators:

- equal to ==
- not equal to !=

The precedence of the equality operators is lower than that of the relational operators. The relational and equality operators are used to form logical expressions, which represent conditions that can be either true or false. The resulting expressions are of type *int*, because true is represented by the integer value 1 and false by the integer value 0. For example, the expression i < j is true (value 1) if the value of i is less than the value of j, and false (value 0) otherwise. Likewise, the expression j == 3 is true if the value of j is equal to 3, and false otherwise.

C also possess two *logical operators*. These are:

- && (logical AND)
- || (logical OR)

The logical operators act on operands which are themselves logical expressions. The net effect is to combine the individual logical expressions into more complex expressions that are either true or false. The result of a logical AND operation is only true if both operands are true, whereas the result of a logical OR operation is only false if both operands are false. For instance, the expression (i >= 5) && (j == 3) is true if the value of i is greater than or equal to 5 and the value of j is equal to 3, and false otherwise. The precedence of the logical AND operator is higher than that of the logical OR operator, but lower than that of the equality operators. C also includes the unary operator ! that negates the value of a logical expression, i.e., it causes an expression that initially was true to false and vice versa. This operator is referred to as the logical negation or *logical NOT operator*. For instance, the expression !(k == 4) is true if the value of k is not equal to 4, and false otherwise. Note that it is poor programming practice to rely too heavily on operator precedence, since such reliance tends to makes C programs very hard for other people to follow. For instance, instead of writing

```
i + j == 3 && i * l >= 5
```

and relying on the fact that arithmetic operators have precedence over relational and equality operators, which, in turn, have precedence over logical operators, it is better to write

```
((i + j) == 3) && (i * l >= 5)
```

whose meaning is fairly unambiguous, even to people who cannot remember the order of precedence of the various operators in C. The most common assignment operator in C is =. For instance, the expression f = 3.4 causes the floating point value 3.4 to be assigned to the variable f. Note that, in C, the assignment operator = and the equality operator == perform completely different functions and should not be confused. In C, multiple assignments are permissible. For example, i = j = k = 4 causes the integer value 4 to be assigned to i, j, and k simultaneously. Note, again, that it is poor programming practice to mix datatypes in assignment expressions. Thus, the datatypes of the constants or variables on either side of the = sign should always match. C contains four additional assignment operators: + =, − =, * =, and / =. The expression

```
i += 6
```

is equivalent to i = i + 6. Likewise, the expression

```
i -= 6
```

is equivalent to i = i - 6. The expression

```
i *= 6
```

is equivalent to i = i * 6. Finally, the expression

```
i /= 6
```

is equivalent to i = i / 6. Note that the precedence of assignment operators is below that of all the operators discussed previously.

Simple *conditional operations* can be carried out with the conditional operator (? :). An expression that makes use of the conditional operator is called a conditional expression. Such an expression takes the general form

```
expression 1 ? expression 2 : expression 3
```

If expression 1 is true (i.e., if its value is nonzero) then expression 2 is evaluated and becomes the value of the conditional expression. On the other hand, if expression 1 is false (i.e., if its value is zero) then expression 3 is evaluated and becomes the value of the conditional expression. For instance, the expression

```
(j < 5) ? 12 : -6
```

takes the value 12 if the value of j is less than 5, and the value −6 otherwise. The assignment statement

```
k = (i < 0) ? x : y
```

causes the value of x to be assigned to the variable k if the value of i is less than zero, and the value of y to be assigned to k otherwise. The precedence of the conditional operator is just above that of the assignment operators.

Scientific programs tend to be extremely resource intensive. Scientific programmers should, therefore, always be on the lookout for methods of speeding up the execution of their code. It is important to realize that multiplication $*$ and division $/$ operations consume considerably more CPU time than addition $+$, subtraction $-$, comparison, or assignment operations. Thus, a simple rule of thumb for writing efficient code is to try to avoid redundant multiplication and division operations. This is particularly important for sections of code which are executed repeatedly, e.g., code located within control loops. The classic illustration of this point is the evaluation of a polynomial. The most straightforward method of evaluating, say, a fourth-order polynomial would be to write something like:

```
p = a0 + a1 * x + a2 * x * x + a3 * x * x * x + a4 * x * x * x * x
```

Note that the above expression employs ten expensive multiplication operations. However, this number can be reduced to four via a simple algebraic rearrangement:

```
p = a0 + x * (a1 + x * (a2 + x * (a3 + x * a4)))
```

Clearly, the latter expression is computationally much more efficient than the former.

2.2.4 Control structures

Control structures are available in any high-level language, because they are needed to control the flow of a program in various directions, depending on the result of certain decisions. We will see these control structures in many programs in later sections. Here, we just introduce the basic syntax and control flow of four main control structures of C, namely:

- *if-else*-statement
- *for* loops
- *while* loops
- *do-while* loops

Decisions: if-else-statements

We first look at the *if*-statement as a control structure. It has the following syntax:

```
if ( expression ) {
    command1;
}
command2;
```

First, the logical condition[16] for `expression` within the brackets () is evaluated. Depending on whether the expression is true (!=0) the command `command1` in the

---

[16] The brackets () denote a logical expression here. "Logical" in C always means integer! Thus, the expression in the *if*-statement can have any numerical integer value. However, *only* 0 is interpreted as false and any other value as true.

Section 2.2 First steps in C

command block between the curly brackets is executed. Note, that command1 can also consist of many individual commands, even many function calls – it doesn't have to be only one single command. Then, the program continues with command2. In case the expression is not true (==0), the command block between the brackets is not executed. Rather, the program directly continues with the execution of command2.

Decisions: the else branch

In a program, after one or several *if*-statements, there will usually be an optional branch of execution. This branch will be executed if no expressions are satisfied. Such a branch is specified in an alternative *else*-branch[17] with the following syntax:

```
if ( expression ) {
    command1;
}
else {
    command2;
}
command3;
```

Note, that this time the expression if (ival != 0) is used, which is the same as if (ival). In case an *if*-statement has only one command within the command block, the parentheses can be left out and the single command can be written directly in the next line:

```
if ( ival != 0)
  printf(``The number is not equal to zero\n'');
else
  printf(``The number is equal to zero\n'');
printf(``Now, the program continues...\n'');
```

Figure 2.5 depicts the control flow of the *if-else* control statement as a flow chart.

Loops: *for*-statement for repeating program parts

When you want to repeat a group of commands, C provides several loops, so called iteration commands. We first discuss *for* loops which have the following syntax:

```
for ( expression1; expression2; expression3) {
command(s);
}
```

---

[17] Note, that an *else*-branch can only be used with a precedent *if*-statement.

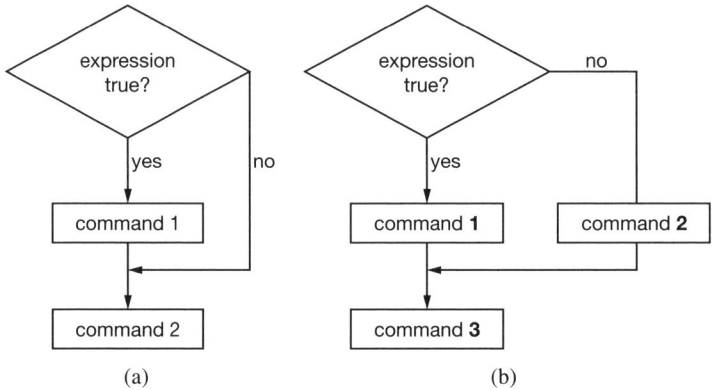

Figure 2.5. Flow chart of the *if-else* control statement. (a) The *if*-statement. (b) The *else*-branch of an *if*-statement.

Before entering the *for* loop, expression1 is executed *once*. Usually, expression1 is simply the initialization of the loop variable, but it could be any valid C command! expression2 is the logical condition which regulates the loop, i.e. it determines the exit-condition of the loop: As long as expression2 is true, the loop is executed. Once expression2 returns 0 (not true) the loop ends and the program is continued with the command(s) after the loop. As a rule, expression3 is used for re-initializing of the loop variable (but, again, in principle, expression3 can be any valid C command!) The simplest and standard usage of for loops is demonstrated in the following program Listing 7: The *#include* command and the function *printf()* are explained in Section 2.2.5.

Listing 7. Simple example of a *for*-loop.

```
1  /* Simple example of a for-loop */
2  #include <stdio.h>
3
4  int main( void ){
5    int counter;
6
7    for (counter = 1; counter <= 5; counter ++) {
8      printf(``%d. Number of for loops\n'',counter);
9    }
10   return 0;
11 }
```

Section 2.2 First steps in C   111

Loops: *while*-statement with entry condition

The *while*-loop has an entry condition `expression` that is checked first. As long as `expression` is true, the while loop is executed. The basic syntax is:

```
while ( expression ) {
   command(s)
}

otherCommand(s)
```

Before execution of the loop is started, `expression` is checked. If `expression` is true (i.e. != 0) the `command(s)` within the while loop are executed. As soon as `expression` is not true, the program continues execution with the `otherCommand(s)` following the *while*-loop.

Loops: *do-while*-statement with exit condition

The counterpart of the *while* loop is the *do-while*-loop, with exit condition. The basic syntax is:

```
do {
   command(s)
} while ( expression );

otherCommands
```

When executing a *do-while*-loop, the `command(s)` are first executed. Hence, this loop type is guaranteed to execute `command(s)` at least once. Next, `expression` is evaluated and if it is true the `command(s)` after the key word *do* are executed again. If `expression` is not true the program continues execution with the `otherCommand(s)` after the loop.

2.2.5 Scientific "Hello world!"

Our next program demonstrates the obligatory "Hello World!" program, presented in a more scientific disguise. I first present it in the C version and then show a C++ version, which uses the *cout* stream for printing.

Think!
What happens when you try to compile Program Listing 8 using the command "gcc -Wall main.c", if the code has been saved as "main.c"?

Listing 8. Scientific "Hello World!" C program. (*)

```
1  /* Scientific Hello World! in C */
2  #include <stdio.h>      /* includes the printf() function */
3  #include <math.h>       /* includes the sqrt() function */
4
5  int main(int argc, char *argv[])
6  {
7    /* declare variables, initialize the first one */
8    double squareRoot = 0.0, s;
9
10   s = argc + 9;
11   squareRoot = sqrt(s);
12
13   printf("Hello World! This program: %s\n", argv[0]);
14   printf("argc = %i, Result: sqrt(%g)=%g\n", argc, s,
         squareRoot);
15
16   return 0;           /* return value of main() upon success */
17 }
```

When Listing 8 is executed, using the command

`./MyBinarayName 1 2 test test2`

the following output is printed on the screen:

```
Hello World! This program: ./MyBinarayName
argc = 5, Result: sqrt(14)=3.74166
```

> Assuming that Listing 8 has been stored as "main.c", depending on the compiler version, using the usual command "gcc -o MyBinaryName -Wall main.c" will most likely result in an error message from the linker:
>
> `main.c:(.text+0xb4): undefined reference to 'sqrt'`
> `collect2: ld returned 1 exit status`
>
> The reason for this is that the "#include <math.h>" command only tells the compiler that *sqrt()* function is defined *somewhere* else, i.e. not in module main.c. Actually, the function *sqrt()* is defined in the math library "libmath.a". This library has to be included using the additional option "-lm" in the above command line.

Let's examine what the program in Listing 8 does. Line 1 is simply a commentary that is removed by the preprocessor before the compilation. The *#include* commands

Section 2.2 First steps in C

in lines 2 and 3 are preprocessor commands. The preprocessor performs temporary changes in the program's source code, before it is compiled. For example, #include <file.c> makes sure that the corresponding text file "file.c" is inserted into the source code, in place of the *#include* command. Using the command #include <...> with angular brackets makes sure that the file is searched for in the directories that are known to the compiler, with the default on most systems being /usr/include/. If the file name is provided in quotes as in #include "...", as we will see it in later code examples, the file is searched for in the current working directory. You can also provide complete absolute or relative paths to the include file in parentheses, for example

#include ''/home/users/testuser/myheader/header.h''

will make the compiler search for the file "header.h" in the corresponding directory /home/users/testuser/myheader/. The compiler must find a declaration[18] of a function before you can call it. The declaration of library functions appears in so called "header files". The header file informs the compiler of the name, type, and number and type of arguments, for all functions contained in the library in question. For example, the <*math.h*> library, included with the command #include <math.h> contains a number of useful predefined functions, as listed in Table 2.1.

In line 5 the *main()*-function is called with its return value set to an integer, returning 0 if successful, i.e. if the program runs until the end. As stated in the previous section, the function *main()* is like an ordinary function, with the only difference being that it is always called *first*. So the question arises how to pass an argument list to *main()*, as we do not explicitly call *main()*. We may also wonder what happens with the return value of *main()*. In Listing 8, *main()* is called with two arguments, "int argc" and "char* argv [ ]"; the former is of type integer and the latter is of type "pointer to a field of characters", i.e. a pointer to string[19]. The variable names "argc" and "argv" in the parameter list of *main()* are merely by convention. One can use any variable names for the arguments passed to *main()*, for example,

int main(int argumentNumber, char **myArguments)

would also be a valid choice[20].

Let's look a little closer at the parameter list of *main()*. As the start of a C program is equal to starting the *main()*-function, the parameters have to be passed from outside of *main()*. When you start your program from a shell, the parameters have to be passed on the command line, after the program name. The integer argc contains the number of passed arguments to *main()*, where the name of the program itself also counts. Thus, argc can never be smaller than 1. The second parameter in the example

---

[18] The compiler checks the argument type and return type of the declaration.
[19] We will introduce strings and pointers in Sections 2.4.4 and 2.4.7.
[20] The equivalence of char *argv[] with char **argv will become clear in Section 2.4.4 when we discuss pointers.

Table 2.1. Useful functions of the math library, defined in header file *math.h*.

function	type	purpose
acos(x)	*double*	return arc cosine of x (in range 0 to π)
asin(x)	*double*	return arc sine of x (in range $-\pi/2$ to $\pi/2$)
atan(d)	*double*	return arc tangent of x (in range $-\pi/2$ to $\pi/2$)
atan2(x1, x2)	*double*	return arc tangent of $x1/x2$ (in range $-\pi$ to π)
cbrt(x)	*double*	return cube root of x
cos(x)	*double*	return cosine of x
cosh(x)	*double*	return hyperbolic cosine of x
exp(x)	*double*	return exponential of x
fabs(x)	*double*	return absolute value of x
hypot(x1, x2)	*double*	return $sqrt(x1 * x1 + x2 * x2)$
log(x)	*double*	return natural logarithm of x
log10(x)	*double*	return logarithm (base 10) of x
pow(x1, x2)	*double*	return $x1$ raised to the power $x2$
sin(x)	*double*	return sine of x
sinh(x)	*double*	return hyperbolic sine of x
sqrt(x)	*double*	return square root of x
tan(x)	*double*	return tangent of x
tanh(x)	*double*	return hyperbolic tangent of x

above, `argv`, is a pointer to a *char* field, i.e. `argv` is a field that contains the addresses of the strings that have been passed to *main()* in the command line. For example, with `argv[0]` one can access the address of the first string, which contains the first parameter, i.e. the name of the program. Hence, the `printf`-statement in line 13 prints the program name next to "Hello World!". *printf()* is a predefined function from the standard C-library defined in the header file <stdio.h>. To use it, its prototype has to be included in the program, which is done in line 2.

> What is the use of a return value of the function *main()*?

Section 2.2 First steps in C   115

> The return value of *main()* is mostly interesting in a UNIX/Linux shell environment[a], where certain control structures of the shell or the operating system can be used to react accordingly to certain return values of *main()*. The general convention is, that a return value of 0 means success and any value other than 0 means that some error occurred. This way, one can assign certain error values to the possible errors (e.g. "File not found", "Out of memory", etc.) that may occur in a program, and have the batch file written in a way that the operating system or the shell can react to accordingly.
>
> ---
> [a] For example, when running the program in "batch" mode, which is often done when submitting programs to the queue of a supercomputer.

In line 8 we declare two variables as *double*. This line shows how to initialize variables upon their declaration. In line 10 we add 9 to the number of arguments passed and store the result in *s*. In line 11 we call the function *sqrt(s)* which returns the square root of its argument (here: *s*), and we store the result in the *double* `squareRoot`. Note, that the integer `argc+9` is automatically converted by the compiler[21] into a *double* by the assignment in line 11. Line 14 again prints `argc` together with the result of our calculation. Finally, in line 16, we return 0. Generally, a function that has a return value must always end with the command *return*, and return a value of the specified type.

In C++, this program can be written as shown in Listing 9, where we have replaced the call to *printf()* with the standard C++ function *cout*. Here, the header file `iostream` is needed. In addition, in C++, we don't need to declare variables like s and `squareRoot` at the beginning of the program. However, I recommend to declare all variables at the beginning of a function, as this provides much better code readability. Note that we have used the declaration using `namespace std`. Namespace is a way to collect all functions defined in C++ libraries. If we omit this declaration on top of the program we would have to add the declaration *std* in front of *cout* or *cin*. Our program would then read as shown in Listing 10.

Another feature worth noting is that we have skipped *exception handling* here. In later examples we discuss cases that test our command line input. Here, it is very easy to add such a feature, as shown in Listing 11 on Page 118. In this version of "Hello World!" we check in line 10 whether more than one argument is passed to *main()*. If not, the program stops with the command `exit(1)` (declared in header file *<stdlib.h>*) in line 14 and writes an error message to the screen.

---
[21] This automatic type conversion of variables is called a "cast" and can also be invoked explicitly by writing `s = (double)(argc+9)`.

Listing 9. Scientific C++ "Hello World!". (*)

```
// Scientific Hello World! in C++
using namespace std;
#include <iostream>

int main(int argc, char *argv[])
{
    // declare variables, initialize the first one
    double squareRoot = 0.0, s;

    s = argc + 9;
    squareRoot = s;

    cout << "Hello World! This program: " << argv[0] << endl;
    cout << "argc = " << argc << " Result: sqrt(" << s << ") =" \
         << squareRoot << endl;

    return 0;         // return value of main() upon success
}
```

> **Exercise 2.1.** As an exercise, I suggest to include the exception handling from Listing 11 in Listing 8, using C syntax with *printf()*-statements.

2.2.6 Streams – input/output functionality

Let's talk about one of the most important general functions of any computer program, namely data input/output[22] functionality.

When developing source code you need to be able to read and write data not only from and to the screen (called "standard output" or "standard error"), but also from and to files in the file system. Data I/O in C and C++ is realized with *data streams*. When opening a file, a new stream is created and removed again when closing the file. The single streams are administered by the operating system. In C there are *text streams*, *binary streams* and *standard streams*.

Text streams

A text stream reads and writes single characters of a text from or to a file. Usually, the text is separated in single lines. The internal representation of text is independent of the operating system on which the program is executed. Text streams can use all printable ASCII-characters and several control codes, e.g. new line "\n" or tab "\t". On Windows systems, the end of line is denoted with the control sign "\r\n", but UNIX/Linux systems use only "\n" for that. Thus, the compiler here performs an

---
[22] This is generally abbreviated as "I/O".

Section 2.2 First steps in C 117

Listing 10. Scientific C++ "Hello World!" without namespace. (*)

```
1  // Hello World! in C++ without using namespace std
2  #include <iostream>
3
4  int main(int argc, char *argv[])
5  {
6      // declare variables, initialize the first one
7      double squareRoot = 0.0, s;
8
9      s = argc + 9;
10     squareRoot = s;
11
12     std::cout << "Hello World! This program: " << argv[0] << std
           ::endl;
13     std::cout << "argc = " << argc << " Result: sqrt(" << s << ")
           =" \
14         << squareRoot << std::endl;
15
16     return 0;         // return value of main() upon success
17 }
```

---

automatic conversion. The end of a text is usually denoted by "^z" (ASCII-code 26), which can also be sent to the screen by "Ctrl+z" or "Ctrl+d" under UNIX/Linux.

Binary streams

With binary streams the content of a file is processed byte by byte and not character by character. Thus, data which has been processed in a binary stream is available for reading in exactly the same way. No automatic conversions are done by the compiler.

Standard streams

Three streams are always available in any C program, namely the three *standard streams*. Standard streams are pointers[23] to a FILE object. These are the standard streams:

- stdin: The standard input, which is usually connected to the keyboard.

- stdout: The standard output is connected to the screen. The output is buffered line by line.

- stderr: The standard error output is connected to the screen, just like stdout, but the output is not buffered.

---

[23] Pointers store addresses of computer memory and are explained later in this chapter.

Listing 11. Scientific C++ "Hello World!" with exception handling. (*)

```
 1 using namespace std;
 2 #include <iostream>
 3 #include <stdlib.h>
 4
 5 int main(int argc, char *argv[])
 6 {
 7
 8    // read an output file and abort if there are too few
 9    // command line options
10    if ( argc <= 1 ){
11       cout << "Bad Usage of: " << argv[0] <<
12          " must read a number on the same command line, e.g.: \
13          prog 0.1234" << endl;
14       exit(1);   // The program stops here
15    }
16
17    // declare variables, initialize the first one
18    double squareRoot = 0.0, s;
19    s = argc + 9;
20    squareRoot = s;
21
22    cout << "Hello World! This program: " << argv[0] << endl;
23    cout << "argc = " << argc << " Result: sqrt(" << s << ") =" \
24       << squareRoot << endl;
25
26    return 0;            // return value of main() upon success
27 }
```

When a file is opened in C with the command *fopen()*, a memory object of type FILE is generated and initialized. A successfully opened file returns a pointer to a FILE memory object which is connected to the stream. Whether a binary or text stream is used can be determined with an additional option in the function *fopen()*. The FILE object is a structure which is declared in the header file *<stdio.h>*. Using the returned FILE pointer, one can read or change data of the stream using the standard function *printf()*. The prototype of the function *fopen()* is the following:

FILE *fopen (const char* filename, const char* mode)

fopen() opens a file with name filename. If a file could not be opened, this function returns the NULL pointer[24]. The second argument mode determines the access mode, which must be one of the strings of the following Table 2.2. In Appendix D.1 some simple sample code is provided in Listing 74, which shows a complete working

---

[24] A NULL pointer is a special pointer value that does not point anywhere. This means that no other valid pointer, to any other variable or array cell or anything else, will ever be evaluated as equal to a NULL pointer. This is further explained in section 2.4.4.

Section 2.2 First steps in C

Table 2.2. Modes for processing text files in C using the command *fopen()*.

mode	meaning
"r"	Opens a file for reading
"w"	Opens a file for writing
"a"	Like "w" but appends data, instead of overwriting existing data in the file
"r+"	Open the file for reading and writing
"w+"	Like "r+", but if the file does not exist, a new file is created
"a+"	Open the file for reading and writing. If it doesn't exist, a new file is created

example of how to use the *fopen()* function to open, and *fclose()* to close a binary or ASCII text file.

2.2.7 The preprocessor and symbolic constants

Before the compiler translates source code, the preprocessor parses the code and performs the following tasks:

- concatenation of string literals

- removal of line breaks having a backslash at the beginning of the line

- substitution of commentaries with blanks

- deletion of space characters between tokens

Additionally, there are several preprocessor tasks that can be controlled by the programmer, using special preprocessor commands:

- including header and source files into the code (*#include*)

- including symbolic constants (*#define*)

- performing conditional compilation (*#ifdef*, *#elseif #endif*).

> Preprocessor commands always have a "#" at the beginning of the line and only *one* preprocessor command per command line is allowed. Hence, the command line
>
> ```
> #include <stdio.h>  #define MAX_VALUE 255
> ```
>
> is *not* valid C syntax, whereas commentaries in the same command line are allowed:
>
> ```
> #include <stdio.h> // Header for standard functions
> ```

Symbolic constants and macros

The preprocessor command *#define* is used for defining *symbolic constants* and *macros*, as in the following code snippet:

```
#define NAME text
#define VALUE 123
#define NOT_EQUAL(x,y)   (x != y)
```

Here, NAME represents a symbolic name, typically written in uppercase letters and text represents the sequence of characters that is associated with that name. Note that text does *not* end with a semicolon, since a symbolic constant definition is not a true C statement, but rather a preprocessor command. In the same way, the symbolic constant VALUE is associated with the number 123. Besides symbolic constants the *#define* command can also be used for parametrized macros, where the macro name is followed by parentheses, as in the third statement above. In fact, the resolution of symbolic names is performed by the C preprocessor prior to compilation. For instance, suppose that a C program contains the following symbolic constant definition:

```
#define PI 3.141593
```

Moreover, suppose that the program contains the statement

```
area = PI * radius * radius;
```

During the compilation process, the preprocessor replaces each occurrence of the symbolic constant PI by its corresponding definition. Hence, the above statement actually becomes

```
area = 3.141593 * radius * radius;
```

Symbolic constants are particularly useful in scientific programs for representing mathematical or physical constants.

> Preprocessor commands start with a #.

> There is no type checking for parametrized macros! Therefore, macros are more error prone compared to functions. The compiler performs return type checking of functions, but it never checks parametrized preprocessor macros, as they are replaced before compilation with their corresponding constants. Thus, macros cannot be changed at runtime.

Listing 12. Use of user defined macros. (*)

```c
#define NOT_EQUAL(x, y)    (x != y)
#define XCHANGE(x, y) { \
   int j; j=x; x=y; y=j; }

int main(void) {
   int intValue1, intValue2;

   printf("intValue1: ");
   scanf("%d", &intValue1);
   printf("intValue2: ");
   scanf("%d", &intValue2);

   if( NOT_EQUAL(intValue1, intValue2) ) {
      XCHANGE(intValue1, intValue2);
   }
   printf("intValue1: %d | intValue2: %d\n", intValue1,
       intValue2);
   return EXIT_SUCCESS;
}
```

The main advantage of using *#define* constants is that the program becomes more readable and maintainable. Imagine that, for example, in a 10.000 lines program, you have used the symbolic constants

```
#define ROW 15
#define COLUMN 80
#define NAME ''My super program 0.1''
```

and you want to change these values. Using macros, you only have to change one value at one spot in your program, as opposed to searching the whole source code (which is usually split into many smaller files, so-called *modules*), for the values 15, 80 or the string "My super program 0.1" and changing them. A complete example of a typical use of macros can be seen in the following Listing 12. It should be an easy exercise for the reader to figure out what this piece of code does. Further typical examples of macros for functions are provided in the following lines of code:

```
#define MIN(a,b) ( ((a) < (b))    ?  (a) : (b) )
#define MAX(a,b) (( (a) > (b))    ?  (a) : (b) )
#define ABS(a)     (( (a) <  0)     ? -(a) : (a) )
#define EVEN(a)   (  (a) \% 2 == 0 ?   1 : 0 )
#define TOASCII  (  (a) & 0 x 7f )
```

In C++ we would replace such function definition by employing so called *inline functions*. Three of the above functions could then read, e.g.,

```
inline double MIN(double a, double b) (return(((a) < (b)) ? (a):(b));)
```

```
inline double MAX(double a, double b) (return(((a) > (b)) ? (a):(b));)
inline double ABS(double a) (return ( ((a) < 0) ? - (a) : (a) );)
```

where we have defined the transferred variables to be of type *double*. The functions also return a variable of type *double*[25]. Inline functions are very useful, especially if the overhead for calling a function implies a significant fraction of the total function call cost. When such function call overhead is significant, a function definition can be preceded by the keyword *inline*. When this function is called, we expect the compiler to generate inline code without function call overhead. However, although inline functions eliminate function call overhead, they can introduce other overheads. When a function is inlined, its code is duplicated for each call. Excessive use of inline may thus generate large programs. Large programs can in turn cause excessive paging in virtual memory systems. Too many inline functions can also lengthen compile and link times. On the other hand, not inlining small functions that do only small computations, like the one above, can make programs bigger and slower. However, most modern compilers know better than the programmer which functions to inline or not. When doing this, you should also test various compiler options. With the compiler option -O3 inlining is done automatically by basically all modern compilers.

A good strategy for C++ coding is to first write your program without inline functions, since you should first of all write any code as simple and clear as possible, without strong emphasis on computational speed. Later, when profiling the program, one can spot small functions which are called many times. These functions can then be candidates for inlining. If the overall running time is reduced due to inlining specific functions, one can proceed to other sections of the program which could be sped up. Another problem with inlined functions is that on some systems debugging an inline function is difficult, because the function does not exist at runtime.

Further preprocessor commands

There are several predefined preprocessor macros, which are listed in Table 2.3. Some of them are quite useful for structuring or debugging programs. Finally, in Listing 13 we show how to use several of these predefined macros in real source code.

2.2.8 The function *scanf()*

For you to be able to work on some useful exercises, we introduce the C-function *scanf()* here, which is used for formatted input from the command line. In a way, this function is the complement of *printf()*, see Section 2.2.6 on Page 117. The data stream is read from (stdin), which usually is the keyboard. *scanf()* – like *printf()* – is declared in the header file *<stdio.h>*, which is why you have to include this header file, as in Listing 14. When you run this program, you are asked in line 7 to enter an integer.

---

[25] In pure C++ these functions could easily be generalized to return whatever types of *real*, *complex*, or *integer* variables through the use of classes and templates.

Section 2.2 First steps in C

Listing 13. Use of predefined macros. (*)

```c
#include <stdio.h>
#include <stdlib.h>

#if defined (__STDC_VERSION__)&&__STDC__VERSION >= 199901L
void function(void) {
  printf("Name of this function: %s\n", __func__)
}
#else
void function(void) {
  printf("No C99 standard compiler\n");
}
#endif

int main(void) {
#ifdef __STDC__
  printf("Using an ANSI-C compiler\n");
#endif
  printf("Date of compilation: %s\n", __DATE__);
  printf("Time of compilation: %s\n", __TIME__);

  printf("Line: %3d | File: %s\n", __LINE__, __FILE__);
#line 99 "file.c" /* sets the line number and the file name */
  printf("Line: %3d | File: %s\n", __LINE__, __FILE__);
  function();
  return EXIT_SUCCESS;
}
```

Listing 14. Example for the use of the C *scanf()* function.

```c
/* Usage of scanf() */
#include <stdio.h>

int main(int argc, char *argv[])
{
  int var;
  printf("Please enter an integer: ");
  scanf("%d", &var);
  printf("You entered %d\n", var);
  return (0);
}
```

scanf() in line 8 awaits your input, which you have to verify by pressing the return key. Then the input value is assigned to the integer variable var in line 8 and printed in line 9. The screen output when running Listing 14 is:

```
Please enter an integer: 34567
You entered 34567
```

Table 2.3. Useful predefined preprocessor macros.

macro	description
__LINE__	Returns the current line number of the program.
__FILE__	Returns the name of the program file as a string literal.
__DATE__	Returns the compilation date of the program as a string literal.
__TIME__	Returns the compilation time of the program as a string literal.
__STDC__	If this *int* constant is 1, the compiler is a ANSI-C compiler.
__func__	Returns the name of the function in which this macro is used (only *C99*).
__STD_VERSION__	Contains an integer constant of type *long*, if the compiler supports the *C99* standard (only *C99*).

scanf works similar to *scanf()*, by using two parentheses and brackets, which means that the input read is formatted. The formatting symbol %d stands for an *int* datatype. What do you think the symbol & in line 8 stands for?

> The ampersand & is an address operator in C, which is used to denote a memory address.

In the above example & denotes the *address* of the variable var of type *int* and assigns this address the value typed on the keyboard. Without the address operator, the input value of *scanf()* cannot be assigned to an address in computer memory. Using the address operator, you can directly access an address in computer memory and print it, using *printf()*. The format symbol for an address when printing with *printf()* is %p. So, if you wish, e.g. to plot the actual address of a variable, you can realize this with the following piece of code:

Listing 15. Plotting of an address.

```
1  /* Plot the address of a variable*/
2  #include <stdio.h>
3
4  int main(void)
5  {
6      int var;
7      printf("The address of var is: %p\n", &var);
8      return (0);
9  }
```

The output of Listing 15 could be, for example,

`The address of var is: 001AFC80`

2.3 Programming examples of rounding errors and loss of precision

As we have seen in Section 1.5.2, real numbers are stored with a decimal precision (or mantissa) and the decimal exponent range. The mantissa contains the significant figures of the number (and thereby the precision of the number). A number such as $(9.90625)_{10}$ in decimal representation is given in binary representation by

$$(1001.11101)_2 = 1 \times 2^3 + 0 \times 2^2 + 0 \times 2^1 + 1 \times 2^0 + \\ + 1 \times 2^{-1} + 1 \times 2^{-2} + 1 \times 2^{-3} + 0 \times 2^{-4} + 1 \times 2^{-5},$$

which has an exact machine number representation, since we need a finite number of bits to represent this number. This representation is, however, not very practical. Rather, we prefer to use a scientific notation. In the decimal system, we would write a number like 9.90625 in what is called the normalized scientific notation. This simply means that the decimal point is shifted and appropriate powers of 10 are specified. Our number could then be written as

$$9.90625 = 0.990625 \times 10^1,$$

and a real nonzero number could be generalized as

$$x = \pm r \times 10^n,$$

where r is a number in the range $1/10 \leq r < 1$. In a similar way, we can represent a binary number in scientific notation as

$$x = \pm q \times 2^m,$$

where q is a number in the range $1/2 \pm q < 1$. This means that the mantissa of a binary number can be represented by the general formula

$$(0.a_{-1}a_{-2}\cdots a_{-n})_2 = a_{-1} \times 2^{-1} + a_2 \times 2^{-2} + \cdots + a_{-n} \times 2^{-n}.$$

In a typical computer, floating point numbers are represented in the way described above, but with certain restrictions on q and m, imposed by the available word length. Following our explanation in Section 1.5.2, our number x is represented internally as

$$x = (-1)^s \times \text{mantissa} \times 2^{\text{exponent}}, \tag{2.1}$$

where s is the sign bit, and the exponent specifies the available range. With a single precision word of 32 bits, 8 bits would typically be reserved for the exponent, 1 bit for the sign and 23 for the mantissa. This means that if we define a variable as

```
float valueOfPressure;
```

we actually reserve 4 bytes in memory, with 8 bits for the exponent, 1 for the sign and 23 bits for the mantissa, implying a numerical precision up to the sixth or seventh digit, since the least significant digit is given by $1/2^{23} \approx 10^{-7}$. The exponent ranges from $2^{-128} = 2.9 \times 10^{-39}$ to $2^{127} = 3.4 \times 10^{38}$, where 128 stems from the fact that 8 bits are reserved for the exponent.

A modification of the scientific notation for binary numbers is to require that the leading binary digit 1 appears to the left of the binary point. In this case, the representation of the mantissa q would be $(1.f)_2$ and $1 \leq q < 2$. This form is rather useful when storing binary numbers in a computer word, since we can always assume that the leading bit 1 is present. One bit of space can then be saved meaning that a 23 bits mantissa has actually 24 bits. Explicitly, this means that a binary number with 23 bits for the mantissa reads as

$$(1.a_{-1}a_{-2}\cdots a_{-23})_2 = 1 \times 2^0 + a_{-1} \times 2^{-1} + a_{-2} \times 2^{-2} + \cdots + a_{-n} \times 2^{-23}.$$

As an example, consider the following 32 bits binary number

$$(10111110 \mid 11110100 \mid 00000000 \mid 00000000)_2,$$

where the first bit is reserved for the sign, in this case 1, yielding a negative sign. The exponent m is given by the next 8 binary numbers $(01111101)_2$, resulting in 125 in the decimal system. However, since the exponent has eight bits, this means it has $2^8 - 1 = 255$ possible numbers in the interval $(-128 \leq m \leq 127)$, our final exponent is $125 - 127 = -2$ resulting in 2^{-2}. Inserting the sign and the mantissa yields the final number in decimal representation as

$$-2^{-2}(1 \times 2^0 + 1 \times 2^{-1} + 1 \times 2^{-2} + 1 \times 2^{-3} + 0 \times 2^{-4} + 1 \times 2^{-5}) = (-0.4765625)_{10}.$$

Section 2.3 Programming examples of rounding errors and loss of precision

In this case we have an exact machine representation with 32 bits (actually, we need less than 23 bits for the mantissa).

If our number x can be exactly represented in the machine, we call x a machine number. Unfortunately, most numbers can not, and are thereby only approximated inside the machine. When such a number occurs as the result of a computation or of reading some input data, an error will inevitably arise when representing it by a machine number as accurately as possible. A floating point number x, labeled float(x) will therefore always be represented as

$$\text{float}(x) = x(1 \pm \epsilon_x), \tag{2.2}$$

where x is the exact number and the error $|\epsilon_x| \leq |\epsilon_P|$, where ϵ_P is the precision assigned. A number such as $1/10$ has no exact binary representation with single or double precision. Since, due to its limited number of bits, the mantissa

$$1.(a_{-1}a_{-2}\cdots a_{-n})_2$$

is always truncated at some stage n, there is only a limited number of real binary numbers. The spacing between every real binary number is given by the chosen machine precision. For a 32 bit word this number is approximately $\epsilon_P \sim 10^{-7}$ and for double precision (64 bits), we have $\epsilon_P \sim 10^{-16}$, or in terms of a binary base as 2^{-23} and 2^{-52} for single and double precision, respectively.

To understand that a given floating point number can be written as in equation (2.2), we assume for the sake of simplicity that we work with real numbers, with a word length of 32 bits, or four bytes. In this case, a given number x in binary representation can be represented as

$$x = (1.a_{-1}a_{-2}\cdots a_{-23}a_{-24}\cdots)_2 \times 2^n,$$

or in a more compact form

$$x = r \times 2^n,$$

with $1 \leq r < 2$ and $(-126 \leq n \leq 127)$, since our exponent is defined by eight bits. In most cases, there will not be an exact machine representation of the number x. Our number will be placed between two exact 32 bits machine numbers x_- and x_+. Following the discussion of Kincaid and Cheney [243], these numbers are given by

$$x_- = (1.a_{-1}a_{-2}\cdots a_{-23})_2 \times 2^n,$$

and

$$x_+ = \left[(1.a_{-1}a_{-2}\cdots a_{-23})_2 + 2^{-23}\right] \times 2^n.$$

If we assume that our number x is closer to x_-, the absolute error is constrained by the relation

$$|x - x_-| \leq \frac{1}{2}|x_+ - x_-| = \frac{1}{2} \times 2^{n-23} = 2^{n-24}.$$

A similar expression can be obtained if x is closer to x_+. The absolute error conveys one type of information. However, we may have cases where two equal absolute errors arise from different numbers. Consider for example the decimal numbers $a = 2.0$ and $\hat{a} = 2.001$. The absolute error between these two numbers is 0.001. In a similar way, the two decimal numbers $b = 2000.0$ and $\hat{b} = 2000.001$ produce exactly the same absolute error[26]. When we compare the relative errors

$$\frac{|a - \hat{a}|}{|a|} = 1.0 \times 10^{-3}, \quad \frac{|b - \hat{b}|}{|b|} = 1.0 \times 10^{-6},$$

we see that the relative error in b is much smaller than the relative error in a. We will see below that the relative error is intimately connected with the number of leading digits in the way we approximate a real number. The relative error is therefore the quantity of interest in scientific work. Information about the absolute error is normally of little use in the absence of the magnitude of the quantity being measured. We continue to define the relative error for x as

$$\frac{|x - x_-|}{|x|} \leq \frac{2^{n-24}}{r \times 2n} = \frac{1}{q} \times 2^{-24} \leq 2^{-24}.$$

Instead of using x_- and x_+ as the machine numbers closest to x, we introduce the relative error

$$\frac{|x - \hat{x}|}{|x|} \leq 2^{n-24},$$

with x being the machine number closest to x. Defining

$$\epsilon_x = \frac{\hat{x} - x}{x},$$

we can write the previous inequality

$$\text{float}(x) = x(1 + \epsilon_x),$$

where $|\epsilon_x| \leq \epsilon_M = 2^{-24}$ for 32 bits variables. The notation $\text{float}(x)$ stands for the machine approximation of the number x. The number ϵ_x is given by the specified machine precision, approximately 10^{-7} for single, and 10^{-16} for double precision, respectively.

---

[26] We note here that $b = 2000.001$ has more leading digits than b.

Section 2.3 Programming examples of rounding errors and loss of precision

There are several mathematical operations where an eventual loss of precision may appear. A subtraction, which is especially important in the calculation of numerical derivatives, is one important operation. In the computation of derivatives, we end up subtracting two nearly equal quantities. In case of such a subtraction, say $a = b - c$, we have

$$\text{float}(a) = \text{float}(b) - \text{float}(c) = a(1 + \epsilon_a),$$

or

$$\text{float}(a) = b(1 + \epsilon_b) - c(1 + \epsilon_c),$$

meaning that

$$\frac{\text{float}(a)}{a} = 1 + \epsilon_b \frac{b}{a} - \epsilon_c \frac{c}{a},$$

and if $b \approx c$, we see that there is a potential for an increased error in the machine representation of float(a). This is because we are subtracting two numbers of equal size, and what remains is only the least significant part of these numbers. This part is prone to rounding errors and if a is small we see that (with $b \approx c$)

$$\epsilon_a \approx \frac{b}{a}(\epsilon_b - \epsilon_c),$$

can become very large. The latter equation represents the relative error of this calculation. To see this, we first define the absolute error as

$$|\text{float}(a) - a|,$$

whereas the relative error is

$$\frac{|\text{float}(a) - a|}{a} \leq \epsilon_a.$$

The above subtraction is thus

$$\frac{|\text{float}(a) - a|}{a} = \frac{|\text{float}(b) - \text{float}(c) - (b - c)|}{a},$$

yielding

$$\frac{|\text{float}(a) - a|}{a} = \frac{|b\epsilon_b - c\epsilon_c|}{a}.$$

An interesting question then is how many *significant bits* are lost in a subtraction $a = b - c$ when we have $b \approx c$. The *loss of precision theorem* [243] for a subtraction

$a = b - c$ states that if b and c are positive, normalized floating point binary machine numbers with $b > c$ and

$$2^{-r} \leq 1 - \frac{c}{b} \leq 2^{-s}, \qquad (2.3)$$

then at most r and at least s significant binary bits are lost in the subtraction $b - c$. For a proof of this statement see, for example, Ref. [243].

We note here, that even additions can be troublesome, in particular if the numbers are very different in magnitude. Consider for example the seemingly trivial addition $1 + 10^{-8}$ with 32 bits used to represent the various variables. In this case, the information contained in 10^{-8} is completely lost during the addition. When we perform the addition, the computer first equates the exponents of the two numbers to be added. For 10^{-8}, this has, however, catastrophic consequences, since in order to obtain an exponent equal to 100, bits in the mantissa are shifted to the right. At the end, all bits in the mantissa are zeros. This means in turn that for calculations involving real numbers (if we omit the discussion on overflow and underflow) we need to carefully understand the behavior of our algorithm, and test all possible cases where rounding errors and loss of precision can arise. Other cases which may cause serious problems are singularities of the type $0/0$, which may arise from functions like $\sin(x)/x$ as $x \to 0$. Such problems may also need restructuring of the algorithm.

2.3.1 Algorithms for calculating e^{-x}

In order to illustrate the problems discussed above, we here discuss a couple of examples, including also a discussion of specific programming features. We start by considering three possible algorithms for computing the function e^{-x}:

1. Simply calculate

$$e^{-x} = \sum_{n=0}^{\infty} (-1)^n \frac{x^n}{n!}, \qquad (2.4)$$

or

2. use a recursion relation (a recurrence)[27] of the form

$$e^{-x} = \sum_{n=0}^{\infty} s_n = \sum_{n=0}^{\infty} (-1)^n \frac{x^n}{n!}, \qquad (2.5)$$

using

$$s_n = -s_{n-1} \frac{x}{n}, \qquad (2.6)$$

or

---

[27] Recurrence formulas, in various disguises, as ways to represent either series or continued fractions, are among the most commonly used forms of function approximations. Examples are Bessel functions, Hermite and Laguerre polynomials.

Section 2.3 Programming examples of rounding errors and loss of precision

3. calculate first

$$e^x = \sum_{n=0}^{\infty} s_n, \qquad (2.7)$$

using partial sums and then take the inverse

$$e^{-x} = \frac{1}{e^x}. \qquad (2.8)$$

In Listing 17 below, I have included a small C++ program which calculates equation (2.4) for values of x ranging from 0 to 100 in steps of 10. When performing the summation, one can always define a desired precision, given below by the fixed value of the variable TRUNCATION = 1.0E-10, so that for a certain value of $x > 0$, there is always a value of $n = N$ for which the loss of precision due to terminating the series at $n = N$ is always smaller than the next term in the series $\frac{x^N}{N!}$. The latter is implemented through the *while{...}* control statement. The output of the implemented brute force algorithm for calculating $\exp(-x)$ on my system is:

```
x = 0  exp = 1            series = 1            number of terms = 2
x = 10 exp = 4.53999e-05  series = 4.53999e-05  number of terms = 45
x = 20 exp = 2.06115e-09  series = 4.69008e-10  number of terms = 73
x = 30 exp = 9.35762e-14  series = -6.58635e-05 number of terms = 101
x = 40 exp = 4.24835e-18  series = 4.8818       number of terms = 128
x = 50 exp = 1.92875e-22  series = 32259.7      number of terms = 156
x = 60 exp = 8.75651e-27  series = -1.02777e+09 number of terms = 172
x = 70 exp = 3.97545e-31  series = -nan         number of terms = 172
x = 80 exp = 1.80485e-35  series = -nan         number of terms = 172
x = 90 exp = 8.19401e-40  series = -nan         number of terms = 172
```

I note that different compilers may produce different messages and deal with overflow problems in different ways. When executing Listing 17, we notice that for low values of x, the agreement is good. However, for larger values of x, we see a significant loss of precision. Secondly, for $x = 70$ we have an overflow problem, represented (by my specific compiler) as nan (not a number). The latter is easy to understand, since the calculation of a factorial of the size of 172! is beyond the limit set for the double precision variable factorial. The message nan appears because the computer sets the factorial of 172 equal to zero, and we end up having a division by zero in our expression for e^{-x}. The overflow problem can be dealt with via the recurrence formula of equation (2.5) for the terms in the sum, so that we avoid calculating factorials and instead calculate products according to equation (2.6). In this case (see Listing 16), we do not get an overflow problem, but the results do not make much sense for larger values of x, see Table 2.4. For a change, I have written Listing 16 as C code, using *printf()* as the stream, instead of *cout*.

When checking the numbers in Table 2.4 one realizes that the results do not make much sense for larger values of x. Decreasing the truncation threshold will not help,

Listing 16. Improved calculation of e^{-x}. (*)

```
1  /* Program to compute exp(-x) without factorials */
2
3  #include <stdio.h>
4  #include <math.h>
5  #define TRUNCATION 1.0E-10
6
7  int main(void)
8  {
9      int loop, n;
10     double x, term, sum;
11     for( loop = 0; loop <= 100; loop += 10 ){
12         /* initialization */
13         x = (double) (loop);
14         sum = 1.0;
15         term = 1;
16         n = 1;
17
18         while( fabs(term) > TRUNCATION){
19             term *= -x/ ( (double)n );
20             sum += term;
21             n++;
22         }/* end while loop */
23
24         printf(" x = %.60f    exp = %.60f    series = %.60f number
                  of terms = %i\n", x, exp(-x), sum, n-1);
25     } /* end for loop */
26 }
```

because this is a more serious problem. In order to understand this problem, let us consider the case of $x = 20$, which already differs from the exact result. Writing out each term of the summation shows that the largest term in the sum appears at $n = 19$, with a value that equals $-43\,099\,804$. However, for $n = 20$ we have almost the same value, but with opposite sign. This means that we have an error, relative to the largest term in the summation, of the order of $43\,099\,804 \times 10^{-10} \approx 4 \times 10^{-2}$. This is much larger than the exact value of 0.21×10^{-8}. The large contributions which may appear at a given order in the sum, lead to strong rounding errors, which in turn is reflected in the loss of precision. We can state the above in the following way: Since $\exp(-20)$ is a very small number and each term in the series can be quite large (of the order of 10^8), it is clear that other terms as large as 10^8, but negative, must cancel the figures in front of the decimal point as well as some behind it. Since a computer can only hold a fixed number of significant figures, all those in front of the decimal point are not only useless but are crowding out needed figures to the right side of the number. Unless we are very careful we will find ourselves summing a series that finally consists entirely of rounding errors. An analysis of the contribution to the sum from various terms

Section 2.3 Programming examples of rounding errors and loss of precision

Table 2.4. Output of the improved algorithm for calculating $\exp(-x)$.

x	$\exp(-x)$	series	number of terms in series
0	0.1000000000E + 01	0.1000000000E + 01	0
10	0.4539992976E − 04	0.4539993656E − 04	44
20	0.2061153622E − 08	0.6056246259E − 09	72
30	0.9357622968E − 13	0.9757624848E − 04	100
40	0.4248354255E − 17	0.3955571379E + 01	126
50	0.1928749847E − 21	−0.566764235 E + 05	155
60	0.8756510762E − 26	−0.823949784 E + 09	182
70	0.3975449735E − 32	0.1032197436E + 14	209
80	0.1804851400E − 34	0.2206643222E + 18	237
90	0.8190000000E − 39	0.4537847360E + 22	264
100	0.0000000000	−0.741812245 E + 26	291

shows that the relative error can be huge. This results in an unstable computation, since small errors made at one stage are magnified in subsequent stages. However, in this specific case there is a simple cure. Noting that $\exp(x)$ is the reciprocal of $\exp(-x)$, one may use the series for $\exp(x)$ and simply take the inverse. One has to be aware however, that $\exp(x)$ may quickly exceed the representable range of a double variable.

2.3.2 Algorithm for summing $1/n$

Let us consider another rounding example, which may be surprising. Consider the series

$$s_1 = \sum_{i=1}^{N} \frac{1}{i}, \qquad (2.9)$$

which is finite when N is finite. Then, consider an alternative way of writing this sum as

$$s_2 = \sum_{i=N}^{1} \frac{1}{i}, \qquad (2.10)$$

i.e we sum the terms, starting with large i and finally ending with $i = 1$. In this case, when summed analytically, the result – of course – will be $s_2 = s_1$. Because of rounding errors however, one will *numerically* end up with $s_2 \neq s_1$. Computing these sums with *single precision*, i.e. using floats for $N = 1,000,000$ results in $s_1 = 14.35736$, while $s_2 = 14.39265$. Note that these numbers are machine and compiler dependent. With double precision, the results agree exactly. However, for larger values of N, differences may appear even for double precision. For example, if we choose $N = 108$ and employ double precision, we get $s_1 = 18.9978964829915355$, while $s_2 = 18.9978964794618506$. This example demonstrates two important things. First, that the chosen precision is important for rounding errors (and – as we've mentioned before – single precision is generally not appropriate for scientific computing. One should *always* use double precision). Second, that the choice of an appropriate algorithm, can be of paramount importance for the result of the calculation, as seen in the example of calculating $\exp(-x)$. There are no standard rules that cover all situations. Only experience, and awareness of the pitfalls can help you avoid these errors.

Calculating the standard deviation

In many computer experiments in science and engineering, the results of computations are analyzed in essentially the same manner as a "classical" experiment is analyzed, namely many measurements N of the same value in many repetitions of the same experiment are summed, and divided by the number of trials, to produce a final numerical value, the so-called average $\langle X \rangle$ of some observable X:

$$\langle X \rangle = \frac{\sum_{i=1}^{N} X(i)}{N}, \tag{2.11}$$

where $X(i)$ is the numerical value of the observable X in the ith experiment measuring this quantity. The confidence that one may have in this value is usually associated with the standard deviation σ_X, calculated in the same way in computer experiments:

$$\sigma_X = \sqrt{\frac{\sum_{i=1}^{N} X(i)^2 - \langle X \rangle \sum_{i=1}^{N} X(i)}{N-1}}. \tag{2.12}$$

As an example illustrating the kind of problems which can arise when the standard deviation is small compared to the mean value $\langle X \rangle$, we assume that $X(i) = i + 105$, with $N = 104$, i.e., 104 repetitions of the same experiment measuring the value of X. The standard algorithm computes the two contributions to σ_X separately, i.e. it first takes the sum $\sum_{i=1}^{N} X(i)^2$ and then subtracts $\langle X \rangle \sum_{i=1}^{N} X(i)$. However, the problem that arises here, is that these two numbers can become nearly equal and large, and so we may end up in a situation with a potential loss of precision. A possible second algorithm first computes $X(i) - \langle X \rangle$ and then squares it when summing. With this second strategy, one may avoid having nearly equal numbers that cancel each other. Both algorithms are implemented in Listing 18.

Using single precision results in a standard deviation of $\sigma = 22.2436$ for the first and most frequently used algorithm, while the exact answer is $\sigma = 27.2743$, a number which results from the second algorithm in Listing 18. With double precision, the two

Section 2.3 Programming examples of rounding errors and loss of precision

Listing 17. Calculation of the exponential function e^{-x}. (*)

```
1  // Program to calculate function exp(-x)
2  // using straightforward summation with differing precision
3
4  using namespace std;
5  #include <iostream>
6  #include <cmath>     // declarations of fabs(), pow() and exp()
7  #include <cstdlib>   // declaration of abs() function
8  // type float:  32 bits precision
9  // type double: 64 bits precision
10
11 #define TYPE double
12 #define PHASE(a) (1-2*(abs(a) % 2))
13 #define TRUNCATION 1.0E-10
14
15 //function declaration
16 TYPE Factorial(int);
17
18 int main()
19 {
20   int n;
21   TYPE x, term, sum;
22   for(x = 0.0; x < 100.0; x += 10.0){
23     sum =0.0;
24     n = 0;
25     term = 1;
26     while (fabs(term) > TRUNCATION) {
27       term = PHASE(n) * (TYPE) pow((TYPE) x, (TYPE) n) /
             Factorial(n);
28       sum += term;
29       n++;
30     }
31     cout << " x = " << x << " exp = " << exp(-x) << " series =
              " << sum;
32     cout << " number of terms = " << n << endl;
33   } // end of for loop
34 } // end of function main()
35
36 // Function Factorial() calculates and returns n!
37 TYPE Factorial(int n)
38 {
39   int loop;
40   TYPE fac;
41   for (loop = 1, fac = 1.0; loop <= n; loop++){
42     fac *= loop;
43   }
44   return fac;
45 } // END function Factorial()
```

Listing 18. Calculation of the standard deviation. (*)

```
//This is a program to calculate the mean and standard
    deviation of
//a user created data set stored in array x[]
using namespace std;

#include <iostream>
#include <cmath>

#define NUM_TRIALS 104
int main()
{
  int i;
  float sum, sumSquared, xAverage, sigma1, sigma2;

  //Array declaration with fixed dimension
  float x[NUM_TRIALS];
  //Initialize data set
  for( i = 0; i < NUM_TRIALS; i++){
    x[i] = i + 100000.;
  }
  //The variable sum is just the sum over all elements
  //The variable sumSquared is the sum over x^2
  sum = 0.;
  sumSquared =0.;
  //The first algorithm for the summation
  for( i = 0; i < NUM_TRIALS; i++){
    sum += x[i];
    sumSquared += pow( (double)x[i],2.);
  }

  //Calculate the average and sigma
  xAverage = sum / NUM_TRIALS;
  sigma1=sqrt((sumSquared - sum * xAverage) / 126.);

  // The second algorithm for the summation, where we calculate
  // separately first the average, and then the
  // sum which defines the standard deviation. The average
  // has already been evaluated through xAverage
  sumSquared = 0.;
  for(i = 0; i < NUM_TRIALS; i++){
    sumSquared += pow((double)(x[i]-xAverage),2.);
  }
  sigma2 = sqrt(sumSquared / 126.);
  cout << "xAverage = " << xAverage << " sigma1= " << sigma1 <<
        " sigma2 = " << sigma2;
  cout << endl;
  return (0);
}
```

algorithms produce the same answer. The reason for this difference lies in the fact that the first algorithm includes the subtraction of two large numbers, which are squared. Since the average value for this example is $\langle X \rangle = 100\,052.0$, it is easy to see that computing $\sum_{i=1}^{N} X(i)^2 - \langle X \rangle \sum_{i=1}^{N} X(i)$ can give rise to very large numbers, with a possible loss of precision when the subtraction is performed. To see this, consider the case where $i = 64$. Then, we have $X_{64}^2 - \langle X \rangle X_{64} = 100\,352$, while the exact answer is $X_{64}^2 - \langle X \rangle X_{64} = 100\,064$. You can even check this by calculating it by hand. The second algorithm first computes the difference between $X(i)$ and the average value. This difference is then squared. However, when using the second algorithm, we have for $X_{64} - \langle X \rangle = 1$, so there is no potential for loss of precision.

2.4 Details on C-Arrays

In this section, we consider an important composite datatype: *arrays*. Arrays are used to store individual elements of a certain elementary numeric datatype (*short, int, float, double, long*), as a sequence of values[28]. Arrays are also called *vectors, fields,* or *sequences*, and their use simplifies the management of elementary numeric data in a way similar to the use of matrices in engineering and in the numerical sciences.

The basic syntax for the definition of an array of type datatype is:

datatype ArrayName[NumerOfElements];

The datatype denotes the datatype of the elements that are stored in the array. ArrayName can be chosen freely, according to the rules for the choice of names.[29]

An array is essentially a matrix, containing variables of a given type. For example,

double myArray[5] = {1.2, 2.1, 3.7, 4.3, 5.0};

declares a variable myArray with five members of type double, and at the same time assigns these member the values in the curly brackets, separated by commas. This simple form of assigning values to the elements of an array can only be done in the definition. If some elements of an array are to be changed within the program after its definition, one has to do this by single assignments like myArray[4]=4.56 or within a loop like in the following code snippet:

```
int x;
for (x = 0; x < 10; x++)
{
```

---

[28] There are also arrays of characters, which are called *strings*, see Section 2.4.7.
[29] See Section 2.2.1.

```
    myArray[x] = x * 1.2;
}
```

Members of an array can be accessed by indices and manipulated in the same way as other variables. Indexing of an array in C always begins at zero. For example,

`d=myArray[0];`

assigns d the value of the first element in myArray, which is 1.2. The array myArray has five elements of type *double*. Hence, internally, the operating system allocates memory for five arrays of type *double*. One can find out how much memory this corresponds to by using the C operator sizeof. Listing 19 provides an example of how the operator sizeof works. The output of Listing 19 on my system is:

```
sizeof(ivalue)       : 4
sizeof(dvalue)       : 8
sizeof(myArray)      : 40
sizeof(myArray[0])   : 8
sizeof(&myArray)     : 4
myArray[3]           : 5.000000
```

The output of this small program shows, that on my system, the word length is 4 bytes. Hence, this is also the size of type int and the size of memory needed to store an address (compare line 16 in Listing 19).

Listing 19. The use of the `sizeof` operator.

```
1  /* Determine the size of a type in bytes
2     and manipulate an element of an array
3  */
4  #include <stdio.h>
5
6  int main(void) {
7      int iValue;
8      double dValue;
9      double myArray[5] = {1.2, 2.1, 3.7, 4.3, 5.0};
10     myArray[3] += .7;
11
12     printf("sizeof(iValue)     : %d\n", sizeof(iValue));
13     printf("sizeof(dValue)     : %d\n", sizeof(dValue));
14     printf("sizeof(myArray)    : %d\n", sizeof(myArray));
15     printf("sizeof(myArray[0]) : %d\n", sizeof(myArray[0]));
16     printf("sizeof(&myArray)   : %d\n", sizeof(&myArray));
17     printf("myArray[3]         : %f\n", myArray[3]);
18     return 0;
19 }
```

> In standard C, *arrays have a fixed number of elements*, as they are defined with a constant number in the brackets [ ]. This means that the size of the array has to be known when compiling the code. Variable array lengths are only possible with *dynamic memory allocation*, discussed in Section 2.4.6, or in the *C99* standard.

As we have seen, an array in C is formed by laying out all the elements in memory contiguously from low to high. The array as a whole is referred to by the address of its first element. For example, the variable name `intArray` below is synonymous with the address of the first element `&intArray[0]` and can be used in expressions like "int *", i.e. like a pointer. Hence, the following statement

```
int intArray[6];
```

can be graphically represented as shown in Figure 2.6.

| intArray[0] | intArray[1] | intArray[2] | intArray[3] | intArray[4] | intArray[5] |

Figure 2.6. Representation of an *int*-array in memory. Note that arrays in C always start at index 0, so the 6 elements of intArray[6] are indexed from 0 to 5.

> **The first element of an array in C has index 0!** This is a common source of *access violations* when using *for* loops for the initialization of arrays, as in the following code snippet
>
> ```
> long lArray[10];
> ...
> for (i = 0; i <= 10; i++){
>     lArray[i]=i;
> }
> ```
>
> Here, the use of the "<=" operator instead of "<" leads to initializing 11 instead of only 10 arrays.

The programmer can refer to elements in the array with the simple [ ] syntax such as `intArray[1]`. This scheme works by combining the base address of the array with the index to compute the base address of the desired element in the array, via some simple arithmetic. Each element takes up a fixed number of bytes, known at compile time. So, the address of the nth element in the array (0-based indexing!) will be at an offset of ($n \times$ elementSize) bytes from the base address of the whole array, see Figure 2.7:

address of nth element = address of 0th element + ($n \times$ elementSizeInBytes)

The square bracket syntax [ ] takes care of this address arithmetic for you, but it's useful to know what it's doing. The [ ] multiplies the integer index by the element size, adds the resulting offset to the array base address, and finally dereferences the resulting pointer in order to obtain the desired element. As an example, let's look at the following command

```
intArray[3] = 48;
```

and see what it does in memory, see Figure 2.7.

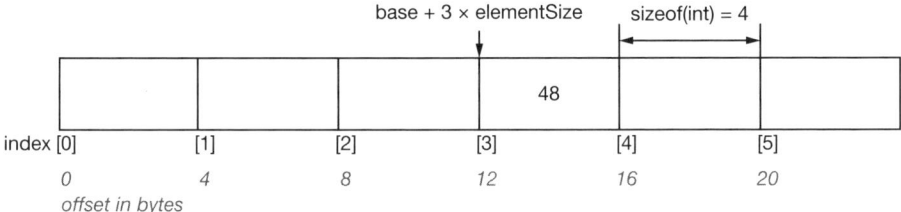

Figure 2.7. Calculation of the memory offset in an int-array. The assignment "intArray[3] = 48;" actually stores the number 48 at the address "(base + 3× elementSize)".

> On many systems there is a compiler (or debugging) option, which checks out of bounds arrays at runtime. However, it is not recommended to use this option for the compilation of the final program, because it leads to very bad runtime behavior. In the future *C1x* standard, there will be a *bounds checking interface*.

In a closely related piece of syntax, the "+ syntax", adding an integer to a pointer does the same offset computation, but leaves the result as a pointer. The square bracket syntax dereferences that pointer to access the nth element while the + syntax just computes the pointer to the nth element. So the expression (intArray + 3) is a pointer to the integer intArray[3]. (intArray + 3) is of type (*int **) while intArray[3] is of type *int*. The two expressions only differ in whether the pointer is dereferenced or not. (For an introduction to pointer arithmetic see the next Section 2.4.4).

> The expression (intArray + 3) is exactly equivalent to the expression (&(intArray[3])). In fact those two probably compile to exactly the same code. They both represent a pointer to the element at index 3.

Any [ ] expression can be written with the + syntax instead. We just need to add the pointer dereference. So intArray[3] is exactly equivalent to *(intArray + 3). For most purposes, the easiest and most readable option is the [ ] syntax. Every once in a while the + syntax is convenient, if you need a pointer to the element instead of the element itself.

2.4.1 Direct initialization of certain array elements (C99)

The *C99* standard allows for initializing a particular element of an array. To do this, one has to write the particular element in brackets in the initialization list of the definition of the array. Here is an example:

```
int iArray[5] = {112, 435, [MAX-1]=969};
```

Here, the first two elements were initialized and then the last item in the initialization list was assigned a value. After this definition and initialization, the individual elements of iArray have the following values:

```
iArray[0] = 112
iArray[1] = 435
iArray[2] = 0
iArray[3] = 0
iArray[4] = 969
```

If one wants to make sure that the values of the elements of an array are *not* changed at runtime, one can precede the keyword *const* in front of the array definition, like in this example:

```
const double dArray[3] = {1.0, 7.345, 9.01};
```

In this case, commands like dArray[0] = 0.8 will lead to a compile error.

2.4.2 Arrays with variable length (C99)

The *C99* standard can also handle arrays of variable length. Provided the array has been defined in a *local scope* (i.e. not outside a command block) and has not been marked with the keyword *const*, it is possible to define an array for which the number of elements is not fixed at compile time. Here is an explicit code example of how to do this: line 12 shows the definition of an *int*-array, where the number of elements is

Listing 20. An array of variable size (only *C99*).

```c
/* Example for an array of variable size */
#include <stdio.h>
#include <stdlib.h>

int main(void) {
  int value;

  printf("Please Input the number of elements: ");
  scanf("%d", &value);

  if(value > 0) {
    int iArray[value];
    int i;

    for(i = 0; i < value; i++) {
      iArray[i] = i;
    }
    for(i = 0; i < value; i++) {
      printf("Assigned array values: %d\n", iArray[i]);
    }
  }
  return EXIT_SUCCESS;
}
```

not fixed at the start of the program. In fact, it is determined by the input of the user in line 9 using the *scanf()* function. You can see that this actually works in line 16, where values are assigned to the elements. Line 19 outputs the values assigned. In order for this to work, it is essential that the definition in line 12 is inside of a command block between lines 11 and 21. `iArray` is only defined within this scope.

2.4.3 Arrays as function parameters

In order to discuss how to pass an array to a function in C we have to anticipate some of our discussion on pointers in the next Section 2.4.4. As we discussed above, the name of an array (without square brackets) stands for the start address of this array in memory, i.e. the name of an array is a *vector* (a *constant* address in memory)[30]. Thus, arrays are passed to functions as *call-by-reference* and not as *call-by-value*. This means, that in the case of an array, no copy of the array is passed to the function, but only the base address of the array. As a consequence, the function can directly access the values of the elements of the original array. Let's look at an example to make things clear.

---

[30] In contrast to vectors, *pointers* are *variable* addresses in computer memory.

Listing 21. An array as function parameter. (*)

```
1  /* Example of an array as function parameter */
2  #include <stdio.h>
3  #define MAX_ELEMENT 10
4
5  void Input(int *array, int nElements)
6  {
7    int i;
8    for(i = 0; i < nElements; i++)
9      {
10       printf("Please input the number for element %i: ", i+1);
11       scanf("%i", &array[i]);
12      }
13 }
14
15 void Calculate(int array[], int nElements)
16 {
17   int i;
18   for(i = 0; i < nElements; i++)
19     array[i] *= 2;
20 }
21
22 void Output(int array[10])
23 {
24   int y;
25   for(y = 0; y < 10; y++)
26     printf("The %i. element of the array is %i\n", y+1, array[y
           ]);
27 }
28
29 int main(void)
30 {
31   int iArray[MAX_ELEMENT];
32   Input(iArray, MAX_ELEMENT);
33   Calculate(&iArray[0], MAX_ELEMENT);
34   Output(iArray);
35
36   return (0);
37 }
```

Example 2.2. Assume, you want to define an *int*-array with 10 elements. We ask the user to assign 10 integer values, each of which we then want to multiply with 2 and output the 10 results as a list on the screen. Write a C program that performs this task.

In Listing 21 we provide a complete solution to the task of Example 2.2, using functions. The listing shows the different ways that arrays can be passed as arguments to functions. Listings with several functions are most easily read by starting with the *main()*-function. Hence, let's analyze what the code in Listing 21 does. In line 31 an *int*-array named "iArray" is defined with 10 entries. In line 32 the address of the array[31], defined in line 31, and the number of elements (MAX_ELEMENT) is passed to the function *Input()*. This function is defined in lines 5–13. The function initializes the individual elements of the array with some values. In line 33 the base address of iArray is passed to the function *Calculate()* (lines 15–20), along with the number of elements in the array. In *Calculate()*, each single value of the elements of iArray is multiplied by 2. The syntax of passing the array iArray to the functions in lines 32 and 33 is equivalent. In line 34 only the array is passed to the function *Output()* and no information on the number of elements is passed. In this case, the information on the number of elements is contained in the definition of the function *Output()* in line 22. Here, the parameter list indicates that a pointer to a field of type *int*, with 10 elements, is passed. The slight disadvantage here is that the "magic number" 10 here appears in the loop in line 25, which makes the code harder to read and should actually be avoided[32]. In this case, it would be better to replace the value 10 in lines 22 and 25 with the macro MAX_ELEMENT.

> With pointers to arrays one needn't use a dereferencing operator (*) when passing the array to a function, because the name of the array represents its base address, just like the name of a pointer represents an address. When passing an array such as int a[17] to a function of the form function(a), then both definitions void function(int *x) and function(int x[17]) are equivalent. In the latter definition one can explicitly see that a pointer to a static array with 17 elements is passed and this function can only be used for *int*-arrays with 17 elements. The former definition is equivalent, but can be used for *int*-arrays of *any* size, because here the compiler only knows that a pointer to an address is passed, the content of which is to be interpreted as *int*. When adhering to the *ANSI*-C standard, only constant fields may be used. In this case, it is better to use the latter definition of a function because it makes the code more readable.

2.4.4 Pointers

In this section, we discuss a very important and powerful feature of C programming, namely pointers. Pointers are used abundantly in scientific programming for dynamic memory allocation and efficient data management, using so-called *abstract*, or *complex* datatypes such as linked lists, trees, or LIFO[33] and FIFO[34] structures. This topic

---

[31] Remember: The name of an array without square brackets is a synonym for its base address, see our remarks on Page 141.
[32] Compare our discussion of "Good software development practices" in Section 2.7.1.
[33] Last *I*n, First *O*ut.
[34] First *I*n, First *O*ut.

is important, because the efficiency of algorithms often relies on a particular, advantageous organization of data. Pointers are needed for:

- dynamic allocation, management and deletion of memory during runtime of a program
- passing data objects to functions via call by reference
- passing functions as argument to functions
- creating abstract datatypes such as lists and trees
- defining typeless pointers (void *) for managing objects of any type.

One of the main characteristics of a scientific program is that large amounts of numerical information are exchanged between the various functions which make up the program. It is generally most convenient to pass this information via the argument lists, rather than via the names of the these functions. After all, only one number can be passed via a function name (which is called *call by value*), whereas scientific programs generally require far more than one number to be passed in a function call. Hence, the functions employed in scientific programs generally return no values through their names (i.e., they often tend to be of datatype void) but have large strings of arguments. There is one obvious problem with this approach. Namely, a void function, which passes all of its arguments by value, is incapable of returning any information to the program segment from which it was called. Fortunately, there is a way of getting around this difficulty: we can pass the arguments of a function *by reference*, rather than by value, by using pointers. This allows the two way communication of information, during function calls, via arguments. Pointer variables, like all other variables, must be declared before they can appear in executable statements. A pointer declaration takes the general form

```
datatype *pointerName;
```

where `pointerName` is the name of the pointer variable, and `datatype` is the datatype of the data item that the pointer points to. Note that an asterisk must always precede the name of a pointer variable in a pointer declaration. Referring to Section 2.2.6, we can now fully appreciate the mysterious asterisk which appears in the declaration of an input/output stream, e.g.,

```
FILE *fopen(...);
```

and which is there because *fopen()* is actually a pointer variable (pointing towards an object of the special datatype `FILE`). In fact, *fopen()* points towards the beginning of the associated I/O stream in memory.

Suppose now that `velocity` is a variable in a C program which represents some particular data item, say, an integer. Of course, the program stores this data item at some particular location in the computer's memory. The data item can thus be accessed if we know its location, or address, in computer memory. The address of velocity's memory location is determined by the expression &velocity, where & is a *unary operator* known as the *address operator*. Suppose now that we assign the address of velocity to another variable pv, in other words, pv = &velocity. This new variable is called a pointer to velocity, since it points to the location where velocity is stored in memory. Remember, however, that pv represents velocity's *address*, and not its *value*. The data item represented by velocity (i.e., the data item stored at velocity's memory location) can be accessed via the expression *pv, where "*" is a unary operator, called the *indirection*, or *dereferencing* operator, which only operates on pointer variables. Thus, *pv and velocity both represent the same data item. Furthermore, if we write pv = &velocity and u = *pv then both u and velocity represent the same value. The simple program Listing 22 below illustrates some of the points made above.

> For the novice in programing, it is strongly recommended to study this code example thoroughly.

Listing 22. Simple illustration of the use of pointers in C. (*)

```c
/* Example of the use of pointers */
#include <stdio.h>
int main(void)
{
    int u = 5;
    int velocity;
    int *pu = NULL;  // Declare a pointer to an integer variable
    int *pv = NULL;  // Declare a pointer to an integer variable
    pu = &u;         // Assign the address of u to pu
    velocity = *pu;  // Assign the value of pu to velocity
    pv = &velocity;  // Assign the address of velocity to pv
    printf("\n        u = %d        &u = %p   pu = %X  *pu = %d",\
            u, &u, pu, *pu);
    printf("\nvelocity = %d &velocity = %p   pv = %X  *pv = %d\n",\
            velocity, &velocity, pv, *pv);
    return 0;
}
```

Note that pu is a pointer to u, whereas pv is a pointer to velocity. Incidentally, the conversion character "X", which appears in the control strings of the *printf()* function calls in lines 13 and 14, indicates that the associated data item should be output as a hexadecimal number. We have seen in Section 1.5 that this is the conventional method of representing an address in computer memory. The corresponding control string "%p" indicates that the data item should be output as an address, which is the same hexadecimal number prefixed with "0x". Lines 7 and 8 initialize the pointers *pu and *pv with the NULL pointer[35]. Execution of the above program yields the following output:

```
u = 5         &u = 0x7fff50b7458c pu = 50B7458C *pu = 5
velocity = 5 &velocity = 0x7fff50b74588 pv = 50B74588 *pv = 5
```

Note that on your own system, the addresses in memory will be different. In the first line, we see that u represents the value 5, as specified in its declaration statement. The address of u is determined automatically by the compiler to be 7fff50b7458c (hexadecimal). The pointer pu is assigned this value. Finally, the value to which pu points is 5, as expected. Similarly, the second line shows that velocity also represents the value 5. This also is to be expected, since we assigned the value *pu to velocity. The address of velocity is 7fff50b74588. Of course, u and velocity have different addresses in memory. The unary operators & and * are members of the same precedence group as the other unary operators (e.g., ++ and –). The address operator (&) can only act upon operands that possess a unique address, such as ordinary variables. Thus, the address operator cannot act upon arithmetic expressions, such as 2 * (u + velocity). The indirection operator (*) can only act upon operands which are pointers.

Pointer arithmetic

If p is a pointer to an element in an array, then (p+1) points to the next element. C-Code can exploit this using the construct p++ (i.e the increment operator) to step a pointer over the elements in an array. It doesn't help readability, so I can't really recommend this technique, but one may see it in code written by others. *Pointer arithmetic* in C/C++ is a very powerful tool to write extremely concise and cryptic code using pointers. Provided that a pointer p has been assigned a certain datatype, C/C++ allows to perform the following arithmetic operations:

- **Comparison of pointers.** (pointer1 op pointer2) with the following allowed operators op $\in \{==, !=, <, <=, >, >=\}$. The use of comparison operators is only useful if the pointers point to addresses of array elements. In this case, a comparison "pointer1 > pointer 2" returns TRUE, if the memory address of pointer1 is larger than the address of pointer2.

- **Subtraction of pointers.** (pointer1 op pointer2), where op $\in \{-, -=\}$. As a result of the operation "pointer1 - pointer 2" the number of elements between the pointers is returned. For this, the header file <stddef.h> defines the datatype ptrdiff_t which is usually declared to be of type *int*.

---

[35] "NULL" is the macro definition of a pointer with value 0, i.e. which points to the address 0x0. This is the reason why many functions in C (for example *fopen()*) return NULL upon failure. The assignment int *pu = NULL is equivalent to int *pu = (void *)0.

> Pointers defined within a block {} that have not been initialized with a valid address, remain uninitialized even when they are accessed later in the program. This can lead to completely undefined behavior as an uninitialized pointer can access any memory address at random. Pointers defined globally or with the keyword `static` are automatically initialized to be a NULL pointer. Wrong usage of pointers is *the* number one source of segmentation faults in C programs. Often, such errors are very hard to detect in an advanced stage of programming. Pointers provide the C programmer with a lot of power but they can also be dangerous, because they can literally point to *any* memory address in the heap of computer memory. Usually, there is neither an automatic memory protection provided by the operating system nor any out-of-bounds pointer checking by C, so this remains entirely the responsibility of the scientific programmer. This is often where the novice distinguishes himself from the experienced programmer.

- **Addition or subtraction of a pointer and an integer.** (pointer1 op integer) where op $\in \{+, -, +=, -=, --, ++\}$. For example, if pointer1 points to the array array[i], then the assignment pointer2=pointer1+3 has pointer2 pointing to the array element array[i+2].

In the following, as an example, I present a sequence of versions of the C-function *strcpy()*, written in the order from most verbose to most cryptic. This also provides an example of the infamous way you can write concise code in C. In the first one, the straightforward *for* loop needs a little follow-up to ensure that the terminating NULL character is copied. Note the short version of the for loop in C. This is possible if you execute only one command within the for loop.

```
void strcpy1(char dest[], const char source[])
{
  int i;
  for (i = 0; source[i] != '\0'; i++)
    dest[i] = source[i];
  /* Don't forget to NULL-terminate the char field (a string)! */
  dest[i] = '\0';
}
```

The second version removes that opportunity for introducing an error by rearranging it into a *while* loop and moving the assignment into the test. So here you have removed one line of code making everything more concise.

Section 2.4 Details on C-Arrays

```
void strcpy2(char dest[ ], const char source[ ])
{
  int i = 0;
  while ((dest[i] = source[i]) != '\0')
    i++;
}
```

In the last example you achieve the same result with only a single line of code (demonstrating the use of "++" on pointers), but this is not really the sort of code you want to maintain. It is a fancy version of C code, where you get rid of i and just move the pointers. This relies on the precedence of * and ++.

```
void strcpy3(char dest[ ], const char source[ ])
{
  while ((*dest++ = *source++) != '\0') ;
}
```

Among the three versions, I think the second one is the best stylistically. With a smart compiler, all three will compile to basically the same machine code with the same efficiency, so one does not necessarily gain speed by writing super-fancy concise C code, but just makes one's code harder to understand for others (and for oneself, too!). As a daunting example, a super-fancy – even shorter – version of this function is provided in Section 2.7 on Page 182.

Pointer type effects

Both [] and + implicitly use the compile time type of the pointer to compute the element size which effects the offset arithmetic. When looking at code, it's easy to assume that everything is in units of bytes.

```
int *p;
p = p + 12;   /* At run-time, what does this add to p? 12? */
```

The above code snippet does not add the number 12 to the address in p, which would increment p by 12 bytes. The code above increments p by 12 *ints*. Each *int* takes 4 bytes, so at runtime the code will effectively increment the address in p by 48 bytes. The compiler figures all of this out, based on the type of the pointer. You can manipulate this by using casts. For example, the following code does just add 12 to the address in the pointer p. It works by telling the compiler that the pointer points to a char instead of an int. The size of char is defined to be exactly 1 byte (or whatever the smallest addressable unit is on the computer). In other words, sizeof(char) is always 1. We then cast the resulting (*char**) back to an (*int**). You can use casting like this to change the code the compiler generates; the compiler just blindly follows your orders.

```
p = (int*) ((char*)p + 12);
```

Arithmetic on a void pointer

For a (*void**) pointer, array subscripting and pointer arithmetic don't quite make sense. These manipulations include implicit multiplication by the size of the element type. What is sizeof(void)? Unknown! Some compilers assume that it should be treated like a (*char**), but if you were to depend on this you would be creating non-portable code. To be precise and correct, you should cast the (*void**) to (*char**) before doing any math, in order to make clear that all arithmetic is done in one byte increments and any necessary multiplication will be done explicitly.

```
void *ptr;
p = (char*)ptr + 4;    /* Increments ptr by exactly 4 bytes */
```

Note that you do not need to cast the result back to (*void**). A (*void**) is the "universal recipient" of pointer types and can be freely assigned any type of pointer. It is best to use casts only when you absolutely must.

Arrays and pointers

One effect of the C array scheme is that the compiler does not make any meaningful distinction between arrays and pointers – they both simply look like pointers. In the following example, the value of intArray is a pointer to the first element in the array, so it's an (*int**). The value of the variable intPtr is also (*int**) and it is set to point to a single integer i. So what's the difference between intArray and intPtr? Not much as far as the compiler is concerned. They are both just (*int**) pointers, and the compiler is perfectly happy to apply the [] or + syntax to either. It's the programmer's responsibility to ensure that the elements referred to by a [] or + operation are actually present. It's just the same old habit of C not to do any bounds checking. C thinks of the single integer i as just a sort of degenerate array of size 1.

```
int intArray[6];
int *intPtr;       /* Declaration of a pointer to int */
int i;
intPtr = &i;       /* 'intPtr' now points to the address \
                      of the 4 byte integer i */
intArray[3] = 31;  /* ok */
intPtr[0] = 21;    /* odd, but ok. Changes i, because \
                      inPtr = intPtr[0] */
intPtr[3] = 31;    /* BAD! There is no integer reserved here!
                      intPtr is NOT a field */
```

Figure 2.8 shows a graphical representation of intArray and intPtr.

Section 2.4 Details on C-Arrays

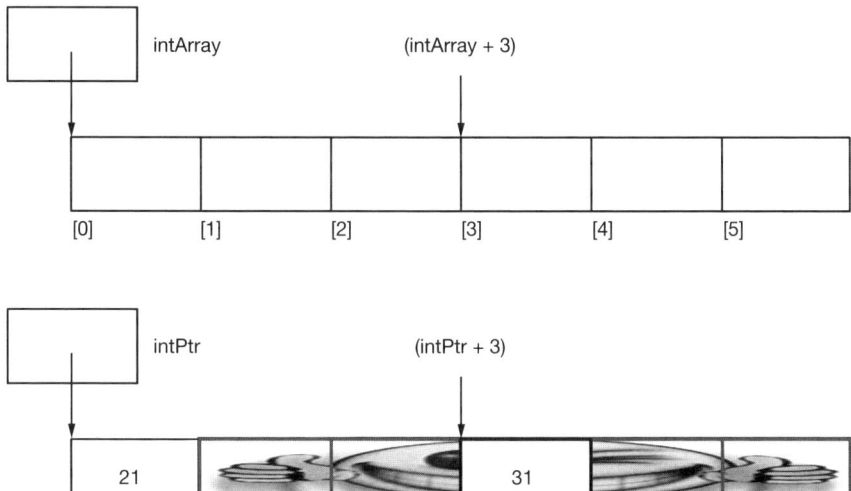

Figure 2.8. The bytes with the thumbs-up smiley face in the lower part of the figure do exist, but they have not been explicitly reserved. They are the bytes which happen to be adjacent to the memory of i. They are probably already being used to store something different, such as the thumbs-up smiley face. The 31 just is blindly written over the smiley face, i.e. over whatever bit pattern happens to be residing at this location in memory. This error will only be apparent later, when the program tries to read the smiley face data.

Array names are *const*

One subtle distinction between an array and a pointer is that the pointer which represents the base address of an array cannot be changed in the code. Technically, the array base address is a *const* pointer. The constraint applies to the name of the array where it is declared in the code – the variable intArray in the examples below.

```
int intArray[100]
int *intPtr;
int i;
intArray = NULL; /* no, cannot change the base address
                    of a pointer */
intArray = &i;          /* no */
intArray = intArray + 1; /* no */
intArray++;             /* no */
intPtr = intArray;      /* ok, intPtr is a pointer
                           which can be changed. */
intPtr++;               /* ok, intPtr can still be changed
                           (and intArray cannot) */
intPtr = NULL;          /* no */
```

```
intPtr = &i;              /* no */
Function (intArray);      /* ok */
Function (intPtr);        /* ok */
```

Array parameters are passed as pointers. The arguments of the two definitions of Function look different, but to the compiler they mean exactly the same thing. It's preferable to use whichever syntax best aids readability. If the pointer coming in will be treated as the base address of a whole array, then use []. If it is merely a pointer to one integer, the * notation is more appropriate.

```
void Function (int arrayParam[])
{
  arrayParam = NULL; /* Silly but valid. Simply changes the local pointer */
}

void Function (int *arrayParam)
{
  arrayParam = NULL; /* ditto */
}
```

Note that either `intArray` or `intPtr` above could be passed to either version of Function. In each case, the address of an integer is passed. For `intArray` it is a copy of base address of the array, for `intPtr`, it is a copy of its current value. Either way, once in the function Function, either pointer can be changed to point elsewhere without affecting the value of the pointers in the calling function.

2.4.5 Pointers as function parameters

I mentioned above that pointers are often passed to a function as arguments. This allows the function to access data items within the calling part of the program, which can then be altered within the function, and passed back to the calling portion of the program in altered form. This use of pointers is referred to as passing arguments *by reference*, rather than by value. When an argument is passed by value, the associated data item is simply copied to the function. Thus, any alteration to the data item within the function is not passed back to the calling routine. When an argument is passed by reference, however, the address of the associated data item is passed to the function. The contents of this address can be freely accessed by both the function and the calling routine. Furthermore, any changes made to the data item stored at this address are recognized by both the function and the calling routine. Thus, the use of a pointer as an argument allows the two-way communication of information between a function and its calling routine. The program Listing 23 below uses a pointer to pass information from a function back to its calling routine:

Think! What happens if you remove the prototype of *Factorial()* in line 8 of Listing 23?

Listing 23. The use of pointers as function parameters. (*)

```c
/* Program to print factorials of all integers between $0$ and
    $30$ */
#include <stdio.h>
#include <stdlib.h>

/* Prototype for function Factorial(); this would usually be
    put
   in a header file
*/
void Factorial(int, double *);

int main()
{
   int j;
   double fact;
   /* Print factorials of all integers between 0 and 20 */
   for (j = 0; j <= 30; ++j)
    {
       Factorial(j, &fact);
       printf("j = %3d Factorial(j) = %12.3e\n", j, fact);
    }
   return 0;
}

/* The definition of factorial() */
void Factorial(int n, double *fact)
{
  *fact = 1.;

  /* Abort if n is negative integer */
  if (n < 0)
   {
      printf("\nError: factorial of negative integer not
          defined\n");
      exit(1);
   }
  /* Calculate factorial */
  for (; n > 0;~--n) *fact *= (double) n;
    return;
}
```

The argument list of the function *Factorial()* in line 24 has two arguments. The first argument is the value of the positive integer n whose factorial is to be evaluated by the function. The second argument, `fact`, is a pointer which passes back the factorial of n (in the form of a floating point number) to the main part of the program. Incidentally, the compiler knows that `fact` is a pointer, because its name is proceeded by an asterisk in the argument declaration for *Factorial()*. Of course, in the body of the function, reference is made to `*fact` (i.e., the *value* of the data item stored in the memory location towards which fact points), rather than fact (i.e., the *address* of the memory location towards which fact points). Note that a *void* function, which returns no value, can only be called via a statement consisting of the function name followed by a list of its arguments (in parentheses and separated by commas). Thus, the function *Factorial()* is called in the main part of the program (line 17) via the statement

`Factorial(j, &fact);`

This statement passes the integer value j to *Factorial()*, which, in turn, passes back the value of the factorial of j via its second argument. Note that since the second argument is passed by reference, rather than by value, it is written `&fact` (i.e., the address of the memory location where the floating point value fact is stored) rather than `fact` (i.e., the value of the floating point variable `fact`). Note, finally, that the function prototype for *Factorial()* (line 8) takes the form

`void Factorial(int, double *);`

Here, the asterisk after *double* indicates that the second argument is a pointer to a floating point data item. We can now appreciate the ampersands which must precede variable names in *scanf()* calls: e.g.,

`scanf("%d %lf %lf", &k, &x, &y);`

scanf() is a function which returns data to its calling routine via its arguments (excluding its first argument, which is a control string). Hence, these arguments must be passed to *scanf()* by reference, rather than by value, otherwise they would be unable to pass information back to the calling routine. It follows that we must pass the addresses of variables (e.g., &k) to *scanf()*, rather than the values of these variables (e.g., k). Note that, since the *printf()* function does not return any information to its calling routine via its arguments, there is no need to pass these arguments by reference. This explains why there are no ampersands in the argument list of a *printf()* function.

2.4.6 Pointers to functions as function parameters

A pointer to a function can be passed to another function as an argument. This allows one function to be transferred to another, as though the first function were a variable. This is very useful in scientific programming. Imagine that we have a routine which

Section 2.4 Details on C-Arrays

numerically integrates a general one-dimensional function. Ideally, we would like to use this routine to integrate more than one specific function. We can achieve this by passing the name of the function to be integrated as an argument to the routine. Thus, we can use the same routine to, for example, integrate a polynomial, a trigonometric function, or a logarithmic function[36]. Let us refer to the function whose name is passed as an argument as the guest function. Likewise, the function to which this name is passed is called the host function. A pointer to a guest function is identified in the host function definition by an entry of the form

```
datatype (*functionName)(type1, type2, ...)
```

in the host function's argument declaration[37]. Here, datatype is the datatype of the guest function, functionName is the local name of the guest function in the host function definition, and type1, type2, ... are the datatypes of the guest function's arguments. The pointer to the guest function also requires an entry of the form

```
datatype (*)(type1, type2, ...)
```

in the argument declaration of the host function's prototype. The guest function can be accessed within the host function definition by means of the indirection operator. To achieve this, the indirection operator must precede the guest function name, and both the indirection operator and the guest function name must be enclosed in parentheses: i.e.,

```
(*functionName)(arg1, arg2, ...)
```

Here, arg1, arg2,... are the arguments passed to the guest function. Finally, the name of a guest function is passed to the host function, during a call to the latter function, via an entry like functionName in the host function's argument list. Program Listing 24 below is an example which illustrates the above points on passing function names as arguments to another function. In the listing, the function *cube()* accepts the name of a guest function (with one argument) as its first argument, evaluates this function at x (the value of its second argument, which is ultimately specified by the user), cubes the result, and then passes the final result back to the main part of the program via its third argument (which is, of course, a pointer). The two guest functions, *Fun1()* and *Fun2()*, whose names are passed to *Cube()*, are both simple polynomials. The output from the above program looks like this:

```
x = 5
x = 5.0000 res1 = 343000.0000 res2 = -221445125.0000
```

---

[36] This aspect of programming is called "reusability".
[37] The parentheses around *functionName are very important: datatype *functionName(type1, type2, ...) is interpreted by the compiler as a reference to a function which returns a pointer to type datatype, rather than a pointer to a function which returns type datatype.

Listing 24. Passing of function names as arguments to other functions via pointers. (*)

```
 1 #include <stdio.h>
 2
 3 /* Function prototypes */
 4 void    Cube(double (*)(double), double, double *);
 5
 6 double Fun1(double);  // Function prototype for first guest
       function
 7 double Fun2(double);  // Function prototype for second guest
       function
 8
 9 int main()
10 {
11 double x, res1, res2;
12
13 /* Input value of x */
14 printf("\nx = ");
15 scanf("%lf", &x);
16
17 /* Evaluate the cube of the value of the first guest function
       at x */
18 Cube(Fun1, x, &res1);
19 /* Evaluate the cube of the value of the second guest function
       at x */
20 Cube(Fun2, x, &res2);
21
22 /* Output results */
23 printf("x = %8.4f res1 = %8.4f res2 = %8.4f\n", x, res1, res2);
24 return 0;
25 }
26
27 void Cube(double (*Fun)(double), double x, double *result)
28 {
29 double y;
30 y = (*Fun)(x);          // Evaluate guest function at x
31 *result = y * y * y;    // Cube the value of the guest function at
       x
32 return;
33 }
34
35 /* First guest function */
36 double Fun1(double z) {
37 return 3.0 * z * z - z;
38 }
39
40 /* Second guest function */
41 double Fun2(double z){
42 return 4.0 * z - 5.0 * z * z * z;
43 }
```

Dynamic arrays

Since arrays are just contiguous areas of bytes, you can allocate your own arrays in the heap using *malloc()*. The following code allocates two arrays of 1 000 *ints* – one from the stack (the usual way), one from the heap using C's *malloc()*. Other than the different allocations, the two are syntactically similar, as can be seen in the following code snippet:

```
int a[1000];
int *b;
b = malloc(sizeof(int) * 1000);
/* malloc() returns the address of the memory that's allocated. So
   now b points to this address.
*/
a[123] = 13;   /* Just use [] to access elements */
b[123] = 13;
free(b);       /* This frees the allocated memory from the heap */
```

The stack and the heap

The computer memory holds the executable code for the different applications currently running on the machine, along with the executable code for the operating system itself. Each application has certain global variables associated with it. These variables also consume memory. Finally, each application uses an area of memory called the *stack*, which holds all local variables and parameters used by any function. The stack also remembers the order in which functions are called so that functions return to the correct place. Each time a function is called, its local variables and parameters are "pushed onto" the stack. When the function returns, these locals and parameters are "popped"[38]. Because of this, the size of a program's stack fluctuates constantly as the program is running, but it has some maximum size. As a program finishes execution, the operating system unloads it, and deallocates its globals and its stack to release the allocated memory. A new program can make use of that space at a later time. In this way, the memory in a computer system is constantly "recycled" and reused by programs as they execute and complete.

---

[38] "Push" and "Pop" are the two elementary operations of the abstract datatype "stack", which is in essence a LIFO (Last In First Out) data structure.

> **Advantages of being in the heap**:
>
> - Size (in the above example 1 000) can be decided *at runtime*. This is not the case for an *array* like a in the example above.
>
> - The array will exist until it is explicitly deallocated with a call to *free()*. This means the function can safely return a pointer to the array – this is not the case with a local stack variable that is deallocated on function *exit()*.
>
> - You can change the size of the malloc-ed array at will at runtime, using *realloc()*. The following code snippet changes the size of the array to 2 000. The *realloc()* function will potentially just stretch the original region to the new size but, if necessary, it will move the region to a new location, copy over the old elements, and free the previous region.
>
> ```
> b = realloc(b, sizeof(int) * 2000);
> ```

In general, perhaps 50 percent of the computer's total memory space might be unused at any given moment. The operating system owns and manages the unused memory, which is collectively known as the heap. The heap is extremely important, because it is available for use by applications during execution using the C functions *malloc()*[39] (memory allocate) and *free()*. The heap allows programs to allocate memory right when they need it during the execution of a program, rather than pre-allocating by declaring an array with a specific size (which corresponds to coding horror Fortran-style).

> **Disadvantages of being in the heap**:
>
> - You have to remember to allocate the array, and you have to get it right.
>
> - You have to remember to de-allocate it exactly once when you are done with it, and you have to get that right.
>
> - The above two disadvantages have the same basic profile: if you get them wrong, your code still looks right. It compiles fine. It even runs for small cases, but for some input it just crashes unexpectedly because random memory is getting overwritten somewhere, similar to the smiley face on Page 151. This sort of "random memory smasher" bug can be REALLY difficult to track down.

---

[39] The functions *malloc()*, *free()* and *realloc()* are declared in the header file *<stdlib.h>*.

Listing 25. Initialization of strings.

```c
/* Initialization of strings */
#include <stdlib.h>   /* Needed for macro EXIT_SUCCESS */

int main(void) {
  char string1[20] = "String";
  char string2[20] = {'S', 't', 'r', 'i', 'n', 'g', '\0'};

  printf("%s\n", string1);
  printf("%s\n", string2);
  return EXIT_SUCCESS;
}
```

2.4.7 Strings

The (C++) Standard Library (STL) implements a powerful string class, which is very useful to handle and manipulate strings of characters. However, because strings are in fact sequences of characters, we can represent them also as plain arrays of char elements. In pure C, there is no elementary "string" datatype. Hence, there is no C-operator that uses strings as operands. From the previous section we know how to define arrays of elementary *numeric* datatypes. This can also be done with the elementary datatype *char*:

char [80];

This command defines a string, i.a. field which can store 80 characters.

> One important thing to remember and a common source of errors with strings is the fact that a consecutive series of characters is always terminated with the NULL character, which has the ASCII symbol "\0". This means that the size of a *char*-array always has to be one item larger than the actual number of relevant characters.

In C, the standard library offers many functions for manipulating strings, in the header file *<string.h>*, which can be found in Appendix F.

Initialization of strings

There are two possible ways to initialize *char*-arrays as shown in Listing 25. One can use literals with single quotation marks as in line 6, or one can use a whole string word in double quotes as in line 5. Both initializations of strings in line 6 and 7 are equivalent. Both define a *char*-array which can contain 19(!) characters. Both strings only contain 6 characters each, so the minimum required length of the array is 7 (and not 6), 6 characters for the string and 1 for the terminating character \0. Listing 25 also shows how a string is printed using the function *printf()* with the format symbol "%s".

> A *char*-array which stores a string always has to be one item larger than the number of relevant characters, because a string has to be terminated with the NULL character \0. If there is no NULL terminator, a function operating on this string will continue after the actual string constant, until somewhere in memory there happens to be a NULL-value. When accessing data in write mode this could potentially overwrite useful data. The string ends as soon as the \0 character is encountered. For example, a string "Str\ring" would only be processed as "Str".

One has to distinguish between a *character constant* and a *string constant*. The following definitions are *not* equivalent.

```
// char constant with one character
char ch = 'X'
// string constant with two characters: 'X' and '\0'
char ch[] = 'X'
```

When initializing strings, one can leave the string length unspecified. Thus, the following three definitions are all equivalent:

```
// Three equivalent definitions for strings
char str[] = {'S', 'T', 'R', 'I', 'N', 'G', '\n', '\0'};
char str[] = ``STRING\n'';
char *str   = ``STRING\n'';
```

The advantage of the last two methods (besides a more concise syntax) is that the compiler calculates the array-length including the terminating character.

Reading strings

Special input functions for strings are the functions *gets()* and *fgets()*[40]. Both functions are defined in the header file *<stdio.h>* and their syntax is:

```
char *fgets(char *str, int nChars, FILE *stream);
```

and

```
char *gets(char *str);
```

The simple example in Listing 26 shows how to read strings using *gets()* and *fgets()*. In line 11, the address of the string (indicated by its name) is passed to the function

---

[40] The function *scanf()* should not be used for reading strings because this function only reads a string until the first blank and all subsequent characters are ignored. Besides, *scanf()* is prone to buffer overflows.

Listing 26. Reading strings.

```
 1  /* Example of the use of gets() and fgets() */
 2  #include <stdio.h>
 3  #include <stdlib.h>
 4  #define MAX 10
 5
 6  int main(void) {
 7    char string1[MAX];
 8    char string2[MAX];
 9
10    printf("Again, input a string (<= 0 characters): ");
11    gets(string1);
12    printf("Your input was: %s\n", string1);
13
14    printf("Input a string (including blanks) : \n");
15    fgets(string2, MAX, stdin);
16    printf("Your input was: %s\n", string2);
17
18    return EXIT_SUCCESS;
19  }
```

gets() which then overwrites it. In line 15, the function *fgets()* reads from stdin (i.e. from the screen) at most MAX characters and stores them in string2. Function *fgets()* guarantees that the NULL terminator is added at the end of string2, so *fputs()* actually writes at most MAX-1 characters into string2. If there is still room, the newline '\n' is added at the end, so if you type a string with exactly 9 characters and press enter (the equivalent to '\n'), no newline is added to the string, as the last character is reserved for \0. Instead of the *printf()* commands in lines 12 and 16 one could have used *puts()* and *fputs()*, the counterparts of *gets()* and *fgets()*. We note here, that Listings 25 and 26 are not very robust and it is very likely that you manage to break the code (i.e. produce a core dump or segmentation fault due to buffer overflow). The purpose of these examples is not to demonstrate fool-proof C code but to illustrate how certain features work in principle. In production code you would have to include *exception handling* routines that check, for example, whether you produced a buffer overflow by typing more characters than there are elements in the *char* array.

2.5 Structures and their representation in computer memory

A very important datatype for data organization in C/C++ is the *struct*. Structs are used to assemble different datatypes which are logically connected, e.g. the properties of particles in an MD simulation or the properties of a finite element. The simplest example is the following:

```
struct student{
char name[30];
int age;
int numClasses;
double grade;
};
```

Note the period at the end of the *struct* declaration. The nice thing about structs is that they can be used like an elementary datatype, i.e. one can now declare variables that are of type "struct student" by typing:

```
struct student student1, student2, student3;
```

Such a declaration reserves an amount of heap memory the size of the *members* that are contained within the *struct*, in our case for each of the student variables a *char*-field of length 30, two integers and a double, is reserved, adding up to a memory space of $30+4+4+8 = 46$ bytes. You can access the members of a struct in the following way using the "." operator:

```
student1.name = ''Lazy Dog'';
student1.age  = 43;
student1.numClasses = 1;
student1.grade = 2.5;
student3.name = ''Bright Guy'';
```

> The address of the whole *struct* always coincides with the address of the first field of the first member.

If you don't want to always use the word *struct* when declaring variables you can declare your own *struct* type using the command *typedef*, for example:

```
typedef struct fraction{
    int numerator;
    int denominator;
} frac;
```

Instead of struct fraction one can now simply use frac in type declarations, e.g. frac pi;. The assignment pi.numerator=22 places a 22 in the lower 4 bytes of the entire struct (which is 8 bytes of memory). Technically, a 22 is stored in the field at an offset 0 from the base address of the entire *struct*. The assignment pi.denominator=17 is stacked on top of numerator, i.e. it is 4 bytes above of the address of the entire *struct* pi, see Figure 2.9 (a).

Section 2.5 Structures and their representation in computer memory

Think!

What is the state of memory after executing the command

`((fraction *) & (pi.denominator)) -> numerator = 14;`

Now, the computer assumes it is addressing an 8 byte structure that overlays the memory space, see Figure 2.9 (b). The arrow operator `->` in the above statement is the way to access the contents of the memory to which the pointer of a *struct* points. One could also use the dereference operator `*`, but then one has to use parentheses, because the point operator has higher priority, i.e.

`(* ((fraction *) & (pi.denominator))).numerator = 14;`

Let's do the same with `denominator`:

`((fraction *) & (pi.denominator)) -> denominator = 57;`

Now, we have the state of memory as indicated in Figure 2.9 (b). Thus, we realize

Figure 2.9. State of memory after declaration of the *struct* `fraction`. (a) The base address of the *struct* coincides with the address of the first field in the struct, i.e. the *int* numerator. (b) After the address of `pi.denominator` is interpreted as a pointer of type fraction, i.e. as an 8 byte segment of memory, this 8 byte structure just overlays the previously occupied memory space as indicated.

that structs allocate contiguous space in memory, which is *interpreted* according to the size of the members of the *struct*.

2.5.1 Blending structs and arrays

Let's consider the following definition of a *struct*:

```
struct student{
char *name;
char studentID[8];
int numClasses;
};
```

This structure reserves 16 bytes in memory, with its base address coinciding with the address of the zeroth character of the string to which name points, see Figure 2.10.

Next, we want to perform a series of commands manipulating this *struct* and understand what this means in terms of memory. The commands we want to execute (in this order) are:

1. `student pupils[4];`

2. `pupils[0].numClasses = 5;`

3. `pupils[2].name = strdup(''Adam'');`

4. `pupils[3].name = pupils[0].studentID + 6;`

5. `strcpy(pupils[1].studentID,''40415xx'');`

6. `strcpy(pupils[3].name,''12345678'');`

7. `pupils[7].studentID[11] = 'A';`

> What is the state of memory after the above commands have been executed?

> The state of memory is graphically addressed in Figure 2.11.

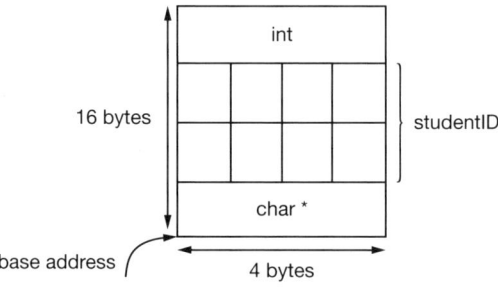

Figure 2.10. Memory allocated by the *struct* student.

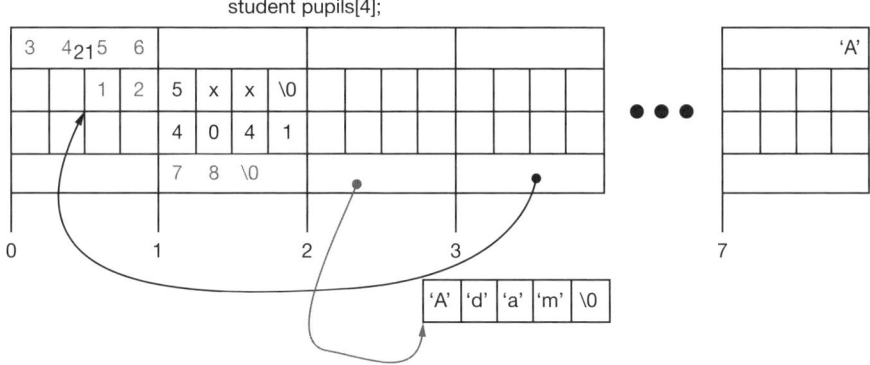

Figure 2.11. Graphical representation of the state of stack and heap memory after executing various commands. A detailed description is provided in the main text.

The first command simply allocates space for an array of type student with 4 entries (numbered from 0 to 3). Altogether, a chunk of memory 64 bytes large is reserved on the stack. The second command writes the byte pattern for 21 – interpreted as an integer – in the upper 4 bytes of pupils[0]. The third command using strdup(''Adam'') *dynamically* allocates just enough memory (in the heap because it's dynamically allocated!) for a *string* "Adam" and returns the base address of this string, i.e. the address of 'A'. Note the null terminating character "\0" at the end of the string. Then, the fourth command does some pointer arithmetic against a char and the result is indicated with the arrow in Figure 2.10: pupils[3].name now points to the base address of pupils[0].studentID plus an offset of 6 bytes, because studentID is of type *char*. The *strcpy()*-function in the fifth command does not allocate any memory, because *strcpy()* assumes that the address where it should copy to, is identified by the first argument. Note again the null terminating character at the end of the string. The next *strcpy()*-command copies the series of numbers "12345678" – interpreted as chars – to the address of pupils[3].name, which points to the address of (pupils[0]-studentID +6). At this address the characters are written as indicated in Figure 2.11, overwriting any memory content that happens to be there – in our case, the 21 is overwritten. Finally, the seventh command writes an 'A' at the location in memory with an offset of $7 \times 64 + 4 + 11 = 463$ bytes from the base address of struct pupils, as indicated in Figure 2.11 and overwrites whatever memory content happened to be at this address on the stack.

2.6 Numerical differentiation and integration

As a first "real" application of what we have so far learned in C/C++, we will deal now with numerical differentiation and integration. The treatment of differential equations is deferred to Chapter 5. Numerical integration and differentiation are some of the most frequently needed methods in computational physics. Quite often one is confronted with the need of numerically evaluating either f' or an integral $\int f(x)dx$.

2.6.1 Numerical differentiation

In this section, I present the most commonly used formulas for computing first and second derivatives, which in turn find their most important applications in the numerical solution of ordinary and partial differential equations, e.g. when solving the *Navier-Stokes equations*, the *diffusion equation* or the *wave equation* to name just a few. Let's start with the mathematical definition of the derivative of a function $f(x)$ which is

$$\frac{df(x)}{dx} = \lim_{h \to \infty} \frac{f(x+h) - f(x)}{h}, \qquad (2.13)$$

where h is the *step size*. If we use a Taylor expansion for $f(x)$ we can write

$$f(x+h) = f(x) + hf'(x) + \frac{h^2 f''(x)}{2} + \cdots \qquad (2.14)$$

We can then set the computed derivative $f_c'(x)$ as

$$f_c'(x) \approx \frac{f(x+h) - f(x)}{h} \approx f'(x) + \frac{hf''(x)}{2} + \cdots \qquad (2.15)$$

Assume now that we use *two points* to represent the function f as a straight line between x and $x+h$, as well as between the points at the boundaries of all other intervals, as indicated in Figure 2.12. This means that we can represent the derivative with

$$f_2'(x) = \frac{f(x+h) - f(x)}{h} + O(h), \qquad (2.16)$$

where the suffix 2 refers to the fact that we use two points to define the derivative. The predominant error of this approximation is of order $O(h)$. This is the *forward two-point formula*. Alternatively, we could use the *backward two-point formula*, which is simply:

$$f_2'(x) = \frac{f(x) - f(x-h)}{h} + O(h). \qquad (2.17)$$

If the second derivative is close to zero, the simple two point formula can be used to approximate the derivative. If, however, a function like $f(x) = a + bx^2$ is to be approximated, we see that the approximated derivative becomes

$$f_2'(x) = 2bx + bh, \qquad (2.18)$$

Section 2.6 Numerical differentiation and integration

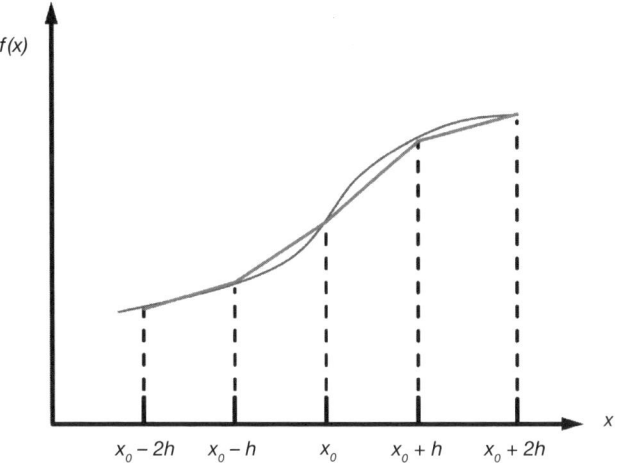

Figure 2.12. One-dimensional grid of equidistant points, for approximating functions and derivatives. The x-axis is subdivided into small intervals h by which the value of x is incremented. When the points x_0 and $x_0 + h$ are used, one can approximate the function by a straight line and use the slope of this line as approximation of the first derivative.

while the exact answer is $2bx$. Unless h is made very small, and b is not too large, we could approach the exact answer by choosing smaller and smaller values for h. However, in this case, the subtraction in the numerator, $f(x+h) - f(x)$ can give rise to rounding errors and eventually to loss of precision. A better approach in case of a quadratic expression for $f(x)$ is to use a symmetric *two-point formula*, where the derivative is calculated on both sides of a chosen point x_0, using the two-point formulas in equations (2.16) and (2.17), and taking the average. In this case, one simply performs a Taylor expansion about $x_0 \pm h$:

$$f(x)|_{x_0 \pm h} = f(x_0) \pm hf' + \frac{2hf''}{2} \pm \frac{h^3 f'''}{6} + O(h^4). \tag{2.19}$$

We rewrite equation (2.19) as

$$f_{\pm h} = f_0 \pm hf' + \frac{h_2 f''}{2} \pm \frac{h^3 f'''}{6} + O(h^4). \tag{2.20}$$

Calculating both $f_{\pm h}$ and subtracting, one obtains the *three-point formula*

$$f_3' = \frac{f_h - f_{-h}}{2h} - \frac{h^2 f'''}{6} + O(h^3). \tag{2.21}$$

Hence, the predominant error in this case is of order $O(h^2)$, if we truncate at the second derivative. The term $h^2 f'''/6$ is called the truncation error, which is the error that arises because of the truncation of the Taylor series. Truncation errors and rounding errors play an important role in the numerical determination of derivatives.

For the above expression with a quadratic function $f(x) = a + bx^2$ we see that the three-point formula f_3' for the derivative produces the exact answer $2bx$. Thus, if our function has quadratic behavior in x in a certain region of space, the three-point formula will result in reliable first derivatives in the interval $[-h, h]$. Using the relation

$$f_h - 2f_0 + f_{-h} = h^2 f'' + O(h^4), \tag{2.22}$$

one can define the second derivative as

$$f'' = \frac{f_h - 2f_0 + f_{-h}}{h^2} + O(h^2). \tag{2.23}$$

One can continue in this manner and also define five-points formulas by expanding up to two steps on each side of x_0. Using a Taylor expansion around x_0 in a region $[-2h, 2h]$, one obtains

$$f_{\pm 2h} = f_0 \pm 2hf' + 2h^2 f'' \pm \frac{4h^3 f'''}{3} + O(h^4). \tag{2.24}$$

Using equations (2.19) and (2.24), multiplying f_h and f_{-h} by a factor of 8, and subtracting $(8f_h - f_{2h}) - (8f_{-h} - f_{-2h})$, we obtain for the first derivative

$$f'_{5c} = \frac{f_{-2h} - 8f_{-h} + 8f_h - f_{2h}}{12h} + O(h^4), \tag{2.25}$$

which has a predominant error of the order $O(h^4)$ at the computational price of two additional function evaluations. Such a formula might be useful, e.g. when f is a fourth-order polynomial in x in the region $[-2h, 2h]$. The two additional subtractions in equation (2.25) can lead to a larger risk of loss of numerical precision when h becomes small. Solving, for example, a differential equation which involves the first derivative, one needs to always strike a balance between numerical accuracy and the time needed to achieve a given result. It can be shown[41] that the widely used formulas for the first and second derivatives of a function can be written in concise form as

$$\frac{f_h - f_{-h}}{2h} = f_0' + \sum_{j=1}^{\infty} \frac{f_0^{(2j+1)}}{(2j+1)!}, \tag{2.26}$$

and

$$\frac{f_h - 2f_0 + f_{-h}}{h^2} = f_0'' + 2\sum_{j=1}^{\infty} \frac{f_0^{(2j+2)}}{(2j+2)!} h^{2j}. \tag{2.27}$$

The error is in both cases of $O(h^{2j})$. These expressions will also be used when we evaluate integrals in the next section.

---

[41] I suggest this as an exercise.

Section 2.6 Numerical differentiation and integration

> **Time discretization in one dimension.** For a differentiable function $f : \mathbb{R} \to \mathbb{R}$, one can approximate the continuous differential operator df/dx at a grid point x_i by the *discrete difference operators*
>
> $$\left[\frac{df}{dx}\right]_i^f := \frac{f(x_{i+1}) - f(x_i)}{\delta x} + O(\delta x) \quad \text{forward difference,} \quad (2.28)$$
>
> by not taking the limit. Here, $x_{i+1} = x_i + \delta x$ is the neighboring grid point on a linear mesh of equidistant points δx. Besides the forward difference, one can also use the backward form of the difference operator
>
> $$\left[\frac{df}{dx}\right]_i^b := \frac{f(x_i) - f(x_{i-1})}{\delta x} + O(\delta x) \quad \text{backward difference,} \quad (2.29)$$
>
> or the central difference
>
> $$\left[\frac{df}{dx}\right]_i^c := \frac{f(x_{i+1}) - f(x_{i-1})}{2\delta x} + O(\delta x^2) \quad \text{central difference.} \quad (2.30)$$
>
> The second total derivative of a function $x : \mathbb{R} \to \mathbb{R}$ may be approximated at a grid point t_n by the difference operator
>
> $$\left[\frac{d^2 x}{dt^2}\right]_n := \frac{1}{\delta t} \left[x(t_{n+1}) - 2x(t_n) + x(t_{n-1})\right], \quad (2.31)$$
>
> which implies a discretization error of the order $O(\delta t^2)$.

2.6.2 Case study: the second derivative of e^x

As an example, we calculate the second derivative of $\exp(x)$ for various values of x. In addition, in this case study, we introduce by way of example, how a program (with non-trivial complexity) should be organized in principle, namely in functions organized in modules – so-called header files – that are included in the main module, which is where the *main()*-function is located. Usually, it is a good strategy, to start with the *main()* function, from where other functions are called and then to deal with the various functions which themselves might call various other functions.

The *main()*-function

The programming style introduced here starts with the *main()*-function, followed thereafter by the detailed tasks performed by each function called in *main()*. The different functions are organized in modules, which are included at the beginning of main.c. Another possibility is to include the function definitions *directly*, before the main program in main.c, meaning that the main program appears at the very end. However, I find this programming style much less readable since I prefer to read a program from top to bottom. In addition, this leads to very large files, where you have to read hundreds or maybe thousands of lines of code in one single file, which is very cumbersome. A division of different tasks into specialized functions, as shown here, is very useful, particularly for larger projects and programs. In the first version of this program, we use a more C-like style for writing and reading to a file. At the end of this section we also include a C++ version for I/O handling, which sometimes is a little bit easier than in C, because of the property of operator overloading. This C++ feature takes the necessity to think about the datatype to be read or output away from the programmer. By overloading the operator *cout*, C++ automatically determines the datatype. While this is a nice feature, it is not really a big deal, but I present it here in a direct comparison. Listing 27 presents the C *main()*-function, stored under the name main.c.

In the *main()*-function, one can see all the basic elements of good program organization. First of all, function definitions are included in the header files at the beginning. This leads to a clear program structure that allows us to focus on the *main()*-function and its functionality. The various functions are called in lines 16, 20 and 23. For the function *Output()*, I have included a return value that is checked within *main()* by the variable check. This variable is assigned the return value of *Output()* in line 26 and then, in line 28, checks whether the function successfully wrote the results to a file, in which case 1 is returned and a short message is printed on screen. If not, i.e. if 0 was returned by *Output()*, the program stops, using the function *exit()* with pre-defined argument EXIT_FAILURE, which is defined in the library <*stdlib.h*>. In the end, we free the memory using the command *free()* in lines 34 and 35, as these two variables have been allocated dynamically (in the heap of the computer's memory) in lines 19 and 20. In this case – since it is the end of the program anyway – we don't have to do that, because a program's memory is freed by the operating system anyway when the program comes to an end.

The header functions

Function *Initialize()*. The first header function is *Initialize()* which is shown in Listing 28 and defined in the module named Initialization.c. This function is very simple and just reads the three basic variables initStep, x and numSteps from std-out using the *scanf()*-function.

Function *SecondDerivative()*. The function *SecondDerivative()* is shown in Listing 29. The second derivative is computed, within the loop over the number of steps, for different increasingly smaller values of h, which is itself halved at each iteration. The step values are stored in the arrays hStep and computedDerivatives. If the reader wonders, why these variables can be addressed as arrays, seeing as they have been, after all, defined as type double *, she should reread the section about point-

Section 2.6 Numerical differentiation and integration

Listing 27. *main()*-function for calculating the second derivative of exp(x). (*)

```
1  /* Program that computes the second derivative of exp(x). */
2  #include <stdlib.h>
3  #include <stdio.h>
4
5  #include "Initialization.h"
6  #include "Derivatives.h"
7  #include "IoHandling.h"
8
9  int main(){
10    /* Declarations of variables */
11    int numberOfSteps, check;
12    double x, initialStep;
13    double *hStep, *computedDerivative;
14
15    /*Read input data from screen */
16    Initialize(&initialStep, &x, &numberOfSteps);
17
18    /* Allocate space in memory dynamically for the 1-dimensional
         arrays */
19    hStep = (double *) ( malloc (numberOfSteps * sizeof(double))
         );
20    computedDerivative = (double *) ( malloc (numberOfSteps *
         sizeof(double)) );;
21
22    /* Compute the second derivative of exp(x) */
23    SecondDerivative(numberOfSteps, x, initialStep, hStep,
         computedDerivative);
24
25    /* Print the results to a file */
26    check = Output(hStep, computedDerivative, x, numberOfSteps);
27
28    if (check == 0)
29      exit(EXIT_FAILURE);
30    else
31      printf("Output successfully written to file\n");
32
33    /* free memory - not really necessary, as this is the end of
         the program */
34    free(hStep);
35    free(computedDerivative);
36    return (0);
37  }
```

Listing 28. The *Initialize()*-function. (*)

```c
/* Function that reads basic variables */
#include <stdio.h>

#include "Initialization.h"

void Initialize(double initStep, double *x, int *numSteps){
  printf("Please read the intial step, x, and the number of steps from screen\n");
  scanf("%lf %lf %d", initStep, x, numSteps);
  return;
}
```

Listing 29. The function *SecondDerivative()*. (*)

```c
/* This function computes the second derivative */

#include <stdio.h>
#include <math.h>

#include "Derivatives.h"

void SecondDerivative(int numberOfSteps, double x, double initialStep, \
        double *hStep, double *computedDerivative)
{
  int counter;
  double h;

  /* calculate the step size, initialize the derivative, y and
      x and the iteration counter */
  h = initialStep;

  /* Start computing for different step sizes */
  for(counter = 0; counter < numberOfSteps; counter++)
    {
      /* Setup arrays with derivatives and step sizes */
      hStep[counter] = h;
      computedDerivative[counter] = (exp(x+h)-2.*exp(x)+ exp(x-h))/ (h*h);
      h *= .5;
    }
  return;
}
```

ers in this chapter. The dynamic memory allocation in lines 19 and 20 in function *main()* allocates two contiguous blocks of memory *numberOfSteps* times 8 bytes large and the content is interpreted as *numberofSteps* doubles, so they can be as-

Section 2.6 Numerical differentiation and integration

signed by using an array index, attached to the pointer storing the addresses of the doubles within this block of memory.

Function *InputOutput()*. Finally, function *InputOutput()* in Listing 30 is a very simple realization of a function that writes the computed data into a file. Note the exception handling in lines 14 to 17 for the case that for some reason, a text file could not be opened for writing.

Listing 30. The function *InputOutput()*. (*)

```
1  /* This function outputs the results of the computation */
2  #include <stdio.h>
3  #include <math.h>
4
5  #include "IoHandling.h"
6
7  int Output (double *step, double *Derivative, double x, int
       numSteps)
8  {
9    FILE *out;
10   int counter;
11   counter = numSteps;
12
13   out = fopen("OutFile.data","w");
14   if (out == NULL){
15     printf("Cannot open file for some reason\n");
16     return (0);
17   }
18
19   for (counter = 0; counter <= numSteps; counter++)
20     fprintf(out,"%i  %f %f %f\n",counter, x, Derivative[counter
         ], step[counter]);
21   return (1);
22 }
```

The three header files that are included in the main file, which contains *main()*, just contain the declaration of the respective functions with the keyword extern, which means that the definition of the function can be written somewhere in some other module. In Listing 31, I show the header file of the function *Initialize()*. All other headers are defined analogously.

Please note that, in the header declaration of a function, it is enough to just provide the *type declarations* – you don't have to provide any variable *names* and in the function you can even choose them as you like. The function argument list in the header file is also called the *interface* of a function because it tells you how to call it, i.e. what type of variables you have to provide when you want to call it. In the calling function and in the called function, you can pick whatever valid names you like. For example, the function *Initialize()* in *main()* is called with the first argument initialStep

Listing 31. The header file *Initialization.h* with the prototype definition of *Initialize()*. (*)

```
1  /* Header file of Initialize() */
2
3  #ifndef __INITIALIZE_H
4  #define __INITIALIZE_H
5
6  extern void Initialize(double *, double *, int *);
7
8  #endif
```

in line 16, but the name of the first argument in the definition of *Initialize()* in file *Initialization.c* is initStep.

Altogether, this simple program for calculating the second derivative of $\exp(x)$ consists of *four* *.c files (which I named *main.c*, *Initialization.c*, *Derivatives.c* and *IoHandling.c*), and *three* *.h files, analogously named *Initialization.h*, *Derivatives.h* and *IoHandling.h*, which have to be included in all modules that make use of the functions declared within these header files. Compiling all sources can be done with the command lines

```
gcc -c main.c Initialization.c Derivatives.h IoHandling.h
gcc -o Derivative main.o Initialization.o Derivatives.o \
        IoHandling.o -lm
```

which creates a binary "Derivative". One can summarize all of these commands in one short command line (assuming that there are *only* these source files in the current directory):

```
gcc -o Derivative *.c
```

which does the same.

A C++ version for computing the second derivative of $e^{(x)}$

An alternative implementation of the *main()*, written in C++ syntax and stored under the name main.cc, is provided in Listing 32. Note that we have declared a *char*-field variable char *outfilename in line 4 which is later passed to the C++ version of *Output()*, called *Output2()* in line 30. In order to use this functionality, one needs to include the corresponding function libraries, using the "#include"-statement. One of the disadvantages of C++ is that *formatted output* is not as easy to use as the printf() and *scanf()* functions in C. The *Output2()*-function, using C++ style is included below in Listing 33. The main part of the source code now includes an object declaration ofstream outfile in line 11, which is included in C++ and allows the programmer to open and declare files. This is done via the statement outfile.open(outfilename) in line 17 The file is closed at the end of *Output()*

Section 2.6 Numerical differentiation and integration

Listing 32. The C++ version of the *main()*-function to compute exp(x)". (*)

```
int main(int argc, char * argv[]){

  // Declarations of variables
  char *outfilename;
  int numberOfSteps;

  double x, initialStep;
  double *hStep, *computedDerivative;

  // Read the outputfile, abort if there are too few command-
      line arguments
  if(argc <= 1){
    cout << "BadUsage:" << argv[0] << "also read outputfile on
        same line " << endl;
    exit(1);
  }
  else{
    outfilename = argv[1];
  }

  //read input data from screen
  Initialize(&initialStep, &x, &numberOfSteps);

  //Allocate space in memory for the 1-dimensional arrays
  hStep              = new double[numberOfSteps];
  computedDerivative = new double[numberOfSteps];

  //Compute the second derivative of exp(x)
  SecondDerivative(numberOfSteps, x, initialStep,hStep,
      computedDerivative);

  //Print the results to file
  Output2(hStep, computedDerivative, x, numberOfSteps,
      outfilename);

  //Free memory
  delete [] hStep;
  delete [] computedDerivative;

  return (0);
}
```

Listing 33. The C++ version of the *Output()*-function, called *Output2()*. (*)

```
using namespace std;

#include <iostream>
#include <fstream>
#include <iomanip>
#include <cmath>
#include <cstdlib>

#include "IoHandling.h"

ofstream outfile;

//Function to write the final results to the file system
int Output2(double *hStep, double *computedDerivative, double x
    , int numberOfSteps, char *outfilename){
  int i;

  outfile.open(outfilename);
  outfile << "RESULTS: " << endl;
  outfile << setiosflags(ios::showpoint|ios::uppercase);
  for(i = 0; i < numberOfSteps; i++){
    outfile << setw(15) << setprecision(8) << log10(hStep[i]);
    outfile << setw(15)<< setprecision(8)<< log10(fabs(
        computedDerivative[i]-exp(x))/exp(x)) << endl;
  }
  //Close outputfile
  outfile.close();
  return (0);
}
```

by writing `outfile.close()`. There is also a corresponding object for reading input files (which we haven't used here).

As we are using the same header file *IoHandling.h* for the C++ version of this code, we have to add a line for the function *Outfile2()* to *IoHandling.h*

```
extern void Output2(hStep, computedDerivative, x, \
                    numberOfSteps, outfilename);
```

2.6.3 Numerical integration

In this section, we discuss some of the classic formulas for numerical integration, such as the trapezoidal rule and Simpson's rule for equally spaced abscissas. The emphasis is on methods for evaluating *one*-dimensional integrals. In chapter 6 we show how MC methods can be used to compute multidimensional integrals[42].

---
[42] MC is, in fact, the most efficient method for the calculation of high-dimensional integrals.

Section 2.6 Numerical differentiation and integration

Trapezoidal rule

Our basic tool for numerical integration is the Taylor expansion of the function $f(x)$ around a point x and a set of surrounding neighboring points. The algorithm is rather simple and is provided as pseudocode in Listing 34. In Listing 34, N is the number

Listing 34. Trapezoidal rule. (pseudocode).

```
1. choose a step size h = (b-a)/N
2. stop the Taylor expansion of f(x) at a certain derivative
3. choose how many points around x should be included in the
   evaluation of the derivatives
4. perform the usual integration
```

of steps and a and b are the lower and upper limits of the integration. In the following I illustrate this method with the integral

$$I = \int_a^b f(x). \tag{2.32}$$

I can be written as

$$I = \int_a^{a+2h} f(x)\,dx + \int_{a+2h}^{a+4h} f(x)\,dx + \cdots \int_{b-2h}^{b} f(x)\,dx. \tag{2.33}$$

The strategy is to find a reliable Taylor expansion for $f(x)$ on the smaller sub intervals. For example, consider evaluating

$$\int_{-h}^{+h} f(x)\,dx, \tag{2.34}$$

and expanding $f(x)$ around a point x_0 in the form

$$f(x)|_{x=x_0+h} = f(x_0) \pm hf' + \frac{h^2 f''}{2} \pm \frac{h^3 f'''}{6} + O(h^4). \tag{2.35}$$

Let us suppose that the integral in equation (2.34) is split in two parts, one from $-h$ to x_0 and the other from x_0 to h. Next, we assume that we can use the two-point formula for the derivative, i.e. that in these two regions we can approximate $f(x)$ by a straight line, as indicated in Figure 2.12. This means that every small element underneath the function $f(x)$ looks like a trapezoid. Consequently, this numerical approach to calculating the integral bears the name *trapezoidal rule*. In essence, what is does is approximate $f(x)$ with a first order polynomial, that is, $f(x) = a + bx$. The constant b is the slope given by the first derivative at $x = x_0$

$$f' = \frac{f(x_0 + h) - f(x_0)}{h} + O(h), \tag{2.36}$$

or

$$f' = \frac{f(x_0) - f(x_0 - h)}{h} + O(h), \qquad (2.37)$$

and if we stop the Taylor expansion at that point, f is

$$f(x) = f_0 + \frac{f_h - f_0}{h}x + O(x^2) \qquad (2.38)$$

for $x = x_0$ to $x = x_0 + h$ and

$$f(x) = f_0 + \frac{f_0 - f_{-h}}{h}x + O(x^2) \qquad (2.39)$$

for $x = x_0 - h$ to $x = x_0$. The error is of order $O(x^2)$. Finally, evaluating the integral in equation (2.34) yields:

$$\int_{-h}^{+h} f(x)\,dx = \frac{h}{2}(f_h + 2f_0 + f_{-h}) + O(h^3), \qquad (2.40)$$

which *is* the trapezoidal rule. The error in the approximation, $O(h^3) = O((b-a)^3/N^3)$ is the *local* error. Since we split the integral from a to b in N pieces, we have to perform approximately N such operations. This means that the *global* error is of order $O(h^2)$. To see this, we use the trapezoidal rule to compute the integral of equation (2.32),

$$I = \int_a^b f(x)\,dx = h\Big(\frac{f(a)}{2} + f(a+h) + f(a+2h) + \cdots + f(b-h) + \frac{f_b}{2}\Big), \qquad (2.41)$$

with a global error of the order $O(h^2)$.

The correct mathematical expression for the local error for the trapezoidal rule is

$$\int_a^b f(x)\,dx - \frac{b-a}{2}[f(a) + f(b)] = -\frac{h^3}{12}f^{(2)}(\xi), \qquad (2.42)$$

and the global error reads

$$\int_a^b f(x)\,dx - T_h(f) = -\frac{b-a}{12}h^2 f^{(2)}(\xi), \qquad (2.43)$$

where T_h is the trapezoidal result and $\xi \in [a, b]$. This scheme can easily be implemented numerically with the algorithm shown in Listing 35. A simple function which implements this algorithm is provided in Listing 36.

In Listing 36, the function *TrapezoidalRule()* has a user-defined function as its 4th argument, which returns a *double* and which has as input a *double* value.

Listing 35. Trapezoidal rule detailed. (pseudocode).

```
1. choose the number of mesh points and fix the step
2. calculate f(a) and f(b), and multiply by h/2
3. loop from n=1 to n-1 and sum the terms in equation (2.31)
4. multiply the final result by h and add hf(a)/2 and hf(b)/2
```

Listing 36. Function that implements the trapezoidal rule.

```
double TrapezoidalRule (double a, double b, int n, double
    (*func)(double))
{
    double trapezSum;
    double fa, fb, x, step;
    int    j;
    step = (b-a) / ((double) n);
    fa = (*func)(a)/2.;
    fb = (*func)(b)/2.;
    trapezSum = 0;
    for (j = 1; j <= n -1; j++){
        x = j * step + a;
        trapezSum += (*func)(x);
    }
    trapezSum = (trapezSum + fa + fb) * step;

    return (trapezSum);
}
```

Midpoint or rectangle method

Another very simple approach is the so-called *midpoint* or *rectangle method*. In this case, the integration area is split in a given number of rectangles with length h and height given by the mid-point value of the function f. This produces the following simple rule for approximating an integral

$$I = \int_a^b f(x)\,dx \approx h \sum_{i=1}^{N} f(xi - 1/2), \qquad (2.44)$$

where $f(x_{i-1/2})$ is the mid-point value of f for a given rectangle. It is very easy to implement this algorithm, as is shown in Listing 37 The analytical expression for the local error for the rectangular rule $R_i(h)$ for element i is

$$\int_{-h}^{+h} f(x)\,dx - R_i(h) = -\frac{h^3}{24} f^{(2)}(\xi), \qquad (2.45)$$

Listing 37. Function that implements the rectangle rule.

```
double RectangleRule(double a,double b,int
    n,double (*func)(double))
{
    double rectangleSum;
    double fa, fb, x, step;
    int j;
    step = (b-a) / ((double) n);
    rectangleSum = 0.;
    for (j = 0; j <= n;j ++){
        x = (j + 0.5) * step+;       //Mid-point of a given
            rectangle.
        rectangleSum += (*func)(x); //Add value of function.
    }
    rectangleSum *= step; // Multiply by step length.
    return (rectangleSum);
}
```

and the global error reads

$$\int_a^b f(x)\,dx - R_h(f) = -\frac{b-a}{24}h^2 f^{(2)}(\xi), \qquad (2.46)$$

where R_h is the result obtained with the rectangular rule and $\xi \in [a,b]$.

Simpson's rule

Instead of using the above linear two-point approximations for f, we could use the three-point formula of equation (2.21) for the derivatives. This means that we will choose formulas based on function values which lie symmetrically around the point where we preform the Taylor expansion. This means also that we approximate our function with a second-order polynomial $f(x) = a + bx + cx^2$. The first and second derivatives are given by equations (2.26) and (2.27). With these expressions, we can approximate the function f as

$$f(x) = f_0 + \frac{f_h - f_{-h}}{2h} + \frac{f_h - 2f_0 + f_{-h}}{2h^2}x^2 + O(x^3). \qquad (2.47)$$

Inserting this formula in the integral of equation (2.34), one obtains

$$\int_{-h}^{+h} f(x)\,dx = \frac{h}{3}(f_h + 4f_0 + f_{-h}) + O(h^5). \qquad (2.48)$$

which is Simpson's rule. Note, that the improved accuracy in the evaluation of the derivatives produces a better error approximation, namely $O(h^5)$ vs. $O(h^3)$. But this is just the local error approximation. Using Simpson's rule, we can easily compute the integral of equation (2.32) as

$$I = \int_a^b f(x)\,dx = \frac{h}{3}(f(a) + 4f(a+h) + 2f(a+2h) + \cdots + 4f(b-h) + f_b), \tag{2.49}$$

with a global error, which is of order $O(h^4)$. The outlined method can easily be implemented numerically through the following simple algorithm provided in Listing 38

Listing 38. Simpson's rule. (pseudocode)

```
1. choose the number of mesh points and fix the step
2. calculate f(a) and f(b)
3. loop from n = 1 to n-1 (f(a) and f(b) are known) and
   sum the terms 4f(a + h) + 2f(a + 2h) + 4f(a + 3h) + ...
   + 4f(b - h). Each step of the loop corresponds to a given
        value
   a + nh. Odd values of n produce 4 as factor whilst even
        values
   yield 2 as factor
4. multiply the final result by h
```

I leave it as an exercise for the reader to implement the algorithm of Listing 38 in a C program.

2.7 Remarks on programming and software engineering

Although leading-edge software development and engineering practices have advanced rapidly, common programming practices have not[43]. Researchers in both software industry and academic settings have discovered effective practices that eliminate most of the programming problems that were prevalent in the 1970s and 80s. Because, however, these practices are not often communicated beyond the pages of highly specialized journals, they have not become general standard knowledge.

2.7.1 Good software development practices

Learning a programming language is much like learning a human language. Once you have learned the basic vocabulary and some grammar, you need to go out and apply it. The same goes for programming. You can only master it by practice. So,

---

[43] This is often particularly true in academic settings, with a general lack of good software practices and the common approach that each new PhD student of computational physics or engineering has to start over, instead of re-using code written by her predecessors.

we recommend that you type the provided code examples in the various program listings of this book yourself and try to compile and run them. While the language C/C++ is infamous for its capability of allowing one to write extremely concise (and unreadable) code, it only takes a little discipline to do much better. Just look at the simple function *StringCopyVersion()* in the following example in Listing 39 and see, if you can really figure out what it does: This code snippet for a string copy function

Listing 39. Implementation of the C *strcpy()* function.

```
// Remark: This is a super fancy version of C code
// Remark: Relies on the fact that '\0' is equivalent to FALSE
void StrcpyVersion(char dest[], const char source[])
{
   while (*dest++ = *source++);
}
```

is actually the most fancy (and worst version) of writing C code as we discussed in Section 2.4.4. Hence, it is important that you adopt a good style of programming from the beginning, by adhering to some general rules of software design, which do not only apply to C programming but are valid for the development of any software of more than trivial complexity. These general *good software development practices* as well as more sophisticated *software design rules* are general recommendations to ensure that your written code can be understood, maintained and extended by others (and by yourself!), and we provide some of them here.

Before writing a single line of code, you should have the relevant algorithm(s) clarified and understood. It is crucial to have a logical structure of data organization on a low level, before one starts writing anything. To put it shortly: "Think first, then start writing". So, here I will provide some rules, in the hope that they might assist you in future computational science projects:

- Implement working code with emphasis on designing for extension and maintenance. Focus on the design of your code in the beginning and do not think too much about efficiency before you have verified your program. A rule of thumb is the 20–80 rule[44]: 80% of the CPU time is spent in 20% of the code, and you will experience that, typically, only a small part of your code is responsible for most of the CPU expenditure[45]. Therefore, I recommend spending most of your time in devising a good algorithm to solve your problem.

- The planning of the program should be from top to bottom, trying to keep the flow of commands as linear as possible. Avoid jumping back and forth in any program. First, arrange the major tasks to be done. Then try to break the major tasks into subtasks which can be represented using functions.

---

[44] This is also called the "Pareto principle".
[45] In particle-based scientific computations, about 90% of CPU time goes into the neighbor list search, as we will see in Chapter 5.

Section 2.7 Remarks on programming and software engineering 183

- Write your code in a modular way using different *.c-files which collectively include the definitions of functions which logically belong together, e.g. *InputOutput.c*, *ReadInputFiles.c*, *InitalizeSystem.c*, *Integrate.c*, etc. Then, use appropriate *.h header files to include the declarations of your functions into those .c-files, where these functions are really used. By this way of modular organization of functions, you will always make sure that the definition of your function is in one particular place, which you can easily find by means of the module names that are clear about the type of functionality implemented in the functions of the respective module. Modular programming improves maintainability of code.

- Never use FORTRAN-style *goto* statements! The use of the *goto* statement is considered very poor programming for good reasons. Although all varieties of spaghetti are a culinary temptation (albeit not really considered sophisticated food), spaghetti-coding in FORTRAN-style with *goto*-statements is to be avoided. It is worthwhile reading the three historic articles on this subject by Edsger W. Dijkstra [133, 134] and Knuth [252].

- Always use variables with a meaningful name, i.e. use names that explain the specific meaning – the same goes for function names. For example, do not use v1=1.0 when you can use speedOfLight=1.0.

- For a scientific problem, try to find some cases where an analytical solution exists or where simple test cases can be applied. If possible, devise different algorithms for the same problem. If you get the same answers, you may have coded things correctly (or made the same error twice).

- Avoid awkward pointer-to-pointer-to...-pointer constructs such as ***particleArray[i][j][k], particularly inside loops.

- Avoid global variables whenever possible! Global variables make code harder to understand and to maintain, because global variables can be accessed and changed by any function anywhere in the program. This completely contradicts the idea of writing modularized code that can be easily maintained, because data access is localized in certain modules through specific function calls[46]

- Avoid local dialects of the language you use, if you wish to port your code to other machines (for example on different supercomputers). In C/C++ for example, you can always enforce the *ANSI*-standard with the compiler option -ansi, the *C99*

---

[46] Modularized programming and thus avoiding globals whenever possible adheres to the "data encapsulation" programming paradigm, which can be realized (without much discipline) in object oriented languages such as JAVA or C++. However, also in C you can write programs that encapsulate your data and only make it accessible through certain function calls. For beginner programmers, it is usually very difficult to accept why the use of globals is to be avoided, as by declaring all variables as global, data access is seemingly most simple and one can completely avoid passing any parameter lists to functions.

standard with option -std=C99, or use the compiler option -std=C++98 respectively.

- Always add comment lines describing what your code actually does. This will help you to understand what you did in your own code many months later.

- Use consistent naming conventions for variable and function names, for example:

  1. Function names always start with a capital letter which is followed by lower- and uppercase letters.

  2. To enhance readability and ease of writing, do not use underscores "_" in any names. For example, do not use `Print_all_local_variables()` but rather `PrintAllLocalVariables()`.

  3. Ordinary variable names always start with a lowercase letter, which is followed by lower- and uppercase letters to enhance readability, e.g. instead of `speed_of_light` use `speedOfLight`.

- Preprocessor variables ("*#define*", which are usually global variables for at least the local module in which they are defined) are always written in capital letters, in order to indicate their global character, e.g. `#define MAX_ARRAY_LENGTH 30`. Here, underscores are used to increase readability, as the only exception to the earlier rule.

2.7.2 Reduction of complexity

Scientific computing is probably the only area of expertise in which a single mind is obliged to span the distance from one bit to a few hundred gigabytes, a ratio of 1 to 10^{12}, or twelve orders of magnitude. This gigantic ratio is staggering. Edsger W. Dijkstra put it this way [135]: "Compared to that number of semantic levels, the average mathematical theory is almost flat. By evoking the need for deep conceptual hierarchies, the automatic computer confronts us with a radically new intellectual challenge that has no precedent in our history".

The most challenging part of programming is conceptualizing the problem, and many errors in programming are conceptual errors. Because each program is conceptually unique, it is difficult or impossible to create a general set of directions that provide a solution in all cases. Thus, knowing how to approach complex problems in general is at least as valuable as knowing specific solutions for specific problems.

Section 2.7 Remarks on programming and software engineering

> One principal task of (computational) science is the reduction of the complexity of real-world systems.

The drive to reduce the complexity of a real-world problem (for example, the N-particle problem in physics and engineering) is at the heart of computer science. Although it is tempting to try to be a hero and deal with computer science problems at all levels, nobody's brain is really capable of spanning twelve orders of magnitude of detail. Computer science and software engineering have developed many intellectual tools for handling such complexity, and a discussion of some of them will help the interested reader and student to make better use of these tools in her own scientific projects involving computer simulation.

At the software-architecture level, the complexity of a problem is reduced by dividing the system into subsystems[47]. The more independent the subsystems are, the more you reduce the complexity. Carefully defined modules in your own program design separate concerns, so that you can focus on *one thing at a time*. Packaging code into objects produces much of the same benefit.

Global data is harmful because it weakens your ability to focus on one thing at a time. By the same token, module data is not harmful because it does not shift your attention away from the module you are working on. If you have data leaks between subsystems in the form of shared global data or sloppily defined interfaces, you have a flood of complexity as well, and the result is a dilution of the intellectual benefit of dividing the system into smaller subsystems.

Complexity should be minimized for the sake of good design, and minimizing complexity also is a motive for many code-level improvements. The point of limiting control structures to *for* and *if* statements (or their equivalents) and straightforward code is to reduce the number of different control patterns your brain has to deal with. *goto*s are particularly harmful because they don't necessarily follow a specific pattern. Your brain cannot simplify their operation in any way that reduces their complexity. Keeping routines short, limiting the nesting of loops and *if* statements, and restricting the number of parameters passed to routines, are additional tools to help manage complexity.

When you put a complicated test into a Boolean function and abstract the purpose of the test, you make the code less complex. When you substitute a table lookup for a complicated chain of logic, you do the same thing. When you create access routines for major data structures, you eliminate the need to worry about higher-level implementation details of the data structures and achieve an overall simplification of your job.

To some extent, the point of having coding conventions is also to reduce complexity. When you can standardize decisions about formatting, loops, and variable names,

---

[47] This is often called the "divide and conquer" approach.

you release mental resources that you need, in order to focus on more challenging aspects of the programming problem. One reason coding conventions are so controversial is that choices among the options have some limited aesthetic base and are essentially arbitrary. People have the most heated arguments over the smallest differences. Conventions are the most useful when they spare you the trouble of making and defending arbitrary decisions. They're less valuable when they impose restrictions in more meaningful areas.

Hierarchies and complexity

A hierarchy minimizes complexity by handling different details at different levels. It frees your brain from worrying about all the details at any particular level. The details do not completely go away, but are simply pushed to another level so that you can think about them when you want to, rather than thinking about all details all the time.

Using hierarchies comes naturally to most people. When they draw a complex object such as a house, they draw it hierarchically. First they draw the outline of the house, then the windows and doors, and then more details. They don't draw the house brick by brick, shingle by shingle, or nail by nail.

In a typical computer system, you'll have, at the very least, the ascending levels from machine instructions, through operating system operations and high-level language programming, to user interface operations. As a programmer in a high-level language, you only have to know about high-level programming and user interface operations, which saves you the grief of having to deal with the machine instructions and operating system calls at the lowest levels of the hierarchy.

If you are smart about designing a software system, you'll create your own hierarchical levels in addition to the ones you automatically get by writing a program in a high-level language.

Abstraction and complexity

Abstraction is another means of reducing complexity by handling different details at different levels. Any time you work with an aggregate, you are working with an abstraction. If you refer to an object as a "molecule" rather than a combination of individual atoms of different extensions, you are making an abstraction. If you refer to a collection of molecules as a "compound", you are making another abstraction.

Abstraction is a more general concept than hierarchy. Hierarchy implies a tiered, structured organization in which levels are ordered and have ranks. Abstraction does not necessarily imply a hierarchically structured organization. Abstraction can reduce complexity by spreading details across, for example, a network of components, rather than among the levels of a tiered hierarchy.

Programming largely advances through increasing the abstraction level of program components. Probably the biggest single gain ever made in computer science was in the jump from machine language to higher-level languages, because it freed programmers from worrying about the detailed quirks of individual pieces of hardware, and allowed them to focus on programming. The idea of routines (functions) was another big step.

Abstract datatypes reduce complexity, because they allow you to work with characteristics of a datatype specific to the problem domain, rather than a computer science structure. Object-oriented programming provides another level of abstraction, which applies to algorithms and data at the same time. This is a kind of abstraction that traditional functional decomposition alone does not provide.

In summary, a primary goal of software design and coding is to conquer complexity. The motivation behind many programming practices is to reduce a program's complexity.

> Reducing complexity is key to being an effective scientific programmer.

Use naming conventions in your code

In some ways, the desire to use short variable names is a historical remnant of an earlier era of computing. Old languages like FORTRAN, assembly or BASIC limited variable names to two to eight characters and forced programmers to create short names. In modern languages like C, C++, etc., you have no reason to shorten meaningful names.

However, if – for some reason – you do want to create abbreviated names in your program, there are several traps in which you may fall. Here are several guidelines for creating abbreviations and avoiding pitfalls:

- Don't abbreviate by removing *one* character from a word. Typing one character is little extra work, and one character savings hardly justify the loss of readability. This is like the calendars that have "Jun" and "Jul". With most one-letter deletions, it is hard to remember whether you removed the character. Either remove more than one character or spell out the word.

- Abbreviate *consistently*. Always use the same abbreviation. For example, use "Num" everywhere or "No" everywhere, but don't use both. Similarly, don't abbreviate a word in some names and not in others. For instance, do not use the full word "Number" in some places and the abbreviation "Num" in others.

- Create names that you can pronounce. Use "xPos" rather than "xPstn" and "curTotal" rather than "CrntTtl". A good test as to whether you should come up with easier to pronounce names is to ask yourself whether you could read your code over the phone.

- Remember that names are more important to the reader of the code than to its author. Read code of your own that you haven't seen for at least six months and note where you have to work to understand what the names mean. Resolve to change the practices that cause confusion.

Generally speaking, naming variables to reflect the "what" of the problem rather than the "how" of the computer science solution, increases the level of abstraction. Using named constants rather than literals also increases the level of abstraction.

2.7.3 Designing a program

On a small project, the talents and capabilities of the individual programmer are the biggest influence on the quality of the software. Part of what makes an individual programmer successful is his or her choice of program design.

On projects with more than one programmer, organizational characteristics make a bigger difference than the skills of the individuals involved. Even if you have a great team, its collective ability is not simply the sum of the team members' individual abilities. The way in which people work together determines whether their abilities are added up, or are subtracted from each other.

One example of the way in which designing a program matters is found by looking at the consequences of not specifying stable requirements before you begin coding. If you don't know what you're building, you can't create a superior design for it. If the requirements, and subsequently the design, change while the software is under development, the code must change too, degrading the quality of the system.

You have to lay a solid foundation, before you can begin building on it. If you rush to coding before the foundation is complete, it will be harder to make fundamental changes in the system's architecture. People will have an emotional investment in the design, because they will have already written code for it. It's hard to throw away a bad foundation once you have started to build a house on it.

The main reason why designing a program matters is that, in software, quality must be built in from the first step. This flies in the face of the naive folk wisdom that you can code like hell and then test all the mistakes out of the software. That idea is dead wrong. Testing merely tells you the specific ways in which your software is defective. Testing won't make your program more usable, faster, smaller, more readable, or more extensible.

Premature optimization is another kind of error that is often committed. In an effective process, you make coarse adjustments at the beginning and fine adjustments at the end. If you were a sculptor, you'd rough out the general shape, before you start polishing individual features. Premature optimization wastes time because you spend time polishing sections of code that don't need to be polished. You might polish sections that are small enough and fast enough as they are, you might polish code that you later throw away, or you might fail to throw away bad code because you have already spent time polishing it. Always ask yourself, "Am I doing this in the right order? Would changing the order make a difference?" Consciously follow a good process of software design.

Designing code means "thinking before writing" software. It is time well spent. Saying that "code is what matters, you have to focus on how good the code is, not on some abstract design process" is very shortsighted and ignores mountains of experimental and practical evidence to the contrary. Scientific software development is a creative exercise. If you don't understand the creative process, you're not getting the most out of the primary tool you use to create software – your brain. A bad process wastes your brain cycles. A good process leverages them to maximum advantage.

2.7.4 Readability of a program

The computer does not care whether your code is readable: it is better at reading binary machine instructions than it is at reading statements written in a high-level language. You should write readable code, because it helps other people to read your code. Readability has a positive effect on all of the following aspects of a program

- comprehensibility
- reviewability
- reusability
- error rate
- debugging
- modifiability
- development time – as a consequence of all of the above
- external quality – as a consequence of all of the above

Readable code does not take any longer to write than confusing code does, and the effort to make it readable justifies itself during initial work on the code. It's easier to be sure your code works, if you can easily read what you wrote. This should be a sufficient reason for writing readable code. But code is also read during reviews. It is read when you or someone else fixes an error. It's read when the code is modified. It's read when someone tries to use part of your code in a similar program.

Making code readable is not an optional part of the development process. You should go to the effort of writing good code, which you can do once, rather than the effort of reading bad code, which you'd have to do again and again.

> "Why should I write readable code if I'm not working in a team but just writing scientific code for myself?" [48]

> The answer is, because a week or two from now you are probably going to be working on another program and think that you already wrote this routine last week. Based on this, you just drop in your old, tested, debugged code and save some time. However, if the code is not readable, good luck!

The idea of writing unreadable code because you are the "lone ranger" on a project is disturbing. It is like not stealing because you might get caught. Habits affect all your work: you can't turn them on and off at will, so be sure that you want what you're doing to become a good habit. It's also good to recognize that whether a piece of code ever belongs exclusively to you is debatable. Douglas Corner came up with a useful distinction between private and public programs [104]: "Private programs are programs for a programmer's own use. They are not used by others. They are not modified by others. Others don't even know the programs exist. They are usually trivial, and they are the rare exception. Public programs are programs used or modified by someone other than the author."

Standards for public and for private programs can be different. Private programs can be sloppily written and full of limitations without affecting anyone but the author. Public programs must be written more carefully: their limitations should be documented, they should be reliable, and they should be modifiable. Beware of a private program becoming public, as private programs often do. You need to convert the program to a public program before it goes into general circulation. Part of making a private program public is making it readable. If you think you don't need to make your code readable because no one else ever looks at it, make sure you are not confusing cause and effect.

2.7.5 Focus your attention by using conventions

A set of conventions is one of the intellectual tools used to manage complexity. Many of the details of programming are somewhat arbitrary, such as: how many spaces do you indent a loop? How do you format a comment? How should you order parameters

---

[48] This is probably the typical situation for somebody developing scientific code in a research institute or at university.

used in a routine? Most of these kinds of questions have no unique right answer[49]. The main point here is to be *consistent* in applying your conventions. Conventions save programmers the trouble of answering the same questions – making the same arbitrary decisions – again and again. On projects with many programmers, using conventions prevents the confusion that results when different programmers make arbitrary decisions differently.

A convention conveys important information concisely. In naming conventions, a single character can differentiate among local, module, and global variables. Capitalization can concisely differentiate between types, named constants, and variables. Indentation conventions can concisely show the logical structure of a program. Alignment conventions can concisely indicate that statements are related.

Conventions protect against known hazards. You can establish conventions to eliminate the use of dangerous practices, to restrict such practices to cases in which they're needed, or to compensate for their known hazards. You could eliminate a dangerous practice by, for example, prohibiting global variables or prohibiting multiple statements on a single line. You could compensate for a hazardous practice by requiring parentheses around complicated expressions or requiring pointers to be cleared immediately after they're freed to help prevent dangling pointers.

Conventions also add predictability to low-level tasks. Having conventional ways of handling memory requests, error processing, input/output, and interfaces for routines adds a meaningful structure to your code and makes it easier for another programmer to figure out – as long as the programmer knows your conventions. One of the biggest benefits of eliminating *goto*s is that you eliminate an unconventional control structure. A reader knows roughly what to expect from a *for*, an *if*, or a *case* statement, but it is hard to tell whether a *goto* jumps up or down, over five lines, or through half of the program. The *goto* increases the reader's uncertainty. With good conventions, you and your readers can take more for granted. The amount of detail that has to be assimilated will be reduced, which in turn will improve program comprehension.

Programmers on large projects sometimes go overboard with conventions. They establish so many standards and guidelines that remembering them becomes a fulltime job. But programmers on small projects tend to go "underboard", not realizing the full benefits of intelligently conceived conventions.

2.8 Ways to improve your programs

Programming is neither fully an art nor fully a science, but is somewhere inbetween. It still takes plenty of individual judgment to create working software. And part of having good judgment in computer programming is being sensitive to a wide array of warning signs which are subtle indications of problems in your program.

---

[49] For example, in my own programs, I never use any underscores in names and for function names I always use initial *capital* letters to distinguish them from ordinary variable names, which always start with a lowercase letter.

When you or someone else says "This is a really tricky routine", that's a warning sign, usually indicating poor code. "Tricky routine" is a code phrase for "bad routine." If you think a routine is tricky, think about rewriting it so that it is no longer tricky. A routine that has more errors than average also is a warning sign. A few erroneous routines tend to be the most expensive part of a program. If you have a routine that has thrown more errors than average, it will probably continue to throw more errors than average. Think about rewriting it.

If programming was a science, each warning sign would imply a specific, well-defined corrective action. However, because programming is still a craft, a warning sign merely points to an issue that you should address. You can't necessarily rewrite tricky code or improve an error-prone routine.

Just as an abnormal number of defects in a routine warns you that the routine is of low quality, an abnormal number of defects in a program implies that your whole design is defective. A good design (and sticking to good software practices) wouldn't allow error-prone code to be developed. Lots of debugging on a project is a warning sign that implies substandard code. Writing a lot of code in a day and then spending two weeks debugging is not making smart use of your time. There is general rule of thumb, which says that if a routine is longer than two printed pages, it is poorly designed. This might not always be true, but it's a warning that the routine is complicated. Similarly, more than about 10 decision points in one routine, more than three levels of logical nesting, an unusual number of variables, high coupling with other routines, or low internal cohesion should raise a warning flag. None of these signs necessarily means that a routine is poorly designed, but the presence of any of them should cause you to look at the routine skeptically.

If you find yourself working on repetitious code or making similar modifications in several areas, you should feel "uneasy and dissatisfied", doubting that control has been adequately centralized in routines or macros. If you find it hard to create scaffolding for test cases because you can't easily call an individual routine, you should feel the "irritation of doubt" and ask whether the routine is coupled too tightly with other routines. If you can't reuse code in other programs because some routines are too interdependent, that's another warning sign that the routines are coupled too tightly.

When you're deep into a program, pay attention to warning signs indicating that part of the program design is not defined well enough to code. Difficulties in writing comments, naming variables, and decomposing the problem into small, well-defined routines all indicate that you need to think harder about the design before doing the actual coding. Bad function names (that do not tell you what the function really does) and difficulty in describing routines in one-line comments are other signs of trouble. When the design is clear in your mind, the low-level details come easily.

Be sensitive to indications that your program is hard to understand. Any discomfort is a clue. If it's hard for you, it will be even harder for the next scientific programmers: they'll appreciate the extra effort you make to improve it. If you are figuring out code instead of reading it, it's too complicated. Make it simpler.

If you want to take full advantage of warning signs, program in such a way that you create your own warnings. This is useful, because even after you know what the signs are, it's surprisingly easy to miss them. Make it hard to overlook problems in your program. One example is setting pointers to NULL at initialization, or after they are freed, so that they'll cause ugly problems if you mistakenly use one. A freed pointer might point to a valid memory location even after it's been freed. Setting it to NULL guarantees that it points to an invalid location making the error harder to overlook.

Compiler warnings are literal warning signs that are often overlooked (compare our remark on compiler warnings on Page 101). If your program generates warnings or errors, fix it so that it does no longer generate them. You don't have much chance of noticing subtle warning signs when you're ignoring those that have "WARNING" printed directly on them.

2.9 Exercises

The following exercises provide extensive training in the basics of the C-language and are offered mostly for those readers who had no prior formal education in writing programs. Some of the exercises may seem simple, and some of them are more difficult, but these are *exactly* the type of problems one has to solve, and expose the type of errors that often occur when writing scientific code, so the student is well advised to try to work some of them. Solutions are not provided in this monograph but can be obtained on demand by emailing me.

2.9.1 Questions

1. What is the difference between declaration and definition?

2. What is a cast?

3. Which value does an if-statement return, when the statement between the brackets () is true/false?

4. What is the difference between a local variable and a global variable?

5. How can you save the value of a variable within a function during consecutive function calls?

6. Explain the difference between #include "file.h" and #include <file.h>.

7. What are arrays?

8. What's the difference between an array, a string and a pointer?

9. What is the major source of errors when using strings or arrays?

10. How do you recognize the declaration of a pointer?

11. What is the NULL pointer and what is it used for?

12. What's the difference in memory used by an *int*-pointer, a *float*-pointer, a *double*-pointer and a *char*-pointer?

13. What do "call-by-reference" and "call-by-value" mean?

14. What's the difference between an array of *int*-pointers and a pointer to an *int*-array?

15. What is a *void*-pointer and what is it used for?

16. What is a C structure?

17. What are streams in C?

18. What is the difference between the following two declarations?

    ```
    char *str1  = ''Hello World'';
    char str2[] = ''Hello World!'';
    ```

19. What is a memory leak and how does it usually come about?

20. Describe briefly which functions you can use to assign memory from the heap at runtime.

21. Draw a picture of the state of memory as discussed in the text after the following assignment:

    ```
    ( (((short *)((char *)(&array[1]))) + 8) )[3]= 100;
    ```

2.9.2 Errors in programs

1. Find the errors in the following program listing and compile it.

    ```
    #include <stdio.h>

    void Main(void)
    {
      printf('Anything wrong here?'\n");
      return 0;
    }
    ```

Section 2.9 Exercises

2. What has been forgotten in this program?

   ```
   int main(void)
   {
     printf("Anything wrong here?\n");
     printf("Look closely!\n");
     return (0);
   }
   ```

3. Why does the following program lead to a compile error?

   ```
   #include <stdio.h>
   #include <stdlib.h>

   int main(void) {
     float floatValue = Multiply(3.33);
     printf("%.2f\n", floatValue);
     return EXIT_SUCCESS;
   }

   float Multiply(float f) {
     return (f*f);
   }
   ```

4. What is wrong in the program Listing 40? Hint: compile and run the program with the input values -7 and 0.

5. What is the error in the following code snippet?

   ```
   int *ptr;
   int intValue;
   ptr = intValue;
   *prt = 255;
   ```

6. Which value is printed in the last two *printf()*-statements in the following piece of code.

   ```
   int *prt = NULL;
   int intValue;
   ptr = &intValue;
   intValue = 23456789;
   *ptr = 987654321;

   printf(``%d\n'', *prt);
   printf(``%d\n'', intValue);
   ```

Listing 40. Comparison of two integer values.

```
1  #include <stdio.h>
2
3  int intCompare(int value1, int value2) {
4    if( value1 > value2 ) {
5      return value1;
6    }
7    else if (value1 < value2) {
8      return value2;
9    }
10   return 0; // Both are equal
11 }
12
13 int main(void) {
14   int intValue1, intValue2, cmp;
15
16   printf("Please input integer 1: ");
17   scanf("%d", &intValue1);
18   printf("Please input integer 2: ");
19   scanf("%d", &intValue2);
20
21   cmp = intCompare( intValue1, intValue2 );
22   if(cmp != 0) {
23     printf("%d is the larger number.\n", cmp);
24   }
25   else {
26     printf("Both numbers are the same.\n");
27   }
28   return 0;
29 }
```

7. The following code snippet contains a severe error. What is it?

   ```
   int *intArray1, *intArray2, i;
   intArray1 = malloc( BLK * sizeof(int) );
   intArray2 = malloc( BLK * sizeof(int) );;
   for ( i = 0; i < BLK; i++){
      intArray1[i] = i;
      intArray2[i] = i + i;
   }
   intArray1 = intArray2;
   intArray2 = intArray1;
   ```

8. Listing 41 contains several (typical) errors with regard to accessing C structures. Find the errors, correct them and compile the program.

Listing 41. Structures.

```
#include <stdio.h>
#include <stdlib.h>
#include <string.h>

typedef struct article {
  char headline[255];
  int page;
  int issue;
} article;

void output( article *a ) {
  printf("%s\n", a->headline);
  printf("%d\n", a->page);
  printf("%d\n\n", a->issue);
}

int main(void) {
  article article1 = {244, "The headline per se.", 33};
  article *article2;
  article articleArr[2];

  strncpy( article2->headline, "A headline", 255);
  article2->page = 212;
  article2->issue = 43;

  strncpy( articleArr.headline[0], "Another one", 255);
  articleArr.page[0] = 266;
  articleArr.issue[0] = 67;

  output( &article1 );
  output( article2 );
  output( &articleArr[0] );
  return EXIT_SUCCESS;
}
```

2.9.3 *printf()*-statement

1. Using format symbols, write a C program that generates the following output with one single *printf* statement.

```
    J
u
s
t for your
        pleasure!
```

2. What values are printed after the following lines of code?

```
int i = 1;
printf("i = %d\n", i--);
printf("i = %d\n", ++i);
printf("i = %d\n", i++);
printf("i = %d\n", ++i);
```

2.9.4 Assignments

1. In the following code snippets, char-variables are assigned a character. Which assignments are syntax errors?

```
char ch1 = ' ';
char ch2 = 66;
char ch3 = "X";
char ch4 = '\x42';
char ch5 = ch1;
char ch6 = 0x43;
char ch7 = A;
```

2. Which values are printed on the screen after execution of the following piece of code. I suggest writing a *main()*-function and testing it if you cannot see it by reading the source code.

```
int    intValue;
float  floatValue = 1234.1234;
char   charValue;
double doubleValue;

intValue    = floatValue;
charValue   = intValue;
doubleValue = floatValue;

printf("intValue :%d\n", intValue);
printf("charValue :%d\n", charValue);
printf("doubleValue :%lf\n", doubleValue);
```

2.9.5 Loops

1. What is printed after execution of this code snippet and which error was made?

   ```
   int intValue = 0;
   while ( intValue > 10 ){
      printf(''%d\n'', intValue);
      intValue++;
   }
   ```

2. Examine closely why this loop turns out to be an endless loop. Where is the error and how can you resolve it?

   ```
   float intValue;
   for ( floatValue = 0.0f; floatValue != 1.0f; floatValue += 0.1f ){
      printf(''%f\n'', floatvalue);
   }
   ```

3. In the following loop, 0 is printed once and then the program enters an infinite loop. What was done wrong?

   ```
   int intValue = 0;
   while ( intValue < 20 ) {
      if (intValue % 2){
         continue;
      }
      printf(''%d\n'', intValue);
      intValue++;
   }
   ```

4. Write a C program which asks the user how many floating point values are to be summed. Read the *float* values within a loop and add them. Print the result of the addition at the end of the program.

2.9.6 Recurrence

The recursive approach tries to simplify the given problem by solving the same problem for a simpler case until the trivial solution is reached. In the case of the calculation of factorials, the problem is successively reduced to simpler ones until the trivial case – the factorial of 0 – is reached.

1. Rewrite the program in Listing 23 in such a way, that the factorial is calculated *recursively*. *Hint:* A recursive function is a function that calls itself.

2. Write a *main()* program and a function *FibonacciRecursive()* that is called within main(), which calculates the Fibonacci numbers *recursively*. A Fibonacci number $F(i)$ is a number which is the sum of its two predecessors, i.e.:

$$F(1) = 1, \quad (2.50a)$$
$$F(2) = 1, \quad (2.50b)$$
$$F(3) = 2 = (F(1) + F(2)), \quad (2.50c)$$
$$F(4) = 3 = (F(2) + F(3)), \quad (2.50d)$$

and so on. Here, the trivial cases are $F(1)$ and $F(2)$.

3. Write a recursive program that calculates the *Ackermann function* $a(m, n)$. The Ackermann function maps two integers onto integers and is defined as:

$$a(m, n) = \begin{cases} m + 1 & \text{for } n = 0 \\ a(n - 1, n) & \text{for } m = 0, \\ a(n - 1, a(n, m - 1)) & \text{else.} \end{cases}$$

4. With this program, calculate $a(2, 0)$, $a(2, 1)$, $a(3, 0)$, $a(3, 1)$, $a(4, 0)$, and $a(4, 1)$.

   Hint: Be patient with $a(4, 1)$.

5. Try if you can also calculate $a(4, 2)$ on your computer system. *Remark:* I am sure, you'll be able to go to lunch in the mean time if your computer does not produce a segmentation fault and stack overflow anyway. Try to understand why that is.

2.9.7 Macros

1. The piece of code in Listing 42 is supposed to print the integer numbers from 0 to 9. However, it only prints 0 once. Which (logical) error was made and how do you resolve it?

2. The code in Listing 43 outputs 190 as result of the multiplication. However, the correct value would have been $100(10 \times (20 - 10))$. Correct the code accordingly.

3. How often is the *for*-loop executed in the program of Listing 44 and why?

2.9.8 Strings

1. Write a program that reads a string from the command line and then separates this string into single words, according to the following tokens: ",.:!?\t \n". For example, the string "Hi,this!is:an\texample\n" must be separated into the words: "Hi", "this", "is", "an" and "example". Store the single words in a pointer array. Hint: use the function *strtok()* of the library <string.h> to split the string.

Listing 42. Printing integers.

```
1  #include <stdio.h>
2  #include <stdlib.h>
3  #define DEBUG_ERROR printf("Fatal debug error\n"); \
4                      return EXIT_FAILURE;
5  #define MAX 10
6
7  int main(void) {
8    int i = 0;
9    do {
10     printf("\%d\n", i);
11     /* Much code... */
12     if( ++i >= MAX )
13       DEBUG_ERROR;
14   }while( 1 );
15
16   return EXIT_SUCCESS;
17 }
```

Listing 43. Macros multiplication.

```
1  #include <stdio.h>
2  #include <stdlib.h>
3
4  #define MULTI(a, b) (a*b)
5
6  int main(void) {
7    int val1 = 10, val2 = 20;
8    printf("Multiplication = %d\n", MULTI(val1, val2-10));
9    return EXIT_SUCCESS;
10 }
```

2.9.9 Structs

1. Memory snapshots.

   Let's focus on the following two definitions of structs:

   ```
   typedef struct{
   char *name;
   char suid[8];
   int numC;
   } student;
   ```

   and

Listing 44. Use of macros.

```
1  #include <stdio.h>
2  #include <stdlib.h>
3
4  #define COUNT 10
5
6  int main(void) {
7    int i;
8  #undef COUNT
9  #define COUNT 5
10   for( i = 0; i < CNT; i++) {
11 #undef COUNT
12 #define COUNT 20
13   printf("%d\n", i);
14   }
15   return EXIT_SUCCESS;
16 }
```

```
typedef struct{
int nun;
int denom;
} fraction;
```

Now, let's execute the following 7 commands:

```
student friends[4];
friends[0].name = friends[2].suid + 3;
friends[5].numC = 21;
strcpy(friends[1].suid, "9926534");
strcpy(friends->name, "Albert Einstein");
strcpy((char *) &friends[0].numC, (const char *)\
 &friends[2].numC
*(char ***)(&(((fraction *)friends)[3].denom)) =\
 &friends[0].name + 1;
```

Draw consecutive pictures of what each one of these commands means, in terms of memory allocated on the stack in which the two structures `student` and `fraction` reside. For the drawing, compare Figure 2.11 on Page 165.

Section 2.10 Projects   203

2. Meet the Wizards. Consider the following *struct* definitions:

```
typedef struct hogwarts {
int dumbledore;
char voldemort[4];
struct hogwarts *wand;
} hogwarts;
```

and

```
typedef struct {
short *harry[2];
short ron[2];
hogwarts hermione;
} wizards;
```

Make an accurate diagram of what the computer memory looks like after the following six lines of code have executed:

```
hogwarts *gonagall;
wizards snape[4];
gonagall = &snape[0].hermione;
snape[1].harry[3] = (short *) &gonagall;
strcpy(gonagall[2].voldemort, "Avada Kedavra!");
*(char **)snape[4].ron = gonagall->voldemort + 4;
```

2.10 Projects

2.10.1 Decimal and binary representation

Devise an algorithm which converts a floating point number in decimal representation (see Chapter 1) to binary representation. You may or may not use a scientific representation. Write a C program which implements this algorithm.

2.10.2 Nearest machine number

Write an algorithm and a C program which reads a real number x and finds the two nearest machine numbers x_- and x_+, the corresponding relative errors and the absolute errors.

2.10.3 Calculating e^{-x}

Assume that you do not have access to the library function $\exp(x)$. Write your own algorithm for $\exp(-x)$, with special attention to how to avoid the loss of precision, described in Chapter 2. Write a C program which implements this algorithm.

2.10.4 Loss of precision

Write a program which computes the function

$$f(x) = x - \sin x$$

for a wide range of values of x. Make a careful analysis of this function for values of x close to zero. For $x \approx 0$ you should consider writing out the series expansions of $\sin x$

$$\sin x = x - \frac{x^3}{3!} + \frac{x^5}{5!} - \frac{x^7}{7!} + \cdots$$

Use the loss of precision theorem of equation (2.3) to show that the loss of bits can be limited to at most one by restricting x such that

$$1 - \frac{\sin x}{x} \geq \frac{1}{2}.$$

Convince yourself that x has to be at least 1.9, implying that for $|x| < 1.9$ we need to consider the series expansion carefully. For $|x| \geq 1.9$, we can directly use the expression $x - \sin x$. For $|x| < 1.9$, you should devise a recurrence relation for the terms in the series expansion, in order to avoid having to compute very large factorials.

2.10.5 Summing series

Write a C program which sums

$$s_{up} = \sum_{i=1}^{N} \frac{1}{i},$$

and

$$s_{down} = \sum_{i=N}^{1} \frac{1}{i}.$$

The program should read N from stdout as input and write the final output to the screen. Compare also s_{up} and s_{down} for different N using both single and double precision for N up to $N = 10^{10}$. Which of the above formulas is the more reliable one? Try to provide an explanation of the possible differences. Make a log-log plot of the relative difference as a function of N in steps of 10^i with $i = 1, 2, \ldots, 10$. Hence, you have to compute $\log_{10}(|(s_{up}(N) - s_{down}(N))/s_{down}(N)|)$ as a function of $\log_{10}(N)$.

2.10.6 Recurrence in orthogonal functions

Write a C program that solves the finite Fourier series

$$F(x) = \sum_{i=0}^{N} a_i \cos(ix) \qquad (2.51)$$

recursively. Assume that $a_i = (i+2)/(i+1)$ and set $N = 1\,000$. Try out different values of x as input. *Hint:* Starting from

$$\cos(i-1)x - 2\cos(x)\cos(ix) + \cos(i+1)x = 0, \qquad (2.52)$$

one can express the entire Fourier series of equation (2.51) in terms of $\cos(x)$ and two constants. One recursively defines a new sequence of coefficients b_i as

$$b_i = (2\cos(x))b_{i-1} - b_{i+2} + a_i \quad \text{for } i \in [0, \ldots, N], \qquad (2.53)$$

which defines $b_{i+1} = b_{i+2} = \cdots = 0$ for all $i > N$, the upper limit of the series. One can then determine all coefficients b_i from a_i and the evaluation of $2\cos(x)$. If one replaces a_i with b_i in equation (2.51), one obtains

$$\begin{aligned}F(x) = &\, b_N \left[\cos(Nx) - 2\cos((N-1)x)\cos(x) + \cos((N-2)x)\right] + \\ &+ b_{N-1}\left[\cos((N-1)x) - 2\cos((N-2)x)\cos(x) + \cos((N-3)x)\right] + \cdots \\ &+ b_2\left[\cos(2x) - 2\cos^2(x) + 1\right] + b1\left[\cos(x) - 2\cos(x)\right] + b_0.\end{aligned}$$

Using the recurrence relation of equation (2.52), we finally obtain

$$F(x) = b_0 - b_1 \cos(x).$$

2.10.7 The Towers of Hanoi

The game "Towers of Hanoi" is a game with three sticks A, B, C. On stick A there is a tower of n disks, piled on top of each other. The largest disk is at the very bottom and the disks get smaller from bottom to top. The aim of the game is to transport the disks from stick A to stick C adhering to the following rules:

1. You may only move one disk at a time in each move.

2. You may only move the topmost disk in each move.

3. You may never pile a larger disk on top of a smaller disk. Only smaller disks on top of larger ones are allowed.

Figure 2.13 shows a solution of the game for $n = 3$: Find a strategy for the solution of this game and implement this solution in a C program.

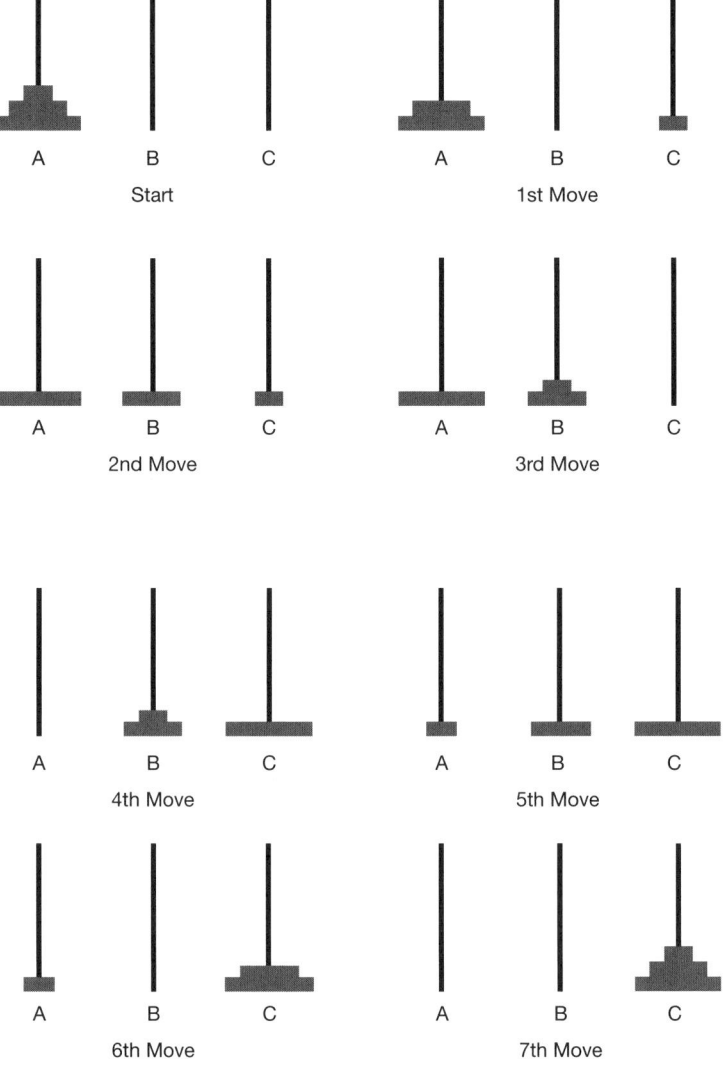

Figure 2.13. Solution of the game "Towers of Hanoi" for $n = 3$.

Hints: Try to use recursion. If you are to find a solution to this game, the first step is to identify the *trivial case* which is, when on stick A there is only one disk and on disk C none. In this case, we can simply move the disk from A to C. The solution of the game for two disks (i.e. two disks on stick A) can be reduced to the trivial case by simply moving the disk on stick A (which lies on top of the lowest disk) to B. In this case, the trivial case where there are no disks on C and only one disk on A is reached. The solution is achieved when finally moving the disk from B to A.

What about three disks? Here, we can apply the same strategy, see Figure 2.13: We start by moving the two upper disks from A to B via C. In the graphical example above we have thus reached the trivial case after the third step, because we can then move the two disks from B to C via A which is the solution.

Let's generalize this strategy: We move the tower of n disks from A to C via B by the following steps:

1. Move $(n-1)$ disks from A to B via C.

2. Move the last disk from A to C (trivial solution).

3. Move the $(n-1)$ tower of disks from B to C via A.

This is a recursive solution strategy.

2.10.8 Spherical harmonics and Legendre polynomials

Many problems in physics utilize spherical harmonics as solutions, e.g., the angular part of the Schrödinger equation for the hydrogen atom, the angular part of the three-dimensional wave equation, or Poisson's equation. The spherical harmonics for a given orbital momentum L, its projection M for $-L \leq M \leq L$, and angles $\theta \in [0, pi]$ and $\phi \in [0, 2\pi]$ are given by

$$Y_L{}^M(\theta, \phi) = \sqrt{\frac{(2L+1)(L-M)!}{4\pi(L+M)!}} P_L{}^M(\cos(\theta)) \exp(iM\phi). \qquad (2.55)$$

The functions $P_L^M(\cos(\theta))$ are the so-called associated Legendre functions, for which the following recursion relations are valid:

$$(L-M)P_L{}^M(x) = x(2L-1)P_{L-1}{}^M(x) - (L+M-1)P_{L-2}{}^M(x),$$
$$P_M{}^M(x) = (-1)^M(2M-1)!!(1-x^2)^{M/2},$$
$$P_{M+1}{}^M(x) = x(2M+1)P_M^M(x).$$

Write a C-function which computes the associated Legendre functions for different values of L and M. Write a program which calculates the real part of the spherical harmonics in equation (2.55). Make plots for various $L = M$ as functions of θ and set $\phi = 0$. Study the behavior of $P_L{}^M$ as L is increased and try to explain why the functions become narrower and narrower as L increases. Also study the behavior of the spherical harmonics when θ is close to 0 and when it approaches 180 degrees. Try to extract a simple explanation for what you see.

2.10.9 Memory diagram of a battle

Trace through the code in Listing 45 and draw a diagram showing the contents of memory (i.e. the content of the members) at the indicated point for the given two structs. Is the memory located on the stack, on the heap, or on both? The two struct definitions are the following:

```
struct harryT {
string name;
string weakness;
int powerLevel;
};

struct snapeT {
   string name;
   string evilPlan;
   int attackLevel;
};
```

Hint: Simply go through the main function (which is executed first) step by step and then trace through the definitions of the datatypes and draw a picture of the memory for the two structs harryT and snapeT. Alternatively, you can try to execute the code and write *printf()* (or *cout*) statements at the indicated point to see what's written in the members of the structs. In order to compile it you will have to include the appropriate library functions.

2.10.10 Computing derivatives numerically

a) Numerically calculate the first derivative of the function, analogous to what was discussed in Section 2.6.1,
$$f(x) = \tan^{-1}(x)$$
for $x = \sqrt{2}$ with step lengths h. The analytic answer is $1/3$. For the calculation, take the forward and the central difference (see Info box on Page 169). b) Find mathematical expressions for the total error due to loss of precision and due to the numerical approximation. Find the step length that produces the smallest value. Perform the analysis with both double and single precision. c) Write a program that computes the first derivative, using the forward and central difference as a function of varying step length h, and let $h \to 0$. Compare with the exact answer. d) Write a makefile for the module files of the examples in Section 2.6.2.

Section 2.10 Projects

Listing 45. Memory diagram of a battle.

```
void Battle(harryT dumbledore, snapeT & voldemort)
{
    int pos = 1;
    int level = dumbledore.powerLevel;
    string name = voldemort.name;
    while(level > 20)
    {
        voldemort.evilPlan[pos--] -= (level / 10);
        level /= 2;
    }

    voldemort.attackLevel -= level;

    pos = name.find(dumbledore.weakness);
    while(pos != string::npos)
    {
        dumbledore.powerLevel /= 2;
        name.replace(pos, 2, "");
        pos = name.find(dumbledore.weakness, pos);
    }

    if(dumbledore.powerLevel > voldemort.attackLevel)
    {
        voldemort.name = "Loser";
    }
    else
    {
        dumbledore.name = "Big Baby";
    }

    /* DRAW THE STATE OF MEMORY HERE */

    return;
}

int main(void)
{
    harryT albus;
    snapeT tom;
    albus.name = "Super Lecturer";
    albus.weakness = "Gr";
    albus.powerLevel = 60;
    tom.name = "Lazy Grad Student";
    tom.evilPlan = "Frowning";
    tom.attackLevel = 30;
    Battle(albus, tom);

    return 0;
}
```

2.11 Chapter literature

There are many books on the C or C++ programming language available, many of which are not really organized in a very pedagogical style and just teach the syntax (e.g. how to declare a *struct*, what to write down in order to pass a pointer to a function as argument to a function, etc.), not covering good programming style or how to organize a more complex program in a useful way. A notable exception from this the book by Darnel and Margolis [115]. Other good books for learning C are Kernighan and Ritchie [240], King [244], Ouallain [322] or Prata [340]. The book by the inventor of C++, Bjarne Stroustrup [405], is very readable, since it explains some of the motivation for certain language elements of C++. Other good books on C++ are, for example, Koenig [254] or Lichner [274]. The Numerical Receipes in C by Press et al. [342] are a very useful source for all sorts of algorithms and can be recommended as a general reference. A good starting point to appreciate the development of algorithms is Lee [268] and Chabert [90]. Cormen [107] also provides a good general introduction to algorithms. Yourdon and Constantine [446] provide a good, albeit very technical, introduction to structured code design. Myers [310] is less technical, and more succinct and to the point. There is a classic article by Parnas [326] describing the gap between how programs are often really designed and how one wishes they were designed. A discussion of creative thought processes in programming can be found in Adams [16] and Simon [384]. The book by Bentley [55] is an expert treatment of code tuning and optimization and Weinberg [435] is a very readable book on software development. More recent books on software engineering practices are e.g. Pohl [332], Otero [321], Capers [86] or Bruegge [79].

Chapter 3

Fundamentals of statistical physics

Summary

The basic assumption of statistical physics is that condensed matter is made of atoms, the dynamics of which is governed by the laws of quantum mechanics and – in very good approximation – by the laws of classical mechanics. For systems with large particle numbers[1] a solution of the dynamic equations of motion is not possible, because one does not know all *microscopic* initial conditions. On the other hand, some fragmentary *macroscopic* knowledge about the systems can always be achieved in terms of measuring e.g. pressure, temperature, energy, volume or particle number. Thus, one tries to solve the equations of motion under the constraint of a reduced knowledge of the system, and is interested only in *average* or *most probable* values. Average quantities are also obtained in measurements that characterize the macroscopic state of a system with enough precision. The axiomatic basis of this averaging process in statistical physics are the axioms of probability theory and classical mechanics. Both form the basis of scientific computations of macroscopic material properties in computer simulations.

Learning targets

✓ Learning the basic features of statistical mechanics.

✓ Understanding the physical axiomatic basis of MD simulations.

✓ Understanding what an ensemble is.

---

[1] For macroscopic systems N is of the order of 10^{23}. This number arises from the astronomical size of Avogadro's constant, $N_A = 6.02204551 \times 10^{23}$ mol^{-1}.

3.1 Introduction and basic ideas

Statistical mechanics (or, equivalently: statistical physics) is often perceived as the theory of how atoms combine to form gases, liquids, solids and even of how they give rise to plasmas or black body radiation, but it is both more and less than that. Statistical mechanics is a useful tool in many areas of science in which a large number of variables have to be dealt with using statistical methods (for example, in the theory of neural networks). It encompasses a very broad set of ideas, and saying that "statistical mechanics is the theory of gases" is rather like saying "calculus is the theory of planetary motion". In fact, statistical mechanics is first of all a *mathematical structure*.

In a nutshell, *statistical mechanics is just probability theory* applied to very specifically prepared large systems composed of many particles (atoms or molecules). Statistical mechanics is a branch of physics that deals with the description of *macroscopic* systems made of very many particles, so-called N-particle systems, with $N \gg 1$, which tries to find answers to the following key questions in modern science:

- What is the microscopic structure of matter?

- How do properties of macroscopic systems follow from their many microscopic constituents?

Statistical physics tries to find answers to these questions by deriving the *macroscopic* properties of matter from models of its *microscopic* properties and subsequently relating these different entities to each other. The understanding of "macroscopic properties" is that these are properties of a system which arise due to the interplay of very many constituents (atoms, molecules, or more general: particles), for example in a gas, a fluid or a solid state. Typical examples of such properties are *pressure* and *temperature* which are only defined for a *many particle system* and which are meaningless when referring to single, individual particles. Such properties – i.e. properties that depend on the amount of substance in a system – are also called *extensive* properties of a system.

In MD computer simulations, one uses statistical mechanics to calculate the relations between different thermodynamic quantities such as temperature, partition functions, correlation functions, distribution functions, and so on[2]. Historically, the development of statistical physics as a branch of science took place in two major areas, namely *thermodynamics* and *statistical mechanics*, which we will discuss briefly.

---

[2] These quantities will be defined properly later in this chapter.

Section 3.1 Introduction and basic ideas 213

Thermodynamics [3] is a purely phenomenological theory of macroscopic systems at *thermal equilibrium*, which harnesses the astonishing (but due to habituation seemingly trivial) fact that macroscopic bodies – despite their complex composition – can be described by a small number of macroscopic observables such as pressure, temperature, magnetization, specific heat, etc. Between these different observables there are certain relationships which become particularly simple at thermal equilibrium. Thermal (or thermodynamic) equilibrium denotes the state of a system in which the macroscopic observables characterizing this system have attained their limiting values of pressure, temperature, etc., and do no longer change with time[4]. Which particular state of equilibrium is attained by a system depends on the type of contact which this system has with its environment. One distinguishes several idealized cases, which are illustrated in Figure 3.1 on Page 214. It turns out that phenomenological thermodynamics can be derived from a set of basic laws, the *laws of thermodynamics*. Thermodynamic theory is very generally applicable but does not allow for calculating material constants such as the specific heat without doing explicit measurements.

Statistical mechanics tries to derive the behavior of macroscopic bodies from the laws of classical mechanics or quantum mechanics, i.e. it tries to solve the so-called N-body problem of statistical physics. The first problem to solve here is to find out those macroscopic variables of the many microscopic variables which are important for the description of a system of macroscopic size. For example, the notions of temperature or entropy are not elements of quantum theory or of the theory of classical mechanics of point systems. Actually, they don't make sense at all for elementary systems such as single point particles or atoms and are only defined meaningfully for systems composed of many subsystems, i.e. for systems with large N. The second important problem that is addressed in statistical mechanics is the observed *irreversibility* of processes in the macroscopic world. For example, a system left to its own devices eventually attains an equilibrium state where the thermodynamic macroscopic variables do not change anymore. The difficulty in explaining this behavior lies in the fact that the fundamental equations of motion of classical and quantum mechanics are *reversible*, i.e. there always is a solution to the equations, when simply changing the sign of the time variable.

Statistical mechanics is often subdivided into the following fields:

- **Classical kinetic gas theory.** This is a special case of statistical mechanics, applicable under the assumption that the interaction between subsystems is small, e.g. for a dilute gas with N atoms or molecules, contained within a volume V. The basic equation of kinetic gas theory is the *Boltzmann equation* [89, 34, 366, 218],

---

[3] Actually, the term "thermodynamics" itself here is really misleading, as this theory has basically *nothing* to say about the *dynamics* of macroscopic systems, but always assumes them to be at thermal equilibrium, so that the phenomenological laws of thermodynamics can be applied.

[4] The ever present Brownian motion of microscopically small particles in a gas or fluid shows that thermal equilibrium is not a static but a *dynamic* equilibrium state with respect to the microscopic movements.

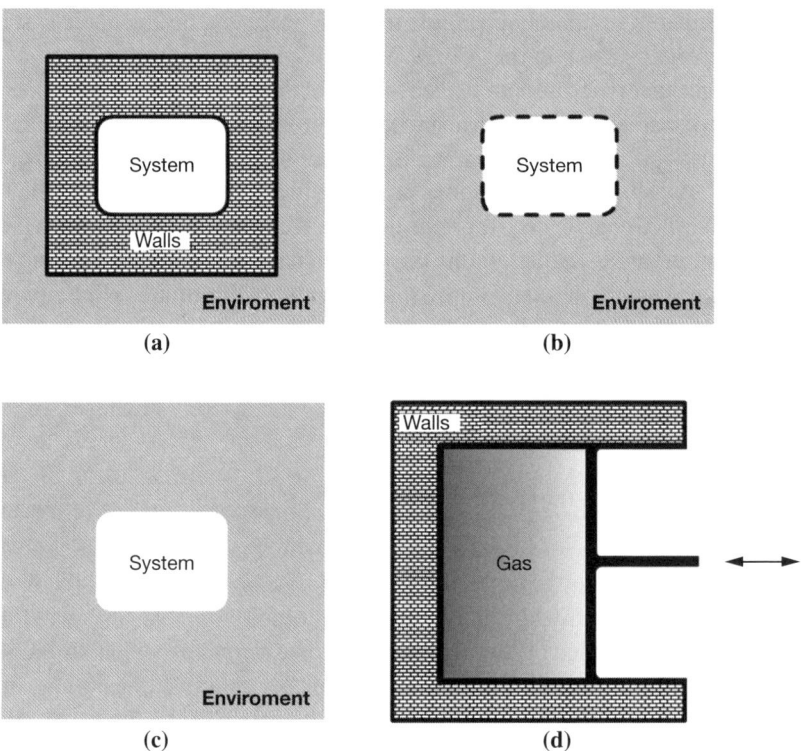

Figure 3.1. Different types of thermodynamic systems in contact with their environment. (a) **Isolated system**. Such a system is isolated against any interaction of the system with its environment. The Walls are impermeable to any form of matter (particles) or energy (heat). For such a system, the total energy E, the particle number N and the volume V are conserved quantities and can be used to characterize the system's macrostate. (b) **Closed system**. This system is only isolated against exchange of energy (heat) but not against exchange of matter, i.e. particles. Hence, energy is not conserved for such a system, but N, V and temperature T can be used to characterize the system's macrostate. (c) **Open system**. Such systems can exchange both particles and energy, in the form of heat, with the environment. Thus, neither of them is a conserved quantity. Instead, T or the chemical potential μ can be used as a definition of the system's macrostate. (d) **Exchange of work**. In the case of a gas, the system can exchange work with the environment, realized either by a plunger that compresses the gas and thereby produces work, or the gas releases mechanical work to the environment by expansion.

Section 3.1 Introduction and basic ideas

which is based on the fundamental assumption of *molecular chaos*[5]. It is required that the thermal de Broglie wavelength λ_{thermal} is small compared to the average distance of particles $\langle d \rangle$ in the system for a classical description to be appropriate, i.e. when

$$\lambda_{\text{thermal}} = \frac{h}{2\pi m k_B T} \ll \langle d \rangle \sqrt[3]{\frac{V}{N}}, \tag{3.1}$$

i.e. the volume of the system and its temperature have to be large enough. For example, for a gas of H_2 molecules at room temperature in a cube with a 1 cm long edge and $N = 10^{19}$, $\langle d \rangle$ can be estimated as $\langle d \rangle = 5 \times 10^{-7}$ cm, and thus, $\lambda_{\text{thermal}} \approx 10^{-8}$ cm.

- **Quantum statistics,** like statistical mechanics, does not rely on either weakly interacting subsystems or on the assumption of molecular chaos. Here, the *ergodic hypothesis* is needed, which states that over long periods of time, the time spent by a system in some region of phase space is proportional to the volume of this region, i.e., that all accessible microstates have equal probability over a long enough period of time. Put differently, the ergodic hypothesis expresses the idea that a system – if it is ergodic – will assume *every* state which is accessible to the system if one observes for long enough time, and that each of these states is attained with equal probability.

The reader might now notice, that there is an obvious problem of getting to grips with an understanding of matter in thermal equilibrium. Let us suppose we are interested in the thermal capacity of copper at $450\,K$. On the one hand we can turn to thermodynamics, but this approach will be of such generality that it is very often difficult to see the point. Relationships between the principal heat capacities, the compressibility, and the thermal expansion are all very well, but they do not really help us understand the particular magnitude and temperature dependence of the actual heat capacity of copper.

On the other hand, you can see that what is needed is a microscopic mechanical picture of what is going on inside the copper. However, this picture quickly becomes unmanageably detailed once one starts to discuss the laws of motion of 10^{19} or so copper atoms[6].

Figure 3.2 shows the relation between different thermodynamic quantities and how they are related to computer simulation and experiments.

---

[5] Molecular chaos is the assumption that the velocities of colliding particles are uncorrelated and independent of position. This assumption, also called in the writings of Boltzmann [67, 372] the "Stosszahlansatz" (collision number hypothesis), makes many calculations tractable.

[6] A simple, rough calculation shows that in a cube of matter with a 1 mm long edge, made only of copper atoms with atomic mass $u(\text{Cu}) = 63.5\,\text{g/mol}$, there are $\approx 7.1 \times 10^{19}$ Cu atoms. If one stores only the positions (x, y, z) (and not the velocities and forces) of these atoms in a *double* on a computer system this would require more than one million petabytes, or 1.7×10^{12} gigabytes, just for the initial configuration of this cube made of Cu atoms.

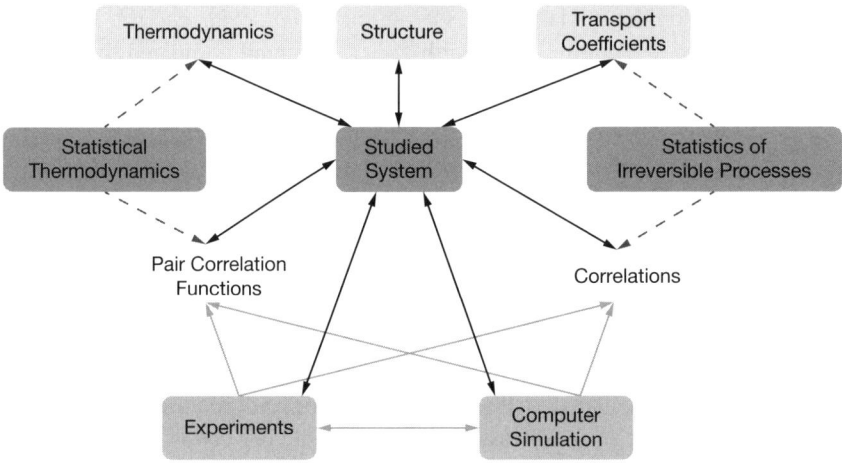

Figure 3.2. Computer simulation and statistical quantities.

3.1.1 The macrostate

The basic task of statistical physics is to take a system which is in a well-defined thermodynamic state[7] and to compute the various thermodynamic properties of that system from an assumed microscopic model. The *macrostate* is another word for the thermodynamic state of a system. It is a specification of a system which contains just enough information for its thermodynamic state to be well-defined, but no more information than that. The macroscopic observables by which the state of a system can be described are called *state variables*. Besides energy E, Volume V, particle number N, entropy S, temperature T, pressure p and chemical potential μ, quantities such as charge, dipole moment, refractive index, viscosity, chemical composition or size of the phase interfaces are also state variables. Not belonging to state variables are *microscopic* properties such as position or momenta of particles. Equations which combine different state variables are called *equations of state*. In thermodynamics, such equations have to be determined empirically. For this, one often uses polynomials of the state variables and experimentally determines their coefficients.

> Empirically determined equations of state are very often only valid within a very limited range of values of the state variables. For example, the very often used model of the *ideal gas* can really only be used for *real* gases at rather low densities. Another typical example is provided by the equations of state of shock waves.

---

[7] This means that the macroscopic thermodynamic variables (temperature, pressure, heat capacity, etc.) are well-determined.

Section 3.1 Introduction and basic ideas

One distinguishes *extensive* and *intensive* state variables. The former are state variables proportional to the amount of substance in a system, like volume or energy. The extensive state variable most characteristic for statistical mechanics and thermodynamics is the entropy $S = \rho \ln \rho$, which is closely connected with the microscopic probability of a system's states. The latter do not depend on the amount of substance in the system and can attain different values in different phases. It is characteristic for intensive state variables that they can be defined *locally* and might change spatially. Typical examples are refractive index, density, temperature or pressure. Often, one simply transforms extensive state variables into intensive ones that describe very similar properties. Energy, volume or number of particles are all examples of extensive quantities, but energy per volume, energy per particle or volume per particle are intensive state variables. We note however, that when enlarging a system (without changing intensive properties) the extensive state variables are enlarged correspondingly, without gaining any new insights into the system.

For the case of a simple, pure substance, specifying its macrostate involves determining the following general information:

- the nature of the substance – e.g. a two-atomic gas (H_2), or a noble gas (Ne), or a solid body made of Fe.

- the amount of substance – e.g. 2.7 moles.

- some pair of thermodynamic state variables for the system – e.g. pressure P and volume V.

Gibbs phase rule answers the question as to how to know how many state variables are actually needed to characterize the macroscopic state of a system uniquely. Gibbs rule states that for a system consisting of P phases[a] and C components[b] one has F degrees of freedom in the system which have to be described using F independent *intensive* state variables:

$$F = K + 2 - P \tag{3.2}$$

---
[a] For example fluid, solid or gaseous.
[b] The number of different kinds of particles, i.e. chemical components of a system.

Each of these pairs of state variables is associated with a way of doing work on the system. For example, for many systems, $P - V$ work is relevant, but the variables generally will be specified in accordance to the external conditions. For instance, a lump of Fe might be at specific pressure $P = 10^5$ Pa and temperature $T = 450$ K. In this case the macrostate would be defined by P and T. Other parameters such as the volume V and internal energy E would then be determined in principle from

P and T. In most examples in this book, if not stated otherwise, we will make an important restriction as to the type of macrostates we are looking at, namely that it be the one *appropriate to an isolated system*, cf. Figure 3.1. In such a system, the macrostate is determined by the *nature* of the substance, the *amount* of substance, and by E and V. An isolated system is in an energy and particle-proof enclosure, so that the internal energy is a fixed constant. V is also a constant as no work is done on the system. The fixed amount of substance can be characterized by the particle number N of "microscopic" particles (or atoms) making up the system.

In practice, this limitation is not too severe, since for an isolated system in which N is reasonably large, the *fluctuations* e.g. in T are very small[8] and thus T is determined rather precisely by (N, E, V). Hence, one can also use results based on the (N, E, V) macrostate, in order to discuss the behavior of the system in any other macrostate.

3.1.2 The microstate

Let us now consider the mechanical microscopic properties of the system of interest, which we are assuming to be an assembly of N identical microscopic particles. For a given (N, E, V) macrostate there are an enormous number of possible microstates. "Microstate" means the most detailed specification of the particle assembly that is possible. For instance, in the classical kinetic theory of gases one would have to specify the position $\vec{r} = (x, y, z)$ and momentum $\vec{p} = (p_x, p_y, p_z)$ of each of the N gas particles, a total of $6N$ coordinates[9]. Of course, just *one* microstate exceeds the storage capacities of all computer systems available today, cf. our remark in the footnote on Page 215. Even worse, any system changes its microstate extremely quickly – for example one mole of gas will typically change its microstate about 10^{32} times a second. Clearly, some sort of averaging over microstates is needed. And here is one of those rare occasions where quantum mechanics turns out to be a lot easier than classical mechanics. The conceptual problem with the microstates in a system treated classically, as outlined above for a typical gas, is that they are infinite in number, because the movement of a particle is thought to be continuous.

In quantum mechanics a microstate is by definition a *quantum state of the whole assembly of particles*. It can thus be described by a single N-particle wave function, containing all possible information about the state of the system. The important point here is that, in principle, quantum states are discrete. Thus, although the macrostate (N, E, V) has an enormous number of possible microstates that are consistent with it, the number is none the less definite and finite. This number is often denoted by the Greek letter Ω and we will see later that Ω plays a central role in the statistical treatment of physical systems.

---

[8] In fact, the fluctuations of thermodynamic variables about their equilibrium values are smaller, the larger the system is, i.e. the more particles are considered. The large fluctuations occurring in very small systems are sometimes called *finite size effects*, see e.g. [112].

[9] Note, that here we assume point particles, i.e. particles with no extension in space (structureless points), with no internal degrees of freedom such as vibration or rotation).

3.1.3 Information conservation in statistical physics

To get a quick start in understanding the parts of statistical physics that are relevant to computer simulations, we first want to establish the important *principle of information conservation* and of *a priory probability* in statistical physics. Let us consider some rather simple examples.

We start with coin flipping. Under ordinary circumstances, one would say that the probabilities of obtaining heads or tails are both $1/2$ – they are equal and sum up to one. But what right do we have to claim that? The most striking (and convincing) argument we can put forward is that fair coins are *symmetric*. That is, the differences between the two sides (e.g. painting, carving and so on) do not influence the actual outcome of a coin flipping experiment. Hence, if the two sides are the same, the probabilities have to be the same. This is the notion of an a priori probability – in this case due to symmetry[10].

Let us take another example and consider a die with six colored sides, say *R*ed, *Y*ellow, *B*lue, *G*reen, *O*range and *P*urple, which we summarize as (R,Y,B,G,O,P). If the six sides of the die are symmetrical, the probability of throwing any one of the six colors is $P_i = 1/6$. With the die there is also a symmetry operation (turning the die by 90 degrees about various axes), which takes any configuration of the die into some other configuration. That symmetry of the die – like in the example above with the coin – again ensures equal probabilities of the different colors of the die. However, with the complex systems considered in most practical situations there are no such symmetries. Hence, the question arises here, whether there is a starting point from which one can back up the notion of an a priori probability no such symmetries are present in a system.

Assume the following example: Let us re-color the die by changing P into R, i.e. $(R,Y,B,G,O,P) \to (R,Y,B,G,O,R)$. What is the probability $P_i(R)$ of rolling a red (R) in this case? By the same logic, by insisting on the symmetry of the colors, one would answer $P_i(R) = 1/5$ which – of course – is wrong, the real answer being $P_i(R) = 2/6 = 1/3$, and $P_i = 1/6$ for all other colors. Hence, the *real* symmetry in this system acts on the faces of the die, not on the five colors. It is easy to imagine an unfair die that does not have any symmetries, for example if it has some extra weight in one corner. Hence, where would you get your a priori probability from in this case?

One idea would be to simply roll it a vast number of times and to derive the probabilities from that experiment – this is, however, intractable, so let us do something else. Let us add a dynamic element to our discussion of rolling dice and think of the die as a *dynamic system* – a system which changes in time under some law of motion. A law of motion is simply an updating process that provides the information about which state a system is in, at consecutive points in time. With respect to the motion of particles, their motion is smooth and continuous and one could subdivide the time into infinitesimal elements when tracking their motion. In the case of a die, our no-

---

[10] Compare our discussion of the importance of symmetries in laws of nature in Section 1.7.1 on Page 73.

tion of motion is connected to the operations the die performs, one per elementary time interval. In other words, we imagine dividing time into equal discrete and finite segments (not taking the limit!). In each instant the die rearranges itself in a way that *only* depends on what is showing on the top of the die. In this case, the motion of the die can be written as a set of rules which constitute the dynamic laws of motion.

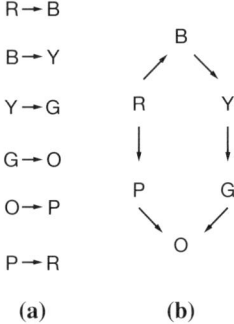

(a)   (b)

Figure 3.3. The dynamical theory of motion of a die. In (a) the equations of motion are written from top to bottom as a set of rules and in (b) these laws of motion are represented as a diagram.

Figure 3.3 represents a complete dynamical theory of a very simple system, in this case, a die. The complete motion of the die is laid down in the above rules. Hence, it does not matter how fast the movement occurs. The decisive point is that the system spends equal amounts of time between successive motions, i.e. it is assumed that the die spends equal amounts of time in each of its states. Because of this, the result of randomly sampling this system will produce equal probability of getting any one of the six different colors, namely $P_i = 1/6$. There are many possible laws like the one stated in this example, all of which lead to the same probability. So this procedure is fairly robust.

Next, let us consider a counterexample of a law that does not give an answer in terms of a priori equal probabilities at all, see Figure 3.4

In the case of the laws of motion laid out in Figure 3.4 there are *two* orbits (or cycles), and there is no way of knowing the probability of being on either cycle a priori. Suppose you are on the (R-G-B) cycle. In that case, the probability is $P_i = 1/3$ to be red, blue or green and $P_i = 0$ to be yellow, purple or orange. Thus, this is a counterexample to the statement that there are equal probabilities for all configurations. However, in both cycles there is a conserved quantity, let's call it C and assume that C = 0 for the cycle (R-G-B) and C = 1 for the cycle (Y-P-O): Two distinct orbits with two distinct values of C. Hence, if a conservation law exists in a system, this means

Section 3.1 Introduction and basic ideas 221

```
R → B
B → G
G → R

Y → P
P → O
O → Y
```

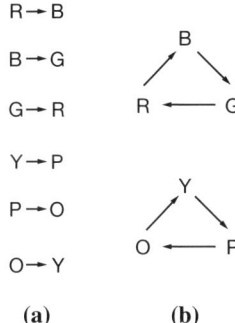

(a)   (b)

Figure 3.4. The motion of a die with two distinct orbits. In (a) the laws of motion are written down and in (b) they are graphically displayed in two separate cycles.

that the system breaks up into different orbits for different values of the conserved quantity[11].

Think!

What are the most important conserved quantities in physics?

Energy conservation is by far the most important conserved quantity[12]. Conservation of momentum does not play as big a role as – in statistical mechanics – we usually think of systems, i.e. gases or liquids, which are *contained within containers*. Hence, when a molecule in a container hits the wall, it bounces off and transfers a little bit of its momentum to the box. Assuming perfect container walls, we ignore the recoil of the box (and also the properties of the box itself). Thus, strictly speaking, when we ignore the recoil of walls, the momentum is *not* conserved, by virtue of molecules hitting the container walls. Hence, in statistical physics, one takes all of the conserved quantities, fixes them, i.e. one measures them, and then studies the system, subject to the constraint that the conserved quantities have certain values.

The structure of statistical mechanics is based on calculating probabilities of things which are subject to constraints, usually taking the form of some (one or more) conserved quantities having fixed values.

---

[11] Note, that there may be many conserved quantities in a complex system.

[12] Of course, there are other conserved quantities, depending on the system under consideration. For example, in Chemistry, the number of atoms for each element is conserved, which is *not* the case in nuclear physics, where it is the total number of protons and neutrons that are conserved.

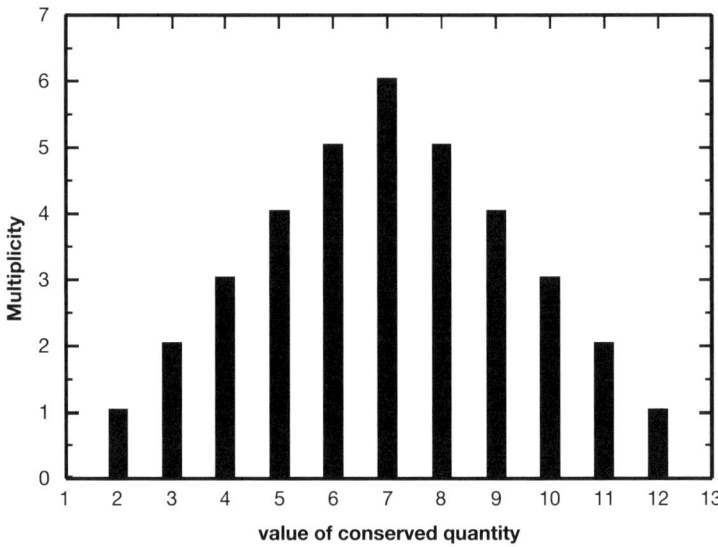

Figure 3.5. Graphical representation of the distribution of the numbers of two dies and their multiplicity.

Consider, as an example, a system of two dice, which interact with each other in a way such that, when one die lands, the other one also lands, in such a way that the sum of the numbers on the two of them does not change. Hence, for this system, the total number on the two dice is a conserved quantity. For each total number on the two dice there is another closed cycle.

> How many possibilities are there for realizing the different possible values of the conserved quantity of the two dice?

Table 3.1 lists all possible combinations of dice numbers pertaining to the different values of the conserved quantity $C \in [1, \ldots, 6]$. These values are graphically represented in Figure 3.5. As a result, for the example of the two dice we have several disconnected orbits are characterized by the different values of the conserved quantity, each with different multiplicity.

As a last example we again consider our colored dice, with a law of motion as displayed in Figure 3.6. This is a perfectly deterministic law, since it tells you what happens next. However, no matter where you start, you always get red. Real, dynamical, physical systems do not behave like this, but rather always conserve information, which means that distinct starting points always result in distinct outputs.

Section 3.1 Introduction and basic ideas

Table 3.1. All combinations of dice numbers with possible values of the conserved quantity and their multiplicity. The distribution of the multiplicities is displayed in Figure 3.5.

die one	die two	value of the conserved quantity	multiplicity
1	1	2	1
2	1	3	2
1	2	3	
1	3	4	
2	2	4	3
3	1	4	
1	4	5	
2	3	5	4
3	2	5	
4	1	5	
1	5	6	
2	4	6	
3	3	6	5
4	2	6	
5	1	6	
1	6	7	
2	5	7	
3	4	7	6
4	3	7	
5	2	7	
6	1	7	
2	6	8	
3	5	8	
4	4	8	5
5	3	8	
6	2	8	
3	6	9	
4	5	9	4
5	4	9	
6	3	9	
4	6	10	
5	5	10	3
6	4	10	
5	6	11	2
6	5	11	
6	6	12	1

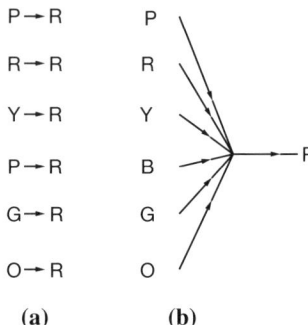

Figure 3.6. Deterministic law of motion with no conserved quantity. No matter where you start, you always obtain red. The trajectories merge so that the initial distinction is lost. This is a violation of the principle of information conservation – real systems do *not* behave like this. In real systems, the mapping between initial and end states is always one to one.

> Real systems always conserve information, which means that distinct starting points do not merge. The rules of thermodynamics are based fundamentally on the idea that the microscopic laws of physics, at the deepest level, are consistent with the *conservation of information*[a].
>
> ---
> [a] This is *not* to be confused with *thermodynamic irreversibility*, which means that initial distinctions between configurations – i.e. starting points – do not disappear.

The principle of conservation of information is a consequence of the basic principles of classical mechanics, which are also fundamentally based on the same principle, consolidated mathematically in *Liouville's Theorem*[13]. Laws of physics are reversible[14] and deterministically lead to output (here: to thermodynamic equilibrium) from a given input (the initial conditions).

> The classical world is only *apparently* statistical. The reason for this is because systems are typically coupled to a much larger system – a so-called "heat bath", about which we know very little. At least, we do not know enough about the heat bath to specify its details. Hence, in classical physics things are "random" not because there is some intrinsic "a priori randomness" in the basic laws of physics, but because we do not know enough about the system to be able to specify everything. This is the idea on which the notion of "entropy" is based.

---
[13] In quantum mechanics, the same principle is called *unitarity*.
[14] Reversibility means you can not only tell where you are going, but also where you came from.

Section 3.1 Introduction and basic ideas

The principle of conservation of information discussed above is rarely mentioned in standard textbooks on thermodynamics or statistical physics, as it is so deeply implicitly assumed by everybody. It is the most central fact, that must be true for thermodynamics to make sense and for it to predict the consequences of thermodynamics, *the theory of heat*.

3.1.4 Equations of motion in classical mechanics

The coin and dice examples in the previous section were just illustrations of a certain concept in statistical mechanics, namely the conservation of information. In these examples, the possible states that the systems may assume, were *discrete* (two sides of a coin and six dice surfaces, either colored or numbered). Statistical mechanics is concerned with the description of the behavior of N-particle systems (with N very large) and our notion of the motion of classical particles is associated with *continuous* positions $\vec{r} = (x, y, z)$ and momenta[15] $\vec{p} = (p_x, p_y, p_z)$ in three-dimensional space.

Think!

How is the state of a mechanical system – say, a point particle – characterized?

In the case of one die, its state was just a color. In the case of the two dice the state of the system was labeled by a pair of colors. In classical mechanics, a single point particle is characterized by specifying its position and momentum with three components each, and thus a total of $f = 6$ degrees of freedom. In the general case of a mechanical system composed of N particles, the system is fully characterized by specifying $fN = 6N$ quantities, $3N$ coordinate components and $3N$ components of the momentum. For reasons of generality these $6N$ quantities can be specified by the *generalized coordinates* q_i

$$\vec{r} = \{q_1, q_2, q_3, \ldots, q_i\} \quad (i = 1, \ldots, 3N), \tag{3.3}$$

and the *generalized velocities* $\dot{q}_i$

$$\vec{v} = \{\dot{q}_1, \dot{q}_2, \dot{q}_3, \ldots, \dot{q}_i\} \quad (i = 1, \ldots, 3N). \tag{3.4}$$

With this, the accelerations $\ddot{q}_i$ of the particles are also determined. The $3N$ quantities $\dot{q}_i$ can also be expressed using generalized momenta, defined as

$$p_i = \frac{\partial L}{\partial \dot{q}_i}, \tag{3.5}$$

---

[15] The momentum $\vec{p} = m \cdot \vec{v}$, where m is the mass and $\vec{v}$ the velocity vector of a particle.

where $L = L(q_i, \dot{q}_i, t)$ is the Lagrangian from which the Hamiltonian can be obtained, using a *Legendre transformation* which is of the form

$$H(p_i, q_i, t) = \sum_i p_i \dot{q}_i - L(\dot{q}_i, q_i, t), \qquad (3.6)$$

and which depends on q_i and p_i.

The Lagrange function is given by

$$L = L(\dot{q}_i, q_i, t) = K(\dot{q}_i, q_i) - U(q_i) = \frac{1}{2} \sum_{i,k} a_{ik}(q_i) \dot{q}_i \dot{q}_k - U, \qquad (3.7)$$

with kinetic energy K and potential energy U. The tensor coefficients a_{ik} only depend on the q_i. To find the equations of motion, i.e. the relation between coordinates, velocities and, accelerations one can use *Hamilton's principle of least action* according to which the dynamics of a system is such that the integral of action takes on an extreme value (normally a minimum), i.e:

$$W = \int_{t_1}^{t_2} L(\dot{q}_i, q_i, t) \, dt = 0. \qquad (3.8)$$

The boundaries of the integral in equation (3.8) are the configurations $q_i(t)$ of the system at times $t = t_1$ and $t = t_2$.

Solution of the Lagrangian variational problem

In order to find a solution of the variational problem of equation (3.8) we re-formulate it in the following form: We are looking for a function $L = L(\dot{q}_i, q, t)$ that minimizes or maximizes the path expressed in the functional I, i.e.:

$$I = \int_{t_0}^{t_1} L(\dot{q}_i, q_i, t) \, dt = 0, \qquad (3.9)$$

with $q_i(t_0) = q_0$ and $\dot{q}_i(t_1) = q_1$, i.e. with fixed boundaries. We find a solution via the following sequence of steps:

1. Assume, that $q(t)$ is the solution of equation (3.9).

    If the function $q(t)$ is the sought-after solution, then all other functions $\bar{q}_i(t) = q_i(t) + \epsilon \xi(t)$ with, fixed endpoints, should deteriorate the extremum. $\xi(t)$ is an arbitrary function, which vanishes at the endpoints t_0, t_1 of I – i.e. $\xi(t_0) = \xi(t_1) = 0$ – and $\epsilon \in \mathbb{R}$ is an arbitrary (but fixed) number.

Section 3.1 Introduction and basic ideas

2. Reduce the variational problem to an ordinary extremum problem.

Inserting the above expression for $\bar{q}_i$ into equation (3.9) with $\xi(t)$ fixed, the integral I becomes a function of parameter ϵ:

$$I = I(\epsilon) = \int_{t_0}^{t_1} L(\dot{\bar{q}}_i, \bar{q}_i, t)\, dt = \int_{t_0}^{t_1} L(\dot{q}_i + \epsilon\dot{\xi}, q_i + \epsilon\xi, t)\, dt. \qquad (3.10)$$

$q_i(t)$ can only be the solution if $I(\epsilon)$ takes on an extremum for $\epsilon = 0$. In this case $\bar{q}_i = q_i$ and $\dot{\bar{q}}_i = \dot{q}_i$. A necessary condition for this is

$$\frac{dI}{d\epsilon} = \int_{t_0}^{t_1} \left(\frac{\partial L}{\partial \bar{q}_i} \underbrace{\frac{\partial \bar{q}_i}{\partial \epsilon}}_{\xi} + \frac{\partial L}{\partial \dot{\bar{q}}_i} \underbrace{\frac{\partial \dot{\bar{q}}_i}{\partial \epsilon}}_{\dot{\xi}} \right) dt = \int_{t_0}^{t_1} \left(\frac{\partial L}{\partial q_i}\xi + \frac{\partial L}{\partial \dot{q}_i}\dot{\xi} \right) dt. \qquad (3.11)$$

3. Rearranging terms using integration by parts.

Using integration by parts we can express $\dot{\xi}$ in equation (3.11) by ξ. The basic trick that simplifies things here, is that the integral vanishes at the endpoints.

$$\int_{t_0}^{t_1} \frac{\partial L}{\partial \dot{q}_i}\dot{\xi}\, dt = \underbrace{\left[\frac{\partial L}{\partial \dot{q}_i}\xi(t) \right]_{t_0}^{t_1}}_{=0 \text{ due to } \xi(t_1)=\xi(t_0)=0} - \int_{t_0}^{t_1} \frac{d}{dt}\left(\frac{\partial L}{\partial \dot{q}_i} \right)\xi(t)\, dt. \qquad (3.12)$$

Inserting this result back into equation (3.11) one obtains:

$$\int_{t_0}^{t_1} \left(\frac{d}{dt}\frac{\partial L}{\partial \dot{q}_i} - \frac{\partial L}{\partial \dot{q}_i} \right)\xi(t)\, dt. \qquad (3.13)$$

4. Application of the fundamental lemma of calculus of variations.

This lemma simply states that – assuming $\int_{t_0}^{t_1} f(t)\xi(t) = 0$ – $f(t)$ is identically zero on the interval $[t_0, t_1]$, if the function ξ is arbitrary and vanishes at the endpoints t_0, t_1. Applied to our integral in equation (3.13) the lemma yields:

$$\frac{d}{dt}\frac{\partial L}{\partial \dot{q}_i} - \frac{\partial L}{\partial \dot{q}_i} = 0 \qquad (3.14)$$

Equation (3.14) is called Lagrange's equation in classical mechanics or Euler equations in variational calculus and solves the original problem of equation (3.8).

Equation (3.14) provides one possible form of the classical equations of motion in the form of N ordinary differential equations for the generalized coordinates q_i of second order in time. In statistical physics however – for theoretical reasons – one considers the phase space of particles based on the generalized coordinates q_i

and momenta p_i. The appropriate form of representing the degrees of freedom of a system this way is given by the Hamiltonian, cf. equation (3.6). In the Hamiltonian formulation of the equations of motion, coordinates q_i and generalized momenta

$$p_i = \frac{\partial L}{\partial \dot{q}_i} \qquad (3.15)$$

are treated equally.

Let us quickly determine the equations of motion of a system in Hamiltonian formulation. Starting from equation (3.6) and taking the total differential of the Hamiltonian yields:

$$dH = \sum_i p_i d\dot{q}_i + \sum_i dp_i \dot{q}_i - dL. \qquad (3.16)$$

The total differential of the Lagrangian is:

$$dL = \sum_i \frac{\partial L}{\partial q_i} dq_i + \sum_i \frac{\partial L}{\partial \dot{q}_i} d\dot{q}_i + \frac{\partial L}{\partial t} dt. \qquad (3.17)$$

Taking the definition of the generalized momenta in equation (3.15) and Lagrange's equation (3.14) in the form

$$\frac{d}{dt} p_i - \frac{\partial L}{\partial \dot{q}_i} = 0, \qquad (3.18)$$

and inserting both into equation (3.17) yields:

$$dL = \sum_i \dot{p}_i dq_i + \sum_i p_i d\dot{q}_i + \frac{\partial L}{\partial t} dt. \qquad (3.19)$$

Finally, inserting equation (3.19) into equation (3.16), after rearrangement of terms, yields

$$dH = \sum_i \frac{\partial H}{\partial q_i} dq_i + \sum_i \frac{\partial H}{\partial p_i} dp_i + \frac{\partial H}{\partial t} dt = \sum_i \dot{q}_i dp_i - \sum_i \dot{p}_i dq_i - \frac{\partial L}{\partial t} dt, \qquad (3.20)$$

and – if L and H are assumed not to be time dependent – the final equations of motion are

$$\dot{q}_i = \frac{\partial H}{\partial p_i} \qquad \dot{p}_i = -\frac{\partial H}{\partial q_i}. \qquad (3.21)$$

Due to their high degree of symmetry in p_i and q_i, they are also called *canonical equations*. They are $2N$ coupled ordinary differential equations of first order for the coordinates and momenta.

Section 3.1 Introduction and basic ideas

The important starting point for MD simulations, however, consists of the simple *Newtonian equations of motion*. Newton's second axiom provides these equations for N particles, indexed with the letter i ($i \in [1, \ldots, 3N]$) as

$$m_i \ddot{q}_i = F_i = -\frac{\partial U}{\partial q_i}. \tag{3.22}$$

The main advantage of the Lagrangian[16] and Hamiltonian formulation of mechanics is that no particular coordinate system is preferred, and – particularly with the latter formulation – a deeper understanding of the mathematical structure of the theory can be obtained.

3.1.5 Statistical physics in phase space

The phase space of a system is the (abstract) space that is spanned by all the quantities that are needed to completely specify the state of a system. Specifying or preparing systems in a particular state, so as to have controlled environmental conditions of a system, is the general way of modeling in physics and engineering. In fact, this is at the core of the scientific method, based on a description with differential equations for which initial conditions and/or boundary conditions have to be specified.

The examples in Figures 3.7 and 3.8 show two very different examples of phase spaces from engineering and physics applications. In the former, the Von Mises cylinder and Tresca hexagonal prism yield surfaces in *principal stress space* – spanned by the three principal stresses ($\sigma_1, \sigma_2, \sigma_3$) – are shown, see e.g. [210, 227, 280, 93] for a thorough reading on this topic.

This phase space is used to illustrate and compare different engineering models describing plastic behavior of a material, described as a continuous medium. In the latter Figure 3.8, the phase space of an undamped harmonic oscillator is shown which, is described by an ellipse given by the equation

$$1 = \frac{1}{2mE_i} p_i^2 + \frac{m\omega^2}{2E_i} x_i^2, \tag{3.23}$$

where ω is the frequency of oscillation.

> **Exercise 3.1.** Derive equation (3.23), starting from the Hamiltonian of the undamped harmonic oscillator.

---

[16] The Lagrangian formulation of the mechanical equations of motion can be obtained using D'Alembert's principle [261, 158], which is a dynamic version of the principle of virtual work. Here, we again see an example of the power of general principles in the formulation of physical theories, see our discussion of this topic in Section 1.7 on Page 59.

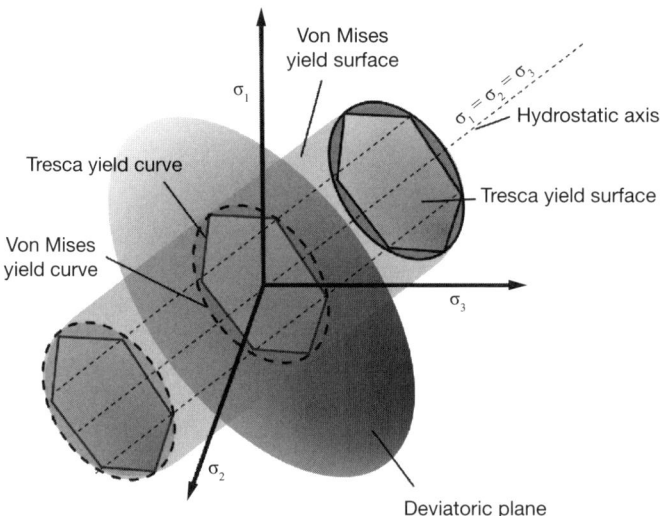

Figure 3.7. Principal stress phase space in Engineering denoting the Von Mises and Tresca models of plasticity.

The configuration space

The mechanical state of a single particle at a certain point in time can be characterized by specifying the (geometric) coordinates $q_i = \{x, y, z\}$ of the particle's position at some point in time. The equations of motion for this particle allow you to follow its trajectory through its phase space spanned solely by its coordinates, which is called the *configuration space*. When the particle moves, its position in configuration space will change with time, so by just specifying the spatial coordinates, the movement of the particle is not yet fully characterized. One also needs to specify the velocities or – more general – the momenta $p_i = \{p_x, p_y, p_z\}$ of the particle, in order to know where it will be next. Hence, the mechanical state of a point particle is fully specified by 6 quantities, 3 coordinates and 3 momenta, spanning another phase space, which we will discuss in the next section.

The molecular phase space (μ-space)

Let us now assume Cartesian coordinates, in which the location of N freely moving mass points are labeled at time $t = t_1$. A snapshot of the system at this moment yields a distribution of points in this 6-dimensional phase space, which is called *molecular phase space*, or μ-space. In Figure 3.9 (a) we consider a system composed of only *one* particle, which is restricted in its movement to one dimension. In this case, its orbit is described in a two-dimensional phase space spanned by one spatial coordinate (x)

Section 3.1 Introduction and basic ideas

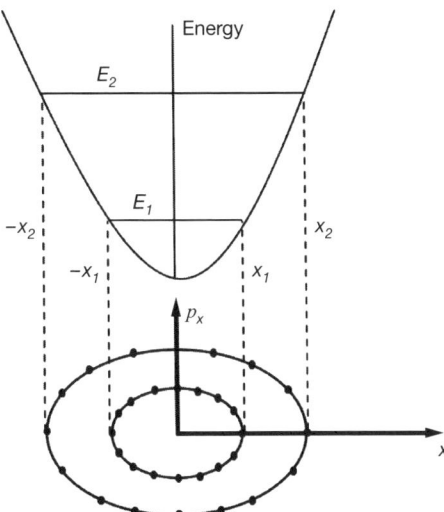

Figure 3.8. Phase space of an undamped one-dimensional harmonic oscillator. The points on the two trajectories mark points equidistant in time. Depending on the fixed value of the energy E_i (which is conserved and does not change along the trajectory), the particle moves on a closed trajectory in its phase space, according to equation (3.23).

and one coordinate of momentum (p_x). The system starts its dynamic development at the initial Point P_1, at time t_1. The laws of motion[17] underlying the dynamics of this particle determine its trajectory (or orbit) through phase space. The little tick marks along the particle's orbit indicate the time intervals by which the particle's path is tracked through phase space until it reaches point P_2 at a time t_2. As we have discussed above, a conservation law leads to a closed loop of a system's path through phase space. In Figure 3.9 (a) the particle's path might continue beyond point P_2 towards infinity, which then describes the trajectory of a free particle in phase space.

The systems that are of interest when performing computer simulations are usually contained in finite boxes with container walls[18] (just like the systems considered in statistical physics), with some quantity that is conserved, so they usually will have finite trajectories in phase space and will not drift to infinity. The properties of the container walls surrounding the system are implemented in computer simulations as *boundary conditions* of the equations of motion that govern the dynamics of the systems and which are expressed in the form of ordinary or partial differential equations.

---

[17] For example, Newton's equations, the Coulomb force law if the particle is electrically charged or whatever equations that govern the system's dynamic behavior.

[18] Periodic boundary conditions can overcome the finite restrictions of systems considered in computer simulations, and are discussed in Chapter 5.

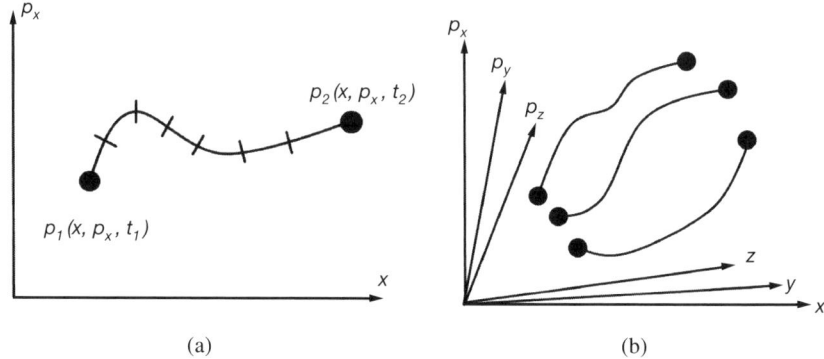

Figure 3.9. Molecule phase space (μ-space). Each point in μ-space represents one particle of the total system. Hence, a complete N-particle system is described by a cloud of N non-interacting points in this phase space. The points have to be independent so that we can apply the methods of statistics to the distribution function of the points. In (a) we display a one-dimensional system in its two-dimensional phase space (spanned by p_x and x) composed of only one particle. In (b) three particles and their orbits, i.e. their history through phase space, are shown.

Figure 3.9 (b) essentially shows the same situation, except that it is not restricted to one dimension. Three particles are shown, together with their orbits through μ-space.

In 6-dimensional μ-space, each of the N particles of a system is represented by one point and the whole system is represented by a scatter plot. If the points of a system are dense enough, one can introduce a distribution function $\rho(\{p_i\},\{q_i\},t)$ for this ensemble of points. The expression

$$\rho(\{p_i\},\{q_i\},t)\, d^{3N}p\, d^{3N}q \tag{3.24}$$

then denotes the number of points in μ-space at time t in the volume element $\rho(\{p_i\}, \{q_i\}, t)$ around the point $(\{p\}, \{q\})$.

This method of using μ-space for the description of systems is also called the *Boltzmann method*. One subdivides μ-space into smaller intervals and calculates the occupation numbers for these cells. However, in order to be able to apply statistical methods, here the different points have to be independent of each other. In other words, statistical methods can only be applied to the concept of μ-space if the different systems do not interact with each other. This restricts the class of systems that can be treated to rather ideal systems with very weak interactions, e.g. ideal gases or ideal crystals.

Section 3.1 Introduction and basic ideas

The gas phase space (Γ-space)

Another way to describe systems in phase space is to use *gas phase space*, or Γ-space where a *complete* N-particle system is represented by one single point, the trajectory of which represents the orbit of a complete system. Γ-space has $6N$ dimensions, spanned by the $3N$ coordinate axes and $3N$ momenta axes, see Figure 3.10.

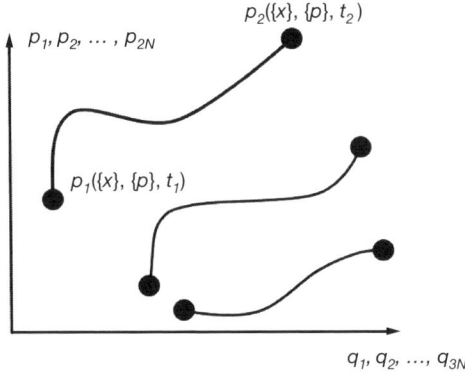

Figure 3.10. Gibbs ensemble in Γ space. Each of the equally prepared systems is represented by a single point in phase space.

We here consider a large number of equally prepared systems, a so-called *Gibbs-ensemble*. Each of these systems is represented by *one* point in Γ-space. For example, consider a large number of systems, prepared in such a way that their energy is fixed at a certain value. In this case, each of these different systems is represented by one single point in phase space and is thought to have a different microstate consistent with the prepared macrostate (fixed value of energy). This procedure then allows for applying the methods of statistics to the phase space density of points.

We will wrap up this section by mentioning three important properties of systems in phase space, which are theorems built into classical mechanics and which are illustrated in Figure 3.11:

- Trajectories do not merge, not even asymptotically.

  This is the same as saying that information, in this case the information about the future and past positions of a system in phase space, does not get lost. One could also say that this information is a conserved quantity. The deeper reason for this is the principal property of all classical mechanical laws of motion that they are completely time-reversible and deterministic. That means that, if you know, where you are in phase space, you know not only where you came from but also where you are going next. This information is provided by the governing equations of motion – differential equations with initial conditions, or formulated as boundary

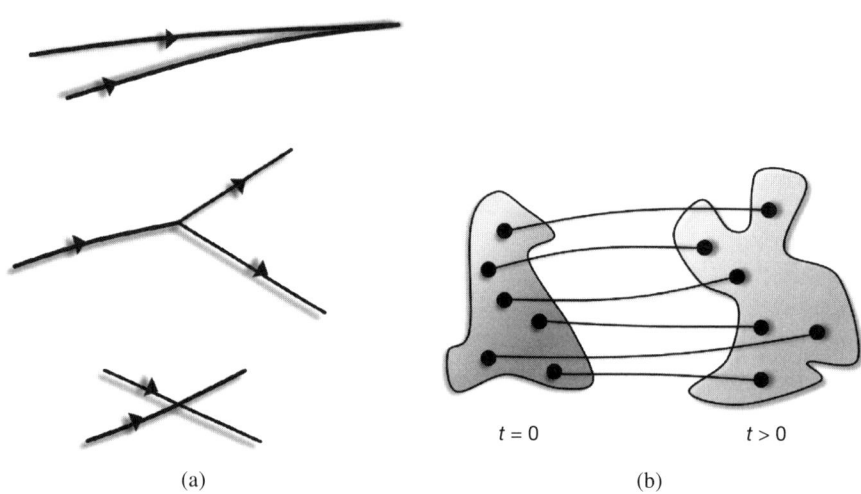

(a) (b)

Figure 3.11. Properties of phase space trajectories. (a) What you will never see in phase space, due to the deterministic mathematical structure of classical mechanics. From top to bottom: merging, splitting or crossing phase space trajectories. (b) Conservation of phase space volume. A fundamental property of the volume in phase space occupied by the points representing the systems is that it is conserved. This is expressed in Liouville' theorem discussed on Page 243.

value problems which endow you with the *complete* information about the system. This strictly deterministic idea of the dynamics of a system is also called *Laplace's demon*. This demon represents the idea, formulated for the first time by Laplace, that the world is completely determined by the laws of motion and the fixed initial conditions.

- Trajectories do not split.

  This is the opposite of the above property, which is also true. With a splitting trajectory, one would not know where to go next: such a system would not any longer be deterministic.

- Trajectories do not cross.

  This is an obvious property, also due to the deterministic properties of the dynamics of classical point systems

Think! What counterarguments against the idea of Laplace's demon can you think of?

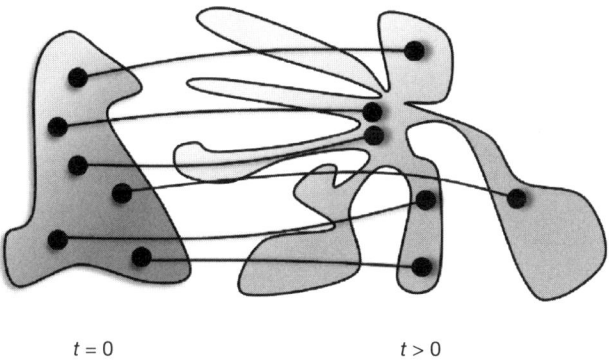

$t = 0$ $t > 0$

Figure 3.12. A chaotic system in phase space.

For almost all systems of interest in computer simulations, the volume in phase space branches out in a very fractulated way

The averaging postulate

We now state the assumption, which forms the whole basis of statistical physics.

> All microstates accessible to a system are equally probable.

This averaging postulate is an assumption, but it is of interest to observe that it is nevertheless a reasonable one.

3.2 Elementary statistics

In this section, we want to discuss elementary *mathematical* aspects and basic terminology of statistical considerations, which one has to understand in order to grasp the statistical description of systems in physics.

As we have seen in Section 3.1.2 on Page 218, when describing systems from a statistical point of view, one usually considers *ensembles*, i.e. a very large number N of identically prepared systems. In that case, the probability of a certain event to be true is always given with respect to such an ensemble, which means that this event is represented by all those systems of the ensemble that are characterized by the event to be true. For example, when throwing a pair of dice, a statistical description takes into account a very large number N of equally prepared pairs of dice. An alternative description of such a system would be to consider the *same* pair of dice thrown N

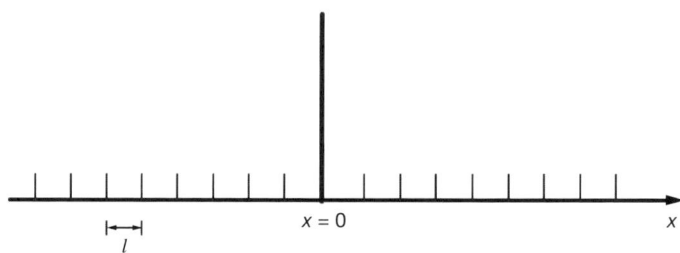

Figure 3.13. One-dimensional random walk. The lamppost in the middle marks the origin of the x-axis. Each step of length l either to the right or the left is *statistically independent* of its previous step.

times under exactly the same conditions, i.e. the same material and dimensions of the dice, the same room temperature, pressure and humidity, etc.

It is also important to keep in mind that the probability of an event to occur is strongly dependent on the considered ensemble of systems underlying the definition of the particular probability under consideration. For example, there is no point in raising the question of the probability of a single seed to become a plant with red blossoms. In contrast, one may usefully ask for the probability of a seed, which is a member of an ensemble of equal seeds of a certain type – for example seeds of the same plant – to become a plant with red blossoms. Hence, the probability of a certain considered event depends on the chosen ensemble.

In the following, we consider an especially simple but important prototype example of how to deal with probabilities, namely the problem of a *random walk*.

3.2.1 Random Walk

In its simplest form, a random walk in one dimension can be described as follows: Imagine a drunken person starting to walk at a lamppost on the street, see Figure 3.13.

Each of his steps – either to the left or to the right – has the same length l and each consecutive step is completely independent of its previous step. All one can say is that each time the man makes a step, the probability of moving to the right is p and the probability of stepping to the left is $q = 1 - p$. In the simplest case, $p = q$, but in general $p \neq q$, because, if for example the street has a slight slope to one direction, a step downhill has a higher probability than one move uphill. If one takes the lamppost as the origin of the x-axis, the spatial location of the drunken man's movement must be of the form $x = ml$, where $m \in \mathbb{N}$. The interesting question here is:

Think!

> What is the probability of the man to be at position $x = ml$ after N random steps, with $m \in \mathbb{N}$?

The statistical formulation of this problem implies considering the movement of a very large number $N \gg 1$, i.e. an *ensemble* of equally drunken persons. In this case, at each step a fraction p of all persons moves to the right. The question is now, what fraction of persons is positioned at $x = ml$ after N steps.

The one-dimensional random walk problem can be easily extended to two dimensions, as shown in Figure 3.14.

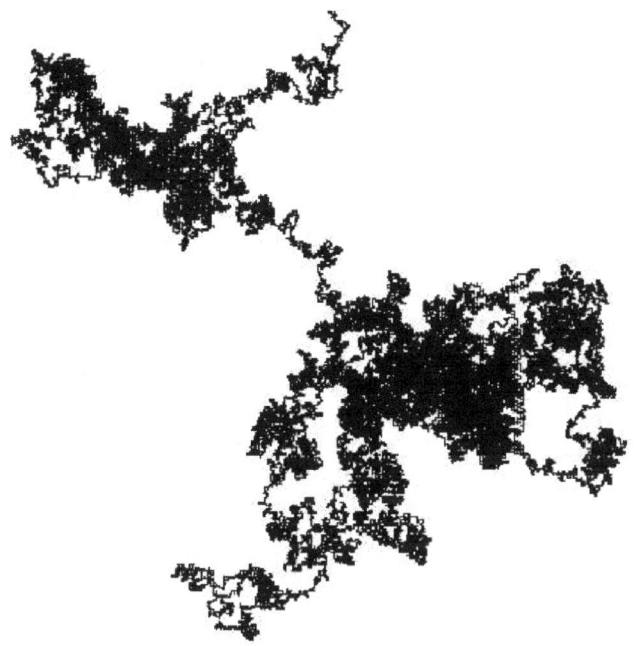

Figure 3.14. Computer-generated two-dimensional random walk on a simple quadratic lattice.

One asks for the probability of the person to be at a certain distance from the origin after N steps. This problem is nothing other than adding N vectors of equal length but with random directions and then asking for the probability that the resulting final vector has a certain length and direction in space. Such a problem occurs, for example, when describing the diffusion of a molecule in a gas: on average, between collisions with other molecules, a given molecule ballistically moves a distance of l in space. One then wants to know what distance the particle has propagated after N collisions. The study of random walks teaches some fundamental aspects of probability theory, which can be directly applied in computational physics. We will see examples of this in Chapter 6 on Page 357.

Mathematical treatment of the random walk in one dimension

As we have seen, the decisive property of a random walk is the statistical independence of consecutive steps. Each step of the random walk depends only on the current position (present) and not on the previously visited sites (past). Previously occupied positions can be occupied again (no *self-avoiding* random walk). In one dimension, the problem can be easily solved. To do so, we abstract from the person moving to the left or right and consider a particle on a one-dimensional lattice, with l being the lattice constant and $P(x,t)$ being the probability to find the particle at position x at time t. The time development of $P(x,t)$ is as follows: each step from position $(x-a)$ or $(x+a)$ increases the probability and each step from x to one of the neighboring lattice sites decreases it, i.e. we have:

$$\frac{\partial}{\partial t}P(x,t) = P(x+a,t)W(x+a \to x) + P(x-a,t)W(x-a \to x)$$
$$- P(x,t)W(x \to x+a) - P(x,t)W(x \to x-a), \qquad (3.25)$$

with

$$W(x, x \pm a) = W(x \pm a \to x) = \frac{\Gamma}{2}, \qquad (3.26)$$

where Γ is the number of steps per unit of time. The above equation (3.25) is called *master equation*.

The solution of the partial differential equation (3.25) can be obtained via a Taylor expansion of the probability $P(x,t)$ about point x:

$$P(x \pm a, t) = P(x,t) \pm a\frac{\partial}{\partial x}P(x,t) + \frac{a^2}{2}\frac{\partial^2}{\partial x^2}P(x,t) + O(a^3). \qquad (3.27)$$

Inserting (3.27) into equation (3.25) and keeping only quadratic terms in a, it follows:

$$\frac{\partial}{\partial t}P(x,t) = \Gamma\frac{a^2}{2}\frac{\partial^2}{\partial x^2}P(x,t). \qquad (3.28)$$

We now set

$$D = \Gamma\frac{a^2}{2}. \qquad (3.29)$$

Generalized to three dimensions $\vec{x} = (x,y,z)$, we obtain

$$\frac{\partial}{\partial t}P(\vec{x},t) = \frac{\Gamma}{6}a^2\Delta P(\vec{x},t), \qquad (3.30)$$

where $\Delta = \left(\frac{\partial^2}{\partial x^2}, \frac{\partial^2}{\partial y^2}, \frac{\partial^2}{\partial z^2}\right)$ is the Laplace operator in Cartesian coordinates.

Section 3.2 Elementary statistics

Let's consider the second moment of the probability distribution $P(x,t)$, which provides the average distance of the random walk from the origin, or the *mean square displacement*:

$$\langle x^2 \rangle = \int_{-\infty}^{\infty} dx\, x^2 P(x,t) \tag{3.31}$$

$$\frac{\partial}{\partial t}\langle x^2 \rangle = \int_{-\infty}^{\infty} dx\, x^2 \frac{\partial}{\partial t} P(x,t) \tag{3.32}$$

$$= \frac{a^2}{2}\Gamma \int_{-\infty}^{\infty} dx\, x^2 \frac{\partial^2}{\partial x^2} P(x,t) \tag{3.33}$$

$$= a^2 \Gamma \int_{-\infty}^{\infty} dx\, P(x) \tag{3.34}$$

$$= 2D. \tag{3.35}$$

In equation (3.33) we have used integration by parts together with the fact that the distribution P very quickly approaches zero for large x. As a result, we see that the mean square displacement is *proportional to time*, i.e.

$$\langle x^2 \rangle = 2Dt. \tag{3.36}$$

Diffusive motion of particles is characterized by the equation

$$\langle x^2 \rangle = 2Dt,$$

where $\langle x^2 \rangle$ is the mean square displacement, t is time and D is the diffusion coefficient.

Factor 2 in equation (3.36) is replaced by a factor 6 in three dimensions and this equation is also valid in more than three spatial dimensions. Figure 3.15 shows that particles in a matrix also show *diffusive behavior*, for which $\langle x^2 \rangle \propto t$.

Now, let us look again at the derivation of the probability distribution of our one-dimensional random walk of a particle. After a total of N steps – each of length l – the particle is at position $x = ml$, where m is a whole number and $-N \leq n \leq N$. Let's calculate the probability $P_N(m)$ of finding the particle at position $x = ml$ after N displacements. Let n_p be the number of displacements to the right and n_q the number of displacements to the left. In this case, the total number of displacements obviously is

$$N = n_p + n_q. \tag{3.37}$$

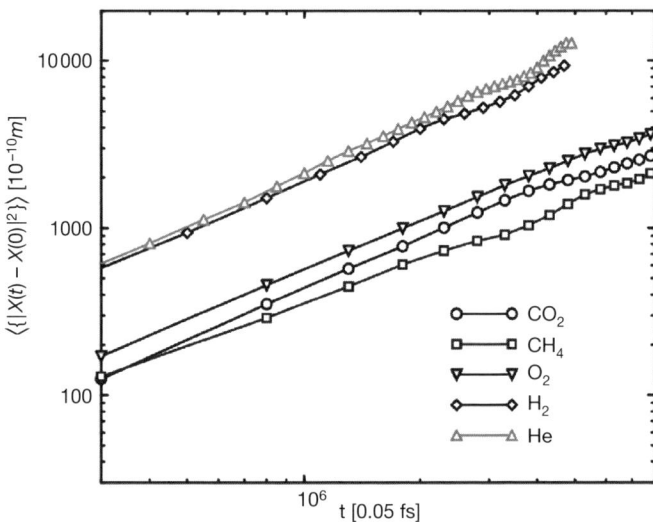

Figure 3.15. Mean square displacement in an atomistic MD simulation of different molecules in a polystyrene matrix with $C = 50$, with degree of polymerization $N = 20$, density $\rho = 0.99 \, \text{g/cm}^2$ and temperature $T = 450 \, \text{K}$.

The resulting displacement (measured in positive units of l, i.e., to the right) is given by

$$m = n_p - n_q. \tag{3.38}$$

Thus, it follows from Equations (3.37) and (3.38) that

$$m = n_p - n_q = n_p - (N - n_p) = 2n_p - N. \tag{3.39}$$

Equation (3.39) shows that the possible values of m are even if N is uneven and the vice versa.

Using our fundamental requirement that individual, consecutive displacements are statistically independent of each other, we realize that – regardless of what happened before – the probabilities for each single displacement are given by p (displacement to the right) and $q = 1 - p$ (displacement to the left). From mathematics we know that, if events in a series $A_1, \ldots, A_N$ are mutually independent, the probability $P\left(\bigcap_{i=1}^{N} A_i\right)$ for all of these events to occur is simply the product of the probabilities of the individual events [128], i.e.

$$P\left(\bigcap_{i=1}^{N} A_i\right) = \prod_{i=1}^{N} P(A_i). \tag{3.40}$$

Thus, the probability of a certain sequence of displacements n_p and n_q is given by the product of the respective probabilities, which is

$$p \cdot p \cdot p \cdots pq \cdot q \cdot q \cdots q = p^{n_p} q^{n_q}. \tag{3.41}$$

For the calculation of the probability $P_N(m)$ we need to consider all possible combinations of N displacements, with n_p displacements to the right and n_q displacements to the left. This number is given by

$$\frac{N!}{n_p! n_q!}. \tag{3.42}$$

Thus, the probability $P_N(m)$ is given by

$$P_N(m) = \frac{N!}{n_p! n_q!} p^{n_p} q^{n_q}. \tag{3.43}$$

The probability distribution in equation (3.43) is called *binomial distribution*[19]. This naming convention is due to the occurrence of equation (3.43) as a typical term in the Taylor expansion of $(p+q)^N$ in the form

$$(p+q)^N = \sum_{n=0}^{N} \frac{N!}{n!(N-n)!} p^n q^{N-n}. \tag{3.44}$$

3.2.2 Discrete and continuous probability distributions

Let us consider a Gibbs ensemble of m systems with f degrees of freedom each[20]. The density function ρ defined in equation (3.24) must be normalized, like any other density distribution function:

$$\int_\Gamma \rho_m(\{q\},\{p\},t) \, dq^{mf} \, dp^{mf} = m. \tag{3.45}$$

We abbreviate the set of generalized momenta $\{p_1, p_2, \ldots, p_{mf}\}$ with p and do the same for the positions, i.e. $q = \{q_1, q_2, \ldots, q_{mf}\}$. We then introduce a normed quantity $\rho(q, p, t)$

$$\rho(q, p, t) = \frac{1}{m} \rho_m(q, p, t), \tag{3.46}$$

called the N-particle distribution function, for which we have:

$$\int_\Gamma \rho(q, p, t) \, dq \, dp = 1. \tag{3.47}$$

$\rho(q, p, t) \, dq \, dp$ is the probability to find the system – consisting of N particles – in the region $q \ldots q + dq, p, \ldots p + dp$ of phase space.

---

[19] For large N, equation (3.43) becomes the Gaussian distribution
[20] Hence, the total number F of degrees of freedom is $F = mf$.

3.2.3 Reduced probability distributions

Often it is enough to know the probability of finding $k < N$ particles. For this, one defines a *reduced probability distribution* $\rho^{(k)}(q, p, t)$, which is obtained by integrating the N-particle probability distribution over all non-interesting $(N - k)$ particles:

$$\rho^{(k)}(q, p, t) = \int \rho^{(N)}(q, p, t) \, dq^{(N-k)} dp^{(N-k)}. \tag{3.48}$$

In particular, one obtains the two-particle and one-particle distribution functions by integrating over $(N - 2)$ or $(N - 1)$ particles:

$$\rho^{(2)}(q, p, t) = \int \rho^{(N)}(q, p, t) \, dq^{(N-2)} dp^{(N-2)}, \tag{3.49}$$

$$\rho^{(1)}(q, p, t) = \int \rho^{(N)}(q, p, t) \, dq^{(N-1)} dp^{(N-1)}. \tag{3.50}$$

From equation (3.49) one can derive the *pair correlation function* and the *radial distribution function* [211]:

- Integration over momentum space yields the probability of finding a system in configuration space in the interval $(q \ldots q + dq)$. This is the density number n of the system:

$$n(q) = \int \rho(q, p, t) \, dp. \tag{3.51}$$

- Integration of the two-particle distribution function of equation (3.49) over momentum space yields:

$$n^{(2)}(q) = \int \rho^{(2)}(q, p, t) \, dp. \tag{3.52}$$

The quantity $n^{(2)}(q_i, q_j) dq_i dq_j$ is the probability of molecule i to be found within the region $(q_i, \ldots, q_i + dq_i)$ and for molecule j to be found in $(q_j, \ldots, q_j + dq_j)$

- Likewise, one obtains the probability $n^{(1)}$ for a molecule i to be found in region $(q_i, \ldots, q_i + dq_i)$:

$$n^{(1)}(q) = \int \rho^{(1)}(q, p, t) \, dp. \tag{3.53}$$

- If $N \gg 1$ one can define a distribution function $g(q_i, q_{ij})$ in the following way:

$$n^{(2)}(q_i, q_j) = n^{(1)}(q_i) n^{(1)}(q_j) g(q_i, q_{ij}). \tag{3.54}$$

The function $g(q_i, q_{ij})$ approaches 1 for $q_{ij} = |\vec{q}_i - \vec{q}_j| \to \infty$. The deviation from 1 is a measure for the degree of correlation of the position of particle pairs. That is why this function is called *pair distribution function*.

Section 3.2 Elementary statistics

- In isotropic systems – for example in ordinary liquids – $g(q_i, q_{ij})$ is a function of distance alone ($g(q_i, q_{ij}) = g(q_{ij} = r)$). The function $g(r)$ is also called *radial distribution function* and is important for characterizing molecular structures and for comparison with experiments. In MD simulations, we can directly calculate $g(r)$.

Liouville's theorem

Let us consider the properties of the volume occupied by the points in phase space that represent an ensemble of systems. The point density in phase space is described by the density function $\rho(q_i, p_i, t)$ of equation (3.24). The points describing a Gibbs ensemble in phase space change their position with time. Thus, the density function $\rho = \rho(q_i, p_i, t)$ also changes with time. This change is given by *Liouville's theorem*.

Theorem 3.2 (Liouville's theorem).

$$\frac{\partial \rho}{\partial t} + \sum_{i=1}^{3N} \left(\frac{\partial \rho}{\partial q_i} \frac{\partial H}{\partial p_i} - \frac{\partial \rho}{\partial p_i} \frac{\partial H}{\partial q_i} \right) = 0. \tag{3.55}$$

H is the classical Hamiltonian, with the equations of motion according to equation (3.21).

Proof. The total number of all systems in Γ-space remains constant. Thus, the only way to change the number of points in a volume Ω of Γ-space is by a flow of points into or out of the volume, through its surface σ. Mathematically stated, this means

$$\frac{\partial}{\partial t} \int_\Omega \rho \, dp^{3N} dq^{3N} = - \int_{\sigma(\Omega)} \rho(\vec{v}\vec{n}) d\sigma, \tag{3.56}$$

where $\vec{v} = \dot{\vec{r}} = (\dot{p}_1, \ldots, \dot{p}_{3N}, \dot{q}_1, \ldots, \dot{p}_{3N})$ is the flow velocity of points in Γ-space, which specifies the speed at which points flow out of Ω. Using Gauss's theorem, equation (3.56) can be written as

$$\int_\Omega dp^{3N} dq^{3N} \left(\frac{\partial}{\partial t} \rho + \text{div}(\rho \vec{v}) \right) = 0, \tag{3.57}$$

and thus, as the volume Ω is arbitrary:

$$\frac{\partial \rho}{\partial t} + \text{div}(\rho \vec{v}) = 0. \tag{3.58}$$

For the divergence we have:

$$\text{div}(\rho \vec{v}) = \sum_{i=1}^{3N} \left[\frac{\partial}{\partial q_i}(\rho \dot{q}_i) + \frac{\partial}{\partial p_i}(\rho \dot{p}_i) \right] \qquad (3.59)$$

$$= \sum_{i=1}^{3N} \left[\frac{\partial \rho}{\partial q_i} \dot{q}_i + \frac{\partial \rho}{\partial p_i} \dot{p}_i + \rho \left(\frac{\partial \dot{q}_i}{\partial q_i} + \frac{\partial \dot{p}_i}{\partial p_i} \right) \right] \qquad (3.60)$$

$$= \sum_{i=1}^{3N} \left[\frac{\partial \rho}{\partial q_i} \frac{\partial H}{\partial p_i} - \frac{\partial \rho}{\partial p_i} \frac{\partial H}{\partial q_i} + \rho \left(\frac{\partial^2 H}{\partial q_i \partial p_i} - \frac{\partial^2 H}{\partial p_i \partial q_i} \right) \right] \qquad (3.61)$$

Since the last term vanishes, we are left with the result:

$$\frac{d\rho}{dt} = \frac{\partial \rho}{\partial t} + \text{div}(\rho \vec{v}) = \frac{\partial \rho}{\partial t} + \sum_{i=1}^{3N} \left(\frac{\partial \rho}{\partial q_i} \frac{\partial H}{\partial p_i} - \frac{\partial \rho}{\partial p_i} \frac{\partial H}{\partial q_i} \right) = 0. \qquad (3.62)$$

□

Thus, the phase space density of points in Γ-space (and with this also the volume) is a conserved quantity.

3.2.4 Important distributions in physics and engineering

Let $P(x)\,dx$ represent the probability of finding the random variable x in the interval x to $x+dx$. Here, $P(x)$ is termed a probability density. Note that $P = 0$ corresponds to *zero chance*, whereas $P = 1$ corresponds to *absolute certainty*. Since it is certain that the value of x lies in the range $-\infty$ to $+\infty$, probability densities are subject to the normalizing constraint

$$P(x) = \int_{-\infty}^{+\infty} = 1. \qquad (3.63)$$

Suppose that we wish to construct a random variable x which is uniformly distributed over the range $R \in [x_1, x_2]$. In other words, the probability density of x is

$$P(x)\,dx = \begin{cases} 1/(x_2 - x_1) & \text{if } x_1 \leq x \leq x_2, \\ 0 & \text{otherwise}, \end{cases} \qquad (3.64)$$

In C, such a variable is constructed as follows:

```
x = x1 + (x2 - x1) * double (Random ()) / double (RAND_MAX);
```

For a detailed explanation and a listing of the function *Random()* and of RAND_MAX, see our discussion of random number generation in Chapter 6 on Page 375. There are two basic methods of constructing non-uniformly distributed random variables –

Section 3.2 Elementary statistics

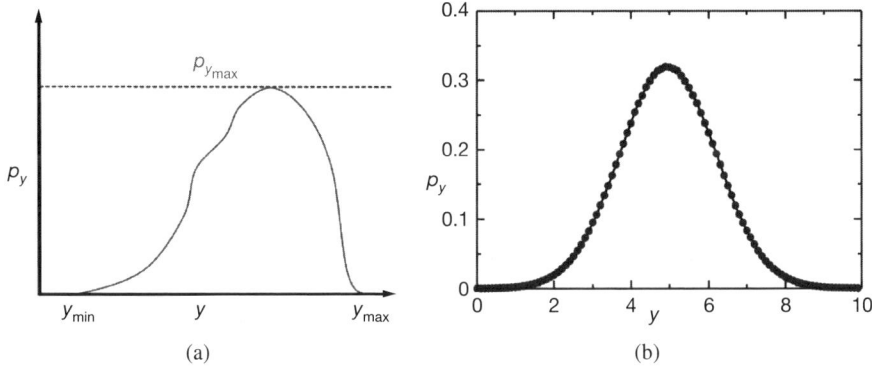

Figure 3.16. The Gaussian distribution. (a) The rejection method. (b) One million values returned by the function *Gaussian()* with mean = 5 and sigma = 1.25. The values are gathered in 100 bins of width 0.1. The figure shows the number of points in each bin, divided by a suitable normalization factor. A Gaussian curve is shown for comparison.

the *transformation method* and the *rejection method*. Let us first consider the former method. Let $y = f(x)$, where f is a known function, and x is a random variable. Suppose that the probability density of x is $P_x(x)$. What is the probability density, $P_y(y)$, of y? The basic law is the conservation of probability:

$$|P_x(x)\,dx| = |P_y(y)\,dy|. \qquad (3.65)$$

Hence, the probability of finding x in the interval $I_x = [x, x + dx]$ is the same as the probability of finding y in the interval $I_y = [y, y + dy]$. From this, it follows that

$$P_y(y) = \frac{P_x(x)}{|f'(x)|}, \qquad (3.66)$$

where $f' = df/dx$. For example, consider the *Poisson distribution*:

$$P_y(y) = \begin{cases} e^{-y} & \text{if } 0 \le y \le \infty, \\ 0 & \text{otherwise.} \end{cases} \qquad (3.67)$$

Let $y = f(x) = ?\ln x$, so that $|f'| = 1/x$. Suppose further that

$$P_x(x) = \begin{cases} 1 & \text{if } 0 \le x \le 1, \\ 0 & \text{otherwise.} \end{cases} \qquad (3.68)$$

It follows that

$$P_y(y) = \frac{1}{|f'|} = x = e^{-y}, \qquad (3.69)$$

with $x = 0$ corresponding to $y \to \infty$, and $x = 1$ corresponding to $y = 0$. Finally, we conclude that if we set

```
x = double (Random ()) / double (RAND_MAX);
y = -log (x);
```

then y is distributed according to the Poisson distribution.

The transformation method requires a differentiable probability distribution function. This is not always very practical. In such cases, preference is given to the rejection method. Suppose one needs a random variable y, distributed with density $P_y(y)$ in the range $R = [y_{\min}, y_{\max}]$. Let $P_{y_{\max}}$ be the maximum value of $P(y)$ in this range, cf. Figure 3.16 (a). The rejection method, as a computational approach, is as follows. The variable y is randomly sampled in the range from y_{min} to y_{max}. For each value of y we first evaluate $P_y(y)$. Next, we generate a random number x, which is uniformly distributed over the range $R \in [0, P_{y_{\max}}]$. Finally, if $P_y(y) < x$ then the y value is rejected, otherwise it is accepted, i.e. it is kept. If this prescription is followed, y will be distributed according to $P_y(y)$. As an example of this, consider the *Gaussian distribution*:

$$P_y(y) = \frac{\exp\left[\frac{(y-\langle y \rangle)^2}{2\sigma^2}\right]}{\sqrt{2\pi}\sigma}, \tag{3.70}$$

where $\langle y \rangle$ is the mean value of y, and σ is the standard deviation. Let

$$y_{\min} = \langle y \rangle - 4\sigma, \tag{3.71a}$$
$$y_{\max} = \langle y \rangle + 4\sigma, \tag{3.71b}$$
$$\tag{3.71c}$$

since there is a small chance that y is more than 4 standard deviations away from its mean value. It follows that

$$P_{y_{\max}} = \frac{1}{\sqrt{2\pi}\sigma}, \tag{3.72}$$

with the maximum occurring at $y = \langle y \rangle$. The function in Listing 46 below employs the rejection method to return a random variable, distributed according to a Gaussian distribution with mean `mean` and standard deviation `sigma`. Also shown is a corresponding *main()*-function in Listing 46, which sorts the numbers into 100 bins. The function *Random()* which is called in function *Gaussian()* is provided in Listing 59 in Section 6.2.1. Figure 3.16 (b) illustrates the performance of the above function. It can be seen that the function *Gaussian()* successfully returns a random value, distributed according to the Gaussian distribution.

Listing 46. Part 1 – Generation of random numbers according to the Gaussian distribution. (*)

```
#include <stdio.h>
#include <math.h>
#include ''Random.h''

#define RAND_MAX 2147483646
#define MAX_TRIALS 10E06

double Gaussian (double mean, double sigma)
{
  double ymin = mean - 4. * sigma;
  double ymax = mean + 4. * sigma;
  double Pymax = 1. / sqrt (2. * M_PI) / sigma;

  /* Calculate a random value uniformly distributed over range
     ymin to ymax */
  double y = ymin + (ymax - ymin) * (double) (Random(0)) / (
     double) (RAND_MAX);

  /* Calculate Py */
  double Py = exp (- (y - mean) * (y - mean) / 2. / sigma /
     sigma) / sqrt (2. * M_PI) / sigma;

  /* Calculate a random value uniformly distributed over range
     0 to Pymax */
  double x = Pymax * (double) (Random(0)) / (double) (RAND_MAX)
     ;

  /* If x > Py, reject value and recalculate */
  if (x > Py) return Gaussian (mean, sigma);
  else return y;
}
```

Listing 46. Part 2 – The main file. (*)

```
27  int main(void)
28  {
29    int x;
30    int bin[100];
31    double number = 0.;
32    double mean   = 5.;
33    double sigma  = 1.25;
34    FILE *fp = NULL;
35
36    for (x = 1; x < 100; x++){
37      bin[x] = 0;
38    }
39
40    for (x = 1; x < MAX_TRIALS; x++){
41      number = Gaussian(mean, sigma);
42      bin[(int)(number * 10)]++;
43    }
44
45    fp = fopen("binOutput.dat","w");
46    for (x = 0; x < 100; x++){
47      fprintf(fp,"%f %f\n",double(x)/10,double(bin[x])/double(
          MAX_TRIALS)*10);
48    }
49    fclose(fp);
50
51    return (0);
52  }
```

3.3 Equilibrium distribution

In this section, we apply the statistical method to realistic thermodynamic systems. This means addressing the properties of an assembly of a large number N of weakly interacting identical particles. There are two types of assembly that fulfill the requirements. One type is a *gaseous assembly*, in which the identical particles are the gas molecules themselves. In quantum mechanics one recognizes that the molecules are not only identical, but are also indistinguishable[21]. It is not possible to put a label on one particular molecule or atom of a gas and to follow its history. Hence the microstate description must take full account of the indistinguishability of the particles.

In this section we shall treat the other type of assembly, in which the particles are considered to be distinguishable. The physical example is that of a solid rather than a gas. Consider a simple solid, made up of N identical atoms. It remains true that the atoms themselves are indistinguishable. However, a good description of our assembly is to think about the solid as a set of N lattice sites, in which each lattice site contains an atom. A particle of the assembly then becomes "the atom at lattice site 7 345" (or whatever)[22]. The particle is distinguished not by the identity of the atom, but by the distinct location of each lattice site. A solid is an assembly of *localized particles*, and it is this locality which makes the particles distinguishable. We will now develop the statistical description of an ideal solid, in which the particles (atoms) are weakly interacting. The main results of this section is the derivation of the thermal equilibrium distribution – the Boltzmann distribution – together with methods for the calculation of thermodynamic quantities using the partition function.

For the macrostate of our system we consider an assembly of N identical, distinguishable (localized) particles contained in a fixed volume V with a fixed internal energy U. The system is mechanically and thermally isolated, but we shall be considering sufficiently large assemblies so that the other thermodynamic quantities (T, S, etc.) are well defined.

The one-particle states will be specified by a state label $j \in [0, 1, 2, \ldots]$. The corresponding energies E_j may or may not be all different. These states will be dependent upon the volume per particle (V/N) for our localized assembly. We use a distribution of states $\{n_j\}$ expressed as a set of numbers $(n_1, n_2, \ldots, n_j, \ldots)$, where the typical distribution number n_j is defined as the number of particles in state j which has energy ε_j. Often, this distribution will be an infinite set. The label j must run over all possible states for one particle. The distribution numbers must satisfy two

---

[21] This is valid in principle as well as in practice
[22] Which one of the atoms is at this site is not specified.

conditions, imposed by the macrostate:

$$\sum_j n_j = N, \tag{3.73a}$$

$$\sum_j n_j \varepsilon_j = U. \tag{3.73b}$$

Equation (3.73)(a) ensures that the distribution contains the correct number of particles and (3.73)(b) guarantees that the distribution corresponds to the correct values of U. All conditions of an (N, V, U) macrostate are now taken care of.

When counting the number of microstates, the fact that we have distinguishable particles raises its head. This means that the particles can be counted just as macroscopic objects. A microstate will specify the state (i.e. the label j) for each distinct particle. We wish to count how many such microstates there are in an allowable distribution $\{n_j\}$. The problem is essentially the same as that discussed in Appendix G, namely the possible arrangements of N objects into piles with n_i objects within a typical pile. The answer is

$$t(\{n_j\}) = \frac{N!}{\prod_j n_j!} \tag{3.74}$$

According to the postulate of the equal probability of all microstates, the thermal distribution should now be obtained by evaluating the average distribution $\{n_j\}_{av}$. This task involves a weighted average of all possible distributions as allowed by equation (3.73) using the values of t in equation (3.74) as statistical weights. This task can be performed but fortunately we are saved from the necessity of having to do anything that is so complicated due to the large numbers involved. Some of the simplifications of large numbers are explored briefly in Appendix G.4.

The crucial point to understand is that it turns out that one distribution, say $\{n_j^*\}$, is overwhelmingly more probable than all others. In other words, the function $t(\{n_j\})$ indeed has a very sharp peak around $\{n_j^*\}$. Consider here, as a comparison, Figure 3.5 in the context of the die combinations, which is also peaked around a mean value. Hence, rather than averaging over all possible distributions, one can obtain essentially the same result by picking out the *most probable distribution alone*. This then reduces to the mathematical problem of maximizing $t(\{n_j\})$ from equation (3.74), subject to the conditions in Equations (3.73). Another, even stronger way of looking at the sharp peaking of t is to consider the relation between the total number of microstates Ω and t. Since Ω is defined as the total number of microstates contained by the macrostate, it follows that

$$\Omega = \sum t(\{n_j\}), \tag{3.75}$$

where the sum is over all distributions. What is now suggested is that this sum can in practice be replaced by its maximum term, i.e. we have

$$\Omega \approx t(\{n_j\}) = t^* \tag{3.76}$$

Section 3.3 Equilibrium distribution

3.3.1 The most probable distribution

To find the thermal equilibrium distribution, we need to find the maximum t^* and identify the distribution $\{n_j^*\}$ at this maximum. Actually, it is a lot less complicated to work with $\ln t$, rather than with t itself. Since $\ln t$ is a monotonically increasing function of t, this does not change the problem, but just makes the solution a lot more straightforward. Taking logarithms of equation (3.74) we obtain

$$\ln t = \ln N! - \sum_j \ln n_j! \qquad (3.77)$$

Here the large numbers come to our aid. Assuming that all ns are large enough for Stirling's approximation to be used, we can eliminate the factorials and obtain

$$\ln t = (N \ln N - N) - \sum_j (n_j \ln n_j - n_j). \qquad (3.78)$$

To find the maximum value of $\ln t$ from (2.6), we express changes in the distribution numbers as differentials (which again are large numbers!) so that the maximum will be obtained by differentiating in point t and setting the result equal to zero. Using the fact that N is constant, and noting the cancellation of two out of the three terms arising from the sum in equation (3.78), this simply gives

$$d(\ln t) = 0 - \sum_j dn_j (\ln n_j + n_j/n_j - 1) = -\sum_j \ln n_j^* dn_j = 0, \qquad (3.79)$$

where the dn_js represent any allowable changes in the distribution numbers from the required distribution $\{n^*\}$. Of course, not all changes are allowed. Only changes which maintain the correct values of N and U – equation (3.73) may be used. This lack of independence of the dn_js produces two restrictive conditions, obtained by differentiating equation (3.73) to obtain:

$$d(N) = \sum_j dn_j = 0, \qquad (3.80a)$$

$$d(U) = \sum_j \varepsilon dn_j = 0. \qquad (3.80b)$$

A convenient way of dealing with the mathematics of a restricted maximum of this type is to use the Lagrange method of undetermined multipliers. The argument in our case runs as follows. First, we note that we can self-evidently add any arbitrary multiples of equation (3.80) to equation (3.79), and still achieve a zero result. Thus

$$\sum_j (-\ln n_j + \alpha + \beta \varepsilon_j) dn_j = 0 \qquad (3.81)$$

for *any* values of the constants α and β. Then, the second, and clever step is to recognize that it will always be possible to write the solution in such a way that each individual term in the sum of equation (3.81) equals zero, as long as specific values of α and β are chosen. In other words, the most probable distribution $\{n^*\}$ will be given by

$$(-\ln n_j^* + \alpha + \beta \varepsilon_j) dn_j = 0, \tag{3.82}$$

with α and β each having a specific (but as yet undetermined) value. This equation can then be written as

$$n_j^* = \exp(\alpha + \beta \varepsilon_j) \tag{3.83}$$

and this is the *Boltzmann distribution*. Before we can appreciate the significance of this central result, we need to explore the meanings of these constants α and β.

α was introduced as a multiplier for the number condition in equation (3.73)(a), and β for the energy condition in equation (3.73)(b). Hence, it follows that α is determined from the fixed number N of particles, and can be thought of as a "potential for particle number". Similarly, β is determined by ensuring that the distribution describes an assembly with the correct energy U and can be interpreted as a "potential for energy".

We determine α by applying the condition in equation (3.73), which caused its introduction in the first place. Substituting equation (3.82) back into (3.73), we obtain

$$N = \sum_j n_j = \exp(\alpha) \exp(\alpha \varepsilon_j), \tag{3.84}$$

since $\exp(\alpha)$ ($= A$, say) is a factor in each term of the distribution. In other words, A is a normalization constant for the distribution, chosen such that the distribution describes the thermal properties of the correct number of particles N. Another way of writing equation (3.84) is: $A = N/Z$, with the "partition function" Z, defined by $Z = \sum_j \exp(\beta \varepsilon_j)$. We may then write the Boltzmann distribution of equation (3.83) as

$$n_j = A \exp(\beta \varepsilon_j) = (N/Z) \exp(\beta \varepsilon_j). \tag{3.85}$$

In contrast to α, the way in which β enters the Boltzmann distribution is more subtle. Nevertheless, the formal statements are easily made. We substitute the thermal distribution equation (3.85) back into the relevant condition in equation (3.73) in order to obtain

$$U = \sum_j n_j \varepsilon_j = (N/Z) \sum_j \varepsilon_j \exp(\beta \varepsilon_j). \tag{3.86}$$

Section 3.3 Equilibrium distribution

This can be written as

$$U/N = \frac{\sum_j \varepsilon_j \exp(\beta \varepsilon_j)}{\sum_j \exp(\beta \varepsilon_j)}. \tag{3.87}$$

The appropriate value of β then is the one which, when put into this equation, precisely produces the internal energy per particle (U/N), specified by the macrostate. Unfortunately, this is not a very tidy result, but it is as far as we can go explicitly, since in general, one cannot invert equation (3.87) to produce an explicit formula for β as a function of (U, V, N). Nevertheless, for a given (U, V, N) macrostate, β is fully specified by (3.87), and one can indeed describe it as a "potential for energy", in that the equation gives a very direct correlation between (U/N) and β. It turns out that this untidy (but absolutely specific) function β has a very clear physical meaning in terms of thermodynamic functions other than (U, V, N). In fact we shall see that it must be related to temperature only. This is a sufficiently important point to justify devoting the following section to it.

3.3.2 A statistical definition of temperature

To show that there is a necessary relation between β and the temperature T, we consider the thermodynamic and statistical treatment of two systems in thermal equilibrium. The thermodynamic treatment is obvious. Two systems in thermal equilibrium have, effectively by definition, the same temperature. This statement is based on the *zeroth law of thermodynamics*, which states that there is some common function of state shared by all systems in mutual thermal equilibrium – and this function of state is what is meant by (empirical) temperature.

The statistical treatment of T can follow directly, along the lines of Section 3.3. The problem can be set up as follows. Consider two systems A and B, which are in thermal contact with each other, but are together isolated from the rest of the universe. We suppose that system A consists of a fixed number N_A of localized particles, which each have states with energy ε_j, as before. The system B need not be of the same type, so we take it as containing N_B particles, whose energy states are ε'_k. The corresponding distributions are $\{n_i\}$ for system A and $\{n'_l\}$ for system B. The restrictions on the distributions are as follows:

$$\sum_j = N_A, \tag{3.88a}$$

$$\sum_k n'_k = N_B, \tag{3.88b}$$

$$U = \sum_j n_j \varepsilon_j + \sum_k n'_k \varepsilon'_k. \tag{3.88c}$$

where U is the total energy (i.e. $U_A + U_B$) of the two systems together. The counting of microstates is easy when we realize that we may write

$$\Omega = \Omega_A \times \Omega_B \quad \text{or} \quad t = t_A \times t_B. \tag{3.89}$$

Finding the most probable distribution may again be achieved by the Lagrange method. The problem is to maximize $\ln t$ with t as given in equation (3.89), subject now to the three conditions in equation (3.88). Using multipliers α_A, α_B and β, respectively for the three conditions, the result on differentiation, compare equation (3.83), is

$$\sum_j (-\ln n_j^* + \alpha_A + \beta \varepsilon_j) dn_j + \sum_k (-\ln n_k'^* + \alpha_B + \beta \varepsilon_k') dn_k = 0. \tag{3.90}$$

The Lagrange method then enables one to see that, for the appropriate values of the three multipliers, each term in the two sums is separately equal to zero, so that the final result for system A is

$$n_j^* = \exp(\alpha_A + \beta \varepsilon_j), \tag{3.91}$$

and for system B:

$$n_k'^* = \exp(\alpha_B + \beta \varepsilon_k'). \tag{3.92}$$

What this shows, is that both system A and system B have their thermal equilibrium distributions of the Boltzmann type, compare equation (3.83). The distributions have their own private values of α, and we can see from the derivation that this follows from the introduction of the separate conditions for particle conservation (3.88a) and (3.88b). However, the two distributions have the same value for β. In the derivation, this arises from the single energy condition (3.88c) – in other words, from the thermal contact or energy interchange between the systems. So, the important conclusion is that two systems in mutual thermal equilibrium and distributions share the same β. From thermodynamics we know that they necessarily have the same empiric temperature and thus the same thermodynamic temperature T. Therefore, it follows that β simply *is a function of T only*, and the relation (which we do not prove here) is:

$$\beta = -\frac{1}{k_B T}. \tag{3.93}$$

Finally, we want to consider how the internal energy U of a system can be changed. From a macroscopic viewpoint, this can be done by adding heat and/or work, i.e. change in U = heat input + work input. The laws of thermodynamics for a differential change in a simple p-V system tell us that

$$dU = TdS - pdV, \tag{3.94}$$

Section 3.3 Equilibrium distribution

where, for reversible processes only the first TdS term can be identified as the heat input, and the second term $-pdV$ as the work input. Now let us consider the microscopic picture. The internal energy is simply the sum of energies of all particles of the system, i.e. $U = \sum n_j \varepsilon_j$. Taking again a differential change in U, we obtain

$$dU = \sum_j \varepsilon_j dn_j + \sum_j n_j d\varepsilon_j, \qquad (3.95)$$

where the first term allows for changes in the occupation numbers n_j and the second term for changes in the energy levels ε_j. It is not hard to convince oneself that the respective first and second terms of (3.94) and (3.95) match. The energy levels only depend on V, so that the work input $-pdV$ can only address the second term of (3.95). And, bearing in mind the correlation between S and Ω (and hence t^*, and $\{n^*\}$), it is equally clear that occupation number changes are directly related to entropy changes. This explains the fact that the first terms match. These ideas turn out to be both interesting and useful. The relation $-pdV = \sum n_j d\varepsilon_j$ gives a direct and physical way of calculating the pressure from a microscopic model. And the other relation bears directly on the topic of this section. The argument can be outlined as follows: start by considering how a change in $\ln \Omega$ can be brought about

$$d(\ln \Omega) = d(\ln t^*) \qquad \text{for a large system} \qquad (3.96)$$

$$= -\sum_j \ln n_j^* dn_j \qquad \text{as in equation (3.79)}$$

$$= -\sum_j (\alpha + \beta \varepsilon_j) dn_j \qquad \text{using the Boltzmann distribution}$$

$$= -\beta \sum_j \varepsilon_j dn_j \qquad \text{as } N \text{ is fixed}$$

$$v = -\beta (dU)_{\text{no work}}, \qquad \text{first term of (3.95)} \qquad (3.97)$$

$$= -\beta (TdS) \qquad \text{first term of (3.94)}.$$

Hence, we see that $S = k_B \ln \Omega$ is consistent with $\beta = -1/k_B T$, because in both definitions the same constant (k_B, Boltzmann's constant) appears.

3.3.3 The Boltzmann distribution and the partition function

We have seen that the Boltzmann distribution is the appropriate one to describe the thermal equilibrium properties of an assembly of N identical localized (distinguishable), weakly interacting particles. We have derived it for an isolated assembly, having a fixed volume V and a fixed internal energy U. An important part in the result is played by the parameter β which is a function of the macrostate (U, V, N). However, the result of the previous section points out that the Boltzmann distribution is most easily written and understood in terms of (T, V, N) rather than (U, V, N). This is no inconvenience, since it frequently happens in practice that T is known, rather than U. It is also no embarrassment from a fundamental point of view, as long as we are dealing with a large enough system, so that fluctuations are unimportant. Therefore, although our method logically determines T (and other thermodynamic quantities) as a function of (U, V, N) for an isolated system, one usually uses the results to describe the behavior of U (and other thermodynamic quantities) as a function of (T, V, N). Therefore, we can write the Boltzmann distribution as

$$n_j = (N/Z) \exp(-\varepsilon_j/k_B T) \qquad (3.98)$$

with the partition function Z defined as

$$Z = \sum_j \exp(-\varepsilon_j/k_B T) \qquad (3.99)$$

The common symbol for the partition function Z is derived from the German word for sum over states[23], for that is all the partition function is, the sum over all one-particle states of the Boltzmann factors $\exp(-\varepsilon_j/k_B T)$. The English name "partition function" comes from the fact that Z is proportional to the corresponding term in the sum. In other words the N particles are partitioned into their possible states, i.e. they are labeled by j, in just the same ratios as Z is split up into the Boltzmann factor terms. This becomes obvious when one rewrites (3.98) as

$$n_j/N = \exp(-\varepsilon_j/k_B T)/Z \qquad (3.100)$$

or, equivalently

$$n_j/n_k = \exp\left[-\frac{(\varepsilon_j - \varepsilon_k)}{k_B T}\right]. \qquad (3.101)$$

Equations of the type $\exp(-\Delta \varepsilon/k_B T)$ turn up in all sorts of different physical situations and we will encounter it again in Chapter 6 when discussing the Monte Carlo method.

---

[23] Zustandssumme.

Section 3.3 Equilibrium distribution

Calculating thermodynamic functions. There are at least three different approaches to using the fundamental Boltzmann distribution derived in statistical mechanics to calculate thermodynamic functions.

1. A method that never fails, uses the Helmholtz free energy F, defined as $F = U - TS$. The reason for the importance of the method is twofold. First, the statistical calculation of F from the Boltzmann distribution turns out to be extremely simple. The calculation is as follows:

$$F = U - TS = \sum n_j \varepsilon_j - k_B T \ln t^* \quad (3.102)$$

$$= \sum n_j \varepsilon_j - k_B T (N \ln N - \sum n_j \ln n_j) \quad (3.103)$$

$$= -N k_B T \ln Z. \quad (3.104)$$

In equation (3.102) we have used the definition of F and U and S, equation (3.78) with $\sum n_j = N$, and equation (3.98). In the last step one takes the logarithm to obtain $\ln n_j = \ln N \ln Z - \varepsilon_j / k_B T$. Everything but the $\ln Z$ term cancels, giving the memorable and simple result in equation (3.104). The second reason for using this approach is that an expression for F in terms of (T, V, N) is of immediate use in thermodynamics, since (T, V, N) are the natural coordinates for F, since $dF = -SdT - pdV + \mu dN$, so that simple differentiation can produce S, p and the chemical potential μ.

2. Use $S = k_B \ln \Omega$. This method is often the shortest to use if *only the entropy is required*. The point is that one can substitute the Boltzmann distribution numbers from equation (3.102) back into (3.74), in order to produce t^* and hence Ω (equation (3.76)), and hence S. Thus S is obtained from a knowledge of the $\varepsilon_j s$ (which depend on V), of N and of T (as it enters the Boltzmann distribution).

3. Use the definition of Z. There is a direct shortcut from the partition function to U. This is particularly useful if only U and perhaps $(dU/dT = C_V$, the heat capacity at constant volume) are needed. In fact, U can be worked out from $U = \sum n_j \varepsilon_j$ at once, but this sum can be neatly performed by looking back at (3.87), which can be re-expressed as $(U/N) = (1/Z) dZ/d\beta = d(\ln Z)/d\beta$. Note that here it is usually convenient to retain β as the variable, rather than to use T.

3.4 The canonical ensemble

We have seen above that the fundamental basis of statistical mechanics is the postulate of *equal a priori probabilities* in the equilibrium state of a completely isolated system. In such a system, referred to as a *microcanonical ensemble*, the total energy E is fixed. Such an ensemble can, in principle, be used to determine both microscopic probabilities and thermodynamic information such as the energy function and the equations of state. In all but the simplest systems, however, the process is too difficult to carry out.

There is another approach, which arrives at the same information in a much simpler manner. It deals with a system which is in thermal equilibrium with a large bath. Since energy can flow to and from the bath, the system is described by the temperature of the bath T, rather than by a fixed energy E. Such a system, and the statistical method based on it, are referred to as a *canonical ensemble*.

Imagine a large, closed system represented by a microcanonical ensemble. Within it is a small subsystem, 1, which may be so small that it has only one microscopic degree of freedom, or so large that a set of thermodynamic variables can be defined for it. The rest of the large system is referred to as the remainder, 2. The subsystem is not closed (isolated) but can interact with the remainder. A phase space can be made up for 1 and another for 2. The spaces do not overlap and the phase space for the whole system is the tensor product of the two. Even though 1 and 2 will not be represented separately by microcanonical ensembles (their energies E_1 and E_2 are not fixed), one can define a phase space volume for each in the same way as one would when dealing with a microcanonical ensemble. Thus, one can find (or at least imagine) $\Omega_1(E_1,$ and other variables) and $\Omega_2(E_2,$ and other variables).

The microcanonical ensemble produces the probability density for the microscopic variables of a closed system (with E and N fixed). The canonical ensemble produces the probability density for the microscopic variables of a system in thermal equilibrium, with a fixed reservoir at temperature T. In the case of the situation that we have constructed, the system is 1, with N_1 fixed and E_1 freely variable, and the remainder 2 is the reservoir which is assumed to be so large that its temperature is insensitive to the state of 1. Let $\{p_1, q_1\}$ indicate the set of microscopic variables associated with 1. We wish to find the joint probability density $p(\{p_1, q_1\})$. For the entire system (microcanonical), one has

$$p(\text{system in state } X) = \frac{\text{volume of accessible phase space consistent with } X}{\Omega(E)} \quad (3.105)$$

In particular, for our case

$$p(\{p1, q1\}) = p(\text{subsystem at } p_1, q_1; \text{ remainder undetermined}) \quad (3.106)$$
$$= \Omega_1(\{p1, q1\})\Omega_2(E - E_1) = \Omega(E).$$

Section 3.4 The canonical ensemble

Note that assigning fixed values to all the microscopic variables in 1, $\{p_1, q_1\}$, means that it is constrained to a single state (a point) in its phase space. Therefore, $\Omega_1(\{p_1, q_1\}) = 1$.

$$k \ln p(\{p1, q1\}) = \underbrace{k \ln \Omega_1}_{=k \ln 1 = 0} + \underbrace{k \ln \Omega_2(E - E_1)}_{S_2(E - E_1)} - \underbrace{k \ln \Omega(E)}_{S(E)}. \quad (3.107)$$

We rewrite the second term on the right by expanding the function S_2 about the point where the energy of the remainder equals E.[24]

$$S_2(E - E_1) \approx S_2(E) - \left.\frac{\partial S_2(E_2)}{\partial E_2}\right|_{E_2 = E} E_1 \quad (3.108)$$

The partial derivative $(\partial S_2 / \partial E_2)$ is, by definition, $1/T_2$ when evaluated at the equilibrium value of E_2. But 2 is so large that the derivative changes little if evaluated at E instead of at $\langle E_2 \rangle$. T_2 is now simply referred to as T, the temperature of the "reservoir". The specific details of the microscopic state of the subsystem enter the expression through the energy E_1, which is the Hamiltonian of the subsystem evaluated at $\{p_1, q_1\}$: $E_1 = H_1(\{p_1, q_1\})$. Using these considerations in the above expression produces

$$k \ln p(\{p1, q1\}) = -\frac{H_1(\{p1, q1\})}{T} + S_2(E) - S(E). \quad (3.109)$$

The first term in the sum depends on the specific state of the subsystem and the second term depends on the reservoir and average properties of the subsystem. It may be that 1 is so small that thermodynamics does not apply to it. That is, the number of particles is too small for macroscopic variables, such as pressure and temperature, to be defined for the subsystem[25]. Under these circumstances, all that we can do is find the probability density of the microscopic variables of the subsystem 1:

$$p(\{p1, q1\}) \propto \exp\left[-\frac{H_1(\{p_1, q_1\})}{k_B T}\right] \quad (3.110)$$

$$= \frac{\exp\left[-\frac{H_1(\{p_1, q_1\})}{k_B T}\right]}{\int \exp\left[-\frac{H_1(\{p_1, q_1\})}{k_B T}\right] \{dp_1, dq_1\}} \quad (3.111)$$

If thermodynamics does apply to 1, one can proceed to get all of the thermodynamic information about the subsystem from the normalization constant, associated with the microscopic probability density. The additive nature of entropy requires

$$S(E) = S_1(\langle E_1 \rangle) + S_2(\langle E_2 \rangle). \quad (3.112)$$

---

[24] We are not physically putting all energy of the system into the remainder, but we are simply imagining evaluating the function S_2 at a point somewhat removed from its most probable value.

[25] Note, however, that the bath has a temperature and it is that temperature which enters into the first term on the right above.

$E_1 + E_2 = E$, due to the way we have divided the subsystem and the remainder within a microcanonical whole. It follows that $\langle E_1 \rangle + \langle E_2 \rangle = E$. We can now simplify the reservoir-dependent part of the expression for $\ln p$.

$$S_2(E) - S(E) = \underbrace{S_2(E) - S_2(\langle E_2 \rangle)}_{\approx (\partial S_2(E_2)/\partial E_2)\langle E_1 \rangle = \langle E_1 \rangle / T} - S_1(\langle E_1 \rangle), \qquad (3.113a)$$

$$k \ln p(\{p_1, q_1\}) = -\frac{H_1(\{p_1, q_1\})}{T} + \frac{\langle E_1 \rangle}{T} - S_1, \qquad (3.113b)$$

$$p(\{p_1, q_1\}) = \underbrace{\exp\left[\frac{(\langle E_1 \rangle - T S_1)}{k_B T}\right]}_{=\frac{1}{Z}} \exp\left[-\frac{H_1(\{p_1, q_1\})}{k_B T}\right]. \qquad (3.113c)$$

Note that $\langle E_1 \rangle$ is what is meant by the thermodynamic internal energy of the system. Also, $T = T_1$ since 1 and 2 are in thermal equilibrium. Thus

$$\langle E_1 \rangle - T S_1 = U_1 - T_1 S_1 = F_1, \qquad (3.114)$$

where F_1 is the *Helmholtz free energy*. From now on, only the subsystem needs to be considered and the subscripts are dropped.

$$p(\{p, q\}) = Z^{-1} \exp\left[-\frac{H(\{p, q\})}{k_B T}\right]. \qquad (3.115)$$

For the partition function we have

$$Z_N(T, V) = \int \exp\left[\frac{-H(\{p, q\})}{k_B T}\right] \{dp, dq\}$$
$$= \exp\left[\frac{(E - TS)}{k_B T}\right] = \exp\left[-\frac{F(T, V, N)}{k_B T}\right], \qquad (3.116)$$

and the thermodynamic relations for the free energy, entropy and pressure are given by

$$F(T, V, N) = -k_B T \ln Z_N(T, V), \qquad (3.117)$$

$$S(T, V, N) = -\left(\frac{\partial F}{\partial T}\right)_{V, N}, \qquad (3.118)$$

$$P(T, V, N) = -\left(\frac{\partial F}{\partial V}\right)_{T, N}. \qquad (3.119)$$

> In the case of the canonical ensemble, the connection between statistical mechanics and thermodynamics is given through the *partition function*.

3.5 Exercises

3.5.1 Trajectories of the one-dimensional harmonic oscillator in phase space

Derive the ellipsoid equation of the harmonic oscillator in equation (3.23).

3.5.2 Important integrals of statistical physics

a) Calculate the integral

$$I = \int_0^\infty e^{-x^2} dx$$

analytically and show that $I = \frac{1}{2}\sqrt{\pi}$.

Hint: Consider I^2 and then use a coordinate transformation to polar coordinates $x \to (r, \theta)$ to calculate the resulting trivial integral.

b) Show that

$$n! = \int_0^\infty e^{-x} x^n dx. \tag{3.120}$$

Hint: Integrate by parts and thereby obtain a recursive formula for the considered integral.

c) Show that the integral

$$\int_0^\infty e^{-\alpha x^2} x^n dx = \frac{\Gamma\left(\frac{n+1}{2}\right)}{2\alpha^{\frac{n+1}{2}}},$$

where $\Gamma(n) = \int_0^\infty e^{-x} x^{n-1} dx = (n-1)$ for $n \in \mathbb{N}$.

Hint: Transform the integral to the form $\int_0^\infty e^{-y^2} dy$ by making the substitution $x = \alpha^{-1/2} y$.

3.5.3 Probability, example from playing cards

Consider a standard deck of playing cards. There are 52 cards in total. These cards are divided among 4 suits (clubs, diamonds, hearts, spades). And within each suit there are 13 types of cards (ace, 2, 3,...,10, jack, queen, king). Suppose you are dealt a 5-card hand from a full, well-shuffled deck.

a) How many different hands are possible? (Here, the ordering of the cards within the hand is not important.)

b) What is the probability of being dealt a flush (all 5 cards have the same suit)?

c) What is the probability of being dealt one pair (two cards of the same type) and only one pair?

d) What is the probability of not getting one pair (and only one pair) until the nth hand? Here, each successive hand is dealt from a full, well shuffled deck.

3.5.4 Rolling dice

Assume that you have a collection of 100 six-sided dice. For a single die, the integers 1 through 6 are all equally likely to result from a roll. Suppose you roll all 100 dice at the same time. What is the approximate probability density for the sum s (the summation of the values on all of the dice)? Even though s is discrete, writing an expression for the continuous envelope function is acceptable. (Don't worry about reducing messy fractions to a simplified form. Just make sure that all of the parameters in your probability density are clearly defined.)

3.5.5 Problems, using the Poisson density

This monograph contains roughly 550 pages. Suppose each page contains roughly the same number of words. Suppose further, that there are also 550 typos randomly scattered throughout this book[26]. Estimate the probability that

a) The first page contains no typos.

b) The first page contains at least 3 typos.

Now, suppose 30 students in a hallway of a dormitory share a bathroom with 4 showers. They all enter the bathroom to shower between 8:00 AM and 9:00 AM, and they arrive at random during this time interval. Assume each student's shower lasts 6 minutes.

a) Estimate the probability that a student who enters the bathroom at 8:06 AM will have to wait (that is, all showers are occupied).

b) Recalculate this probability if an additional shower is built.

3.5.6 Particle inside a sphere

Suppose a particle has a uniform probability density for being located anywhere inside a sphere with unit radius. The joint probability density in the position variables x, y,

---

[26] I hope there are fewer!

and z is given by:

$$p(x,y,z) = \begin{cases} \frac{3}{4\pi} & \text{for } x^2 + y^2 + z^2 \leq 1, \\ 0 & \text{otherwise.} \end{cases} \quad (3.121)$$

Calculate $p(x)$, the probability density for the single random variable x. Make a plot of $p(x)$ with carefully labeled axes.

Chapter 4

Inter- and intramolecular potentials

Summary

In materials science, all macroscopic properties of condensed matter in the gaseous, fluid, or solid state are determined by the intermolecular forces that act between the constituents of matter. In classical MD simulations, matter is modeled as being composed of particles. These particles might be interpreted as atoms when performing simulations on the nanoscale, but sometimes the "particles" are also understood to be more general constituents, capturing the salient features of the system under consideration through the choice of appropriate interaction laws. Hence, in this chapter we provide a short overview about the nature of intermolecular interactions and introduce some of the most commonly used potentials in MD simulations. Of the four known fundamental interactions, the strong force, the weak force, the gravitational force and the electromagnetic force, *only* the latter is really relevant to the description of interactions between particles in materials science. That is, for all practical purposes, *all* particle interactions are of *electromagnetic* origin.

Learning targets

✓ Understanding the quantum mechanical origin of molecular interactions.

✓ Learning the most common potentials used in computer simulations.

✓ Understanding the features of the Lennard-Jones (LJ) potential.

✓ Understanding the properties of pair potentials.

4.1 Introduction

Major experimental sources of our knowledge of intermolecular forces are:

- scattering experiments (X-ray, neutron)
- spectroscopy (IR or Raman spectroscopy)
- thermophysical measurements (virial coefficients, specific heats),
- high-frequency spectroscopy using Nuclear Magnetic Resonance (NMR) with fluids and solids.

For the interpretation of experimental data, one uses simple models where theoretical parameters can be fitted to the experimental findings. Quantum mechanics provides a theoretical foundation for describing and understanding molecular interactions. In quantum mechanics, the Schrödinger equation takes the place of Newton's EOM. But the Schrödinger equation is so complex that it can only be solved analytically for a few simple cases. Also, direct numerical solution on computers is limited to very simple systems with very small numbers of particles, because of the high dimension of the (Hilbert) space in which the Schrödinger equation is posed. Therefore, approximation procedures are used to simplify the problem. The most common approximation is the Born-Oppenheimer approximation (BOA). Approximation procedures are based on the fact that the electron mass is much smaller than the mass of the nuclei.

The idea is to split the Schrödinger equation, which describes the state of both electrons and nuclei, using an approach that separates the problem into two coupled equations. In that case, the influence of the electrons on the interaction between the nuclei is described by an *effective potential*. This potential results from the solution of the electronic Schrödinger equation. As a further approximation, the nuclei are moved according to the classical EOM, using either effective potentials which result from quantum mechanical computations (which include the effects of the electrons) or *empirical potentials*, that have been fitted to the results of quantum mechanical computations or to the results of experiments. This approach is an example of a *hierarchy* of approximation procedures and of the use of *effective* quantities. In the following, the derivation of the MD method from the laws of quantum mechanics is outlined. For further details, see the large body of available literature, for example [296, 262, 270, 285].

Think!

Is the BOA always valid and useful?

> The BOA is certainly *not* always justified – it just is hard to think of any other approximation, so it is prevalent in ab initio methods of quantum chemistry and used almost always. What the BOA essentially assumes is that for *every* set of nuclear coordinates, you are *always* in the electronic ground state, i.e. that the electrons immediately relax to their ground state. This is certainly the case in a stationary configuration, after a long time. Thus, the BOA is *not* appropriate when the atoms move very fast and/or the electrons move very slow and there *are* situations like that, for example:
>
> - long-lived excited electronic states in semi-conductors such as $EL2$ in $GaAs$ [428]
>
> - Dynamics of protons, which are lightweight, so that the electrons cannot follow fast
>
> - diffusion of charged Li^+-ions in isolators
>
> - problems specified at non-zero temperature: here, there is a finite probability for the electrons to occupy higher energy (excited) states. This is not a big problem though, as it can be solved by statistically weighting how the different energy levels are populated.

4.2 The quantum mechanical origin of particle interactions

Until the end of the nineteenth century, classical physics could answer the most important questions, based on the laws of classical mechanics [261, 36], in particular Newton's Lex II. *Classical* here means that the quantum nature of the atoms and their electronic structure is completely neglected. The Lagrange formalism and the Hamilton formalism both lead to generalized classical equations of motion that are essentially equivalent. These equations describe how the change in time of the position of particles depends on the forces acting on them. If the initial positions and velocities are known, the positions of the particles are determined uniquely at all later points in time. Observable quantities, such as angular momentum or kinetic energy, can then be represented as functions of the positions and the momenta of the particles.

In the beginning of the twentieth century, the theory of quantum mechanics was developed. This theory describes the dynamics of particles with a new equation of motion, the Schrödinger equation. In contrast to Newton's equation its solution no longer provides unique trajectories, meaning uniquely determined positions and momenta of the particles, but only probabilistic statements about the positions and momenta. Furthermore, due to Heisenberg's uncertainty principle (equation (1.20)), position and momentum of a single particle can no longer be arbitrarily measured at exactly the same time, and certain observables, such as the energies of bound electrons, can only assume certain *discrete* values. All statements that can be made about a quantum mechanical system can be derived from the state function (or wave function) Ψ which is obtained as the solution of the Schrödinger equation. As an example, let us consider a system consisting of N nuclei and K electrons. The time-dependent state function of such a system can be written in the usual form [367] as

$$\Psi = \Psi(\vec{R}_1, \ldots, \vec{R}_N, \vec{r}_1, \ldots, \vec{r}_K, t), \tag{4.1}$$

where $\vec{R}_i$ and $\vec{r}_i$ denote positions in three-dimensional space $\mathbb{R}^3$ associated to the ith nucleus and the ith electron, respectively. The variable t denotes the time dependency of the state function. The vector space (space of configurations) in which the coordinates of the particles are specified is therefore of dimension $3(N + K)$. In the following, the usual abbreviations $\vec{R}$ and $\vec{r}$ for $(\vec{R}_1, \ldots, \vec{R}_N)$ and $(\vec{r}_1, \ldots, \vec{r}_K)$ are used. According to the statistical interpretation of the state function, the expression

$$|\Psi^*(\vec{R}, \vec{r}, t) \Psi(\vec{R}, \vec{r}, t)| \, dV_1 \cdots dV_{N+K} \tag{4.2}$$

describes the probability of finding the system at time t in the volume element $dV_1 \cdots dV_{N+K}$ of the configuration space, centered at the point $(\vec{R}, \vec{r})$. By integrating over a volume element of the configuration space one determines the probability of finding the system in this domain.

Nuclei and electrons are *charged* particles and interact with each other via the electrostatic (Coulomb) potential. The electrostatic Coulomb potential of a point charge immersed in a dielectric medium[1] with permittivity ϵ_r is given by:

$$\Phi(r) = k \cdot \frac{1}{\epsilon} \frac{e}{r}, \tag{4.3}$$

where r is the distance from the position of the charged particle, the constant $k = 1$ in the cgs-system of units, $k = 1/4\pi$ in the SI-system, e is the elementary charge[2]. The factor $\epsilon = \epsilon_0 \epsilon_r$ is the product of the dielectric constant of the vacuum and ϵ_r, i.e. it is the total dielectric constant of the medium, for example $\epsilon_{\text{air}} = 1$ for air, $\epsilon_{\text{prot}} = 4$ for proteins or $\epsilon_{\text{H}_2\text{O}} = 82$ for water. The electrostatic energy $\Phi_{\text{Coulomb}}(r)$ between N particles in a molecule with position vectors $\vec{r}_i$ and $\vec{r}_j$ is given by:

$$\Phi_{\text{Coulomb}}(r) = k \cdot \frac{1}{\epsilon} \sum_{i}^{N} \sum_{j>i}^{N} \frac{z_i z_j e^2}{|\vec{r}_i - \vec{r}_j|}. \tag{4.4}$$

---

[1] If the space between electric charges is filled with a non-conducting medium or an insulator (e.g. air) called "dielectric", it is found that the dielectric reduces the electrostatic force, compared to the vacuum, by a factor ϵ_r.

[2] See Table H.1 in Appendix H for the most recent experimental value of the elementary charge.

The z_i denote the charge of individual monomers in the molecule.

The electrostatic interaction gives rise to the dipole character of water, which is the basic requirement for the existence of life. Water is a dipole due to the higher electronegativity of oxygen, which gives rise to a partial negative charge at the oxygen atom and partial positive charges at the H-atoms in the H_2O-molecule. If the electronegativity of one atom is large enough, it can attract the whole electron from the bonding partner. This is the case with, for example, NaCl, where the initially electrically neutral Cl atom becomes a Cl^--ion and Na accordingly turns into Na^+. Chemical bonds that emerge from Coulomb attraction of ions are called *ionic bonds*. This type of chemical bond plays an important role in the formation of structures of biomolecules. For example, charged side groups may bind to receptors within the cell membrane, and protein structures are stabilized when a positively charged, protonated ammonium group (NH_4^+) and a negatively charged carboxyl group ($COOH^-$) form an ionic bonding.

Neglecting the electron spin and relativistic interactions and assuming that no external forces act on the system, the Hamilton operator $\mathcal{H}$ associated with a system of nuclei and electrons is obtained as the sum of the operators for the kinetic energy and the Coulomb potentials:

$$\mathcal{H}(\vec{R},\vec{r}) = -\frac{\hbar^2}{2m_e}\sum_{i=1}^{K}\Delta_{\vec{r}_K} + \frac{k}{\epsilon}e^2\sum_{i}^{K}\sum_{j>i}^{N}\frac{1}{|\vec{r}_i - \vec{r}_j|}$$
$$-\frac{k}{\epsilon}e^2\sum_{i=1}^{K}\sum_{j=1}^{N}\frac{z_j}{|\vec{r}_i - \vec{R}_j|} + \frac{k}{\epsilon}e^2\sum_{i=1}^{K}\sum_{j=1}^{N}\frac{z_i z_j}{|\vec{R}_i - \vec{R}_j|} \quad (4.5)$$
$$-\frac{\hbar^2}{2}\sum_{i=1}^{N}\frac{1}{M_i}\Delta_{\vec{R}_i}.$$

In equation (4.5), M_j and z_j denote the mass and the atomic number of the jth nucleus, m_e is the mass of an electron and $\hbar = h/2\pi$, with h being Planck's constant. $|\vec{r}_i - \vec{r}_j|$ are the distances between electrons, $|\vec{r}_i - \vec{R}_j|$ are distances between electrons and nuclei and $|\vec{R}_i - \vec{R}_j|$ are distances between nuclei. The operators $\Delta_{\vec{R}_j}$ and $\Delta_{\vec{r}_j}$ are Laplace operators with respect to the nuclear coordinates $\vec{R}_j$ and the electronic coordinates $\vec{r}_j$. We introduce the following abbreviations of the different summands of equation (4.5):

$$\mathcal{H} = T_e + \Phi_{ee} + \Phi_{eK} + \Phi_{KK} + T_K = \mathcal{H}_e + T_K, \quad (4.6)$$

where T_e and T_K are the operators of the kinetic energy of the electrons and of the nuclei, respectively, and Φ_{ee}, Φ_{KK} and Φ_{eK} refer to the potential energy operators of the interactions between the electrons, the nuclei, and between electrons and nuclei,

Section 4.2 The quantum mechanical origin of particle interactions

respectively. The state function Ψ is the solution of the Schrödinger equation

$$i\hbar \frac{\partial \Psi(\vec{R},\vec{r},t)}{\partial t} = \mathcal{H}\Psi(\vec{R},\vec{r},t), \qquad (4.7)$$

where i denotes the imaginary unit. The expression $\Delta_{\vec{R}_i}\Psi(\vec{R},\vec{r},t)$ that occurs in $\mathcal{H}\Psi$ is to be understood as the Laplace operator applied to the function Ψ at point $\vec{R}_i$. The other Laplace operator $\Delta_{\vec{r}_i}$ is to be understood analogously. In the following, we consider the case of the Hamilton operator $\mathcal{H}$ being not explicitly time dependent, as was already assumed in equation (4.6). Then, in order to solve equation (4.7), one uses a separation of variables approach of the form

$$\Psi(\vec{R},\vec{r},t) = \Psi(\vec{R},\vec{r}) \cdot f(t). \qquad (4.8)$$

$\Psi(\vec{R},\vec{r})$ is a function that does not depend on time and $f(t)$ is some general time-dependent function. Inserting this into equation (4.7) yields

$$i\hbar \frac{df(t)}{dt}\Psi(\vec{R},\vec{r}) = f(t)\mathcal{H}\Psi(\vec{R},\vec{r}), \qquad (4.9)$$

since $\mathcal{H}$ does not act on $f(t)$. Dividing both sides by the term $\Psi(\vec{R},\vec{r}) \cdot f(t) \neq 0$ yields

$$i\hbar \frac{1}{f(t)}\frac{df(t)}{dt} = \frac{1}{\Psi(\vec{R},\vec{r})}\mathcal{H}\Psi(\vec{R},\vec{r}). \qquad (4.10)$$

The left-hand side of equation (4.10) contains only the time coordinate t, the right-hand side only the spatial coordinates. Hence, both sides have to be equal, up to a common constant E and equation (4.10) can be separated. As a result one obtains the two equations

$$i\hbar \frac{1}{f(t)}\frac{df(t)}{dt} = E, \qquad (4.11)$$

and

$$\mathcal{H}\Psi(\vec{R},\vec{r}) = E\Psi(\vec{R},\vec{r}). \qquad (4.12)$$

Equation (4.11) is the *time-dependent Schrödinger equation* that yields the evolution of the wave function Ψ in time with the general solution

$$f(t) = ce^{-iEt/\hbar}. \qquad (4.13)$$

Equation (4.12) is an eigenvalue problem for the Hamilton operator $\mathcal{H}$, with energy eigenvalue E. This equation is called the time-independent, or *stationary Schrödinger equation*. To every energy eigenvalue E_n, as a solution of equation (4.12), there is one (or, in the case of degenerated states, several) associated energy eigenfunction Ψ_n and one time dependent term f_n. In this case, the solution of the time dependent Schrödinger equation is obtained as a linear combination of the energy eigenfunctions Ψ_n and the associated time dependent terms f_n of the form

$$\Psi(\vec{R},\vec{r},t) = \sum_n c_n e^{-iE_n t/\hbar} \Psi_n(\vec{R},\vec{r}), \qquad (4.14)$$

with the coefficients $c_n = \int \Psi_n^\star(\vec{R},\vec{r})\Psi(\vec{R},\vec{r},0)\,d^3R\,d^3r$. Similar to the time dependent Schrödinger equation, the stationary equation (4.12) is so complex that analytical solutions can only be obtained for a few very simple systems. The development of approximation procedures is therefore a fundamental area of research in computational quantum mechanics.

4.3 The energy hypersurface and classical approximations

Starting from equations (4.5), (4.11), and (4.12) and performing a series of approximations following [285, 416], which we will not repeat here, one can derive that in the *ground state*, i.e. in the lowest possible energy state of a system, the atomic nuclei are moving on a single hypersurface of potential energy, which is given by

$$\Phi(\vec{R}) = \int \phi_0^\star(\vec{R},\vec{r})\,\mathcal{H}(\vec{R},\vec{r})\,\phi_0(\vec{R},\vec{r})\,d^3r = E_0(\vec{R}). \qquad (4.15)$$

Here, $E_0(\vec{R})$ is the energy of the stationary electronic Schrödinger equation for the ground state, which only depends on the coordinates of the nuclei. The ground state of the electronic wave function is ϕ_0. The approximation in equation (4.15) is justified as long as the energy difference between the ground state ϕ_0 and the first excited state is large enough everywhere, compared to the thermal energy $k_B T$, so that transitions to excited states do not play a role. In order to solve equation (4.15) one has to solve the time independent Schrödinger equation for the electron

$$\mathcal{H}_e(\vec{R},\vec{r})\,\phi(\vec{R},\vec{r}) = E_0(\vec{R})\,\phi(\vec{R},\vec{r}). \qquad (4.16)$$

As a consequence of equation (4.15) the computation of the dynamics of the nuclei can now be separated from the computation of the hypersurface for the potential energy. If one assumes that the stationary electronic Schrödinger equation can be solved for a given nuclear configuration, then one could derive an entirely classical approach via the following steps:

1. Determine the energy $E_0(\vec{R})$ of the ground state from the stationary electronic Schrödinger equation (4.16) for as many representative nuclear configurations $\vec{R}_i$ as possible.

2. Evaluate the function $\Phi(\vec{R})$ of equation (4.15) at a number of points and obtain a number of data points $(\vec{R}_i)$.

3. Approximately reconstruct the global potential energy hypersurface for $\Phi(\vec{R})$ in analytic form, using an expansion of the many-body potentials (equation (4.17)).

4.4 Non-bonded interactions

Various physical properties are determined by different regions of the potential hypersurface of interacting particles. Thus, for a complete determination of potential curves, several different experiments are necessary. For an N–body system without internal degrees of freedom, the total energy Φ_{nb}, i.e. the potential hypersurface of the non-bonded interactions, can be written as

$$\Phi_{nb}(\vec{R}) \approx \sum_{i=1}^{N} \Phi_1(\vec{R}_i) + \sum_{i}^{N} \sum_{j>i}^{N} \Phi_2(\vec{R}_i, \vec{R}_j)$$
$$+ \sum_{i}^{N} \sum_{j>i}^{N} \sum_{k>j>i}^{N} \Phi_3(\vec{R}_i, \vec{R}_j, \vec{R}_k) + \cdots, \quad (4.17)$$

where $\Phi_1, \Phi_2, \Phi_3, \ldots$ are the interaction contributions due to external fields (e.g. the effect of container walls) and due to pair, triple and higher order interactions of particles. In classical MD, one often simplifies the potential by using the hypothesis that all interactions can be described by pairwise additive potentials, i.e. by restricting oneself to Φ_2. Despite this reduction of complexity, the efficiency of an MD algorithm that takes into account only pair interactions of particles is rather low (of order $O(N^2)$) and several optimization techniques are needed in order to improve the runtime behavior to $O(N)$.

The simplest general form for a *non-bonded* potential for spherically symmetric systems, i.e. $\Phi(\vec{R}) = \Phi(R)$ with $R = |R_i - R_j|$, is a potential of the following form:

$$\Phi_{nb}(R) = \Phi_{\text{Coulomb}}(R) + \left(\frac{C_1}{R}\right)^{12} + \left(\frac{C_2}{R}\right)^{6}. \quad (4.18)$$

Parameters C_1 and C_2 are parameters of the attractive and repulsive interaction and the electrostatic energy $\Phi_{\text{Coulomb}}(R)$ between the particles with position vectors $\vec{R}_i$ and $\vec{R}_j$ is given by equation (4.4).

The energy in equation (4.17) is appropriately truncated. With such an expansion the electronic degrees of freedom are, in essence, replaced with interaction potentials Φ_i and are therefore no longer explicit degrees of freedom of the equations of motion. After the pair potentials Φ_i are specified, the original quantum-mechanical problem is completely reduced to a classical problem, i.e. to Newton's equation of motion for the nuclei

$$M_i \ddot{\vec{R}}_i(t) = -\nabla_{\vec{R}_k} \Phi(\vec{R}(t)). \tag{4.19}$$

Here, the gradients can be computed analytically. This method of classical molecular dynamics is feasible for many-body systems, because the global potential energy is decomposed according to equation (4.17). Here, in practice, the same form of the potential is used for the same kind of particles. For instance, if only a two-body potential function

$$\Phi \approx \sum_{i}^{N} \sum_{j>i}^{N} \Phi_2(|\vec{R}_i, \vec{R}_j|) = \Phi_2(r) \tag{4.20}$$

of the distance $r = |\vec{R}_i, \vec{R}_j|$ is used, only *one* one-dimensional function Φ_2 has to be determined. This is certainly a drastic approximation, which has to be justified in many respects and that brings with it a number of problems. It is not obvious how many, and which typical nuclear configurations have to be considered to reconstruct the potential function from the potentials of these configurations with a not too large error. In addition, there is an error introduced by the truncation of the expansion in equation (4.17). The precise form of the analytic potential functions Φ_n and the subsequent fitting of their parameters also have a decisive influence on the size of the approximation error. The assumption that the global potential function is well represented by a sum of simple potentials of a few generic forms, and the transferability of a potential function to other nuclear configurations are further critical issues. Altogether, in this approach, not all approximation errors can be controlled rigorously. Furthermore, quantum mechanical effects and therefore chemical reactions are excluded by its construction. Nevertheless, the MD method based on parametrized potentials has been proved extremely successful in practice, in particular in the computation of the properties of fluids, gases and macromolecules.

Think!

What can potentials be fitted to?

The methods used in practice to determine the interactions in real systems are either based on the approximate solution of the stationary electronic Schrödinger equation (ab initio methods) and subsequent force matching [147], or on fitting (that is, parametrization) given analytic potentials to experimental or quantum mechanical results.

Potentials can be fitted to e.g.:

- lattice constants
- cohesive energy
- bulk modulus
- the equation of state
- other elastic constants
- phonon frequencies
- forces
- stable crystal structures
- surface energy and relaxation
- liquid pair correlation functions

In the first approach, the potential is constructed implicitly, using ab initio methods. There, the electronic energy E_0 and the corresponding forces are computed approximately for a number of chosen example configurations of the nuclei[3]. By extrapolation/interpolation to other configurations, an approximate potential energy hypersurface can be constructed which can in turn be approximated by simple analytic functions. In the second, more empirical approach, one directly chooses an analytic form of the potential that contains certain form functions that depend on geometric quantities such as distances, angles or particle coordinates. Subsequently, this form is fitted to available results from quantum mechanical computations or from actual experiments, by an appropriate determination of its parameters. In this way, one can model interactions that incorporate different kinds of bond forces, possible constraints, conditions on angles, etc. If the results of the simulation are not satisfactory, the potentials have to be improved by the choice of better parameters or by the selection of better forms of the potential functions with other or even extended sets of

---

[3] The wave function in the electronic Schrödinger equation is still defined in a high dimensional space. The coordinates of the electrons are in $\mathbb{R}^{3K}$. An analytic solution or an approximation via a conventional numerical discretization method is, in general, impossible. Therefore, approximation methods have to be used that substantially reduce the dimension of the problem. Over the years, many variants of such approximation methods have been proposed and used, such as the HF method [164, 102], the DFT [214, 255], configuration interaction methods, coupled-cluster methods [382], generalized valence bond techniques [183, 182, 309], the Tight Binding (TB) approach [167, 335], or the Harris functional method [201]. An overview of the different approaches can be found in [359, 360], for example.

parameters. The construction of good potentials is still an art form and requires much skill, work, and intuition. Quantum chemistry programs such as GULP [8] can help in the creation of new forms of potentials and in the fitting of parameters for solids and crystals.

4.5 Pair potentials

When solving the Schrödinger equation (4.7) for simple molecules, where the interaction only depends on the *distance r* of two particles (i.e. molecules without *internal degrees of freedom*, e.g. He_2), one finds that the intermolecular potential energy $\Phi(r)$ always has a specific functional form, as displayed in Figure 4.1. Such potentials are called *pair potentials*.

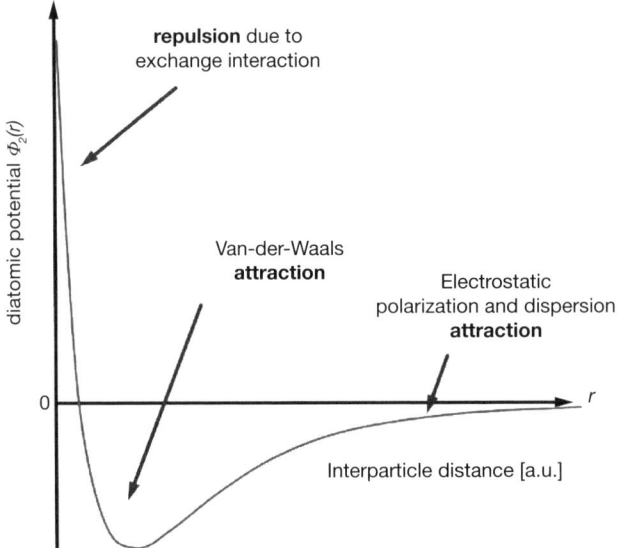

Figure 4.1. Classification of intermolecular interactions. In all intermolecular interactions, one can distinguish three distinct properties: for short distances r, the quantum mechanical exchange energy leads to a strong repulsion, due to the Pauli principle. At intermediate distances, one has electrostatic multipole-multipole interactions, interactions due to polarization, and due to dispersion. Finally, for large particle distances, the interaction is based on electrostatic dispersion effects.

In this figure, we have marked the three distinguishable characteristic regions in the potential:

- At *small distances* (with respect to the molecules' dimension), all potentials are repulsive, due to Pauli's principle. The electrostatic repulsion dominates over attraction.

- At *intermediate distances*, there are both attractive and repulsive interactions, due to electrostatic multipole-multipole interactions, interactions due to polarization[4], and dispersion interactions, which are due to the dynamic polarization of the electron charge distribution in both molecules.

- At *large distances*, there is no electron exchange anymore and the major interaction is an attractive force, due to dispersion.

When using *pair potentials* for modeling, one assumes that the total energy of a number of particles can simply be summed pairwise.

$$E_{\text{total}} = E_0 + \frac{1}{2} \sum_{i=1}^{N} \sum_{j \neq i}^{N} \Phi(|\vec{r}_i - \vec{r}_j|).$$

Pair potentials always have the same principal form shown in Figure 4.1. The analytic form of pair potentials is usually based on some basic physics. However, physical relevance tends to disappear when the potential constants are fitted.

Let's investigate more closely the precise form of the pair potential shown in Figure 4.1. As we have seen, the potential comprises a repulsive part (important for small r) and an attractive part (important for large r). It turns out that there are three major contributions to the attractive part, which we discuss in the next section.

4.5.1 Repulsive Interactions

When two molecular species approach each other so close that their electron clouds overlap, the positively charged nuclei become less well shielded by the negative electrons and so the two species repel each other. The repulsive term is sometimes written as

$$\langle \Phi_{\text{rep}}(r) \rangle = A \exp(-Br) \qquad (4.21)$$

where A and B are specific to the particular molecular pair and have to determined from experiment. The precise form of the repulsive term is not well understood, but results are known from the exchange integral when calculating the interaction between two He-molecules quantum-mechanically. All that is certain is that it must fall off very quickly with distance, and the exponential function is therefore a possibly suitable candidate. In computer simulations, some potential functions are very often

---

[4] Polarization occurs even in electrically neutral molecules, due to the deformation of the electron charge distribution of a molecule under the influence of a different molecule.

used instead. The total interaction is

$$\Phi = \Phi_{\text{rep}} + \Phi_{\text{dipole-dipole}} + \Phi_{\text{induction}} + \Phi_{\text{dispersion}} \quad (4.22)$$

which we can summarize as

$$\Phi = A\exp(-Br) - \frac{C}{r^6}, \quad (4.23)$$

since all attractive forces fall off as $1/r^6$. This is known as the *exp -6 potential*.

The minimum of the non-bonded interaction is attained when the attraction just cancels repulsion. This distance is called Van-der-Waals radius σ_0. The probably most commonly used form of the potential of two neutral atoms which are only bound with Van-der-Waals interactions, is the *Lennard-Jones (LJ), or (a-b) potential* which has the form:

$$\Phi_{a,b}(r) = \alpha\epsilon \left[\left(\frac{\sigma_0}{r}\right)^a + \left(\frac{\sigma_0}{r}\right)^b \right], \quad (4.24)$$

where

$$\alpha = \frac{1}{a-b}\left(\frac{a^a}{b^b}\right)^{\frac{1}{a-b}}, \quad \Phi_{\min} = \epsilon \quad \text{and} \quad \Phi(\sigma) = 0. \quad (4.25)$$

The most often used *LJ-(6-12) potential* for the interaction between two particles at a distance $r = |\vec{r}_i - \vec{r}_j|$ then reads (cf. equation (4.18)):

$$\Phi_{\text{LJ}}(r) = 4\epsilon\left[\left(\frac{\sigma_0}{r}\right)^{12} + \left(\frac{\sigma_0}{r}\right)^6\right]. \quad (4.26)$$

Parameter ϵ is the unit of the energy scale and σ_0 is the unit of the length scale. The $1/r^6$ term can be physically justified – it is caused by fluctuating dipoles. However, the repulsive term is taken as simply $\propto r^{-12}$ for numerical convenience (because it is $(r^{-6})^2$. In simulations one uses dimensionless *reduced units*, which tend to avoid numerical errors when processing very small numbers, arising from physical constants such as the Boltzmann constant $k_B = 1.38 \cdot 10^{-23}$ J/K. In these reduced (simulation) units, one MD time step is measured in units of $\hat{\tau} = (m\sigma^2/\epsilon)^{1/2}$, where m is the mass of a particle and ϵ and σ_0 are often simply set to $\sigma_0 = \epsilon = k_B T = 1$. Applied to real molecules, for example to Argon, with $m = 6.63 \times 10^{-23}$ kg, $\sigma_0 \approx 3.4 \times 10^{-10}$ m and $\epsilon/k_B \approx 120$ K one obtains a typical MD time step $\hat{\tau} \approx 3.1 \times 10^{-13}$ s.

The two LJ parameters σ and ϵ have been determined for a range of atoms. In the literature, the quantity ϵ/k_B (which has dimension of temperature), is usually recorded, rather than ϵ. Examples of atomic parameters are listed in Table 4.1. Over the years, people have extended these ideas to the interaction of simple molecules. Some caution is needed here, because the interaction between two molecules will generally depend on the precise details of their orientation, and the values given in Table 4.2 must be interpreted as some kind of geometrical average.

Section 4.5 Pair potentials

Table 4.1. Representative LJ atomic parameters.

atom	$\epsilon/k_B[K]$	$\sigma[pm]$
He	10.22	258
Ne	35.7	279
Ar	124	342
Xe	229	406

Table 4.2. Representative LJ parameters for simple molecules according to [211].

molecule	$\epsilon/k_B[K]$	$\sigma[pm]$
H_2	33.3	297
O_2	113	343
N_2	91.5	368
Cl_2	357	412
Br_2	520	427
CO_2	190	400
CH_4	137	382
CCl_4	327	588
C_2H_4	205	423
C_6H_6	440	527

Combination rules for LJ parameters

Over the years, a large number of LJ parameters have been deduced, but these relate to pairs of *like* atoms. Rather than try to deduce corresponding parameters for unlike pairs, it is common practice to use so called combination rules, which enable us to relate the (σ^{12}) and (σ^6) parameters for an unlike atom pair A-B to those of A-A and B-B. The use of such combination rules is common in subjects like chemical engineering, and is widely applied to many physical properties. Of the many combination rules that can be found in the literature [286, 186, 132, 408] we here mention three of the most simple ones. The first is

$$\sigma_{ij}^{12} = \left(\frac{r_{0_i}}{2}\frac{r_{0_j}}{2}\right)^{12}\sqrt{\epsilon_i \epsilon_j}, \qquad (4.27a)$$

$$\sigma_{ij}^6 = 2\left(\frac{r_{0_i}}{2}\frac{r_{0_j}}{2}\right)^6\sqrt{\epsilon_i \epsilon_j}, \qquad (4.27b)$$

where r_{0_i} is the minimum energy separation for two atoms of type i, and ϵ_i is the potential well depth. The second is

$$\sigma_{ij}^{12} = 4(\sigma_i \sigma_j)^6 \sqrt{\epsilon_i \epsilon_j}, \tag{4.28a}$$

$$\sigma_{ij}^6 = 4(\sigma_i \sigma_j)^3 \sqrt{\epsilon_i \epsilon_j}. \tag{4.28b}$$

Finally, we mention the rule

$$\sigma_{ij}^{12} = \frac{1}{2}\sigma_{ij}^6 (\sigma_{0_i} + \sigma_{0_j})^6, \tag{4.29a}$$

$$\sigma_{ij}^6 = \frac{\alpha_i \alpha_j}{\sqrt{\frac{\alpha_i}{N_i}} + \sqrt{\frac{\alpha_j}{N_j}}}, \tag{4.29b}$$

where α_i is the dipole polarizability of atom i, N_i the number of valence electrons, and σ_0 the Van der Waals radius.

Comparison with experiment

Generally speaking, one has to determine the parameters in any pair potential by appealing to experiment. There are two kinds of experiments to consider. First there are those that are essentially in the gas phase, where pairs of atoms genuinely interact with each other undisturbed by other species. This means that the total mutual potential energy is given by the sum of the interacting pairs. Second, there are experiments that essentially relate to condensed phases, where the interacting particles are sufficiently close to raise doubts as to the credibility of the pairwise additive assumption.

In gas theory, the deviation of gases from perfect behavior can be expressed in the form of a virial equation of state

$$\frac{pV}{nRT} = 1 + \frac{nB(T)}{V} + \frac{n^2 C(T)}{V^2} + \cdots \tag{4.30}$$

where the virial coefficients $B(T), C(T), \ldots$ depend on the temperature and on the characteristics of the species being studied. Here, n is the amount of substance, p the pressure, V the volume, R the gas constant and T the thermodynamic temperature. $B(T)$ is called the *second virial coefficient*, $C(T)$ is called the *third virial coefficient*, and so on. They have to be determined experimentally by fitting the pVT data of the gas being studied. The virial equation of state has special significance in that the virial coefficients can be directly related to molecular properties. $B(T)$ depends on the pair potential $\Phi(r)$ in the following way:

$$B(R) = 2\pi \int_0^\infty \left(1 - \exp\left(-\frac{\Phi(r)}{k_B T}\right)\right) r^2 dr. \tag{4.31}$$

In a recent publication "What toys can tell us", Philip Ball [43] describes toy experiments using real beads connected with strings, to represent a model of polymer

Section 4.5 Pair potentials 279

chains. This is in essence a reflection of the way chemists have thought about atoms, molecules and liquids ever since John Dalton used wooden balls to represent them, around 1810. Chemists still routinely use plastic models in order to see how molecules fit together.

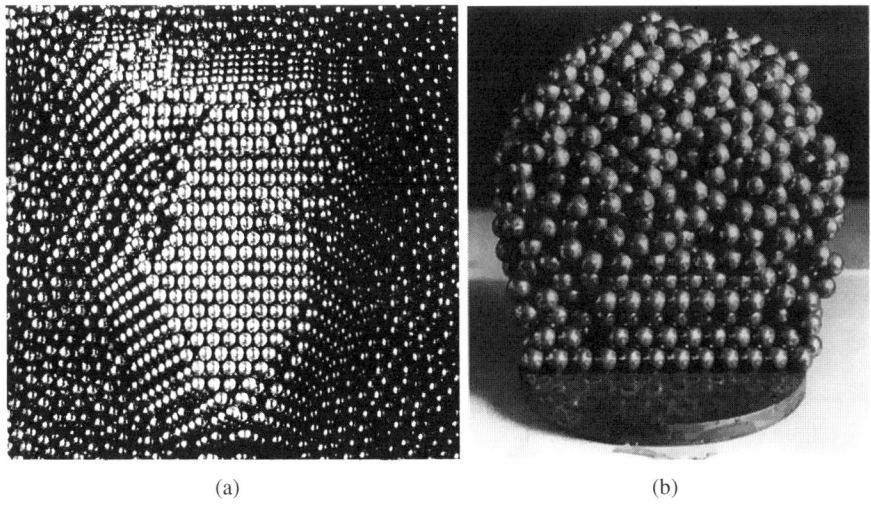

(a) (b)

Figure 4.2. Bernal's ball bearing models of liquids and solids [58]. (a) Face-centered cubic "crystal" surrounded by "liquid" caused by shearing ball bearing mass. (b) Ball bearing assembly showing a transition from random close packing to a regular crystalline array, induced by inserting a flat plate.

In Figure 4.2 we display a well-known study of Bernal on the disorderly packing of atoms in liquids, using ball-bearings, which he presented in the Bakerian Lecture [58] on "The structure of liquids", in 1962. However, ball bearings of course have negligible thermal motion. Real suspensions of mesoscopic colloids actually exhibit Brownian motion and are thermodynamically equivalent to atoms and small molecules [437]. In 1986, Pusey and van Megen [344] showed that a suspension of sterically stabilized PMMA particles showed hard-sphere-like equilibrium phase behavior. Their work led to many experimental studies on the statistical physics of hard spheres, using colloids as models. Since Pusey and van Megen's work, the equation of state of hard sphere colloids has been determined [329], crystal nucleation has been observed [11, 368], and the glass transition has been studied [306]. The body of experimental research that we just reviewed relies on light scattering as structural and dynamical probe.

The advent of single particle tracking in real space with confocal microscopy [420] opened a new dimension in experiments on hard-sphere-like systems, yielding an unprecedented level of detailed information [339]. Confocal microscopy of hard-sphere-like suspensions is thus ideal for studying generic processes where local events are important, such as crystal nucleation [175], melting [26], and dynamic heterogeneity [237]. Another type of experiment is a molecular beam experiment, where a beam of mono-energetic molecules is produced and allowed to collide, either with other molecules in a scattering chamber, or with a similar beam traveling perpendicular to the original beam. Measurements of the amount by which the incident beam is reduced in intensity, or the number of molecules scattered in a particular direction, allow the determination of the parameters in the LJ pair potential.

4.5.2 Electric multipoles and multipole expansion

For many purposes, it is convenient to describe a charge distribution in terms of electric moments $\vec{p}_{\text{el}} = |Q| \cdot \vec{r}$ which, in the case of two point charges $+Q$ and $-Q$, is defined as the vector pointing from the negative charge to the positive charge, multiplied by the magnitude of the charge Q, see Figure 4.3.

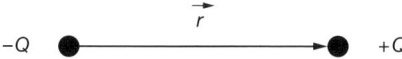

Figure 4.3. Definition of an electric dipole $\vec{p}_{\text{el}}$.

The generalization of the dipole to N point charges is simply:

$$\vec{p}_{\text{el}} = \sum_{i=1}^{N} Q_i \vec{R}_i, \tag{4.32}$$

where the Q_i are to be understood as the absolute positive values of the charges.

Suppose that we have two molecules with centers a distance r apart, see Figure 4.4. The distance r is taken to be large, compared to the molecular dimension. Each molecule consists of a number of charged particles, and in principle we can write down an expression for the mutual potential energy of these two molecules, in terms of the pair potentials between the various point charges. The basic physical idea of the *multipole expansion* is to make use of the fact that several of these particles constitute molecule A and the remainder constitute molecule B, each of which has a distinct chemical identity. We therefore seek to write the mutual potential energy of A and B in terms of properties of the two molecular charge distributions and their separation.

4.5.3 Charge-dipole interaction

Let's consider the mutual potential energy of a simple, small electric dipole and a point charge, see Figure 4.5. Suppose we have a simple dipole consisting of a pair of charges Q_A and Q_B aligned along the horizontal axis and equally separated from the coordinate origin by distance d. We introduce a third charge Q as shown, with the scalar distance R from the origin. This point charge makes an angle θ with the

Section 4.5 Pair potentials 281

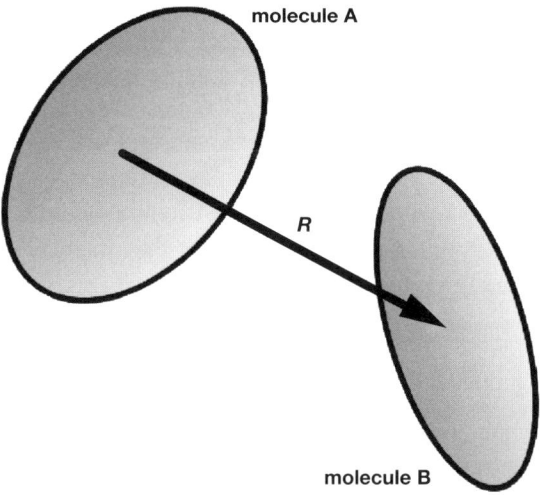

Figure 4.4. Sketch of two interacting molecules.

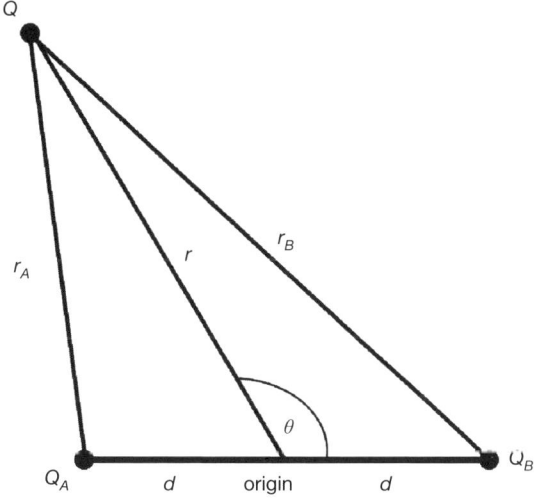

Figure 4.5. Charge-dipole interaction.

electric dipole, as shown. The two point charges Q_A and Q_B have a mutual potential energy of

$$\Phi = \frac{k}{\epsilon} \frac{Q_A Q_B}{2d}. \tag{4.33}$$

> **?** What happens to the mutual potential energy of the system in Figure 4.5 when we change the position vector of Q?

The mutual potential energy Φ(charge-dipole) of the point charge and the electric dipole is given by

$$\Phi_{\text{charge-dipole}} = \frac{k}{\epsilon} Q \left(\frac{Q_A}{r_A} + \frac{Q_B}{r_B} \right), \tag{4.34}$$

This can also be written in terms of R and θ as

$$\Phi_{\text{charge-dipole}} = \frac{k}{\epsilon} Q \left(\frac{Q_A}{\sqrt{r^2 + d^2 + 2dr\cos\theta}} + \frac{Q_B}{\sqrt{r^2 + d^2 - 2dr\cos\theta}} \right) \tag{4.35}$$

and, once again, this is an exact expression. In the case where the point charge Q gets progressively farther away from the coordinate origin, we can usefully expand the two denominators, using the binomial theorem to obtain

$$\Phi_{\text{charge-dipole}} = \frac{k}{\epsilon} Q \left(\frac{Q_A + Q_B}{r} + \frac{(Q_B - Q_A)d}{r^2} \cos\theta \right. \tag{4.36}$$

$$\left. + \frac{(Q_A + Q_B)d^2}{2r^3}(3\cos^2\theta - 1) + \cdots \right). \tag{4.37}$$

The first term on the right-hand side contains the sum of the two charges making up the dipole. Very often, one deals with simple dipoles that carry no overall charge, and this term is zero because $Q_A = -Q_B$. The second term on the right-hand side obviously involves the electric dipole moment, whose magnitude is $(Q_B - Q_A)d$. The third term involves the electric second moment, whose magnitude is $(Q_B + Q_A)d^2$, and so on. Hence, the mutual potential energy is a sum of terms, each a product of a moment of the electric charge, and a function of the inverse distance. As R increases, the magnitude of the successive terms will become less and eventually the mutual potential energy will be dominated by the first few terms of the expansion. In the more general case where one replaces Q_A and Q_B with an arbitrary array of point charges $Q_1, Q_2, \ldots, Q_n$ with position vectors $\vec{r}_1, \vec{r}_2, \ldots, \vec{r}_n$, it turns out [222] that we

Section 4.5 Pair potentials

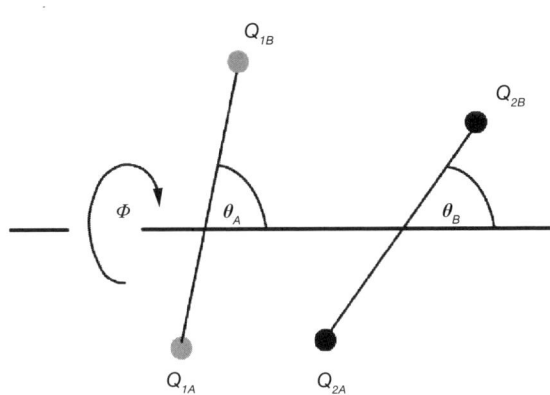

Figure 4.6. Multipole expansion of a dipole-dipole interaction.

can always write the mutual interaction potential with Q as

$$\Phi = \frac{k}{\epsilon}\left(\sum_{i=1}^{N} Q_i\right)\frac{1}{r} - \left(\sum_{i=1}^{N} Q_i \vec{r}\right)\nabla\left(\frac{1}{r}\right) + \text{higher order terms}. \qquad (4.38)$$

The first summation on the right-hand side specifies the overall charge of the charge distribution, the second term involves the electric dipole moment, the third term involves the electric quadrupole moment, and so on.

4.5.4 Dipole-dipole interaction

Next, we consider a slightly more realistic model for the interaction of two simple (diatomic) molecules, see Figure 4.6. Molecule A consists of two point charges, Q_{1A} and Q_{2A}. Molecule B consists of two point charges Q_{1B} and Q_{2B}. The overall charge of molecule A is therefore $Q_A = Q_{1A} + Q_{2A}$ with an analogous expression for molecule B. The electric dipole moments of A and B are written $\vec{p}_A = Q_A \vec{r}_A$, $\vec{p}_B = Q_B \vec{r}_B$ in an obvious notation, and their scalar magnitudes are written p_A and p_B. Because they are linear, the second moments of the two molecules are each determined by a scalar value q_A and q_B (and not by a tensor). Molecule A is centered at the origin, while molecule B has its center a distance r away, along the horizontal axis. The inclinations to the axis are θ_A and θ_B, and ϕ specifies the relative orientation of the two molecules. The distance between the two molecules is much less than their separation, so we can make convenient approximations. After some standard analysis

Table 4.3. Selected dipole-dipole interaction terms for two diatomic molecules.

θ_A	θ_B	relative orientations	$\Phi_{\text{dipole-dipole}}$
0	0	parallel	$-\frac{k}{\epsilon}\frac{2p_A p_B}{r^3}$
0	0	antiparallel	$+\frac{k}{\epsilon}\frac{2p_A p_B}{r^3}$
0	$\pi/2$	perpendicular	0

we find that the mutual potential energy of A and B is

$$\frac{\epsilon}{k}\Phi_{AB}(r) = \frac{Q_A Q_B}{r} + \frac{1}{r^2}(Q_B p_A \cos\theta_A - Q_A p_B \cos\theta_B)$$
$$- \frac{p_A p_B}{r^3}(2\cos\theta_A \cos\theta_B - \sin\theta_A \sin\theta_B \cos\phi)$$
$$+ \frac{1}{2r^3}(Q_A q_B(3\cos^2\Theta_B - 1) + Q_B q_A(3\cos^2\theta_A - 1)) + \cdots .$$
(4.39)

The physical interpretation is as follows: The first term on the right-hand side of equation (4.39) states the mutual potential energy of the two charged molecules A and B. The second term provides a contribution due to each charged molecule with the other dipole. The third term in equation (4.39) is a dipole-dipole contribution. Higher order terms are not written down explicitly in equation (4.39). If A and B correspond to uncharged molecules, then the leading term is seen to be the dipole-dipole interaction:

$$\frac{\epsilon}{k}\Phi_{\text{dipole-dipole}}(r)\frac{p_A p_B}{r^3}(2\cos\theta_A \cos\theta_B - \sin\theta_A \sin\theta_B \cos\phi). \quad (4.40)$$

The sign and magnitude of this term depend critically on the relative orientation of the two molecules. Table 4.3 shows three possible examples, all of which have $\phi = 0$.

4.5.5 Dipole-dipole interaction and temperature

We now imagine that the two molecules are subject to thermal motion. We keep their separation r constant but allow the angles to vary. The aim is to calculate the average dipole-dipole interaction. Some orientations of the two dipoles will be more energetically favorable than others, and we allow for this by including a Boltzmann factor $\exp(-U/k_B T)$, where k_B is the Boltzmann constant and T the thermodynamic temperature. The mean value of the dipole-dipole interaction is formally given by

$$\langle\Phi_{AB}\rangle_{\text{dipole-dipole}} = \frac{\int \Phi_{AB}\exp\left(-\frac{\Phi_{AB}}{k_B T}\right)d\tau}{\exp\left(-\frac{\Phi_{AB}}{k_B T}\right)}. \quad (4.41)$$

Section 4.5 Pair potentials

The integral has to be taken over all possible values of the angles, keeping r fixed. After some standard integration, we find

$$\langle \Phi_{AB} \rangle_{\text{dipole-dipole}} = -\left(\frac{k}{\epsilon}\right)^2 \frac{2p_A^2 p_B^2}{3k_B T} \frac{1}{r^6}. \qquad (4.42)$$

The overall value is therefore *negative*, and the term is inversely dependent on the temperature. It also falls off as $1/r^6$.

4.5.6 Induction energy

The logical next step is to consider the case of two interacting molecules, one of which has a permanent dipole moment and one of which is polarizable but does not have a permanent electric dipole moment. Figure 4.7 shows molecule A, with a permanent

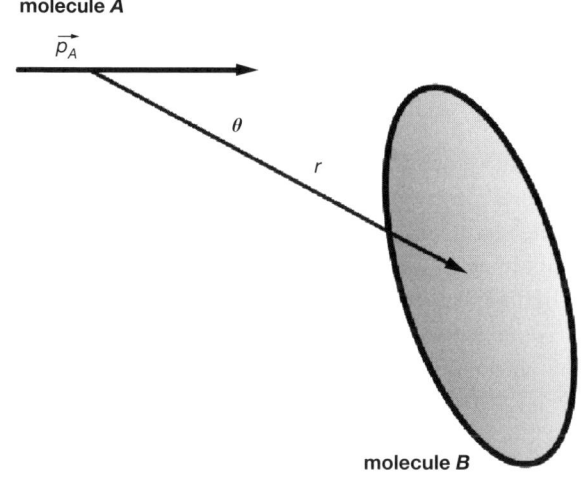

Figure 4.7. Dipole-induced dipole.

dipole moment p_A. The direction of p_A is indicated, pointing from the center of mass of molecule A to the center of mass of molecule B. The dipole p_A is at distance r from point B, and is oriented with an angle ϕ as shown. The molecules are sufficiently far apart for the precise location of the center of mass inside the second molecule to be irrelevant. What happens here physically, is that the electric dipole p_A induces a dipole in molecule B, since B is polarizable. We evaluate the potential energy involved, and finally average over all possible geometrical arrangements, for a fixed value of the intermolecular separation. The steps involved are as follows. The electrostatic potential due to the small dipole p_A is

$$\Phi_A(\vec{r}) = \frac{k}{\epsilon} \frac{\vec{p}\,\vec{r}}{R^3}. \qquad (4.43)$$

Differentiation of equation (4.43) yields the electric field

$$\vec{E}_A(\vec{r}) = -\nabla \Phi(\vec{r}) = -\frac{k}{\epsilon}\left\{\frac{\vec{p}_A}{r^3} - 3\frac{\vec{p}_A\vec{r}}{r^5}\vec{r}\right\}. \tag{4.44}$$

Thus, molecule A generates an electrostatic field in the region of molecule B, according to equation (4.44). The modulus of this vector at a point p in molecule B is

$$E_A = \frac{k}{\epsilon}\frac{p_A}{r^3}\sqrt{1 + 3\cos^2\theta}. \tag{4.45}$$

This electrostatic field induces a dipole in molecule B. For the sake of the argument, It is assumed that the induced dipole points in the direction of the applied field (and so we need not worry about the fact that the polarizability is a tensor property). A calculation of the resulting mutual potential energy Φ_{AB} provides

$$\Phi_{AB} = -\int_0^E \vec{p}_A \cdot d\vec{E}'_A = -\frac{1}{2}\alpha\frac{k}{\epsilon}|\vec{E}|^2 = -\frac{k^2\alpha_B p_A^2}{\epsilon}\frac{1}{r^6}\frac{1}{2}(3\cos^2\theta + 1). \tag{4.46}$$

Polarizabilities are positive quantities. Hence, $\Phi_A B$ is negative for all values of θ at a given intermolecular separation. This is quite different from the dipole-dipole interaction, where some alignments of the dipoles actually make a positive contribution to the mutual potential energy, and some a negative one. Finally, we have to average over all possible alignments, keeping the internuclear separation fixed. This averaging has again to be done using the Boltzmann weightings, and we eventually find an expression for the contribution of the induction to the mutual potential energy of A and B:

$$\langle \Phi_{AB} \rangle_{\text{induction}} = -\frac{k^2}{\epsilon}\frac{p_A^2 \alpha_B}{r^6}. \tag{4.47}$$

Note, that the interaction falls off as $1/r^6$ just as for the dipole-dipole interaction but this time, there is no temperature dependence. For two identical A molecules, each with permanent electric dipole p_A and polarizability α_A, the expression becomes

$$\langle \Phi_{AA} \rangle_{\text{induction}} = -2\frac{k^2}{\epsilon}\frac{p_A^2 \alpha_B}{r^6}. \tag{4.48}$$

This contribution adds to the expression for the dipole-dipole interaction in equation (4.42).

The ability of water molecules to polarize, gives rise to another, directed interaction called *hydrogen bonding*. Hydrogen bonds are not only found in fluid and solid water but also in complex biopolymers and macromolecules, for example in proteins, where hydrogen bonds are responsible for the genesis of tertiary structures, such as α-helices or β-sheets. Despite the directed nature of the hydrogen bond, one often assumes a spherically symmetric, analytic form of the type $(A \cdot r^{-12} - B \cdot r^{-6})$, but a more precise form, which takes into account the nonlinearity of the hydrogen bond by angle θ between N-H-O, has also been proposed [117]:

$$\Phi_{hb} = \sum_{ij} \cos\theta(-A \cdot r_{ij}^{-6} + B \cdot r_{ij}^{-12}) + (1 - \cos\theta)(-C \cdot r_{ij}^{-6} + D \cdot r_{ij}^{-12}).$$
(4.49)

Here, parameters A, B, C, D are constants that depend on the atom pairs under consideration.

4.5.7 Dispersion energy

It is an experimental fact that inert gases can be liquefied. Atoms do not have permanent electric moments, so the dipole-dipole and induction contributions to the mutual potential energy of an array of inert gas atoms must both be zero. There is a third interaction mechanism, referred to as *dispersion*, which is a purely quantum mechanical effect that has no classical analog. Dispersion was first identified by Fritz W. London in 1930 [146].

The two contributions to the mutual potential energy discussed in the previous two sections, can be described by classical electromagnetism. There is no need to invoke the concepts of quantum mechanics. Dispersion interactions can only be correctly described using quantum mechanics. Nevertheless, the following qualitative discussion can be found in all elementary texts. The electrons in an atom or molecule are in continuous motion, even in the ground state. So, although, on average, the dipole moment of a spherically symmetrical system is zero, at any instant a temporary dipole moment can arise. This temporary dipole can induce a further temporary dipole in a neighboring atom or molecule and, as in the case of the inductive interaction, the net effect will be attraction.

Paul K. L. Drude gave a simple quantum mechanical description of the effect, and his theory suggests that the dispersion contribution can be written as

$$\langle \Phi \rangle_{\text{dispersion}} = -\left(\frac{A}{r^6} + \frac{B}{r^8} + \frac{C}{r^{10}} + \cdots\right).$$
(4.50)

The first term is to be identified with the instantaneous dipole-induced dipole mechanism. The higher terms A and B are caused by instantaneous quadrupole-induced quadrupoles, etc. According to Drude's theory, A is given by

$$A = -\left(\frac{k}{\epsilon}\right)^2 \frac{3\alpha^2 \varepsilon_1}{4}.$$
(4.51)

In this expression, ε_1 is the first excitation energy of the atomic or molecular species concerned. The dispersion energy is again seen to be attractive and to fall off as $1/r^6$.

4.5.8 Further remarks on pair potentials

Usually, one wants to mimic at least three material parameters: the energy scale, the length scale, and the elastic properties as the response of a system to perturbations. In order to do this, more parameters than the two LJ parameters are needed. One of the simplest such potentials is the *Morse potential*

$$\Phi(r) = D \exp\left[-2\alpha(r - r_0)\right] - 2D \exp\left[-\alpha(r - r_0)\right], \tag{4.52}$$

that has an additional third parameter α to model the elasticity scale (the curvature of the potential)[5]. Using an exponential function instead of the repulsive r^{-12} term as shown in equation (4.23) and adding an additional attractive r^{-8} from second order Van der Waals perturbation theory, one obtains the *Buckingham potential* [80]:

$$\Phi(r) = b \exp(-ar) - \frac{c}{r^6} - \frac{d}{r^8}. \tag{4.53}$$

This potential is often used in materials science for oxides and has the disadvantage of using a numerically very expensive exponential function, and it is also known to be unrealistic for many substances at small distances r and has to be modified accordingly.

For reasons of efficiency, a classical MD potential should be short-range in order to keep the number of calculations of the force between interacting particles at a minimum. Therefore, instead of using the original form of the potential in equation (4.26), which approaches 0 at infinity, it is common to use a modified form, where the potential is simply cut off at its minimum value $r = r_{\min} = \sqrt[6]{2}$ and shifted to positive values by ϵ, such that it is purely repulsive and smooth at $r = r_{\text{cut}} = \sqrt[6]{2}$:

$$\Phi_{\text{LJ}}^{\text{cut}}(r) = \begin{cases} 4\epsilon \left\{ \left(\frac{\sigma_0}{r}\right)^{12} - \left(\frac{\sigma_0}{r}\right)^6 \right\} + \epsilon & r \leq 2^{1/6}\sigma_0, \\ 0 & \text{otherwise.} \end{cases} \tag{4.54}$$

Another extension of the potential in equation (4.54) is proposed in [396], where a smooth, attractive part is again introduced, in order to allow for including different solvent qualities of the solvent surrounding the polymer:

$$\Phi_{\cos}(r) = \left[\frac{1}{2} \cdot \cos(\alpha r^2 + \beta) + \gamma\right]\epsilon. \tag{4.55}$$

This additional term adds an attractive part to the potential of equation (4.54) and at the same time – by appropriately choosing the parameters α, β and γ – keeps the potential cutoff at r_{cut} smooth. The parameters α, β and γ are determined analytically such that the potential tail of $\Phi_{\cos}$ has a zero derivative at $r = 2^{1/6}$ and at $r = r_{\text{cut}}$, whilst it is zero at $r = r_{\text{cut}}$, and has value γ at $r = 2^{1/6}$, with γ the depth of the

---

[5] For example, the bulk modulus is $B = V \frac{\partial^2 E}{\partial V^2}$, i.e. it is the curvature in the energy minimum.

Section 4.5 Pair potentials

attractive part. Further details can be found in [396]. When setting $r_{\text{cut}} = 1.5$ one sets $\gamma = -1$ and obtains α and β as solutions of the linear set of equations

$$2^{1/3}\alpha + \beta = \pi, \quad (4.56)$$
$$2.25\,\alpha + \beta = 2\pi. \quad (4.57)$$

The total unbounded potential can then be written as:

$$\Phi_{\text{Total}}(r,\lambda) = \begin{cases} \Phi_{\text{LJ}}^{\text{cut}}(r) - \lambda\epsilon & 0 < r < 2^{1/6}\sigma_0, \\ \lambda\Phi_{\cos}(r) & 2^{1/6}\sigma_0 \leq r < r_{\text{cut}}, \\ \infty & \text{otherwise}, \end{cases} \quad (4.58)$$

where λ is a new parameter of the potential, which determines the depth of the attractive part. Instead of varying the solvent quality in the simulation by changing the temperature T directly (and having to equilibrate the particle velocities accordingly), one can achieve a phase transition in polymer behavior by changing λ accordingly, cf. Figure 4.8. Using coarse-grained models in the context of lipids and proteins,

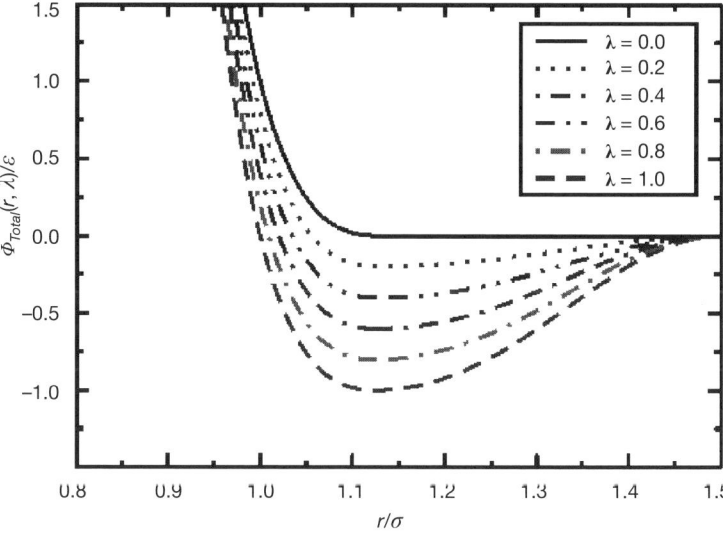

Figure 4.8. Non-bonded potential of equation (4.58) which allows for modeling the effects of different solvent qualities.

where each amino acid of the protein is represented by two coarse-grained beads, it has become possible to simulate lipoprotein assemblies and protein-lipid complexes for several microseconds [380]. The assumption of a short ranged interaction is usually fulfilled very well for all (uncharged) polymeric fluids. However, as soon as charged systems are involved, this assumption breaks down and the calculation of the Coulomb force requires special numerical treatment due to its infinite range.

> **Features and Problems with pair potentials.** In essence, pair potentials just count the energy of the bonds but do not take into account their organization. For example, three atoms, connected by bonds so as to form a triangle, would have exactly the same energy as four atoms connected linearly as a chain. Pair potentials also have the tendency to form *close-packed structures*, as the particles try to maximize their coordination number. So, for example, it would be impossible to stabilize a diamond cubic structure for Si (which is an open structure) using pair potentials. As pair potentials do not have an angular dependence, they have no stability against shear and lack of Cauchy pressure. The most important principal failure of pure pair potentials lies in the fact that they cannot be used to describe metals which form covalent (directed) bonds. The main reason for this is the fact that the cohesive energy of pair potentials scales with Z, the coordination number, i.e $E_b \propto Z$. So, for example, when you calculate the bonding energy of a central particle connected to six other particles, its bonding energy is six times the energy of this particle connected to one particle. This, however, is in contradiction with experiments, which show that for metals the bonding energy $E_b \propto \sqrt{Z}$, see Figure 4.9. That is, in real metals, the bonds get weaker as more atoms are added to a central atom.

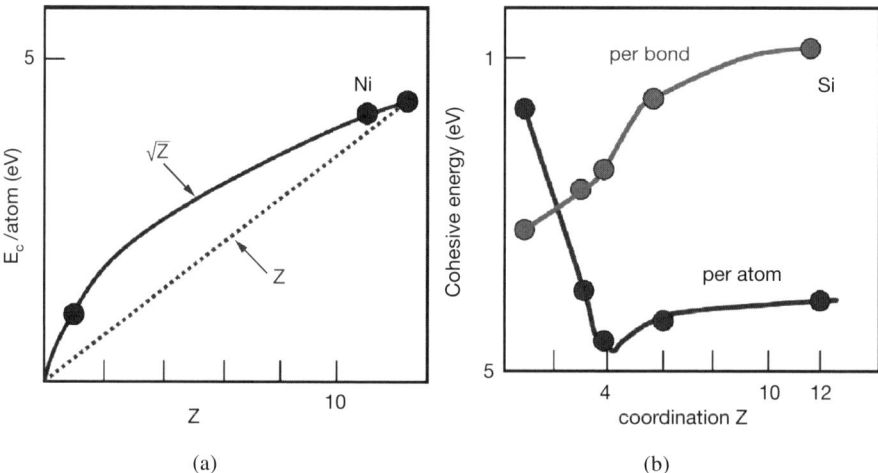

Figure 4.9. Energy as a function of coordination number Z. (a) Experimental data for the cohesive energy of Ni in comparison with the linear relation of a pair potential. (b) Cohesive energy of Si. Adapted from [121].

4.6 Bonded interactions

Using the notion of intermolecular potentials acting between the particles of a system one cannot only model fluids made of simple spherically symmetric particles but also more complex molecules with internal degrees of freedom (due to their specific monomer connectivity). If one intends to incorporate all aspects of the chemical bond in complex molecules, one has to treat the system using quantum chemical methods, see e.g. Cook [105]. Usually, one considers the inner degrees of freedom of polymers and bio-macromolecules by using generic potentials that describe bond lengths l_i, bond angles θ and torsion angles ϕ. When neglecting the fast electronic degrees of freedom, often bond angles and bond lengths can be assumed to be constants. In this case, the potential includes lengths l_0 and the angles θ_0, ϕ_0 about which the molecules are allowed to oscillate at equilibrium, and it includes forces which ensure on average that the system attains these equilibrium values. Hence, the bonded interactions Φ_{bonded} for polymeric macromolecular systems with internal degrees of freedom can be treated by using some or all parts of the following potential term:

$$\Phi_{bonded}(r, \theta, \phi) = \frac{\kappa}{2} \sum_i (|\vec{r}_i - \vec{r}_{i-1}| - l_0|)^2 + \frac{k_\theta}{2} \sum_k (\theta_k - \theta_0)^2 \\ + \frac{\beta}{2} \sum_m (\phi_m - \phi_0)^2. \tag{4.59}$$

Here, the summation indices sum the number of bonds i at positions $\vec{r}_i$, the number of bond angles k between consecutive monomers along a macromolecular chain, and the number of torsion angles m along the polymer chain. A typical value of $\kappa = 5000$ ensures that the fluctuations of bond angles are very small (below 1%). The terms l_0, θ_0 and ϕ_0 are the equilibrium distance, bond angle and torsion angle, respectively.

In particular, in macromolecular physics, very often a Finitely Extensible Non-linear Elastic (FENE) potential is used which – in contrast to a harmonic potential – restricts the maximum bond length of a polymer bond to a prefixed value R_0 [401]:

$$\Phi_{FENE}(r) = \begin{cases} \frac{1}{2}\kappa R_0^2 \ln(1 - \frac{r^2}{R_0^2}) & r < R_0, \\ \infty & \text{otherwise}. \end{cases} \tag{4.60}$$

The FENE potential in equation (4.60) can be used instead of the first term on the right hand side of the bonded potential in equation (4.59). Figure 4.10 illustrates the different parameters used in the description of bonded interactions in equation (4.59). Further details on the use of potentials in macromolecular biology and polymer physics may be found in [369, 156, 387, 395].

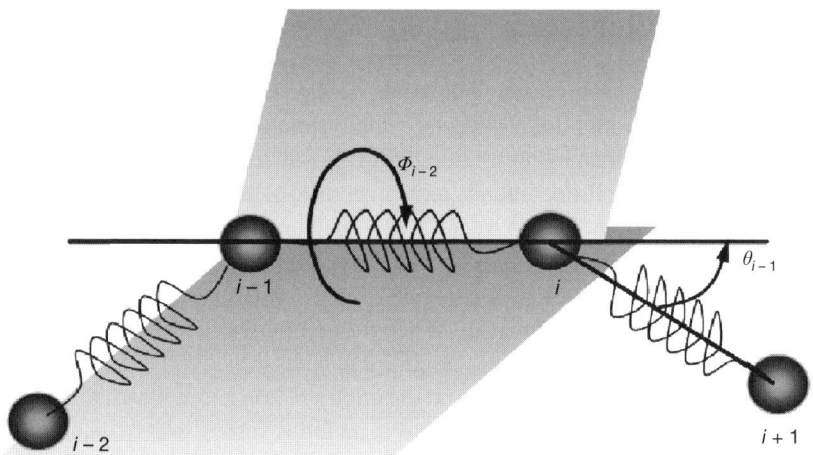

Figure 4.10. Illustration of the potential parameters, used for modeling bonded interactions.

4.7 Chapter literature

General references for potential and functional models for MD simulations of material properties can be found in the book by Torrens [414]. This book is quite old but it is a good book. The specific information on energy models is not up to date, but it may be used as a good general introduction. Oxides are covered in [272, 83]. References to the Embedded Atom Method (EAM) can be found in the excellent review by Daw et al. [121]. Multibody interactions are discussed in Nieminen [315] and a good review paper discussing the need to go beyond pair potentials is [88]. Examples of empirical *bond order potentials* include the Tersoff potential [412, 411], the Brenner potential [76], and the Finnis-Sinclair [161] potentials. They have the advantage over conventional MD force fields that they can describe several different bonding states of an atom using the same parameters, and are thus able to, to some extent, describe chemical reactions correctly. The potentials were partly developed independently of each other, but share the common idea that the strength of a chemical bond depends on the bonding environment, including the number of bonds and possibly also angles and bond length.

Since the early 1970s many-body interactions have been introduced as potential functions. The various approaches involve density and coordination number, respectively, and exploit the idea that bonds are weaker, the higher the local density of the particles. This led to the development of potentials with additional terms that most often consist of two components, a two-body part and a part which takes the coordination number (that is, the local density of particles) into account. Examples of such potentials are the glue model [148], the EAM [120], the Finnis-Sinclair potential [161] and also the so called effective-medium theory [223]. All these approaches differ strongly in the way the coordination number is used in the construction of the potential. Sometimes, different parametrizations are obtained, even for the same material, because of the different constructions. Special many-body potentials have been developed specifically for the study of crack propagation in materials [407]. Still more complex potentials are needed for, e.g., the modeling of semiconductors such as silicon. The potentials developed for these materials also use the concept of coordination number and bond order, meaning that the strength of the bond depends on the local neighborhood. These potentials share a strong connection with the glue models. Stillinger and Weber [403] use a two-body and an additional three-body term in their potential. The family of potentials developed by Tersoff [412] was slightly modified by Brenner [76] and also used, in a similar form, in the modeling of hydrocarbons.

Chapter 5

Molecular Dynamics simulations

Summary

This chapter introduces the Molecular Dynamics (MD) method, which is fundamentally based on the principles of statistical physics at equilibrium and on the concept of particles as the basic constituents of matter. This concept dates back to Presocratic Philosophy. The MD method is typically used on nanoscopic to microscopic length scales, and there are various ab intio variants of this method. This chapter is devoted to *classical* MD – classical in this context means that no electronic degrees of freedom, i.e. those obtained by solving the Schrödinger equation, are taken into account, but rather the classical Newtonian equations for the motion of point particles are considered. Thus, MD is a method which solves the classical N-body problem under certain initial and/or boundary conditions. Conceptually, MD is based on the classical assumption that all particle velocities and positions are known with arbitrary exactness (within the numerical error bounds). Hence, the conceptual consequences of quantum theory (in particular the uncertainty principle) are completely neglected. Also, any relativistic effects arising from the absolute speed limit at which information can be transferred are completely neglected. Despite these fundamental limitations, the MD method is very successful in describing many properties of matter in the gaseous, solid or fluid state.

Learning targets

✓ Learning how to integrate ordinary differential equations.

✓ Understanding different integration methods for solving the N-body problem.

✓ Learning the fundamental linked-list data structure.

✓ Understanding how to analyze the results from MD simulations.

5.1 Introduction

Molecular Dynamics is a meshfree[1] computer simulation technique, where the time evolution of a set of interacting particles (which might be atoms, molecules, or even larger constituents of materials in a corresponding theory) is followed, by integrating their equations of motion. In MD we follow the laws of classical mechanics, most notably Newton's second law, an ordinary DEQ of second order, for each particle i in a system consisting of a total of N particles.:

$$\vec{F}_i = m_i \vec{a}_i = m_i \frac{d\vec{r}_i}{dt} = m_i \frac{d^2\vec{r}_i}{dt^2}. \tag{5.1}$$

Here, m_i is the particle mass, $\vec{a}_i$ its acceleration, and $\vec{F}_i$ is the force acting upon it, due to the interactions with other atoms. Hence, MD is a *deterministic* technique: It assumes that all particle positions and momenta are known *exactly* at all times during the simulation and it implies an infinite speed of force propagation[2], i.e. forces in this theory act instantaneously.

Laplace's demon.

The idea of absolute causal determinism was most prominently formulated by the French mathematician Pierre Simon de Laplace in his 1796 "Exposition du système du monde" [125]. According to determinism, if someone knows the precise location and momentum of every atom in the universe, their past and future values for any given time are fixed, and may be calculated from the laws of classical mechanics. Laplace himself didn't use the term "demon" for an intellect who knows all initial conditions of all atoms, but it was a later embellishment.

Given an initial set of positions and velocities, the subsequent time evolution is in principle completely determined[3], compare Figure 5.1. In a more pictorial description, MD particles move around, if the system is a gas or a fluid, bump into each other, probably oscillate in concert with their neighboring particles, perhaps evaporate out of the system, if there is a free surface, and so forth, in a way pretty similar to what atoms in a real substance would do.

---

[1] Besides the particle methods considered in this book, there are a number of other "meshfree" simulation methods that can be categorized as *particle methods*. Among these are the so-called *Vortex method* [95, 269, 282, 109], which is applied to fluid flow problems, and other *gridless discretization methods* [312, 42, 54, 311].

[2] Classical MD does not involve wave equations, which limit the propagation speed of the interactions between particles.

[3] In practice, the finiteness of the time step in the integration, arithmetic rounding errors, and – most severely – the Liapunov instability, see Section 5.9, will eventually cause the trajectory to deviate exponentially from the "true" trajectory.

> **?** From a modern scientific perspective, can you think of objections against Laplace's demon of 1796?

The algorithm implemented in the computer simulations calculates a particle trajectory in $6N$-dimensional phase space ($3N$ positions $\vec{r}$ and $3N$ momenta $\vec{p}$). However, such trajectories are usually not particularly relevant by themselves. Rather, MD is a statistical mechanics method that provides a way to obtain a set of configurations distributed according to some statistical distribution function, or statistical ensemble. An important example is provided by the microcanonical (NVE) ensemble of an isolated system, corresponding to a probability density in phase space where the total energy E is a constant:

$$\delta(H(\{r\},\{p\}) - E). \tag{5.2}$$

Here, $H(\{r\},\{p\})$ is the Hamiltonian, and $\{r\},\{p\}$ represent the complete set of positions and momenta. The Dirac δ-function in equation (5.2) selects only those states which have a specific energy E. The microcanonical ensemble average of some dynamical observable $A(\{r\},\{p\}) = A(\vec{r}_1, \cdots, \vec{r}_N, \vec{p}_1, \cdots, \vec{p}_N)$ is then given by

$$\langle A \rangle_{NVE} = \frac{\int d\vec{r}_1 \cdots \int d\vec{r}_N \cdots \int d\vec{p}_1 \cdots \int \vec{p}_N\, A(\{r\},\{p\})\, \delta\left(H(\{r\},\{p\}) - E\right)}{\int d\vec{r}_1 \cdots \int d\vec{r}_N \cdots \int d\vec{p}_1 \cdots \int d\vec{p}_N\, \delta\left(H(\{r\},\{p\}) - E\right)}. \tag{5.3}$$

The numerator of equation (5.3) is, of course, the microcanonical partition function (apart from a multiplicative constant), see Chapter 3. Efficiently calculating phase space integrals like in equation (5.3) is the main purpose of the Monte Carlo (MC) method, discussed in Chapter 6. Another example is the canonical ensemble, where the temperature T is constant and the probability density is given by the Boltzmann distribution

$$\exp(-H(\{x\},\{p\})/k_B T) = \exp(-\beta E). \tag{5.4}$$

One can also regard the classical system as a dynamical system. Given initial positions $\{\vec{r}_1(0), \cdots, \vec{r}_N(0)\}$ and initial momenta $\{\vec{p}_1(0), \cdots, \vec{p}_N(0)\}$, the classical equations of motion can be solved to yield the positions $\{\vec{r}_1(t), \cdots, \vec{r}_N(t)\}$ and momenta $\{\vec{p}_1(t), \cdots, \vec{p}_N(t)\}$ at time t, see Figure 5.1. As the system moves along its phase space trajectory, the value of the dynamical quantity

$$A(\{r\},\{p\}) = A(\vec{r}_1, \cdots, \vec{r}_N, \vec{p}_1, \cdots, \vec{p}_N; t) \tag{5.5}$$

now changes with time. The energy E of the system remains constant. The time average $\bar{A}$ is:

$$\bar{A} = \lim_{t \to \infty} \frac{1}{T} \int_0^T dt\, A(\vec{r}_1, \cdots, \vec{r}_N, \vec{p}_1, \cdots, \vec{p}_N; t). \tag{5.6}$$

Section 5.1 Introduction

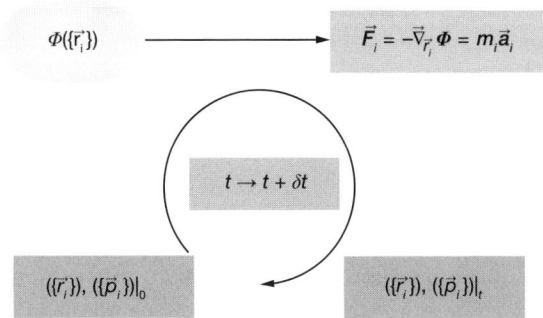

Figure 5.1. The basic integration scheme for solving Newton's equations. Starting from the initial coordinates and momenta at time $t = 0$, one obtains, by explicit time integration, the coordinates and momenta at time $t > 0$. The positions and velocities are determined by the interactions $\phi(\{\vec{r}_i\})$, see Chapter 4, which usually only depend on the positions. The corresponding forces $\vec{F}_i$ on the ith particle that appear in the EOM (equation (5.1)), are obtained by taking the derivative of ϕ with respect to the positions $\vec{r}_i$.

Note that we assume that (for sufficiently long times t) $\bar{A}$ is independent of the initial conditions. We can now write:

$$\bar{A} = \langle A \rangle_{NVE}. \qquad (5.7)$$

Equation (5.7) is referred to as the *ergodic hypothesis*, which appears to be valid for systems in equilibrium. However, the ergodic hypothesis is by no means generally true. Examples of systems for which the ergodic theorem is known to be invalid are metastable systems, such as supercooled liquids and amorphous or vitreous materials. For example, we cannot speak of the equilibrium pressure of a glass. The properties of the glass are partly determined by the manner in which it has been prepared – it is frozen in phase space. For ergodic systems we can determine the equilibrium properties by time-averaging, as in equation (5.6). In order to do so, we determine the phase space trajectory of the system and express the various quantities of interest in terms of the positions and momenta of the system.

Objections against Laplace's demon are

1. The Liapunov instability, see Section 5.9,

2. Quantum theory,

3. Relativity theory.

In Chapter 3 we have seen that, according to statistical physics, physical quantities are represented by averages over configurations distributed according to a certain statistical ensemble. A trajectory obtained by MD provides such a set of configurations. Therefore, a measurement of a physical quantity via simulation is simply an arithmetic average of the various instantaneous values assumed by that quantity during the MD run.

Statistical physics is the link between microscopic behavior and thermodynamics. In the limit of very long simulation times, one could expect the phase space to be fully sampled and in that limit this averaging process would yield the thermodynamic properties. In practice, the runs are always of finite length, and one should exercise caution and estimate when the sampling may be good (i.e. when it constitutes a system at equilibrium) and when not. In this way, MD simulations can be used to measure thermodynamic properties and can therefore evaluate, for example, the phase diagram of a specific material.

Think!

> Why is it reasonable to use the laws of classical mechanics applied to atomic models? Shouldn't the laws of quantum mechanics be used? Should not Schrödinger's equation be employed as equation of motion, instead of Newton's classical laws in equation (5.1)?

When considering a system of interacting atoms consisting of nuclei and electrons, one can *in principle*[4] determine its behavior by solving the Schrödinger equation with the appropriate Hamiltonian operator. However, an analytic or even numerical solution of the Schrödinger equation is only possible for a number of simple special cases. Therefore, even on this fundamental level, approximations have to be made. The most prominent approximation is the BOA [71] which was introduced on theoretical grounds in 1927, i.e. only two years after the famous publication by Heisenberg [204], which laid the foundation for the final matrix formulation of quantum theory, in two successive papers of the same year [70, 69]. The BOA is based on the large difference in mass between the electron (e) and the proton (p), $m_p \approx 1836\, m_e$. This difference in mass allows separation of the equations of motion of the nuclei and the electrons. The intuition behind this approximation is that the significantly smaller mass of the electrons permits them to adapt to the new position of the nuclei almost instantaneously (and adiabatically). The Schrödinger equation for the nuclei is therefore replaced by Newton's law. The nuclei then move according to classical mechanics, but using potentials that result from the solution of the Schrödinger equation for the electrons. For the solution of the electronic Schrödinger equation, approximations have to be employed. Such approximations are derived via, for instance, the *Hartree-Fock* [202, 164] approach, or with density functional theory [214, 255].

---

[4] But not in practice!

Section 5.1 Introduction

These approaches are known as *ab initio MD* (which we do not discuss further in this book – for a basic introduction, see, e.g, [395]). However, the complexity of the model and the resulting algorithms in ab initio approaches enforce a restriction of the system size to a few thousand atoms. A further drastic simplification is the use of parametrized analytical potentials that only depend on the position, i.e. on the energy of the nuclei, known as *classical molecular dynamics*. The potential function itself is then determined by fitting it to the results of quantum mechanical electronic structure computations for a few representative model configurations and subsequent force-matching [147], or by fitting to experimentally measured data. The use of these very crude approximations to the electronic potential hypersurface allows us to treat systems with many billions of atoms. However, in this approach, quantum mechanical effects are lost to a very large extent.

5.1.1 Historical notes on MD

The development of the MD computer simulation technique and the development of particle models for numerical simulation has been closely connected with the development of computers in the 1950s and 60s. In the following, we mention some of the early milestones in the development of the MD method.

- The very first article about MD simulation was written by Alder and Wainwright [21] in 1957. In this work the authors investigated the phase diagram of a two-dimensional hard sphere system, and in particular the solid and liquid regions[5]. In a hard sphere system, particles interact via instantaneous collisions and travel as free particles (on a ballistic curve) between collisions. The calculations were performed on a UNIVAC and on an IBM 704.

- Shortly after, the article "Dynamics of radiation damage", by Gibson, Goland, Milgram and Vineyard, from Brookhaven National Laboratory, appeared in 1960 [179]. This article is interesting, because it is probably the first example of *atomistic modeling of materials*, where the MD method was used with a continuous potential, based on a finite difference time integration method. The calculation of a system with 500 atoms was performed on an IBM 704, and took about a minute per time step[6]. The paper deals with the creation of defects induced by radiation damage – a theme appropriate for the *cold war* era.

- In 1962, Alder and Wainwright continued their studies of phase transitions of two-dimensional hard-disk fluids [22]. In this publication, they studied a 870 particle system and even extended it to three dimensions (with only 500 particles) on a

---

[5] These results were a big surprise at the time, since it was generally assumed that an additional attractive potential was needed to generate such a phase transition.
[6] Today, as a rough estimate and depending on the system, updating 5×10^5 particles in an MD simulation takes about one second on a single CPU.

UNIVAC LARC[7] computer, see also Page 21. Its time needed for addition was 4 microseconds, 8 microseconds for multiplication, 28 microseconds for division. It was the fastest computer in the world in 1960–1961, until the IBM 7030 took this title.

- Aneesur Rahman at Argonne National Laboratory has been a well known pioneer of the MD method. In his landmark paper of 1964 "Correlations in the motion of atoms in liquid argon" [346], he studied the relations in the motion of 864 liquid argon atoms, on a CDC 3600 computer, introducing the Lennard-Jones (LJ) potential to MD simulations.

- In 1967, Loup Verlet introduced an algorithm to efficiently manage data in an MD simulation, using an abstract datatype that is well known among computer scientists, namely *linked lists* or, as he called it, *neighbor-lists* [423]. This paper also introduced a time integration scheme [8], which has ever since become *the* standard integration method[9] used in MD simulations and continues to be until today. In a follow-up paper in 1968 [424], Verlet investigated the correlation function of a fluid made of particles.

- Phase transitions in the same system as the one studied by Verlet, were investigated by Hansen and Verlet in 1969 [200].

- Molecules such as butane were first investigated by Ryckaert in 1975 [364].

- MD simulations with constant pressure or constant temperature were described by H. C. Andersen in the beginning of the 1980s [27, 28]. Other ensembles were considered by Hoover [219] and Nosé [318, 319].

- More complex potentials, with *many-body interactions*, were introduced in MD simulations [45] as early as 1971.

The potentials used in early MD papers of the 1960s and 1970s were mostly *short-range potentials* and the simulations were very limited because of the small capacity of the computers at the time. The simulation of models with *long-range potentials*, compare Figure 5.4, in particular for large numbers of particles, demanded further developments in computer technology and in the algorithms used. One method for the treatment of such potentials relies in its main features on the Ewald summation method of electrostatic forces on a lattice, introduced in 1921 [152]. With this method, the potential is split into its short-range and long-range part, each of which can be computed efficiently with a specific separate approach. The decisive idea is the use of fast Poisson solvers for the long-range part. Hierarchical methods, such as fast Fourier

---

[7] The UNIVAC LARC (Livermore Advanced Research Computer) was a mainframe computer, delivered to Livermore in 1960.
[8] This integration scheme relies on an approach presented by Störmer in 1907 [392].
[9] In fluid dynamics and solid mechanics, one often speaks of "solvers" instead of "integration scheme".

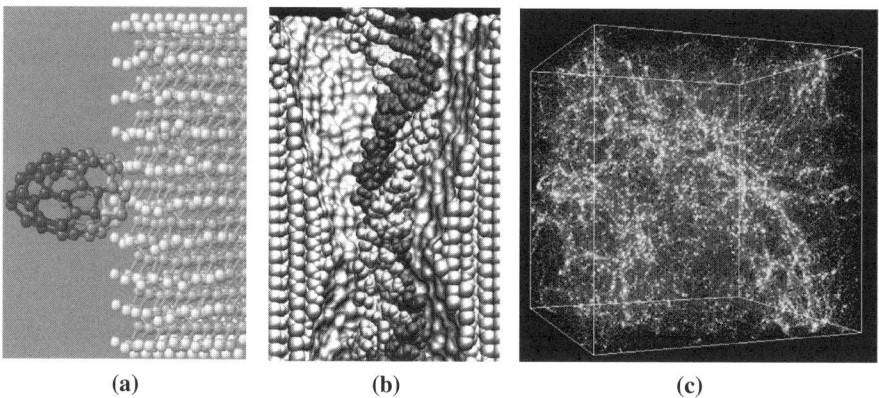

Figure 5.2. Typical applications of MD in solid state and soft matter physics, and astrophysics, spanning at least 26 orders of magnitude. (a) Simulation snapshot of a Bucky ball (C_{60} molecule) bouncing off a surface made of carbon atoms. (b) A high electric field forces DNA through a 2.0 nm diameter pore (a nanotube) which at smaller fields is impermeable. The 6.5 V bias deforms the DNA helix, shifting one of the strands by approximately one nucleotide, whilst preserving the hydrogen bonds between strands. (c) This simulation frame shows the formation of large scale structures of the universe in a 43 million parsecs (or 140 million light years) large simulation box.

transforms or multilevel methods, are applied in these fast Poisson solvers. The variants of this so called P^3M method [141, 212, 150, 390, 334] differ in the selection of algorithms for the single components (interpolation, force evaluation, adaptivity, fast solvers, etc.). A prominent example is the so-called Particle Mesh Ewald (PME) method [114, 151, 267, 415], which uses splines.

Another class of methods for long-range potentials uses a simple Taylor expansion[10] of the potential functions to approximately evaluate particle interactions. To allow efficient data management and computation, the resulting data is stored in another abstract datatype, well known to computer scientists, namely *tree structures*. Some earlier representatives of this class of methods, which are often used in astrophysics, were developed by Appel [32] and by Barnes and Hut [46]. Newer variants by Rokhlin and Greengard [357, 190] use higher moments in the expansions. Since the 1980s ended, developments of algorithms for the parallelization of MD simulations with short-range potentials attracted increasing attention, and expositions of such methods can be found in, for example, [349, 331, 48, 394]. Parallel variants of the P^3M algorithm can be found in [157, 413, 447] and parallel versions of the Barnes-Hut algorithm and the multipole method have been presented for example in [189, 451, 433]. Domain decomposition methods are employed in these versions, as well as in the parallelization of algorithms for short-range potentials. Normally, such

---

[10] In electrodynamics this expansion is called *multipole expansion*.

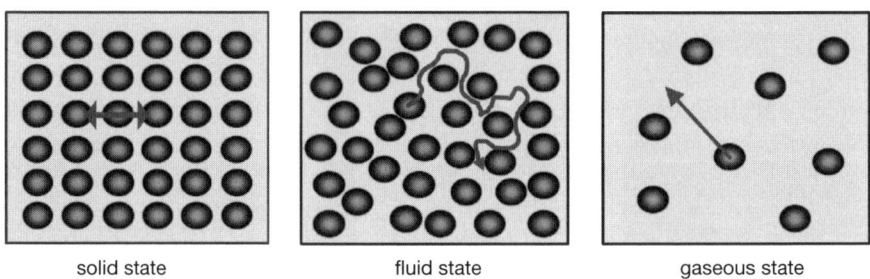

solid state   fluid state   gaseous state

Figure 5.3. Typical applications of MD in fluid, solid state and kinetic gas theory. Depicted are fluctuations of atoms about their lattice positions in a crystal, diffusive behavior of a particle in a fluid and the ballistic motion of a particle in a gas in between collisions.

potentials are used for *large-scale* MD simulations in materials science, which on the largest supercomputers exceed several hundred billion particles [13, 81, 233, 234, 25].

Today's fields of application of MD

MD simulations are abundant in science and engineering today. Whilst the first successful applications of MD were in liquid theory, today, MD is used in a multitude of research areas, comprising all three basic states of matter (liquid, gaseous and solid), see Figure 5.3. In the following, we provide an incomplete list of physical systems which are readily amenable to particle simulations with MD, compare also Figure 5.2:

Liquids and Fluid Dynamics. Particle simulations can serve as a new approach to the study of hydrodynamical instabilities on the microscopic scale, e.g. the Rayleigh-Taylor or Rayleigh-Bénard instability. Furthermore, molecular dynamics simulations allow investigation of complex fluids, such as polymers in solution and fluid mixtures, e.g., emulsions of oil and water, but also of crystallization and of phase transitions on the microscopic level. Through non-equilibrium techniques, transport phenomena such as viscosity and heat flow have been investigated.

Solid State Physics. The simulation (and virtual design) of materials on an atomic scale is primarily used in the analysis of known materials and in the development of new materials with specific desired properties, e.g. in terms of stiffness [393]. Developments in this area of research started at the beginning of the 1990s, when the largest supercomputers in research institutes and universities had become fast enough to simulate hundreds of millions of particles and when *coarse-grained* methods had been developed, which allowed for bridging various length and time scales [120, 121, 160, 165, 118, 119, 429]. Examples of phenomena studied in solid state physics are the structure conversion in metals induced by temperature or shock [216, 215, 74], the formation of cracks initiated by pressure or shear stresses [283, 159, 379, 217, 399], fracture and failure simulations[452], the propagation of sound waves in materials [12], the impact of defects in the structure of materials on their load-bearing capacity, and the analysis of plastic and elastic deformations [292].

Another subject whose investigation with MD started a long time ago, see e.g. [98], is defects in crystals, which are crucial for their mechanical properties and are therefore of technological interest. The focus in this area shifted somewhat from point defects to linear dislocations and planar defects (grain boundaries and stacking faults) [444, 402, 400, 82].

Soft Matter Physics and Biomacromolecules. The dynamics of macromolecules on the atomic level is one of the most prominent applications of MD [436, 230, 108, 78, 248, 386, 422]. With such methods it is possible to simulate molecular fluids, crystals, amorphous polymers, liquid crystals, zeolites, nuclear acids, proteins, membranes and many more biochemical materials.

Astrophysics. In this area, simulations are primarily done to test theoretical models. In a simulation of the formation of the large-scale structure of the universe, particles correspond to entire galaxies. In a simulation of galaxies, particles represent several hundred up to thousand stars. The force (equation (1.19)) acting between these particles results from the gravitational potential.

Beyond these "traditional" applications, MD is nowadays also used for studies of non-equilibrium processes, and as an efficient tool for optimization of structures overcoming local energy minima, e.g. by *simulated annealing* [246].

5.1.2 Limitations of MD

After an initial equilibration phase, the system will usually reach an *equilibrium state*. By averaging over an equilibrium trajectory (coordinates over a function of time) many macroscopic properties can be extracted from the output. While MD is a very powerful technique it also has its limitations. We will briefly examine the most important of them.

Artificial boundary conditions

The system size that can be simulated with MD is very small compared to real molecular systems. Hence, a system of particles will have many unwanted artificial boundaries (surfaces). In order to avoid real boundaries, one introduces periodic boundary conditions (see Section 5.7, which can introduce artificial spatial correlations in too small systems. Therefore, one should always check the influence of system size on the results.

Cut off of long-range interactions

Usually, all non-bonded interactions are cut off at a certain distance, in order to keep the cost of the computation of forces (and the search effort for interacting particles) as small as possible. Due to the minimum image convention (see Section 5.8) the cutoff range may not exceed half the box size. Whilst this is large enough for most systems in practice, problems are to be expected with systems containing charged particles. Here, simulations can go badly wrong and, e.g. lead to an accumulation of the charged particles in one corner of the box. Here, one has to use special algorithms, such as the particle-mesh Ewald method [114, 151].

The simulations are classical

One question that immediately comes to mind is why we can use Newton's law to move atoms, when systems at the atomic level obey quantum laws rather than classical laws, and Schrödinger's equation should be used. Hence, all those material properties connected to the fast electronic degrees of freedom are not correctly described. For example, atomic oscillations (e.g. covalent C-C-bond oscillations in polyethylene molecules, or hydrogen-bonded motion in biopolymers such as DNA, proteins, or biomembranes) are typically of the order 10^{14} Hz. The specific heat is another example that is not correctly described in a classical model, because here, at room temperature, all degrees of freedom are excited, whereas quantum mechanically, the high-frequency bonding oscillations are not excited, thus leading to a smaller (correct) value of the specific heat than in the classical picture. A general solution to this problem is to treat the bond distances and bond angles as constraints in the equations of motion. Thus, the highest frequencies in the molecular motion are removed and one can use a much larger time step in the integration [421].

A simple test of the validity of the classical approximation is based on the de Broglie thermal wavelength λ, see equation (3.1). The classical approximation of the EOM in equation (5.1) is justified if $\lambda \ll a$, where a is the mean nearest neighbor separation. If one considers, for instance, liquids at the triple point, the ratio λ/a is of the order of 0.1 for light elements such as Li and Ar, decreasing further for heavier elements. Hence, the classical approximation is ill-suited for very light systems, such as H_2, He and Ne.

Moreover, quantum effects become important in any system when the temperature T is sufficiently low. The drop below the Debye temperature Θ_D [247] in the specific heat of crystals, or the anomalous behavior of the thermal expansion coefficient, are well known examples of measurable quantum effects in solids. In these regions, classical MD results should be interpreted with caution.

In MD simulations, particles (often interpreted as atoms) interact with each other. These interactions give rise to forces which act upon atoms and atoms move under the action of these instantaneous forces. As the atoms move, their relative positions change and the forces change as well. Hence, the forces form the essential ingredient

Section 5.1 Introduction   305

that contains the physics of a system. an MD simulation is "realistic" when it mimics the behavior of the real system. This is only the case to the extent that interatomic forces are similar to those that real atoms (or, more exactly, nuclei) would experience if arranged in the same configuration. The forces are usually obtained as the gradient of a potential energy function $\phi(\vec{r}_i)$, depending on the positions of the particles. Thus, one needs a realistic energy description of the system, otherwise one does not know whether the simulation will produce something useful. As a result, the quality of an MD simulation depends on the ability of the chosen potential to reproduce the behavior of the system, under the boundary conditions at which the simulation is run.

Restrictions in time and length scales

The ultimate goal of atomic particle calculations in materials science is to virtually "design" materials with predefined properties, by changing their specific atomic bonding and electronic structure. A whole host of theoretical methods have been developed in order to address the above set of problems. Nowadays, chemicals and molecules of interest in research and technology, including drugs, are often first designed on the computer using Molecular Orbital (MO) calculations[11], before the molecules are actually made in the lab. In MO calculations, the Hartree-Fock (HF) approximation is often used to describe electron-electron interaction effects, excluding correlation effects. The *local density* approximation is a much better approximation for ground state properties, but cannot be applied to excited and highly correlated atomic states. The numerical methods for computing structural properties of materials may be divided into two classes:

1. Methods that use empirically determined parameters, and

2. Methods that do not make use of any empirical quantities.

The former methods are often called "empirical" or "semi-empirical", whereas the latter are called "ab initio" or "first principles methods". Several tight-binding schemes fall into the former category. The idea of ab initio methods is to treat many-atom systems as many-body systems composed of electrons and nuclei, and to treat everything on the basis of first principles of quantum mechanics, without introducing any empirical fitting parameters from measurements. Ab initio methods are often useful for predicting the electronic properties of new materials, clusters or surfaces, and for predicting trends across a wide range of materials. We note, however, that even with the most sophisticated methods of modern physics and quantum chemistry, one still has to rely on approximations on which ab initio methods are fundamentally based[12]. One of the main differences between classical (empirical potential) and quantum mechanical methods is that in the former approach the potential energy E between atoms i and j is described by an analytical function of the atom coordinates $E(\vec{R}_{ij})$, whereas in

---

[11] In an industrial setting, this approach is called "Molecular Modeling".
[12] Most prominently, the Born-Oppenheimer approximation [71].

the majority of the latter methods, the energy is calculated by solving the Schrödinger equation for electrons in the field of many atomic cores.

Figure 5.4 exhibits several common methods employed for particle MD simulations, along with their estimated maximum system size and the typical time scales that can be treated. The highest precision and transferability of results of methods is achieved with self-consistent first principles calculations. Self-Consistent Field (SCF) theory is fundamentally based on the Hartree-Fock (HF) method which itself uses the Mean Field Approximation (MFA). Due to the MFA, the energies in HF calculations are always larger than the exact energy values. The second approximation in HF calculations is to express the wave function $\psi(\vec{x}, t)$ in functional form. Such functionals are only known exactly for very few one-electron systems and therefore some approximate functions are usually used instead. The approximate basis wave functions used are either plain waves, Slater Type Orbitals (STO) $\sim \exp(-ax)$, or Gaussian Type Orbitals (GTO) $\sim \exp(-ax^2)$. DFT is an alternative ab initio method to SCF. In DFT, the total energy of a system is not expressed in terms of a wave function but rather in terms of an approximate Hamiltonian and thus an approximate total electron density. This method uses GTO potentials or plane waves as basis sets and correlations are treated with Local Density Approximations (LDA) and corrections.

So called *semi-empirical methods*, such as Tight Binding (TB) (e.g. Porezag et al. or Pettifort), approximate the Hamiltonian $\mathcal{H}$ used in HF calculations by approximating or neglecting several terms (called the Slater-Koster approximation), but re-parameterizing other parts of $\mathcal{H}$ in a way that yields the best possible agreement with experiments or ab initio simulations. In the simplest TB version, the Coulomb repulsion between electrons is neglected. Thus, in this approximation, there exists no correlation problem. However, there also is no self-consistent procedure.

Classical multi-body potentials of Tersoff or Stillinger-Weber type, and two-body potentials, e.g. in the Embedded Atom Method, or generic LJ potentials allow for a very efficient force field calculation, but can hardly be used for systems other than the ones for which the potentials were introduced, e.g. systems with other types of bonding, or atoms. Classical methods are, however, very important in order to overcome the complexities of some materials, and they are the method of choice for coarse-grained models of material behavior.

Despite its fundamental limitations, the MD method is very successful in describing many properties of matter when relativistic and quantum effects may be neglected. Usually, an eigenvolume of the particles is taken into account and in the simplest MD version it is assumed that all forces can be described as arising from pair potentials only. The particles are put into a simulation box (usually having a cubic shape, but other shapes are also used), and the boundary conditions pertaining to a certain statistical ensemble to be simulated, are inserted into the equations of motion. The equations of motion are solved and the system is evolved in its phase space for typically several hundred thousand time steps, using an explicit integration scheme, with the *Verlet velocity* scheme being the most prominent one. After a while, the simulated

Section 5.1 Introduction

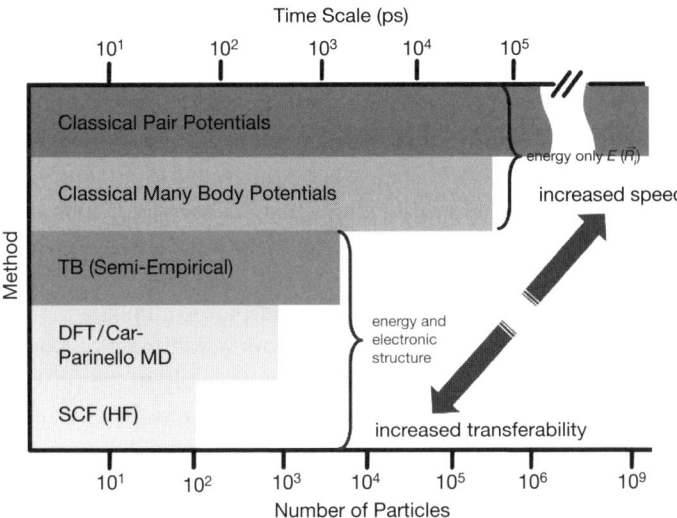

Figure 5.4. Different MD methods used in science and engineering, along with their corresponding level of transferability, and a rough estimate of the number of particles that can be simulated within a couple of days on present day supercomputing systems. Note that the available size of the systems can be considerably increased by using domain decomposition on parallel computing facilities. However, the time scale cannot be "decomposed" and distributed over many CPUs. Therefore, the available time scale with atomic type simulations, using empirical pair potentials, is roughly a few hundred nanoseconds for the longest runs on the largest super computers. Self Consistent Field (SCF), Hartree-Fock (HF) and Density Functional Theory (DFT) calculations yield the electronic structure of atoms and molecules, whereas calculations with Tight Binding and Car-Parinello MD yield the trajectories of a few hundred atoms for a few picoseconds. Semi-empirical methods, such as EAM, bond order type (Tersoff), or cluster type (Stillinger-Weber) potentials, are typically used to obtain structural and energetic properties of clusters, crystals, surfaces or defects. Classical force field MD, based on pair potentials, is often used for calculating phase transitions, diffusion, growth phenomena or surface processes. With these models, the treated particles are usually not atoms but coarse grained structures that neglect chemical details of atomic bonds and only take into account the "large-scale" behavior of these "super particles". Using coarse grained approaches, the available length and time scales can be extended considerably, usually at least to microseconds and micrometers, whilst capturing many microstructural features of the investigated system.

system will have attained its equilibrium state, where the microscopic state variables fluctuate about their macroscopic mean values. At this point one starts performing measurements of observables and the results of these measurements are analyzed *in the same way as in a real experiment*, calculating standard deviations of measured mean values, which characterize the statistical fluctuation of observables.

Generally speaking, an MD simulation is "safe" from the point of view of its duration, when the simulation time is much longer than the relaxation time of the quantities of interest. However, different properties have different relaxation times. In particular, systems tend to become slow in the proximity of phase transitions. A limited system size can also constitute a problem. In this case, one has to compare the size of the MD cell with the correlation lengths of the spatial correlation functions of interest. Correlation lengths may increase or even diverge, particularly in the proximity of phase transitions, and the results are no longer reliable as soon as they become comparable with the box length. This problem can be partially alleviated by a method known as *finite size scaling* [112]. With this method one calculates a physical property P several times, using different simulation box sizes L, and then fits the results to the relation

$$A(L) = A_0 + \frac{c}{L^N}, \qquad (5.8)$$

using A_0, c and N as fitting parameters. A_0 corresponds to $\lim_{L \to \infty} A(L)$ and should therefore be taken as the most reliable estimate for the "true" value of the physical quantity A.

The electrons are in the ground state

Using *conservative force fields* in MD implies that the potential is a function of the atomic positions only. No electronic motions are considered. Thus, the electrons remain in their ground state and are considered to instantaneously follow the movement of the core. This means that electronically excited states, electronic transfer processes, and chemical reactions cannot be treated.

Approximative force fields

Force fields are not really an integral part of the simulation method but are determined from experiments or from a parametrization, using ab initio methods. Also, most often, force fields are pair-additive (except for the long range Coulomb force) and hence cannot incorporate polarizabilities of molecules. However, such force fields exist and there is continuous effort to generate such kind of force fields [375]. However, in most practical applications, e.g. for biomacromolecules in an aqueous solution, pair potentials are quite accurate, mostly because of error cancellation. This does not always work, for example, ab initio predictions of small proteins still yield mixed results and when the proteins fail to fold, it is often unclear whether the failure is due to a deficiency in the underlying force fields or simply a lack of sufficient simulation time [168, 168].

Force fields are pair additive

All *non-bonded* forces result from the sum of non-bonded pair interactions. Non pair-additive interactions, such as the polarizability of molecules and atoms, are represented by averaged *effective pair potentials*. Hence, the pair interactions are not valid for situations that differ considerably from the test systems on which the models were parametrized. The omission of polarizability in the potential implies that the electrons do not provide a dielectric constant, with the consequence that the long-range electrostatic interaction between charges is not reduced (as it should be) and is thus overestimated in simulations.

5.2 Numerical integration of differential equations

In the following sections, a number of techniques is presented that allow one to obtain approximate numerical solutions to initial and boundary value problems, i.e. to numerically integrate systems of ordinary differential equations. Numerical methods to solve complicated initial value and boundary value problems share the discretization of the independent variables (typically time t and space x) and the transformation of the continuous derivative into its discontinuous counterpart, i.e., its finite difference quotient. Using these discretization steps amounts to recasting the continuous problem expressed by differential equations with an infinite number of degrees f or function values, into a discrete algebraic one, with a finite number of unknown parameters, which can be calculated in an approximate fashion.

5.2.1 Ordinary differential equations

By definition, an ordinary differential equation, or ODE, is an equation in which all dependent variables are functions of a *single* independent variable, either a scalar or a tensor quantity [236, 14, 410]. The most general notation of an ODE is the implicit form

$$f(y^{(n)}(x), y^{(n-1)}(x), \ldots, y'(x), y(x), x) = 0. \tag{5.9}$$

Furthermore, an nth order ODE is such that, when it is reduced to its simplest form, the highest order derivative it contains is of nth order.

> The *general form of an ordinary differential equation* (ODE) is:
>
> $$L^{(n)}(x)y(x) = f(x),$$
>
> where $L^{(n)}(x)$ is a linear differential operator
>
> $$L^{(n)}(x) = \sum_{k=0}^{n} f_k(x) \frac{d^k}{dx^k}.$$
>
> - The *order* of the ODE is defined by its highest-order derivative.
>
> - An ODE is *linear* if it is linear in its unknown function $y(x)$ and its derivatives $y'(x)$, else it is called *nonlinear*.
>
> - An ODE is *explicit* if $y^{(n)}$ can be written as a function of $y^{(n-1)}, \ldots, y$ and x, else it is called *implicit*.

According to Newton's classical laws of motion, the motion of any collection of rigid objects can be reduced to a set of second order ODEs in which the time t is the common independent variable. For instance, the equations of motion of a set of n interacting point particles, moving three-dimensionally in the $\vec{r}$ direction, might take the form:

$$\frac{d^2\vec{r}_i(t)}{dt^2} = \frac{\vec{F}_i(\vec{r}_1, \ldots, \vec{r}_n, t)}{m_i}, \qquad (5.10)$$

where $i \in [1, \ldots, n]$ and $\vec{r}_i$ is the position of the ith particle and m_i is its mass. Note that a set of n second order ODEs can always be rewritten as a set of $2n$ first order ODEs. Thus, the above equations of motion can be rewritten in the form:

$$\frac{d\vec{r}_i(t)}{dt} = v_i, \qquad (5.11)$$

and

$$\frac{d\vec{v}_i(t)}{dt} = \frac{\vec{F}_i(\vec{r}_1, \ldots, \vec{r}_n, t)}{m_i}, \qquad (5.12)$$

with $i \in [1, \ldots, n]$.

5.2.2 Finite Difference methods

Finite difference methods fall under the general class of *one-step methods*. The algorithm is rather simple. Suppose we consider the general first order ODE

$$y' = f(x, y), \qquad (5.13)$$

where the symbol "′" denotes d/dx, subject to the general initial-value boundary condition

$$y(x_0) = y_0. \tag{5.14}$$

Clearly, if we can find a method for solving this problem numerically, we should have little difficulty in generalizing it to deal with a system of n simultaneous first order ODEs. It is important to appreciate that the numerical solution to a differential equation is only an approximation to the actual solution. The actual solution of equation (5.13), $y(x)$, is (presumably) a *continuous* function of a continuous variable, x. However, when we solve this equation numerically, the best we can do is to evaluate approximations to the function $y(x)$ at a series of n *discrete* grid points x_n in the interval $[a, b]$, where $n = 0, 1, 2, \ldots$ and $x_0 < x_1 < x_2 \cdots < x_n$. We will restrict our discussion to equally spaced grid points, see Figure 5.5, where

$$x_n = x_0 + nh. \tag{5.15}$$

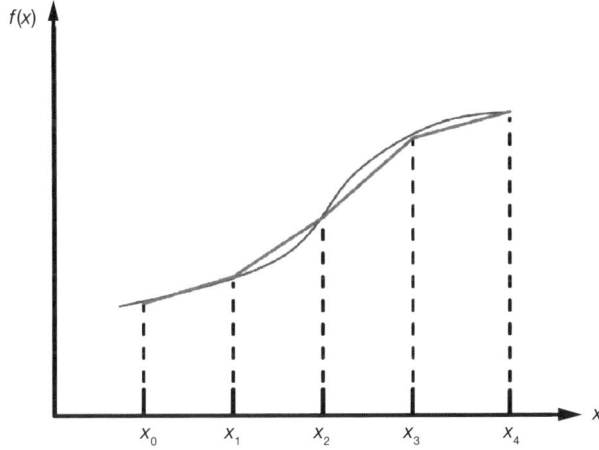

Figure 5.5. Illustration of Euler's method with equally spaced grid points.

Here, the quantity $h = (b - a)/n$ is referred to as the *step length* or *step size*, compare Section 2.6. Let y_n be our approximation of $y(x)$ at the grid point x_n. A numerical integration scheme is essentially a method which somehow employs the information contained in the original ODE of equation (5.13), in order to construct a series of rules interrelating the various y_n. The simplest possible integration scheme was invented by the 18th century Swiss mathematician Leonhard Euler, and is therefore called *Euler's method*. Incidentally, it is interesting to note that virtually all standard methods used in numerical analysis were invented before the advent of electronic computers. People actually performed numerical calculations by hand in a very long and tedious process. Suppose that we have evaluated an approximation y_n to the solution $y(x)$ of equation (5.13) at the grid point x_n. The approximate gradient of $y(x)$ at this point is, therefore, given by

$$y_n' = f(x_n, y_n). \tag{5.16}$$

Let us approximate the curve $y(x)$ as a straight line between the neighboring grid points x_n and x_{n+1}. It follows that

$$y_{n+1} = y_n + y_n'h = y_n + f(x_n, y_n)h. \tag{5.17}$$

Numerical errors

There are two major sources of error associated with a numerical integration scheme for ODEs, namely *truncation errors* and *rounding errors*. Truncation errors arise in Euler's method because the curve $y(x)$ is generally not a straight line between the neighboring grid points x_n and x_{n+1}, as assumed above. The error associated with this approximation can be easily assessed by a Taylor expansion of $y(x)$ about $x = x_n$:

$$\begin{aligned} y(x_n + h) &= y(x_n) + hy'(x_n) + \frac{h^2}{2}y''(x_n) + O(h^3) \\ &= y_n + hf(x_n, y_n) + \frac{h^2}{2}y''(x_n) + O(h^3). \end{aligned} \tag{5.18}$$

A comparison of equations (5.17) and (5.18) yields

$$y_{n+1} = y_n + h\,f(x_n, y_n) + O(h^2). \tag{5.19}$$

Equation (5.18) conveys the essence of *Euler's method*.

> The *Euler method* is based on a simple, truncated Taylor expansion and *cannot be recommended* for computer simulations, because
>
> - it is *not* time reversible
>
> - it does *not* conserve the phase space volume
>
> - it suffers catastrophically from energy drift

Figure 5.6 shows an example of the bad performance of Euler's algorithm, displaying the huge energy drift in the simulation of a one-dimensional harmonic oscillator. Euler's method is *not recommended for computer simulations*, although it is handy

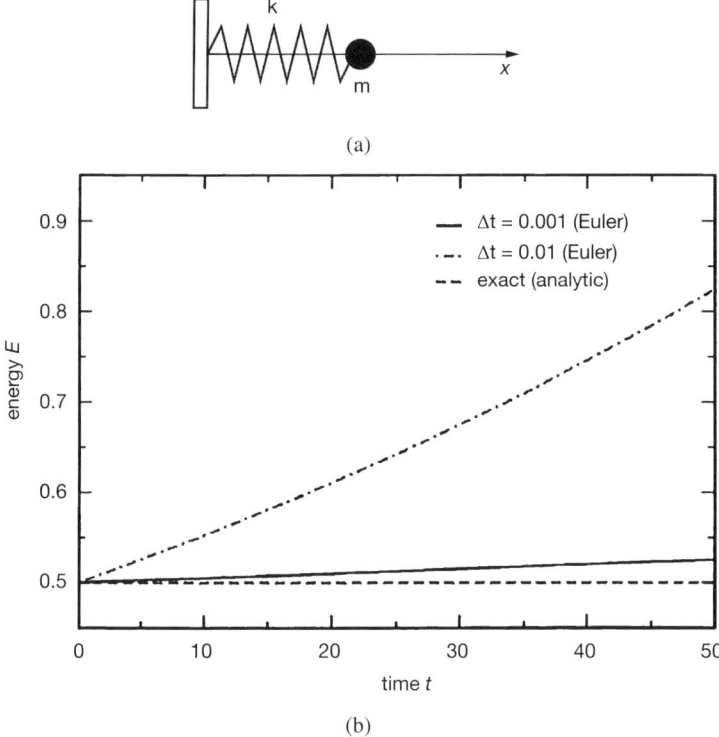

Figure 5.6. Euler's method applied to a one-dimensional oscillator. (a) Oscillator. (b) Energy, calculated for two different discretizations Δt, compared to the analytic solution. The calculation with Euler's algorithm shows a huge energy drift.

to use in order to get a first idea on what a solution may look like. As an example, consider Newton's law in equations (5.11) and (5.12). The first step (omitting index i) would be

$$\vec{r}_{n+1} = \vec{r}_n + h\vec{v}_n + O(h^2) \tag{5.20}$$

and

$$\vec{v}_{n+1} = \vec{v}_n + h\frac{\vec{F}_i}{m_i} = \vec{v}_n h \vec{a}_n + O(h^2), \tag{5.21}$$

where $\vec{a}$ is the acceleration. The Euler method is *asymmetrical in time*, since it uses information about the derivative from the integration point that preceded it. This means that we evaluate the position at time $n + 1$ using the velocity at time n. A simple variation of this calculation would be:

$$\vec{r}_{n+1} = \vec{r}_n + h\vec{v}_{n+1} + O(h^2) \qquad (5.22)$$
$$\vec{v}_{n+1} = \vec{v}_n + h\vec{a}_n + O(h^2).$$

The acceleration $\vec{a}_n$ in equation (5.21) and (5.22) has to be calculated as well. This is called the *Euler–Cromer method*.

When using Euler's method, every time we take a step, we incur a truncation error of $O(h^2)$, where h is the step length. Suppose that we use Euler's method to integrate our ODE over an x-interval of order unity. This requires $O(h^{-1})$ steps. If each step produces an error of $O(h^2)$ and the errors are simply cumulative (a fairly conservative assumption), the net truncation error is $O(h)$. In other words, the error associated with integrating an ODE over a finite interval, using Euler's method, is directly proportional to the step length h. Thus, if we want to keep the relative error in the integration under about 10^{-6}, we need to take about one million steps per unit interval in x. Incidentally, Euler's method is termed a first order integration method, because the truncation error associated with integration over a finite interval scales like h^1. More generally, an integration method is conventionally called n^{th} order, if its truncation error per step is $O(h^{n+1})$.

> Note that truncation errors would arise even if computers performed floating point arithmetic operations with infinite accuracy.

Unfortunately, computers do not perform operations with infinite accuracy. In fact, we have seen in Chapter 1 that, in the bit system, floating point numbers are only represented with an accuracy of a fixed number of decimal places. For every type of computer, there is a characteristic number ξ which is defined as the smallest number which, when added to a number of order unity, gives rise to a new number, i.e., a number which, when taken away from the original number, yields a nonzero result. Every floating point operation produces a rounding error of $O(\xi)$, which arises from the finite accuracy with which floating point numbers are stored by a computer. Suppose that we use Euler's method to integrate our ODE over an x interval of order unity. This entails $O(h^{-1})$ integration steps and, therefore, $O(h^{-1})$ floating point operations. If each floating point operation incurs an error of $O(\xi)$, and the errors are simply cumulative, then the net rounding error is $O(\xi/h)$. The total error ε, associated with integrating an ODE over an x-interval of order unity is (approximately) the sum of the truncation and rounding errors. Thus, for Euler's method, we have

$$\varepsilon \approx \frac{\xi}{h} + h. \qquad (5.23)$$

Clearly, at large step lengths, the error is dominated by truncation errors, whereas rounding errors dominate at small step lengths. The net error attains its minimum

Section 5.2 Numerical integration of differential equations

value, $\varepsilon_0 \simeq xi^{1/2}$, when $h = h_0 \simeq \xi^{1/2}$. There is clearly no point in making the step length h any smaller than h_0, since this increases the number of floating point operations, but does not lead to an increase in overall accuracy. It is also clear that the ultimate accuracy of Euler's method – or any other integration method – is determined by the accuracy ξ with which floating point numbers are stored on the computer that performs the calculation. The value of ξ depends on how many bytes the computer hardware uses to store floating point numbers. Let's assume that the appropriate value for *double* precision floating point numbers is $\xi = 2.22 \times 10^{-16}$.[13] In this case, it follows that the minimum practical step length for Euler's method on such a computer is $h_0 \simeq 10^{-8}$, yielding a minimum relative integration error of $\varepsilon_0 \simeq 10^{-8}$. This level of accuracy is perfectly adequate for most scientific calculations. Note, however, that the corresponding ξ-value for *single* precision floating point numbers is only $\xi = 1.19 \times 10^{-7}$, yielding a minimum practical step length and a minimum relative error for Euler's method of $h_0 \simeq 3 \times 10^{-4}$ and $\varepsilon_0 \simeq 3 \times 10^{-4}$, respectively. This level of accuracy is generally not at all adequate for scientific calculations, which explains why such calculations are invariably performed using *double*, rather than *single* precision floating point numbers.

Numerical instabilities

Consider the following example: Suppose that we investigate the ODE

$$y' = -\alpha y, \tag{5.24}$$

where $\alpha > 0$, subject to the boundary condition

$$y(0) = 1. \tag{5.25}$$

Of course, we can solve this problem analytically to obtain

$$y(x) = \exp(-\alpha x) \tag{5.26}$$

Note that the solution is a monotonically decreasing function of x. We can also solve this problem numerically, using Euler's method. Appropriate grid points (in one dimension) are

$$x_n = nh, \tag{5.27}$$

where $n = 0, 1, 2, \ldots$. Euler's method yields

$$y_{n+1} = (1 - \alpha h) y_n. \tag{5.28}$$

Note one curious fact. If $h > 2/\alpha$, then $|y_n + 1| > |y_n|$. In other words, if the step-length is made too large, the numerical solution becomes an oscillatory function of x

---

[13] This value is actually specified in the system header file float.h.

of monotonically increasing amplitude, i.e., the numerical solution diverges from the actual solution. This type of catastrophic failure of a numerical integration scheme is called a *numerical instability*. All simple integration schemes become unstable if the step-length is made sufficiently large.

5.2.3 Improvements to Euler's algorithm

The most obvious improvements to Euler's and Euler-Cromer's algorithms – avoiding the need for additionally computing a second derivative – is the so-called *midpoint method*. In this case one has

$$\vec{r}_{n+1} = \vec{r}_n + \frac{h}{2}\left(\vec{v}_{n+1} + \vec{v}_n\right) + O(h^2), \tag{5.29}$$

and

$$\vec{v}_{n+1} = \vec{v}_n + h\vec{a}_n + O(h^2). \tag{5.30}$$

Inserting equation (5.29) into (5.30) finally yields

$$\vec{r}_{n+1} = \vec{r}_n + h\vec{v}_n + \frac{h^2}{2}\vec{a}_n + O(h^3). \tag{5.31}$$

This equation implies that the local truncation error in the position is now $O(h^3)$, whereas Euler's or Euler-Cromer's methods have a local error of $O(h^2)$. Thus, the midpoint method yields a global error with second-order accuracy for the position and first-order accuracy for the velocity. However, although these methods yield exact results for constant accelerations, the error in general increases with each time step. One method that avoids this is the so-called half-step method. Here, one defines

$$\vec{v}_{n+1/2} = \vec{v}_{n-1/2} + h\vec{a}_n + O(h^3), \tag{5.32}$$

and

$$\vec{r}_{n+1} = \vec{r}_n + h\vec{v}_{n+1/2} + O(h^2). \tag{5.33}$$

Note that this method needs the calculation of $\vec{v}_{n+1/2}$. This is done using e.g. Euler's method

$$\vec{v}_{n+1/2} = \vec{v}_n + \frac{h^2}{2}\vec{a}_n + O(h^2). \tag{5.34}$$

As this method is a little more numerically stable, it is often used instead of Euler's method.

> All the previous methods, based on higher-order derivatives, are in general not used in numerical computation, since they rely on evaluating derivatives several times. Unless one has analytical expressions for these derivatives, the risk of rounding errors is very large.

5.2.4 Predictor-corrector methods

Consider again the first-order differential equation

$$\frac{dy}{dx} = f(t, y), \tag{5.35}$$

which can be solved, using Euler's algorithm. This results in

$$y_{n+1} \approx y_n + hf(t_n, y_n). \tag{5.36}$$

with $t_{n+1} = t_n + h$. This means geometrically, that we compute the slope at y_n and use it to predict y_{n+1} at a later time t_{n+1}. We introduce $k_1 = f(t_n, y_n)$ and rewrite our prediction for y_{n+1} as

$$y_{n+1} \approx y_n + hk_1. \tag{5.37}$$

We can then use the prediction y_{n+1} to compute a new slope at t_{n+1} by defining $k_2 = f(t_{n+1}, y_{n+1})$. We define the new value of y_{n+1} by taking the average of the two slopes, resulting in

$$y_{n+1} \approx y(t_n) + \frac{h}{2}(k_1 + k_2). \tag{5.38}$$

The algorithm for this strategy of calculating y_{n+1} is very simple and is displayed in Listing 47.

Listing 47. Predictor-corrector algorithm.

```
1. Compute the slope at t_n, that is, define the quantity
   k_1 = f(t_, y_n).
2. Make a prediction for the solution by computing y_(i+1) by
   Euler's method.
3. Use the prediction y_(i+1) to compute a new slope at t_(i+1),
   defining the quantity k_2.
4. Correct the value of y_(i+1) by taking the average of the
   two slopes according to equation (5.38).
```

It can be shown [260] that this procedure results in a mathematical truncation which scales like $O(h^2)$, to be contrasted with Euler's method which scales like $O(h)$.

Thus, one additional function evaluation yields a better error estimate. This simple algorithm in Listing 47 conveys the philosophy of a large class of methods called *predictor-corrector methods*, see Chapter 15 in [342] for additional algorithms.

5.2.5 Runge–Kutta methods

Runge-Kutta methods are based on Taylor expansion formulas, but in general yield better algorithms for solutions of an ODE. The basic philosophy is that it provides an intermediate step in the computation of y_{n+1}. To see this, consider first the following definitions

$$\frac{dy}{dt} = f(t, y), \tag{5.39}$$

and

$$y(t) = \int f(t, y)\, dt, \tag{5.40}$$

and

$$y_{n+1} = y_n + \int_{t_n}^{t_{n+1}} f(t, y)\, dt. \tag{5.41}$$

To demonstrate the strategy behind Runge-Kutta methods, let us consider the second-order Runge-Kutta method, RK2. The first approximation consists of a Taylor expansion of $f(t, y)$ around the center of the integration interval t_n to t_{n+1}, i.e., at $t_n + h/2$, h being the step size. Using the midpoint formula for an integral defining $y(t_n + h/2) = y_{n+1/2}$ and $t_n + h/2 = t_{n+1/2}$, we obtain

$$\int_{t_i}^{t_{i+1}} f(t, y)\, dt \approx hf(t_{n+1/2}, y_{i+1/2}) + O(h^3). \tag{5.42}$$

This in turn means that we have

$$y_{n+1} = y_n + hf(t_{n+1/2}, y_{n+1/2}) + O(h^3). \tag{5.43}$$

However, we do not know the value of $y_{i+1/2}$. Thus, we arrive at the next approximation, where we use Euler's method to approximate $y_{n+1/2}$. We then have

$$y_{n+1/2} = y_n + \frac{h}{2}\frac{dy}{dt} = y(t_n) + \frac{h}{2}f(t_n, y_n). \tag{5.44}$$

This means that we can define the following *algorithm for the second-order Runge–Kutta method, RK2*.

$$k_1 = hf(t_n, y_n), \tag{5.45a}$$
$$k_2 = hf(t_{n+1/2}, y_n + k_{1/2}), \tag{5.45b}$$
$$y_{n+1} \approx y_n + k_2 + (h^3). \tag{5.45c}$$

Section 5.2 Numerical integration of differential equations

The difference with the previous one-step methods is that we now need an intermediate step in our evaluation, namely $t_n + h/2 = t_{n+1/2}$, where we evaluate the derivative of f, see Figure 5.7. This involves more operations, but we gain better stability of the

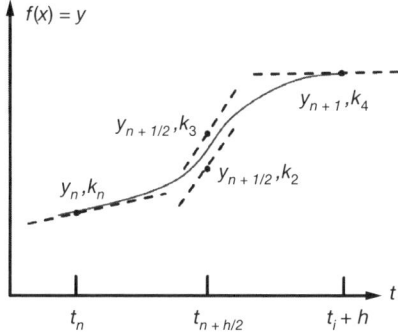

Figure 5.7. Geometrical interpretation of the fourth-order Runge-Kutta method. The derivative is evaluated at four points, once at the initial point, twice at the trial midpoint and once at the trial endpoint. These four derivatives constitute one Runge-Kutta step, resulting in a final value for $y_{n+1} = y_n + 1/6(k_1 + 2k_2 + 2k_3 + k_4)$.

solution. The *fourth-order Runge–Kutta, RK4*, which we will employ in a case study below, is easily derived. The steps are as follows. We start again with the equation

$$y_{n+1} = y_n + \int_{t_n}^{t_{n+1}} f(t, y)\, dt, \tag{5.46}$$

but instead of approximating the integral using the midpoint rule, we now use Simpson's rule at $t_{n+h/2}$, h being the step. Using Simpson's formula for an integral, defining $y(t_n + h/2) = y_{n+1/2}$ and $t_{n+h/2} = t_{n+1/2}$, we obtain

$$\int_{t_n}^{t_{n+1}} f(t, y)\, dt \approx \frac{h}{6}\left[f(t_n, y_n) + 4f(t_{n+1/2}, y_{n+1/2}) + f(t_{n+1}, y_{n+1})\right] + O(h^5). \tag{5.47}$$

This in turn means that we have

$$y_{n+1} = y_n + \frac{h}{6}\bigl[f(t_n, y_n) + 2f(t_{n+1/2}, y_{n+1/2}) + 2f(t_{n+1/2}, y_{n+1/2}) \\ + f(t_{n+1}, y_{n+1})\bigr] + O(h^5), \tag{5.48}$$

since we want to approximate the slope at $y_{n+1/2}$ in two steps. The first two function evaluations are the same as for RK2. *The algorithm for RK4 is as follows:*

1. First compute

$$k_1 = hf(t_n, y_n), \qquad (5.49)$$

which is nothing but the slope at t_n. If one stops here, this is Euler's method.

2. Compute the slope at the midpoint using Euler's method to predict $y_{n+1/2}$, as in RK2. This leads to the computation of

$$k_2 = hf(t_n + h/2, y_n + k_1/2). \qquad (5.50)$$

3. The improved slope at the midpoint is used to further improve the slope of $y_{n+1/2}$ by computing

$$k_3 = hf(t_n + h/2, y_n + k_2/2). \qquad (5.51)$$

4. With the latter slope, we can in turn predict the value of y_{n+1} via the computation of

$$k_4 = hf(t_n + h, y_n + k_3). \qquad (5.52)$$

5. Finally, calculate

$$y_{n+1} = y_n + \frac{1}{6}(k_1 + 2k_2 + 2k_3 + k_4). \qquad (5.53)$$

Thus, the algorithm consists of first calculating k_1 with t_n, y_1 and f as input. Thereafter, we increase the step size by $h/2$ and calculate k_2, then k_3 and finally k_4. With this caveat, we can then obtain the new value for the variable y, which results in four function evaluations, but the accuracy is increased by two orders, compared with the second-order Runge-Kutta method. The fourth order Runge-Kutta method has a global truncation error which scales like $O(h^4)$.

5.3 Integrating Newton's equation of motion: the Verlet algorithm

For discretizing Newton's equation of motion

$$m_i \ddot{\vec{r}} = \vec{F}_i, \qquad (5.54)$$

the integration scheme used most often is the Verlet algorithm [423, 424], which is

Section 5.3 Integrating Newton's equation of motion: the Verlet algorithm

used in computational science in three common variants:

- the original Verlet algorithm [423]
- the Verlet leapfrog algorithm
- the Verlet velocity algorithm [424]

All three methods are of order $\mathcal{O}(\delta t^2)$. From now on, we use δt for the temporal discretization step instead of h. The original method is based on the difference operators introduced in Chapter 2 and is obtained by *directly inserting* the discrete difference operator for the second derivative of equation (2.31) from the Info Box on Page 169 into (5.54)

$$m_i \frac{1}{\delta t^2} \left(\vec{r}_i^{\,n+1} - 2\vec{r}_i^{\,n} + \vec{r}_i^{\,n-1} \right) = \vec{F}_i^{\,n}. \tag{5.55}$$

In (5.55) the abbreviation $\vec{x}_i^{\,n} := \vec{r}_i(t_n)$ has been introduced (and analogously for $\vec{v}_i$ and $\vec{F}_i$). Solving (5.55) for the positions $\vec{r}_i^{\,n+1}$ at time $t_n + \delta t$ one obtains

$$\vec{r}_i^{\,n+1} = 2\vec{r}_i^{\,n} - \vec{r}_i^{\,n-1} + \frac{\delta t^2 \cdot \vec{F}_i^{\,n}}{m_i}, \tag{5.56}$$

which involves calculating the force $\vec{F}_i$ at time t_n.[14]

The *original Verlet method* [423], displayed in (5.56), has the disadvantage that two numbers are added that are very different in size (as the force is multiplied with the small number δt^2)[15], thus possibly leading to large numerical errors. Also, the velocity is not *directly* calculated in this scheme and consequently has to be obtained by averaging, e.g. by using the central difference method

$$\vec{v}_i^{\,n} = \frac{\vec{r}_i^{\,n-1} + \vec{r}_i^{\,n+1}}{2\,\delta t}. \tag{5.57}$$

However, the velocity is needed for example for calculating the energy of the system.

The *Verlet leapfrog algorithm* is algebraically identical to the original method but calculates the velocities at time $t_{n+1/2}$. This method first calculates the velocities $\vec{v}_i^{\,n+1/2}$ from the velocities at time $t_{n-\delta 1/2}$ and the forces at time t_n, according to

$$\vec{v}_i^{\,n+1/2} = \vec{v}_i^{\,n-1/2} + \frac{\delta t}{m_i} \vec{F}_i^{\,n}. \tag{5.58}$$

The positions x_i^{n+1} are then determined as

$$\vec{r}_i^{\,n+1} = \vec{r}_i^{\,n} + \delta t\, \vec{v}_i^{\,n+1/2} \tag{5.59}$$

---

[14] An integration scheme where a function on one side of an equation is calculated at time t_n and a function on the other side at time t_{n+1} is called an *explicit* integration scheme.

[15] The time step δt used in a simulation is chosen as $\ll 1$ in order to obtain a stable integration scheme.

which involves the positions at time t_n and the velocities $v_{n+1/2}$ just computed. The Verlet leapfrog variant avoids adding up terms that are very different in size, but has the disadvantage of calculating the velocities and positions at *different times* (namely at $t_{n+1/2}$ and t_n, respectively). Thus, one cannot readily calculate the total energy[16]

$$E_{\text{tot}} = E_{\text{kin}} + E_{\text{pot}} = \sum_{i=1}^{N} \frac{\vec{p}_i{}^2}{2m_i} + V(\vec{r}_1, \dots, \vec{r}_N) \tag{5.60}$$

of the considered system, because the kinetic part depends on the velocity at time $t_{n+1/2}$ and the potential part depends on the position at time t_n. Therefore, here again, the velocity at time t_n has to be calculated similar to (5.57) by e.g. using the average

$$\vec{v}_i{}^n = \frac{\vec{v}_i{}^{n-1/2} + \vec{v}_i{}^{n+1/2}}{2}. \tag{5.61}$$

Think!

> Show that the total energy E_{tot} is a conserved quantity for a system with a conservative potential.

Finally, the *Verlet velocity algorithm* [424] calculates all quantities at the same time t_n, including the velocities. It also has excellent long-term stability properties in terms of rounding errors, particularly for the energy, and is therefore practically *the* integration scheme used exclusively for any MD simulation involving ODE:

$$\vec{r}_i{}^{n+1} = \vec{r}_i{}^n + \delta t \, \vec{v}_i{}^n + \frac{\vec{F}_i{}^n \delta t^2}{2m_i}, \tag{5.62a}$$

$$\vec{v}_i{}^{n+1} = \vec{v}_i{}^n + \frac{(\vec{F}_i{}^{n+1} + \vec{F}_i{}^n) \delta t^2}{2m_i}. \tag{5.62b}$$

---

[16] If the potential of a system is *conservative*, i.e. not explicitly time-dependent, the total energy E_{tot} is identical with the Hamiltonian $\mathcal{H}$ of the system and a conserved quantity.

Section 5.4 The basic MD algorithm

> **Criteria for good integration algorithms (Solvers):** The most important criteria for the choice of a particular integration algorithm[a] is the long-term stability with respect to the *conservation of energy*. Other important criteria are:
>
> - robustness (with respect to the choice of time step δt)
>
> - time-reversibility
>
> - conservation of the volume of a system in phase space[b]
>
> The Verlet algorithm – and in particular its velocity variant – is robust, time-reversible, conserves the volume in phase space and has excellent properties with respect to energy conservation, even after hundreds of millions of integration steps. Thus, this is the recommended algorithm to be used for MD simulations.
>
> ---
>
> [a] Particularly in the context of engineering applications in fluid dynamics or solid dynamics one likes to speak of "solvers" instead of "integrators".
> [b] An algorithm with this property is called *symplectic*.

Equation (5.62) can be derived as follows: starting from the central difference for the velocity at time t_n in equation (5.57), solving for $\vec{r}_i^{\,n-1}$, substituting the result into the original Verlet algorithm in equation (5.56), and then solving for $\vec{x}_i^{\,n+1}$, one obtains equation (5.62a). Solving equation (5.62a) for $\vec{v}_i{}^n$ yields

$$\vec{v}_i{}^n = \frac{\vec{r}_i^{\,n}}{\delta t} - \frac{\vec{r}_i^{\,n-1}}{\delta t} - \frac{\vec{F}_i^{\,n}\delta t}{2m_i}. \tag{5.63}$$

Adding the corresponding expression for $\vec{v}_i{}^{n+1}$ (by substituting n with $n+1$ in equation (5.63)) yields:

$$\vec{v}_i{}^{n+1} + \vec{v}_i{}^n = \frac{\vec{r}_i^{\,n+1} - \vec{r}_i^{\,n-1}}{\delta t} + \frac{(\vec{F}_i^{\,n+1} + \vec{F}_i^{\,n})\delta t}{2m_i}. \tag{5.64}$$

Finally, using equation (5.63) together with (5.64) yields equation (5.62b).

5.4 The basic MD algorithm

In the MD simulation of particle models, the time evolution of a system of interacting particles is determined by the integration of the EOM. Here, one can follow individual particles, see how they collide, repel each other, attract each other, how several particles are bound to each other, or separate from each other. Distances and angles between several particles, and similar geometric quantities can also be computed and observed over time. Such measurements allow the computation of relevant macroscopic variables such as kinetic or potential energy, pressure, diffusion constants, transport coefficients, structure factors, spectral density functions, distribution functions, and many more. After having chosen appropriate interaction potentials, the four principal components of MD simulations of particles are:

1. generate initial conditions for all particles
2. calculate the forces
3. integrate the EOM
4. consider boundary conditions (e.g. periodic in certain directions)

Listing 48. Basic MD algorithm (pseudocode).

```
Input : Initial configuration of particles
set (t = timeStart) and (timeEnd > timeStart)
loop (i = 1 to N)
   set initial conditions for x_i, v_i and F_i

while (timeStep < timeEnd){
   loop (j = 1 to N){
      calculate forces
      integrate EOM
      consider boundary conditions
   }
   store snapshots of data on filesystem
   set t = t + deltaT
}
Output : Stored data on filesystem
```

If we denote the position, velocity and force of the ith particle with x_i, v_i and F_i, respectively, we can write the four basic MD steps as pseudocode[17] as displayed in Listing 48. Based on the initial values chosen for x_i, v_i and F_i (usually all F_i are simply set to zero), for $i \in \{1, \ldots, N\}$, in each integration step within a loop starting at time $t = 0$, the time is increased by Δt, until the final time step $t = t_{end}$ is reached. The forces on the individual particles and their new positions and velocities are then computed by integrating the EOM over all particles within an inner loop.

In principle, when calculating the forces of N particles one has to consider $O(N^2)$ interactions between them. So, in the most naive approach to calculating the forces one has a double loop of the kind:

```
for (i = 0; i < N; i++){
   for (j = 0; j < N; j++){
      CalculateForces(i, j);
   }
}
```

---

[17] In the pseudocode, in order to be unambiguous, we use brackets along with indentation.

Section 5.4 The basic MD algorithm   325

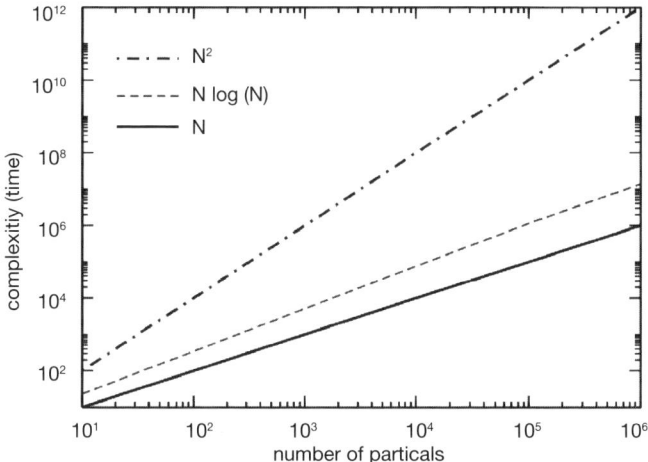

Figure 5.8. Efficiency of force calculations in MD. Comparison of the runtime behavior of a linear, an $N \log N$ and an N^2 algorithm for the calculation of the forces between N particles in an MD simulation. One always tries to come as close as possible to the optimal linear algorithm.

If self-interactions ($i = j$) are excluded, the computational complexity is still $O(N^2)$. Thus, if the number of particles is doubled, the number of operations quadruples. If we further consider that the double loop actually counts all interactions twice[18], which can be avoided due to Newton's Lex III (see the footnote on Page 61), one still has $(N^2 - N)/2$ interactions to compute, but the loop can be changed to

```
for (i = 0; i < N; i++){
    for (j = i + 1; j < N; j++){ // Avoid counting interactions twice
        CalculateForces(i, j);
    }
}
```

Because of the limited performance of computers, this inefficient $O(N^2)$ approach to the computation of the forces is only feasible for relatively small numbers of particles. However, if only an approximation of the forces up to a certain accuracy is required, a substantial reduction of the complexity is possible. The complexity of an approximate evaluation of the forces at a fixed time is obviously at least of order $O(N)$ since every particle has to be processed at least once. Algorithms are called *optimal* if the complexity for the computation, up to a given accuracy is $O(N)$. If the complexity

---

[18] The force $\vec{F}_{ij}$ that particle i exerts on particle j is opposite and equal to the force $\vec{F}_{ji}$ that particle j exerts on particle i, i.e. $\vec{F}_{ij} = -\vec{F}_{ji}$.

of the algorithm differs from the optimal, e.g. by a logarithmic factor[19], meaning it is of the order $O(N \log(N))$, the algorithm is called *quasi-optimal*. Figure 5.8 shows a graphical comparison of $O(N^2)$, $O(N \log(N))$ and $O(N)$ algorithms. As one can see, the time needed by the $O(N^2)$-algorithm to calculate the interaction of only 1000 particles is roughly the same as the optimal linear algorithm needs for the calculation of the interaction of more than one million particles.

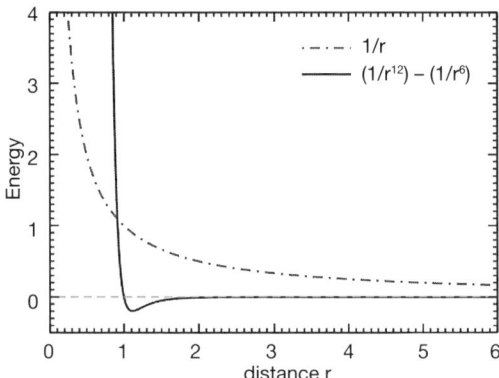

Figure 5.9. Comparison of the energy for a short-range Lennard-Jones (LJ) potential and a long-range Coulomb potential ($\propto 1/r$). As one can see, the $1/r$ potential approaches zero much slower than the LJ potential, which is practically zero at a distance of less than 2 (in reduced simulation units, where the unit of distance is set to one). This justifies the general practice with short-range potentials of introducing a cutoff in the interaction. For the Coulomb potential, a cutoff set e.g. at a distance of $r = 2$ would mean a large error in the energy calculation (but not for the LJ potential), which is why the calculation of forces with long-range potentials needs more sophisticated treatment.

The design of a suitable algorithm for the force calculation necessarily has to be adapted to the kind of interactions modeled and to other "global" simulation parameters, such as changes in the overall density of the particles[20]. It is clear that algorithms which are optimal for some form of interaction potentials may not be suitable for others. This adds some complexity to MD simulations, which can be most easily demonstrated by considering the difference between a potential that decays rapidly (e.g. LJ) and one that decays slowly (e.g. Coulomb), see Figure 5.9. Here, for a fast decaying LJ potential, a particle only exerts a significant force on another particle if

---

[19] The complexity $N \log N$ of an algorithm is almost always an indication of a *divide and conquer* solution of the problem.

[20] The reason for this is – as we will see – that the simulation box is usually subdivided into subcells (smaller boxes) with the particles sorted into these subcells and stored in an abstract datatype, called a linked list, for each subcell. The calculation of the interaction then simply involves a loop over the particles in the *linked lists* of all subcells. However, if the particles are strongly homogeneously distributed among the subcells, many loops over empty cells are done, which is inefficient.

the distance between the two particles is small. The graph also schematically shows a $1/r$ behavior is typical for long-range potentials. For very small values of distance r both potentials are very large and decrease rapidly with increasing distance. Thus, in this case, a small change in the position of the particle has a very strong effect on the resulting potential and the resulting force. In contrast, for large distances r, a small change in the position of the particles only has a very small effect on the resulting potential value and the corresponding force. In particular, in the case of very large r, a small change in position affects the energy value much less than at small distances. Here, for the approximate evaluation of potentials and forces, one does not have to distinguish between two particles close to each other, since the resulting values of potential and force will be approximately the same. This fact is exploited in algorithms for long-range potentials.

5.5 Basic MD: planetary motion

The key issue of the MD method is to solve the (classical) N-body problem, in an attempt to predict and explain the properties of complex systems (solid states, gases, fluids, engineering or biological materials) based on the assumption that these systems are composed of many small constituents, called particles. In this section, I will introduce a basic MD code that uses the Verlet algorithm to solve the problem of planetary motion, i.e. we integrate Newton's law of gravitation:

$$\vec{F} = -\nabla \Phi(\vec{r}) = -\Gamma m_i m_j \frac{\vec{r}_i - \vec{r}_j}{|\vec{r}_i - \vec{r}_j|^3}. \tag{5.65}$$

In the following code listings, all elements needed to write such a program are introduced. The various codes in the listings are stored in *one* big file, named, say "PlanetaryMotion.c" which can then be compiled, using `gcc -o Planets PlanetaryMotion.c`, which creates a binary "Planets". Here, we use this one big file approach on purpose, so that the reader can practice writing modules in our suggested project at the end of this section. The complete source code of Listing 49 is available at the De Gruyter book's website (`www.degruyter.com`).

5.5.1 Preprocessor statements and basic definitions

We first define some variables and macros which make things a little easier.

In Listing 49 Part 1 we have introduced the standard way of commenting source code using the tool "Doxygen". Doxygen is a very useful standard tool in the Unix/Linux environment, which parses the sources and generates documentation in various formats[21] fully automatically. In order to use the tool, when documenting the sources, one has to use certain *tags* for which Doxygen searches in the source code. In the

---

[21] The supported formats are *HTML*, *latex*, *rtf*, *XML* and the Unix man page format.

Listing 49. Part 1 – Planetary motion code. Basic definitions. (*)

```
1  #include <math.h>
2  #include <stdlib.h>
3  #include <stdio.h>
4
5  #define DIM 3
6
7  /**
8      \def VERSION
9      A global variable indicating the current version of the
         application.
10 */
11 #define VERSION 1.0
12
13 /** \def sqr(x)
14     This is the definition of the squared value of a number.
15 */
16 #define sqr(x) ((x) * (x));
```

example of the listing above, we used the definition tag def which has to be written within /** */ to be recognized by Doxygen. You will find good documentation of Doxygen on the webpage www.doxygen.org.

5.5.2 Organization of the data

Generally speaking, data is usually best organized by using structs, as shown in Listing 49 Part 2.

Listing 49. Part 2 – Planetary motion code. Data organization. (*)

```
17 /**
18     \struct Particle
19 */
20 typedef struct {
21     double m;            /* mass */
22     double x[DIM];       /* position */
23     double v[DIM];       /* velocity */
24     double f[DIM];       /* force */
25     double fOld[DIM];    /* Old force - needed for Verlet-velocity
           */
26 } Particle;
```

5.5.3 Function that computes the energy

Listing 49 Part 3 shows the function *computeStatistics()* that calculates the energy of the particles and could be used to print out (or store on disk) information about the system such as the current potential and kinetic energy and their averages. (This is left as an exercise to the reader).

Listing 49. Part 3 – Planetary motion code. Energy calculation. (*)

```
27  void ComputeStatistics(Particle *p, int N, double time){
28    int i, d;
29    double velocity;
30    double energy = 0.;
31
32    for (i = 0; i < N; i++){
33
34      velocity = 0.;
35
36      for (d = 0; d < DIM; d++)
37        velocity += sqr(p[i].v[d]);
38
39      energy += .5 * p[i].m * velocity;
40    }
41    /* Optionally: insert a function here, that prints out the
         kinetic energy
42       at this time step */
43    /* printf("Check: E_kin = %f\n", energy); */
44  }
```

5.5.4 The Verlet velocity algorithm

Listing 49 Part 4 shows a straightforward implementation of the Verlet algorithm that was introduced and discussed in Section 5.3. It is useful to split the individual computations into very small functions, since this eases handling of the source code, makes it clearer to read and has advantages in case one later wants to reuse parts of this code in other modules.

5.5.5 The force calculation

Listing 49 Part 5 introduces the force calculation and should be easy to understand. It is a direct implementation of Newton's law of gravitation. The calculation is again split into two functions, since this has advantages if one wants to reuse parts of the force calculation in other sections of the code. That way, the calculation itself is encapsulated in the function *ForceCalculate()* and separated from the loop over the interacting particles that is performed in *ComputeForces()*. In the latter function, another example of the use of Doxygen tags is shown, namely, tags that allow for displaying formulas in code comments using latex commands.

Listing 49. Part 4 – Planetary motion code. Integration. (*)

```
45  void UpdateX(Particle *p, double deltaTime){
46    int i, d;
47    double a = deltaTime * .5 / p -> m;
48
49    for ( d = 0; d < DIM; d++) {
50      p -> x[d] += deltaTime * (p -> v[d] + a * p -> f[d]);
51      p -> fOld[d] = p -> f[d];
52    }
53  }
54
55  void UpdateV(Particle *p, double timeDelta){
56    int i, d;
57    double a = timeDelta * .5 / p -> m;
58    for ( d = 0; d < DIM; d++)
59      p -> v[d] += a * (p -> f[d] + p -> fOld[d]);
60  }
61
62  void ComputeX(Particle *p, int N, double deltaTime){
63    int i;
64    for (i = 0; i < N; i++)
65      updateX(&p[i], deltaTime);
66  }
67
68  void ComputeV(Particle *p, int N, double deltaTime){
69    int i;
70    for (i = 0; i < N; i++)
71      updateV(&p[i], deltaTime);
72  }
73
74  void TimeIntegration(double time,
75          double deltaTime,
76          double timeEnd,
77          Particle *p,
78          int N){
79    ComputeForces(p, N);
80
81    while (time < timeEnd) {
82      time += deltaTime;
83
84      ComputeX (p, N, deltaTime);
85      ComputeForces (p, N);
86      ComputeV (p, N, deltaTime);
87      ComputeStatistics (p, N, time);
88      OutputResults(p, N, time);
89    }
90  }
```

Section 5.5 Basic MD: planetary motion

Listing 49. Part 5 – Planetary motion code. The $O(N^2)$ force calculation. (*)

```
91  void ForceCalculate(Particle *i, Particle *j){
92    int d;
93    double distance = 0.0;
94    double force = 0.0;
95
96    for (d = 0; d < DIM; d++)
97      distance += sqr(j -> x[d] - i ->x[d]);
98
99    force = i -> m * j -> m / (sqrt(distance) * distance);
100
101   for (d = 0; d < DIM; d++)
102     i -> f[d] += force * (j -> x[d] - i -> x[d]);
103 }
104
105 /**
106    void ComputeForces(Particle *p, int N)
107
108    \brief Computation of the new forces after \em one time step
       ,
109    using the Verlet-velocity method with the all particle
          approach
110
111    \param *p A pointer to the particle information
112    \param N The total number of particles
113
114    In the basic version of the code, the force is calculated
          according to
115    \f[
116    \vec{F}_i=-\nabla_{\vec{x}_i} V(\vec{x}_1,...,\vec{x}_N) =
          -\sum_{j=1,j\neq i}^{N}
117    \nabla_{\vec{x}_i}U(r_{ij})=\sum_{j=1,j\neq i}^{N}\vec{F}_{
          ij}
118    \f], i.e.\ the two forces \f$\vec{F}_{ij}\f$ and   \f$\vec{F}
          _{ji}\f$ are calculated
119    separately.\n\n
120 */
121 void ComputeForces(Particle *p, int N){
122   int d,i,j;
123
124   for (i = 0; i < N; i++) // Comment
125     for (d = 0; d < DIM; d++)
126       p[i].f[d] = 0.0;
127   /* All forces are now set to zero */
128
129   for (i = 0; i < N; i++)
130     for (j = 0; j < N ; j++)
131       if (i != j) ForceCalculate(&p[i], &p[j]);
132 }
```

5.5.6 The initialization and output functions

The initialization function in Listing 49 Part 6 simply initializes the masses, initial positions and velocities of the different particles. As one can see, the masses are scaled, relative to the mass of the sun. The function *OutputResults()* is a simple *printf()*-statement which prints – in this case – only the information about the sun on stdout (the screen). If one wants to output more information, one has to add more *print()*-statements or to put the statement within a loop.

5.5.7 The *main()*-function

Finally, we list the *main()*-function in Listing 49 Part 7, which comes at the end of our file *PlanetaryMotion.c*, because we didn't use header files in this example, i.e. we did not include function prototypes, the definition of which can be found somewhere else in corresponding *.c files. We will, however, reorganize this code in the case study suggested below. This is then left as a very important exercise to the reader, because it shows all essential features of *all* code development and follows best coding practices (see Chapter 2). As one can see, by using function definitions that are included with prototypes, the *main()*-function clearly shows the logical flow of the program. Note that $N = 4$ in line 164, because here we only consider 4 objects of the solar system (2 planets, a comet and the sun). If one wants to add objects, then this number has to be adjusted accordingly. Note, that in Listing 49 Part 6 we have used a *static* declaration before the *fprinf()*-statement in line 154, so as to only print a commentary once. This is an often useful "trick". An example of how this simulation may be visualized is presented in a series of snapshots in Figure 5.10.

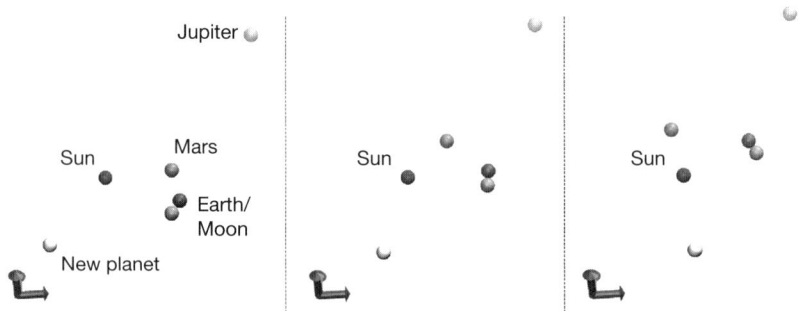

Figure 5.10. Snapshots of planetary motion, generated with the VMD tool, see http://www.ks.uiuc.edu/Research/vmd/. The integration steps are (from left to right). $\delta T = 20, 22, 26$ in reduced (dimensionless) simulation units.

The VMD tool is a very useful, *free tool* for the visualization of scientific data, in particular of proteins and macromolecules. It uses the Protein Data Bank (PDB) format,

Section 5.5   Basic MD: planetary motion   333

Listing 49. Part 6 – Planetary motion code. Initialization and output. (*)

```
133  void InitAllData(Particle *p)
134  {
135      /* The initial data for the case study of planetary motion.
            This study works in
136         2D AND 3D. The particle information is always allocated as
            3D, but in
137         the 2D, case the 3D information is ignored in the
            corresponding loops. */
138
139      /* Sun */
140      p[0].m = 1.; p[0].x[0] = p[0].x[1] = p[0].x[2] = 0.;    p[0].v
            [0] = p[0].v[1] = p[0].v[2] = 0.;
141
142      /* Earth */
143      p[1].m = .000003; p[1].x[0] = 0; p[1].x[1] = 1.; p[1].x[2] =
            0.;   p[1].v[0] = -1.0; p[1].v[1] = 0.; p[1].v[2] = 0.;
144
145      /* Jupiter */
146      p[2].m = .000955;   p[2].x[0] = 0.; p[2].x[1] = 5.36; p[2].x
            [2] = 0.;  p[2].v[0] = -0.425; p[2].v[1] = p[2].v[2] = 0.;
147
148      /* Comet */
149      p[3].m = .00000000000001;   p[3].x[0] = 34.75; p[3].x[1] = p
            [3].x[2] = 0.;   p[3].v[0] = 0; p[3].v[1] = 0.0296; p[3].
            v[2] = 0.;
150  }
151
152  void OutputResults(Particle *p, int N, double time){
153
154      static int once = 1;
155
156      if (once){
157         printf("# Sun:   time X  Y  Z\n");
158         once = 0;
159      }
160
161      printf(" %f %f %f %f\n",time, p[0].x[0], p[0].x[1], p[0].x
            [2]);
162  }
```

Listing 49. Part 7 – Planetary motion code. The *main()*-function. (*)

```
163  int main (void){
164    int      N          = 4;
165    double   timeDelta  = 0.015;
166    double   timeEnd    = 468.5;
167
168    Particle *p = (Particle*) malloc(N * sizeof (*p));
169
170    InitAllData(p);
171
172    TimeIntegration(0, timeDelta, timeEnd, p, N);
173    free(p);
174    return (0);
175  }
```

which is tailored to the detailed description of NMR spectroscopic data of proteins and nucleic acids. It can also be used to visualize much simpler systems, such as planets, represented as spheres, that circle around the sun. VMD can read XYZ information of particle positions and many more formats. Here, I provide a template function *WritePDBOutput()* in Listing 50, with an interface that can be used to produce PDB snapshot files in any particle code. The nice thing about this is that many snapshots are saved in one file, so that VMD can even make a movie of the individual snapshots. This file is also used and is provided in the suggested project at the end of this section.

The function in Listing 50 is an example of the exchange of programs with a given interface. One can directly use this function in any code and call the function with the appropriate data types given in the its argument list. Here, the function expects as its input the number of particles Nparticles as *int*, an address that points to the starting address of an array of *double*s that contains the positions (x, y, z) of the particles, and a pointer to *char* pdbfile[], which is simply a string that contains the name of the file in which the PDB data are to be written.

What will happen if you exchange the two positions of the functions *ForceCalculate()* and *ComputeForces()* of Listing 49 Part 5 in the source code file *PlanetaryMotion.c*?

Listing 50. Function *WritePDBOutput()*.

```
void WritePDBOutput (
   int Nparticles,
   double* positions,
   char pdbfile[])
{
   int i;
   int Countatom=0;
   FILE* ofile;
   ofile = fopen(pdbfile, "a+"); // open file
   fprintf(ofile, "%s %9d\n", "Teilchen",0); // write pdb header

   for ( i = 0; i < 3*N_particles; i+=3) { // for all particles
      Countatom++;
      // print a line in pdb format:
      fprintf(ofile, "ATOM %6d%4s   UNX F%4d    %8.3f%8.3f%8.3f
         0.00  0.00       T%03d\n",
         Countatom,"FE",Countatom,10.*positions[i+0],10.*positions
            [i+1],10.*positions[i+2],i);
   }
   fprintf(ofile, "%s\n", "ENDMDL"); // write pdb file ending
   fclose(ofile);
}
```

5.6 Planetary motion: suggested project

I suggest to perform the following tasks with the above program for planetary motion:

1. Change *OutputResults()* so that the data of *all* objects can be stored in a file.

2. Change the code in such a way that the initial conditions for the simulation run are read from an ASCII-text file, called "InputParams.dat", including the number of total particles N, i.e. provide a text file that includes all necessary parameters for running the code and the initial conditions of the particles. Next, write an appropriate function that reads this file at the beginning of the program and initializes the corresponding local variables. By changing the code design this way, one does not have to hardcode the initial conditions in the source code but can do this in a file, separately from the code. Thus, when changing the initial conditions or the number of particles, the whole code does not have to be recompiled every single time.

3. Change the code in such a way, that you provide a *restart-functionality*, i.e. save all relevant data after, say, half the total number of time steps, in a file called SYSTEM.restart". These files must contain all relevant information about the system, so as to be able to continue the simulation from this system snapshot. Such restart

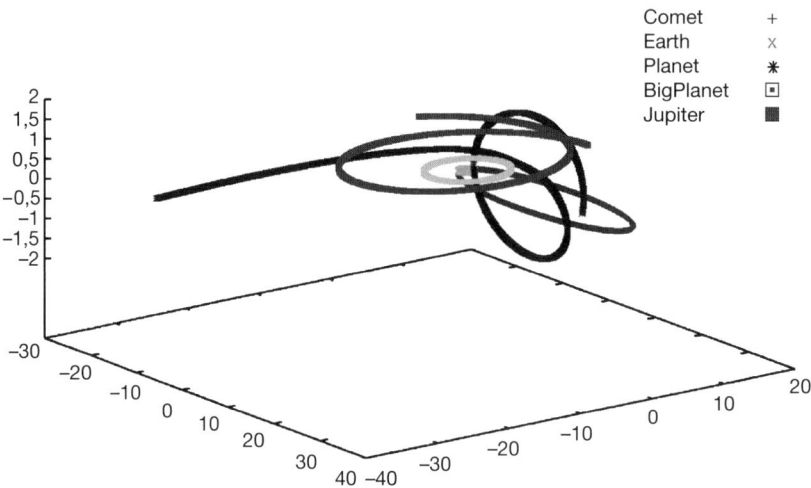

Figure 5.11. Snapshot of planetary motion in 3D. This plot was generated using the Linux/UNIX tool "gnuplot", see www.gnuplot.info.

functionality should always be provided in MD codes as, usually, real production runs may take several days to finish on a computer cluster, so you need this, in case something goes wrong (i.e. an unprecedented shutdown of the cluster, a file system crash, etc.). The restart functionality allows you to simply continue your simulation run from the saved moment in time. Write an appropriate function that implements this functionality.

4. Change the code in such a way that the name of the ASCII inputFile and the name of the System-restart file can be provided as parameters to the function *main()*, i.e. define local variables in the code and initialize them with the command line arguments of *main()* when starting the program in a command shell.

5. Write a makefile for this program, which allows you to conveniently compile the sources by typing the command "make" and which includes the command "make clean", which removes all object files and executables in the current working directory. You can use the sample makefile provided on Page 96.

6. Add two more masses to the program with corresponding initial conditions

```
m_Planet = 0.0005; PosPlanet_0 =(25.0,0,0);\
VeloPlanet0 = (0,0.02,0.02)
m_BigPlanet = 0.1; PosBigPlanet_0 =(0,17.0,0);\
VeloBigPlanet0 = (-0.25,0,0)
```

Run the code and make a print of the funny trajectories of all planets, using a 3D perspective. You should obtain a picture of the trajectories similar to Figure 5.11.

5.7 Periodic boundary conditions

In an MD simulation only a very small number of particles can be considered. To avoid the (usually) undesired artificial effects of surface particles not surrounded by neighboring particles in any direction and thus subject to non-isotropic forces, one introduces *periodic boundary conditions*. Using this technique, one measures the "bulk" properties of the system, due to particles located far away from surfaces. As a rule, one uses a cubic simulation box, in which the particles are located. This cubic box is periodically repeated in all directions. If, during a simulation run, a particle leaves the central simulation box, one of its image particles enters the central box from the opposite direction. Each of the image particles in the neighboring boxes moves in exactly the same way, cf. Figure 5.12 for a two dimensional visualization.

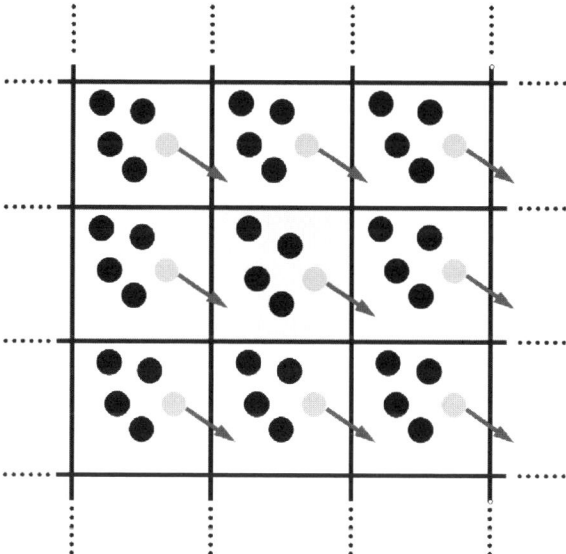

Figure 5.12. Two-dimensional scheme of periodic boundary conditions. The particle trajectories in the central simulation box are copied in every direction.

In simulations with periodic boundaries, the cubic box is almost exclusively used, mainly due to its simplicity, however spherical boundary conditions, where the three-dimensional surface of the sphere induces a non-Euclidean metric, have also been investigated[15, 259]. The use of periodic boundary conditions allows for a simulation of bulk properties of systems with a relatively small number of particles.

5.8 Minimum image convention

The question as to whether the measured properties obtained with a small, periodically extended system are to be regarded as representative for the modeled system depends on the specific observable that is investigated as well as on the range of the intermolecular potential. For a LJ potential (equation (4.26)) with additional cutoff, no particle can interact with one of its images and be thus exposed to the artificial periodic box structure imposed upon the system. For long range forces, interactions of far away particles also have to be included. Thus, for such systems the periodic box structure is superimposed, although they are actually isotropic. Therefore, for long-range forces one only takes into account those contributions to the energy of each of the particles, which is contributed by a particle within a cutoff radius that is at the most $1/2 L_B$, with box length L_B. This procedure is called *minimum image convention*. Using the minimum image convention, each particle interacts with at most $(N-1)$ particles. Particularly for ionic systems, a cutoff must be chosen such that the electroneutrality is not violated.

5.9 Lyapunov instability

The Lyapunov instability of atomic or molecular systems is a consequence of the (basically) convex surface of the particles, which leads to *molecular chaos*. An infinitesimal perturbation of the initial conditions is modified during each interaction event with neighboring particles and grows linearly in time. On average, this leads to *exponential growth* in time, see Figure 5.13, which can be characterized by a finite number of time-averaged rate constants, the so-called *Lyapunov exponents* [277, 441, 278]. The number of exponents is equal to the dimension of the phase space D, and the whole set, $\{\lambda_1, \ldots, \lambda_D\}$, ordered according to size, $\lambda_1 \geq \lambda_l = 1$, is known as the *Lyapunov spectrum*. Thus, *any* small error in the numerical integration of the equations of motion will blow up exponentially, and this is true for *any* algorithm [142, 336].

> What is the point of simulating the dynamics of a system using MD if we cannot trust the resulting time evolution of the system?

In order to see why MD is a useful simulation method that can be trusted despite the Lyapunov instability, we have to go back to the Lagrangian formulation of Newtonian mechanics and apply yet another important principle of physics, the *principle of least action*, which can be states as follows:

$$S = \int_0^t dt' L(\vec{r}(t), \vec{v}(t), t) = \text{extremum}, \tag{5.66}$$

Section 5.9 Lyapunov instability

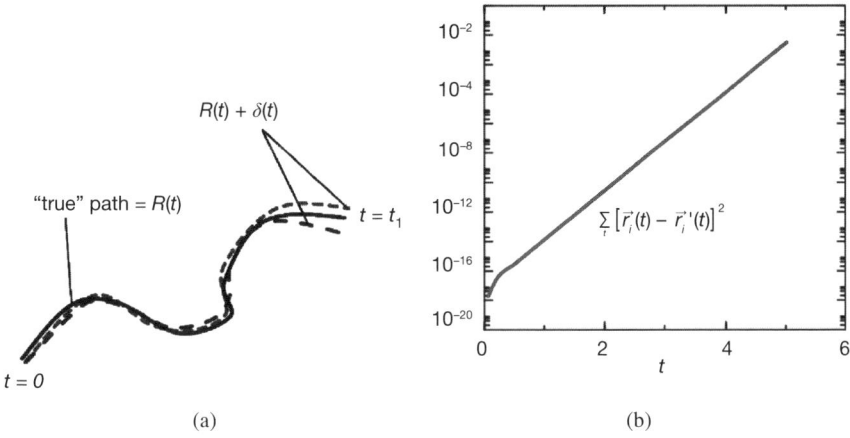

Figure 5.13. Lyapunov instability. (a) Two particle trajectories with only slightly different initial conditions will eventually diverge exponentially. (b) The distance of the trajectories of two particles in a molecular LJ fluid.

where S is the action[22], i.e. a functional of a system's trajectory, expressed by the Lagrangian L, generally a function of position $\vec{r}$, velocity $\vec{v}$ and time t, compare our discussion in Section 3.1.4 on Page 225.

For conservative forces, i.e. forces that do not depend on velocity $L = E - U$, where E is the kinetic energy of the system and U is the potential energy, that is,

$$L(\vec{r}(t)) = E - U(\vec{r}) = \sum_{i=1}^{N} \frac{1}{2} m_i \vec{v}_i^{\,2} - U(\{\vec{r}_i\}), \tag{5.67}$$

where index i in the potential energy runs from 1 to N.

Now, let us consider a variation of the functional in equation (5.66) by looking at a path $\vec{r}(t)$ close to the true path of a particle in phase space, i.e.

$$\vec{r}(t) = \vec{R}(t) + \delta\vec{r}(t). \tag{5.68}$$

Then, the action S takes on an extremum if

$$\frac{\delta S}{\delta \vec{r}(t)} = 0 \quad \forall t. \tag{5.69}$$

The discretized version of equation (5.66) is

$$S_{\text{discrete}} = \Delta t \sum_{i=1}^{i_{\max}} L(t_i). \tag{5.70}$$

---

[22] The dimension of the action is [Energy × time=Js].

For simplicity, we consider only the x-coordinate in one dimension and assume all masses $m_i = m$. Then, the Lagrangian is

$$L(t_i)\Delta t = \frac{1}{2} m \Delta t \frac{(x_{i+1} - x_i)^2}{\Delta t^2} - U(x_i) \Delta_t, \tag{5.71}$$

and thus, the discretized version of the action is:

$$S_{\text{discrete}} = \sum_{i=1}^{i_{\max}} \left[\frac{m(x_{i+1} - x_i)^2}{2\Delta t} - U(x_i)\Delta t \right]. \tag{5.72}$$

Next, we look for small variations along the discretized trajectory, i.e. for small variations in x_i, for which

$$\frac{\partial S_{\text{discrete}}}{\partial x_i} = 0 \quad \forall i. \tag{5.73}$$

The variation yields:

$$\frac{\partial S_{\text{discrete}}}{\partial x_i} = \frac{\partial}{\partial x_i} \sum_{i=1}^{i_{\max}} \left[\frac{m(x_{i+1} - x_i)^2}{2\Delta t} - U(x_i)\Delta t \right] \tag{5.74}$$

$$= \frac{-m(x_{i+1} - x_i) + m(x_i - x_{i-1})}{\Delta t} - \Delta t \frac{U(x_i)}{\partial x_i}, \tag{5.75}$$

and thus:

$$0 = \frac{m}{\Delta t}\left[2x_i - x_{i+1} - x_{i-1} - \frac{\Delta t^2}{m} \frac{\partial U(x_i)}{\partial x_i} \right]$$

$$= \left[2x_i - x_{i+1} - x_{i-1} - \frac{\Delta t^2}{m} \frac{\partial U(x_i)}{\partial x_i} \right]. \tag{5.76}$$

Finally, equation (5.76) can be rewritten as:

$$x_{i+1} = 2x_i - x_{i-1} + \frac{\Delta t^2}{m} F(x_i), \tag{5.77}$$

which is nothing else than the Verlet algorithm.

> The Verlet algorithm generates trajectories that satisfy the boundary conditions of a "real" trajectory at the beginning and at the endpoint, i.e. the Verlet algorithm produces particle trajectories close to "real" trajectories in phase space.

5.10 Case study: static and dynamic properties of a microcanonical LJ fluid

For this case study you will have to download the file *NVE_LJFluid.c* from the book's website `www.degruyter.com`. Let us reconsider just a few of the important elements of any MD program:

1. evaluation of the forces
2. impose periodic boundaries
3. perform integration
4. output the configuration

We will understand these elements by manipulating an existing simulation program that implements the LJ fluid based on the interaction potential of equation (4.26).

Reduced units

When implementing this potential in an MD program, we adopt a *reduced unit system*, in which we measure length in σ and energy in ϵ, i.e. we use dimensionless units. Additionally, for MD, we need to measure mass, and we simply adopt units of particle mass, m. This convention makes the force on a particle numerically equivalent to its acceleration. For simple liquids, composed of particles which interact with the same pairwise additive potential, one can write the potential in general form as:

$$\Phi(r) = \varepsilon f\left(\frac{\sigma}{r}\right). \tag{5.78}$$

In this case, it is possible to define a set of dimensionless, reduced units such as $x^* = x/\sigma$, $V^* = V/\sigma^3$, $T^* = k_B T/\varepsilon$ and $P^*\sigma/\varepsilon$. Within this set of reduced units, all systems interacting with potentials of the form given in equation (5.78) follow *one single universal equation of state*. Therefore, results of simulations of simple fluids are presented in reduced units. The results can then be transferred to a particular temperature and pressure, when the values of ε and σ for a particular fluid are known. One particular implication of this is that we can perform all simulations with $\sigma = 1.0$ and $\varepsilon = 1.0$ and, if desired, transfer the results to other values of the interaction. With these conventions, time is a derived unit:

$$[t] = \sigma\sqrt{m/\epsilon} \tag{5.79}$$

For a system of identical LJ particles, the equipartition theorem of classical physics states that, for each degree of freedom, we have a contribution of $1/2 k_B T$. Thus, for a system of N particles in 3D, we have:

$$\frac{3}{2}NT = E_{\text{kin}} = \frac{1}{2}\sum_{i=1}^{N}|\vec{v}_i|^2 \tag{5.80}$$

Recall that the mass m equals 1 in reduced LJ units. These conventions obviate the need to perform unit conversions in the code.

In our chosen system of units, the force $\vec{F}_{ij}$ exerted on particle i by virtue of its LJ interaction with particle j, is obtained by $-\nabla\Phi(\vec{r})$:

$$\vec{f}_{ij}(r_{ij}) = \frac{\vec{r}_{ij}}{r_{ij}}\left\{48\varepsilon\left[\left(\frac{\sigma}{r_{ij}}\right)^{1}2 - \frac{1}{2}\left(\frac{\sigma}{r_{ij}}\right)^{1}2\right]\right\} \equiv \vec{r}_{ij}F. \tag{5.81}$$

Once we have computed the vector $\vec{F}_{ij}$, we automatically have $\vec{F}_{ji}$, because $\vec{F}_{ij} = -\vec{F}_{ji}$. The scalar f in line 21 in Listing 51 is called a "force factor". If F is negative, the force vector $\vec{F}_{ij}$ points from i to j, meaning that i is attracted to j. Likewise, if f is positive, the force vector $\vec{F}_{ij}$ points from j to i, meaning that i is being forced away from j.

Listing 51 presents a code fragment of the source code that shows how to compute both the total potential energy *and* the interparticle forces:

Notice that the argument list of *Forces()* includes arrays for the forces. Because force is a vector quantity, we have three parallel arrays for a three-dimensional system. These forces must, of course, be initialized, as shown in lines 7–11. The N^2 loop for visiting all unique pairs of particles is opened on lines 12–13. The inside of this loop is a very straightforward implementation of equation (5.81), along with the "force factor" f on line 21, and the subsequent incrementation of force vector components on lines 22–27. Notice as well that, for simplicity, there is no implementation of the periodic boundary conditions in this code fragment. The periodic boundaries are implemented by the piece of code provided in Listing 52 and have to be included within the force calculation after line 13:

Another major aspect of MD is the integrator. As discussed above the Verlet-style (explicit) integrators are used primarily. The most common version is the Verlet velocity algorithm, see equations (5.62) The code fragment in Listing 53 shows how to execute one time step of the integration, for a system of N particles: Notice the update of the positions in lines 1–8, where vx[i] is the x-component of the velocity, fx[i] is the x-component of the force, dt and dt2 are the time step and squared time step, respectively. lines 5–7 are the first half of the velocity update. The force routine computes the new forces working on the currently updated configuration in line 10. Then, lines 12–18 perform the second half of the velocity update. Also note that the ki-

Section 5.10 Case study: static and dynamic properties of a microcanonical LJ fluid

Listing 51. Code fragment for calculating LJ forces.

```
double Forces ( double * rx, double * ry, double * rz,
                double * fx, double * fy, double * fz, int n ){
   int i,j;
   double dx, dy, dz, r2, r6, r12;
   double energy = 0.0, f = 0.0;

   for (i=0;i<n;i++) {
     fx[i] = 0.0;
     fy[i] = 0.0;
     fz[i] = 0.0;
   }
   for (i=0;i<(n-1);i++) {
     for (j=i+1;j<n;j++) {
       dx    = (rx[i]-rx[j]);
       dy    = (ry[i]-ry[j]);
       dz    = (rz[i]-rz[j]);
       r2    = dx * dx + dy * dy + dz * dz;
       r6i   = 1.0/(r2*r2*r2);
       r12i  = r6i*r6i;
       e    += 4*(r12i - r6i);
       f     = 48/r2*(r6i*r6i-0.5*r6i);
       fx[i] += dx*f;
       fx[j] -= dx*f;
       fy[i] += dy*f;
       fy[j] -= dy*f;
       fz[i] += dz*f;
       fz[j] -= dz*f;
     }
   }
   return energy;
}
```

netic energy KE, which is returned by *Forces()*, is computed in this loop. Compilation instructions appear in the header comments on screen when running *NVE_LJFluid.c*:

```
Options.
-N [integer]    number of particles
-rho [real]     number density
-dt [real]      time step
-rc [real]      cutoff radius
-ns [real]      number of integration steps
-T0 [real]      initial temperature
-fs [integer]   sample frequency
-sf [a|w]       append or write config output file
-icf [string]   initial configuration file
-seed [integer] random number generator seed
-h              print this info.
```

Listing 52. Code fragment for implementing periodic boundary conditions.

```
1   dz   = (rz[i]-rz[j]);
2   /* Periodic boundary conditions: apply the minimum image
3      convention. Note that this is *not* used to truncate the
4      potential, as long as there an explicit cutoff. */
5   if (dx>hL)         dx-=L;
6      else if (dx<-hL) dx+=L;
7   if (dy>hL)         dy-=L;
8      else if (dy<-hL) dy+=L;
9   if (dz>hL)         dz-=L;
10     else if (dz<-hL) dz+=L;
11  r2 = dx*dx + dy*dy + dz*dz;
12  if (r2<rc2) {
13    r6i  = 1.0/(r2*r2*r2);
14    e   += 4*(r6i*r6i - r6i) - ecut;
15    f    = 48*(r6i*r6i-0.5*r6i);
16    fx[i] += dx*f/r2;
17    fx[j] -= dx*f/r2;
18    fy[i] += dy*f/r2;
19    fy[j] -= dy*f/r2;
20    fz[i] += dz*f/r2;
21    fz[j] -= dz*f/r2;
22    *vir += f;
23  }
```

Listing 53. Code fragment for calculating LJ forces.

```
1   for (i=0;i<N;i++) {
2     rx[i]+=vx[i]*dt+0.5*dt2*fx[i];
3     ry[i]+=vy[i]*dt+0.5*dt2*fy[i];
4     rz[i]+=vz[i]*dt+0.5*dt2*fz[i];
5     vx[i]+=0.5*dt*fx[i];
6     vy[i]+=0.5*dt*fy[i];
7     vz[i]+=0.5*dt*fz[i];
8   }
9
10  PE = Forces(rx,ry,rz,fx,fy,fz,N,L,rc2,ecor,ecut,&vir);
11
12  KE = 0.0;
13  for (i=0;i<N;i++) {
14    vx[i]+=0.5*dt*fx[i];
15    vy[i]+=0.5*dt*fy[i];
16    vz[i]+=0.5*dt*fz[i];
17    KE+=vx[i]*vx[i]+vy[i]*vy[i]+vz[i]*vz[i];
18  }
19  KE*=0.5;
```

Section 5.10 Case study: static and dynamic properties of a microcanonical LJ fluid

Let us run *NVE_LJFluid.c* for 1 024 particles and 10 000 time steps at a density of $\rho = 0.85$ and an initial temperature of $T = 2.5$. We will pick a relatively conservative (small) time step of $\delta t = 0.001$. Unlike the previous case study on planetary motion, we will not specify an input configuration but instead we allow the code to create initial positions on a cubic lattice. You can see how this is done, in the source code. We run this setup with the command line

```
NVE_LJFluid -N 2024 -fs 500 -ns 10000 -rho 0.85 -T0 2.5 -dt 0.001
```

Here are the first few lines that we see in the terminal:

```
[Steinhauser@lap1] NVE_LJFluid.exe -N 1024 -fs 500
-ns 10000 -rho 0.85 -T0 2.5 -dt 0.001
# NVE MD Simulation of a Lennard-Jones fluid
# L = 10.64046; rho = 0.85000; N = 1024; rc = 1000000000000.00000
# nSteps 10000, seed 23410981, dt 0.00100
HEY!
# step PE KE TE drift T P
0 0.00000 -1622.88736 3838.10690 2215.21954 1.51390e-06 2.498 29.447
1 0.00100 -1611.38324 3826.60469 2215.22145 2.37582e-06 2.491 29.479
2 0.00200 -1590.32358 3805.54588 2215.22230 2.75873e-06 2.477 29.542
3 0.00300 -1559.77135 3774.99403 2215.22268 2.93165e-06 2.457 29.635
4 0.00400 -1519.80921 3735.03266 2215.22345 3.27877e-06 2.431 29.757
5 0.00500 -1470.55217 3685.77795 2215.22577 4.32632e-06 2.399 29.908
6 0.00600 -1412.16496 3627.39615 2215.23119 6.77444e-06 2.361 30.088
7 0.00700 -1344.88492 3560.12664 2215.24172 1.15279e-05 2.317 30.296
8 0.00800 -1269.05130 3484.31116 2215.25986 1.97135e-05 2.268 30.530
9 0.00900 -1185.14120 3400.42975 2215.28854 3.26637e-05 2.213 30.789
10 0.01000 -1093.81119 3309.14220 2215.33101 5.18345e-05 2.159 31.079
```

Each line of output after the header information corresponds to one time step. The first column is the time step, the second the potential energy, the third the kinetic energy, the fourth the total energy, the fifth the energy drift, the sixth the instantaneous temperature, and the seventh the instantaneous pressure.

The drift is output to assess the stability of the explicit integration. As a rule of thumb, we would like to keep the drift below 0.01% of the total energy TE. The drift reported by *NVE_LJFluid.c* is computed as

$$\Delta \text{TE}(t) = \frac{TE(t) - TE(0)}{TE(0)}. \tag{5.82}$$

Figure 5.14 shows the total, kinetic, and potential energies and the drift of the total energy.

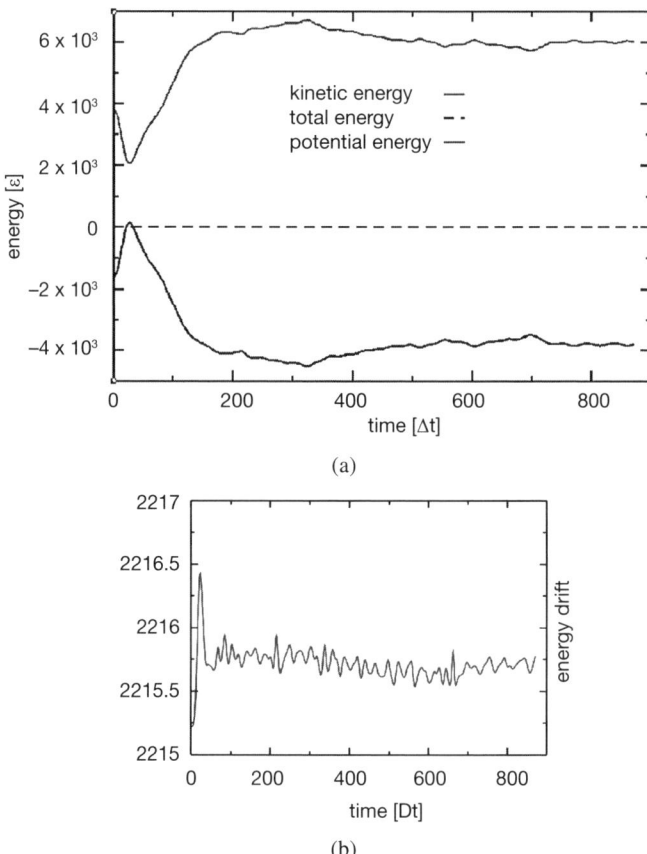

Figure 5.14. Energies and energy drift of an NVE LJ fluid. $\rho = 0.85$, $N = 1\,024$, $\delta t = 0.001$.

5.10.1 Microcanonical LJ fluid: suggested projects

Below, I suggest several tasks with the program *NVE_LJFluid.c* introduced above.

Doing measurements

Once the system has been running for a while and the energy is fluctuating around a mean value, measurements can be done. Sample snapshots of a molecular fluid are shown, both in the initial setup (which is far away from equilibrium) and after a microcanonical simulation run for 4 and 1 LJ time, respectively. In this particular initial setup the particles are arranged at their equilibrium distances and at time step $t = 0$ the full LJ potential acts on the particles, which are assigned random initial velocities. It is not useful to waste too much time on the initial configuration, as it is artificial anyway and the system will very quickly approach a Maxwell (equilibrium) velocity distribution. This is, however, different when simulating macromolecules – such as polymers – due to their particular connectivity and entropic effects that come into play. In Figures 5.15 and 5.16 we show two snapshots of the LJ fluid, one directly at the beginning ($\delta t = 0$) and one after 1 000 integration steps.

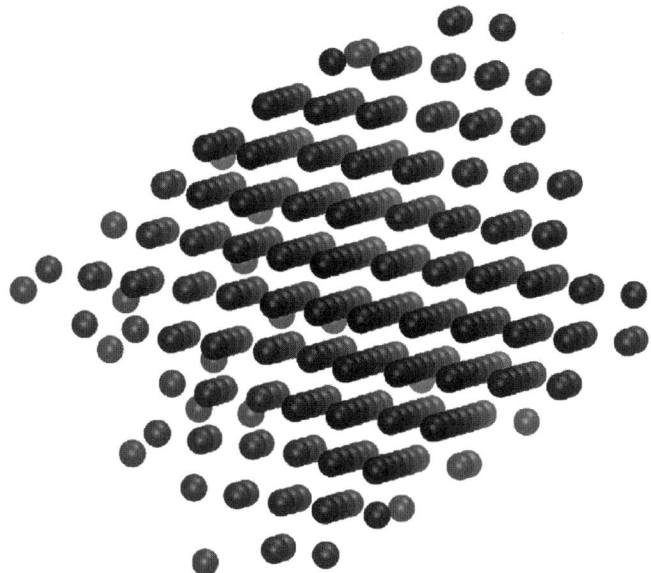

Figure 5.15. Snapshot of the initial setup of particles in an LJ molecular fluid simulation.

Checking equilibration

The decision as to when a system is at equilibrium in a simulation, is not an easy one. Usually, one checks the decay of the correlation of consecutive snapshots as well as typical configuration properties, such as coordination numbers, energy, pressure, or – in the case of molecules – the average extension, e.g. the radius of gyration of the molecules – as an indication of the system having reached an equilibrium state. In polymer melt simulations, one usually lets the system evolve for several times the typical relaxation time τ of the molecules in a melt. The pressure p of a fluid is defined in terms of a virial expansion. In reduced units this writes as:

$$pV = NT + \frac{1}{d}\left\langle \sum_{i=1}^{N} \vec{r}_i \vec{F}_i \right\rangle. \tag{5.83}$$

Figure 5.16. Snapshot of particles in an LJ molecular fluid simulation after 1 000 integration steps.

In Figure 5.17 the energy and the pressure of a Lennard-Jones fluid are displayed. After an initial equilibration phase, both quantities have reached their equilibrium value and eventually fluctuate about their mean value.

Here, we quantify the notion of "equilibration" of the system by assessing correlations in (apparently) randomly fluctuating quantities like the potential energy. Remember that, in order to perform accurate ensemble averaging over an MD trajectory, we have to be sure that, in the properties we are measuring, correlations have "died out." This is another way of saying that the length of the time interval over which we conduct the time average must be much longer than the appropriate correlation time. In this case study, we use "block averaging" to determine the timescale of equilibration of the potential energy U.

First, compute the variance of the L samples:

$$\sigma_0^2(U) \approx \frac{1}{L} \sum_{i=1}^{L} [U_i - \langle U \rangle]^2 \qquad (5.84)$$

This is an approximation, because of so far undetermined time correlations in U. That is, not all L samples are uncorrelated. For example, two samples one time step apart will likely be very close to one another. We now regroup the data by averaging

Section 5.10 Case study: static and dynamic properties of a microcanonical LJ fluid

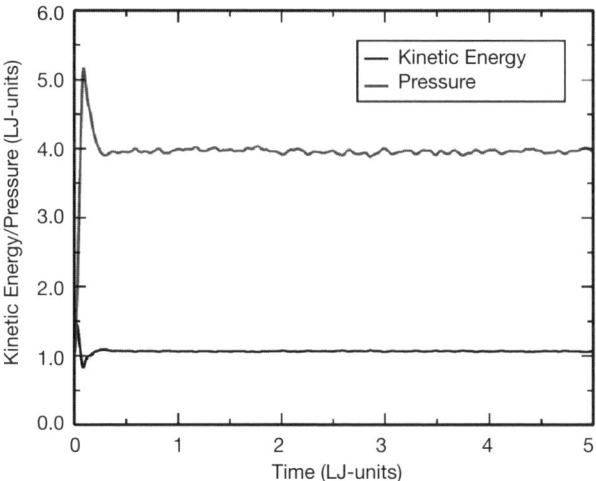

Figure 5.17. Energy and pressure of a Lennard-Jones fluid.

$L/2$ pairs of adjacent values:

$$U_i^{(1)} = \frac{U_{2i-1} + U_{2i}}{2} \tag{5.85}$$

The superscript (1) indicates that this is a "first generation" coarsening of the potential energy trace. The variance of this new set, σ_1^2 is computed. Next, the process is carried out recursively through many subsequent generations. After many blocking operations, the coarsened samples $U_i^{(j)}$ become uncorrelated and the variance (temporarily) saturates. This means that we should observe a plateau in a plot of variance vs generation. I suggest implementing the method outlined in the provided code and check the equilibration by plotting σ vs the number of blocks B.

Scattering experiments with fluids

Scattering experiments are performed with electromagnetic radiation, neutrons, or electrons. In all cases, the principle of a scattering experiment is always the same: a primary beam with frequency ω_0, intensity I_0 and wave vector $\vec{k}$ hits a target specimen and initiates scattering waves. The scattered intensity I of the secondary waves emerging from the target is measured. In general, I depends on the direction of observation and the angle θ under which the direction of observation diverges from the direction of the incoming beam, called the *Bragg angle*. In order to analyze the structure of a specimen, one has to choose a wave length of the incoming radiation or particles comparable to the typical length scale of the investigated structure, e.g. a lattice constant in the case of a solid. Other nanoscopic structures, e.g. molecular fluids, typically have structural dimensions of the order of 1–10 Å. Hence, for high resolution microscopy, electrons can be used, which have may wave lengths below 1 Å. Thermal neutrons, with energies in the range of $\sim 10\ meV$ may also be used. With colloidal systems (soft matter) much larger typical length scales of up to $\mu\ m$ may arise. Thus, for such systems, photons (light) may be used.

Electrons interact with matter via the electromagnetic force. This is the reason why specimens that are investigated by X-ray microscopy have to be small – the electrons interact with the electrons of the atoms in the material and dissipate energy inside of the specimen, which is why the scattered intensity will be rather low. Neutrons, however, can be used with large specimens, as they are electrically neutral and can thus penetrate deeper into a material. Information on the structure of the specimen is obtained by measuring the scattered intensity from different Bragg angles. Differences in intensity are a consequence of the interference of waves emerging from the particles of the specimen. Using elementary electrodynamics, the outgoing primary beam can be described as a plane wave with amplitude E_0:

$$E_0 \exp(-i w_0 t + i \vec{k}\vec{r}). \tag{5.86}$$

This wave hits the specimen and initiates oscillations of atoms and molecules. Due to these oscillations, secondary waves are emitted from the specimen in direction $\vec{k}'$. To simplify calculations one introduces the scattering vector

$$\vec{q} = \vec{k}' - \vec{k}. \tag{5.87}$$

The total amplitude E of the scattered waves is obtained by summing over all contributions of the scatterers in the material, i.e. one has to calculate the sum

$$E = E(\vec{q}) \sim \sum_j f_j \exp(-i\vec{q}\vec{r}_j). \tag{5.88}$$

where parameter f_j is the so-called *form factor*, which is introduced because the waves scattered off the molecules are different from the ones scattered off electrons. Most often however, scattering experiments are done with specimens where the scatterers are made of only one kind of particle. In this case, the amplitude E only depends on the phase factor of the scattered wave, i.e.

$$E = E(\vec{q}) \sim \sum_j \exp(-i\vec{q}\vec{r}_j) = C(\vec{q}). \tag{5.89}$$

No detector can follow the rapidly changing electromagnetic field vector $\vec{E}(\vec{r},t)$. Rather, all measurements are averages of some quantities, in this case the average of the different intensities of the quickly changing field vector. This average is called the *intensity* I, and this is what is measured in a scattering experiment

$$I(\vec{q} \sim \langle |E_0(\vec{q})|^2 \rangle = \langle |C(\vec{q})|^2 \rangle. \tag{5.90}$$

Section 5.10 Case study: static and dynamic properties of a microcanonical LJ fluid

The total intensity measured in a scattering experiment is proportional to the number of scattering particles N. Hence, one introduces a function $S(\vec{q})$ as

$$S(\vec{q}) = \frac{1}{N} \sum_{j,k=1}^{N} \langle \exp[-i\vec{q}(\vec{r}_j - \vec{r}_k)] \rangle . \tag{5.91}$$

$S(\vec{q})$ is called *Structure factor*[23] $S(\vec{q})$ does not depend on the nature of the scatterers, is universally valid and can thus be applied to any condensed matter system. For molecular fluids, the scattering function can be written, using the pair correlation function $g(\vec{r})$, as

$$S(\vec{q}) = \frac{1}{N}\left(N + N \int_V \exp(-i\vec{q}\vec{r}) g(\vec{r}) \, d^3r\right) = 1 + \int_V \exp(-i\vec{q}\vec{r}) g(\vec{r}) \, d^3r . \tag{5.92}$$

The first term on the right hand side of (5.92) is the contribution to the sum by all terms for which $i = j$. The second term is the sum over all contributions from particles with $j \neq k$. The integral in (5.92) is the Fourier transform of $g(\vec{r})$. For $r \to \infty$, the limiting value of $g(r)$ is the particle density ρ. This limiting value has to be subtracted from $S(\vec{q})$ in order to obtain

$$S(\vec{q}) - 1 = \int_V \exp(-i\vec{q}\vec{r}) g(\vec{r} - \rho) \, d^3r + \rho \int_V \exp(-i\vec{q}\vec{r}) d^3\vec{r} . \tag{5.93}$$

The second term is the Fourier transform of a macroscopic volume which only has contributions for $\vec{q} \approx 0$. In practical experiments, this forward scattering has to be avoided in order to avoid detecting primary waves. Hence, this term can usually be neglected. For fluids, $S(\vec{q})$ is isotropic, which means that

$$g(\vec{r}) = g(|\vec{r}| = r) , \tag{5.94}$$

and

$$S(\vec{q}) = S(|\vec{q}|) . \tag{5.95}$$

For isotropic functions, the Fourier transform may be written as

$$S(q) - 1 = \int_{\infty}^{\infty} (g(r) - \rho) 4\pi r^2 \frac{\sin(qr)}{qr} \, dr \tag{5.96}$$

and the inverse transformation is

$$g(r) - \rho = \left(\frac{1}{2\pi}\right)^3 \int_{r=0}^{\infty} (S(q) - 1) 4\pi r^2 \frac{\sin(qr)}{qr} \, dr . \tag{5.97}$$

---

[23] Sometimes $S(\vec{q})$ is also called Scattering function or interference function.

The radial distribution function is an important statistical mechanical function that captures the structure of liquids and amorphous solids. We can express $g(r)$ using the following statement:

$$\rho g(r) = \text{average density of particles at } \vec{r}, \text{ given that a tagged particle is at the origin.} \tag{5.98}$$

We can use $g(r)$ to count particles within a distance r from a central atom:

$$n(r) = \int_0^r \int_0^\pi \int_0^{2\pi} g(r') r' r' \sin\theta\, dr'\, d\theta\, d\phi = 4\pi \int_0^r r' r' g(r')\, dr' \tag{5.99}$$

In principle, any quantity that can be written as a sum of pairwise terms (such as potential energy and pressure) can be written as an integral over $g(r)$. For example, the total potential can be computed by "averaging" the pairwise potential $U(r)$ over $g(r)$:

$$U/N = \frac{1}{2}\rho \int_0^\infty \int_0^\pi \int_0^{2\pi} u(r) g(r) r^2 \sin\theta\, dr\, d\theta\, d\phi \tag{5.100}$$

$$= 2\pi \int_0^\infty r^2 u(r) g(r)\, dr \tag{5.101}$$

Theories of the liquid state have the primary goal of predicting $g(r)$, given the intermolecular potentials and the molecular structure. MD simulation therefore is a useful complement to theoretical investigations. Let us now consider how to compute $g(r)$ from an MD simulation of the Lennard-Jones liquid. The procedure we will follow will be to write a second program (a "post processing program") which will read the configuration files ("samples") produced by the simulation, *NVE_LJFluid.c*.

The general structure of a $g(r)$ post-processing program could look like

1. determine limits: start, stop, and step

2. initialize histogram

3. for each configuration: read it, visit all unique pairs of particles, and update histogram for each visit, if applicable

4. normalize histogram and output

5. end

A suggestion for an implementation of this algorithm, which produces a histogram, can be found in Listing 54.

One can see in Listing 54 that the bin value is computed on line 19 by first dividing the actual distance between members of the pair by the resolution of the histogram δr. This resolution can be specified on the command line, when the program is executed. Also notice that the histogram is updated by 2, which reflects the fact that either of

Listing 54. Calculation of the radial density distribution function $\rho(r)$.

```
1  void update_hist ( double * rx, double * ry, double * rz,
2                     int N, double L, double rc2, double dr, int
                        * H ) {
3      int i,j,bin;
4      double dx, dy, dz, r2, hL = L/2;
5
6      for (i=0;i<(N-1);i++) {
7          for (j=i+1;j<N;j++) {
8              dx = rx[i]-rx[j];
9              dy = ry[i]-ry[j];
10             dz = rz[i]-rz[j];
11             if (dx>hL)        dx-=L;
12             else if (dx<-hL)  dx+=L;
13             if (dy>hL)        dy-=L;
14             else if (dy<-hL)  dy+=L;
15             if (dz>hL)        dz-=L;
16             else if (dz<-hL)  dz+=L;
17             r2 = dx*dx + dy*dy + dz*dz;
18             if (r2<rc2) {
19                 bin=(int)(sqrt(r2)/dr);
20                 H[bin]+=2;
21             }
22         }
23     }
24 }
```

the two particles in the pair can be placed at the origin. Also notice that lines 11–16 implement the minimum image convention. I suggest implementing this algorithm in the program *NVE_LJFluid.c*.

In Figure 5.18, some exemplary measurements of the MD simulation of molecular LJ fluid are displayed for four different densities of the fluids. The maxima of $g(r)$ correspond to different shells of neighbors. Finally, in Figure 5.19, we show more snapshots of LJ systems at different densities.

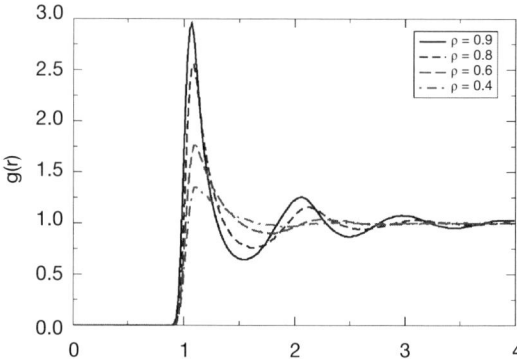

Figure 5.18. The radial distribution function $g(r)$ of a molecular fluid, with different densities as displayed. For a visualization of the structures, see Figure 5.19. The RDF counts the number of particles found in a spherical shell of radius r in the vicinity of a selected particle. The denser the systems, the more structure (pronounced peaks in $g(r)$) can be seen.

Section 5.10 Case study: static and dynamic properties of a microcanonical LJ fluid 355

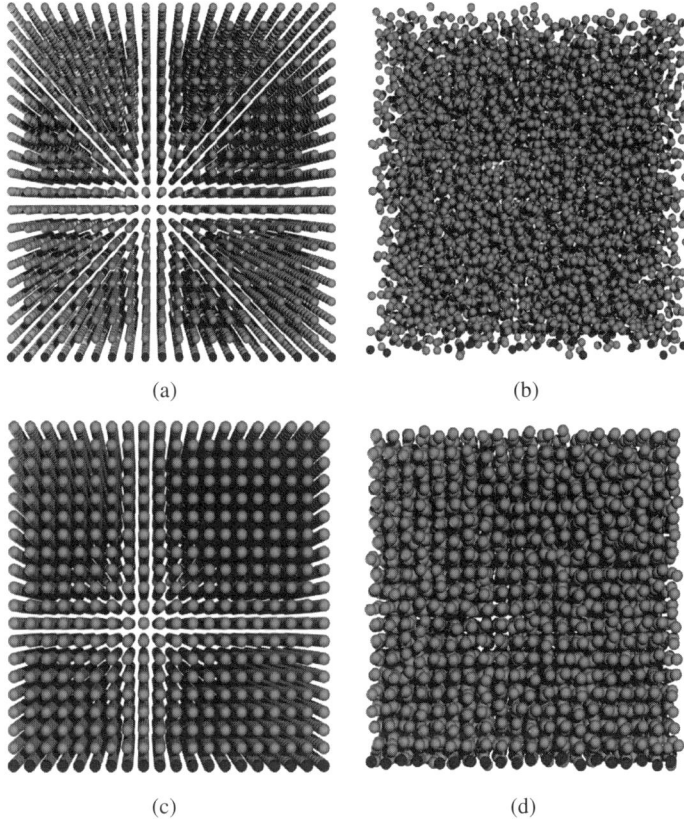

Figure 5.19. Simulation snapshots of a micro-canonical molecular dynamics simulation of a Lennard-Jones fluid, starting from a cubic initial configuration with $N = 8\,000$ particles. **(a)** Initial cubic configuration of fluid particles at density $\rho = 0.4$. **(b)** System snapshot after 4 reduced time units. **(c)** Initial configuration of particles in a simulation box with $\rho = 0.8$. **(d)** System snapshot after 1 reduced time unit. The lowest front row of particles in **(a)** and **(c)** are colored darker, so as to show their diffusion in the fluid.

5.11 Chapter literature

The first and most authoritative monograph on differential equations is probably Courant and Hilbert [111]. Other detailed monographs on differential equations and their numerical solution have been published by Forsythe and Wasow [166], Morton [307] and Dettman [130]. More modern texts are e.g. Golub [184] or Johnson [225].

One of the earliest books that summarizes aspects of the MD method was edited by Ciccotti and Hoover [98]. This book does, however, not provide a very pedagogical preparation of the material, since it is more a collection of many different essays, written by many experts who contributed single chapters. It has a strong emphasis on statistical methods and is not of much help in developing your own simulation programs. The book by Heermann [203] is quite old and by now is outdated. The classic by Allen and Tildesley [24], albeit by now also quite old, not up to date, and only providing code samples in ancient FORTRAN 77, is still readable for historic reasons. This book focuses heavily on liquid theory and applications of MD to liquids. The book edited by Ciccotti, Frenkel and McDonald [98] focuses almost entirely on liquids and touches a little bit on solids, and again is a collection of papers ranging from the field's origins to the state of the art in 1986. No source code at all is provided. The book by Hockney and Eastwood [212] is a nice introduction to general particle simulation methods, but is by now very outdated. The first issue was published in 1981. The book by Haile [198] is also a nice general introduction on a very elementary level. Code examples are provided only in FORTRAN 77. The classic by Rappaport [348] is still a very readable introduction to many standard methods of MD and it provides code samples in C. The C code however, is often very substandard and written in FORTRAN style – obviously this is the language in which the code was originally written. In the code samples, usually no argument lists for functions are used and everything is defined globally – very bad style, but the programs do their job. The more recent book by Frenkel and Smit [170] is obviously written by chemists or chemical physicists, since it has a very strong focus on thermodynamics and typical statistical applications in the field of physical chemistry. It provides a good introduction to the field and is focused almost completely on the MC method. MD is covered only marginally. It provides many (pseudocode) algorithms, but the explicit programming examples are all in FORTRAN.

# Chapter 6

Monte Carlo simulations

"I suggested an obvious name for the statistical method – a suggestion not unrelated to the fact that Stan Ulam had an uncle who would borrow money from relatives because he 'just had to go to Monte Carlo.' The name seems to have endured."

N. Metropolis in "The Beginning of the Monte Carlo Method", 1987

Summary

In this chapter, we look at a simulation method known as "Monte Carlo" and its typical applications in physics and engineering. Numerical methods known as MC methods can be loosely described as statistical simulation methods, where "statistical simulation" is defined in quite general terms to be any method that utilizes sequences of random numbers to perform the simulation. Hence, this method can neither be assigned to particle or mesh-based models, but is based on the theory of stochastic processes, in particular on the law of large numbers and the central limit theorem. We will look at numerical integration, the calculation of phase space integrals and the Metropolis algorithm, to perform importance sampling. This algorithm is considered to be among the "top ten algorithms of the 20th century".

Learning targets

✓ Learning the basics of the Monte Carlo method.

✓ Learning applications of MC.

✓ Understanding the Metropolis algorithm.

6.1 Introduction to MC simulation

Monte Carlo methods are nowadays widely used, from the integration of multi-dimensional integrals to solving ab initio problems in physics, chemistry, medicine, biology, or even Dow-Jones forecasting. Computational finance is one of the novel emerging fields where, since the 1980s, Monte Carlo methods found new applications [404, 221, 391, 181, 290, 425].

One of the major advantages of MC methods is their systematic improvement with the number of samples N, because the error Δ decreases as follows:

$$\Delta \propto \frac{1}{\sqrt{N}}. \tag{6.1}$$

A good example of this is the computation of π in Section 6.2.2. MC methods are also very popular in statistical physics (MC molecular modeling), when an exact solution to a given problem cannot be found with a deterministic algorithm, and in the context of higher-dimensional integration. MC methods are also used in fundamental science such as Quantum Chromodynamics [31, 288]. In the Large Hadron Collider (LHC) at CERN, MC methods were used to simulate signals of Higgs particles, and are used for designing detectors (and to help understand as well as predict their behavior).

MC simulations are generally concerned with large series of computer experiments using uncorrelated random numbers [199, 381, 60, 235, 62, 61, 324, 445]. Depending on the distribution from which the random numbers are selected for numerical integration, one distinguishes *simple (or naive) sampling* and *importance sampling* MC methods. The former method uses an equal distribution of random numbers, whereas the latter employs a distribution tailored to the problem under consideration. Importance sampling means that a large weight is used in those regions where the integrand has large values and a small weight in those regions where it assumes small values. i.e. the sampling is focused on those (most important regions) where the function considered is different from zero.

A central algorithm in MC methods is the *Metropolis algorithm*, ranked as one of the top ten algorithms in the last century, see Section 6.4. Statistical simulation methods may be contrasted with conventional numerical discretization methods, which are typically applied to ordinary or partial differential equations that describe some underlying physical or mathematical system. In many applications of Monte Carlo, the physical process is simulated directly, and there is no need to even write down the differential equations that describe the behavior of the system. The only requirement is that the physical (or mathematical) system be described by a Probability Distribution Function (PDF). Once the PDFs are known, the MC simulation can proceed by random sampling of the PDFs. Many simulations are then performed (multiple "trials" or "histories") and the desired result is taken as an average over the number of observations (which may be a single observation or perhaps millions of observations). In many practical applications, one can predict the statistical error (the "variance") of this average result, and hence an estimate of the number of MC trials that are needed to achieve a given error level. If we assume that the physical system can be described by a given probability density function, the MC simulation can proceed by sampling from these PDFs, which necessitates a fast and effective way to generate random numbers distributed uniformly over the interval $I = [0, 1]$. The results of these random

Section 6.1 Introduction to MC simulation 359

samplings, or trials, must be accumulated in an appropriate manner in order to produce the desired result, but the essential characteristic of MC is the use of random sampling techniques (and perhaps other algebra to manipulate the outcomes) to arrive at a solution of the physical problem. In contrast, a conventional numerical solution approach would start with the mathematical model of the physical system, discretizing the differential equations, and then solving a set of algebraic equations for the unknown state of the system. It should be kept in mind though, that this general description of MC methods may not directly apply to some applications. It is natural to think that MC methods are used to simulate random, or stochastic, processes, since these can be described by PDFs. However, this coupling is actually too restrictive because many MC applications have no apparent stochastic content, such as the evaluation of a definite integral or the inversion of a system of linear equations. However, in these and other cases, one can pose the desired solution in terms of PDFs, and while this transformation may seem artificial, this step allows the system to be treated as a stochastic process for the purpose of simulation and hence Monte Carlo methods can be applied to simulate the system. There are at least four crucial ingredients in a basic MC strategy. These are

1. random variables

2. PDFs

3. moments of a PDF

4. calculation of variances σ

I feel that a brief explanation may be appropriate in order to explain the strategy behind an MC simulation. Let us first demystify the somewhat obscure concept of a *random variable*. The example we choose is the classic one, the tossing of two dice, its outcome and the corresponding probability. In principle, we could imagine being able to determine the motion of the two dice exactly, and determine the outcome of the tossing with given initial conditions. However, in practice this ideal situation never occurs. This does, however, not mean that we do not have certain knowledge about the outcome. Our partial knowledge about this stochastic process is given by the probability of obtaining a certain number when tossing the dice. To be more precise, the tossing of the dice yields the following possible values: $D \in [2, 3, 4, 5, 6, 7, 8, 9, 10, 11, 12]$. These values are called the *domain D*. To this domain we have the corresponding probabilities $P_i \in [1/36, 2/36, /3/36, 4/36, 5/36, 6/36, 5/36, 4/36, 3/36, 2/36, 1/36]$. The numbers in the domain are the outcomes of the physical process of tossing the dice. One cannot tell beforehand whether the outcome will be 2 or 10 or any other number in this domain. This defines the randomness of the outcome. The only thing we can tell beforehand is that, say, the outcome 2 has a certain probability. If one spends some time tossing two dice and registering the sequence of outcomes, one will notice that the numbers in the above domain D appear in a random order. After 12 throws the results may be the following sequence: 6, 11, 8, 10, 6, 9, 11, 7, 12, 4, 5, 1.

Repeating this exercise will most likely never give you the exact same sequence. Random variables are hence characterized by a domain which contains all possible values that the random value may take. This domain has a corresponding PDF.

To provide yet another example of random number generation, consider the radioactive decay of an α-particle from a certain nucleus. Assume that you have a Geiger-counter which registers every 10ms whether an α-particle reaches the counter or not. If we record a hit as 1 and "no observation" as zero, and repeat this experiment many times, the outcome of the experiment is also truly random. We cannot form a specific pattern out of the above observations. The only possibility to say something about the outcome is given by the PDF which, in this case, is the well-known exponential function of radioactive decay

$$\lambda_0 \exp(-\lambda x), \tag{6.2}$$

with λ being proportional to the half-life of the given decaying nucleus.

6.1.1 Historical remarks

The first important contribution to what would be later denoted as the Monte Carlo method was published by Lord Kelvin in 1901 [239], who used this method for a discussion of the Boltzmann equation. The origins of the systematic development of the MC method as a statistical simulation tool dates back to 1944, when S. Ulam and J. von Neumann introduced MC simulation to mimic the uncorrelated spatial diffusion of neutrons in fissile materials[1] during the Manhattan project [297, 300]. They managed to find a probabilistic analogue of this problem, which could be solved by random walk type many-particle simulations or by random sampling of adequate integral formulations. The basic idea of solving state function integrals by randomly sampling a non-uniform distribution of numbers was introduced in the 1950s by N. Metropolis and others [418, 298].

The following Table 6.1 provides a selected overview of MC references in different areas of research and may serve as a starting point for getting acquainted with classic publications and recent original literature in the field of MC applications in physics and engineering.

---

[1] In essence, this is a random walk problem.

Table 6.1. MC applications in various domains of materials science.

area of application	Reference
classics	Metropolis and Ulam (1949) [300]
	Potts (1952) [337]
	Metropolis, Rosenbluth, Rosenbluth, Teller and Teller (1953) [298]
	Kalos and Whitlock (1986) [235]
	Binder and Stauffer (1987) [64]
diffusion	Limoge and Bocquet (1988) [273]
	Gladyszewski and Gladyszewski (1991) [180]
	Frontera, Vives and Planes (1993) [171]
	Mattsson, Engberg and Wahnström [287]
	Uebing and Gomer (1991) [417]
	Chen, Gomez and Freeman (1996) [92]
	Wang, Rickman and Chou (1996) [431]
epitaxial growth	Kew, Wilby and Vvedensky (1993) [242]
grain boundaries	Alba and Whaley (1992) [19]
	Tagwerker, Plotz and Sitter [385]
Ising model	Wansleben and Landau (1991) [432]
	Wang and Young (1993) [430]
	Zang and Yang (1993) [450]
polymers	Baumgärtner (1984) [47]
	Milik and Orszak (1989) [301]
	Cifra, Karasz and MacKnight (1992) [99]
	Haas, Hilfer and Binder (1995) [197]
phase diagrams	Bichara and Inden (1991) [59]
	Farooq and Khwaja (1993) [155]
	Silverman, Zunger, Kalsih and Adler [383]
phase transformation	Zamkova and Zinenko (1994) [448]
	Roland and Grant [358]

6.2 Simple random numbers

No numerical algorithm can generate a truly random sequence of numbers [325]. However, there exist algorithms which generate repeating sequences of m (say) integers which are, to a fairly good approximation, randomly distributed over some range. This type of number sequence is termed *pseudorandom*. C provides a standard function for the generation of pseudorandom numbers called *rand()*. This function is included in the library <*stdlib.h*> and produces integer numbers in the range $R \in [0, \text{RAND\_MAX}]$ where RAND_MAX is a predefined macro in <*stdlib.h*>. Listing 55 shows how to use this function in practice.

Listing 55. Usage of *rand()*. (*)

```
/** This program demonstrates the most primitive form of
    generating random integer numbers in C
*/

#include <stdio.h>
#include <stdlib.h>

void main(void)
{
   int x;

   for(x = 1; x <= 20; x++)
      printf("Random Number %2i :%i\n",x,rand());
}
```

The program of Listing 55 outputs 20 integer pseudorandom numbers:

```
Random Number  1 :1481765933
Random Number  2 :1085377743
Random Number  3 :1270216262
Random Number  4 :1191391529
Random Number  5 :812669700
Random Number  6 :553475508
Random Number  7 :445349752
Random Number  8 :1344887256
Random Number  9 :730417256
Random Number 10 :1812158119
Random Number 11 :147699711
Random Number 12 :880268351
Random Number 13 :1889772843
Random Number 14 :686078705
Random Number 15 :2105754108
```

Section 6.2 Simple random numbers

```
Random Number 16 :182546393
Random Number 17 :1949118330
Random Number 18 :220137366
Random Number 19 :1979932169
Random Number 20 :108995793
```

When you run the program several times you will notice that the produced "random" numbers are always the same. Of course, the computer calculates the numbers in *rand()* by using an algorithm. The numbers produced by this algorithm depend on the initialization value of the function *rand()*, for which there is another function *srand()*, also defined in the library *<stdlib.h>*. So, let's initialize the function *rand()* with several different values. To do this, we insert the following lines of C code just before line 12 in Listing 55:

```
int y;

printf("Please provide an integer as initialization value :");
scanf("%i",&y);
srand(y);
```

After this little change, our program first asks for an initialization value (called a "seed") for *rand()*. Without explicit initialization, the algorithm is started with the default value 1, i.e. it is initialized with *srand(1)*. When testing this version of our Pseudo Random Number Generator (PRNG), we notice that we always get the same numbers when using the same value for the initialization. To avoid user interaction and to ensure that *rand()* is always initialized with a different value, we change Listing 55 once again and introduce the following lines of code just before line 12:

```
 time_t tim;

 srand(time(&tim));
```

and on line 7 we write #include <time.h>. We have now included the additional header file *<time.h>*, which provides functions for time management, one of which is the function *time()*, which expects the address of a variable of type time_t as an argument, a datatype declared in *<time.h>* and which is specifically provided for storing the system time. The time format stored in time_h is the number of seconds which have passed since a certain date[2] in the past. Provided you don't call the PRNG several times per second, the *time()*-function ensures two things: first, it makes sure that at all times a different number is used for the initialization, and second, that the initialization is hidden from the user. The user who runs the code has no influence whatsoever on the initialization of *rand()*. Listing 56 shows a more general *main()*-function for the generation of pseudorandom numbers, which asks the user to limit the random numbers to within a certain range. Listing 56 Part 1 includes a user header

---

[2] 1 January 1970, 00 : 00 : 00 Greenwich Mean Time.

Listing 56. Part 1 – Pseudorandom numbers limited to a certain interval.

```
1  /** This program shows how to generate general pseudorandom
2      numbers that lie within a certain interval.
3  */
4
5  #include <stdio.h>
6  #include <stdlib.h>
7  #include <time.h>
8  #include ''Random.h''
9
10 void main(void)
11 {
12    int x, min, max;
13    time_t tim;
14
15    srand(time(&tim));
16
17    printf("Minimum Integer Value :");
18    scanf("%i",&min);
19    printf("Maximum Integer Value :");
20    scanf("%i",&max);
21
22    for(x = 1; x <= 20; x++)
23       printf("Random Number is %2i :%i\n",x,Randomize(min,max));
24 }
```

file "Random.h" which contains the definition of the function *Randomize()* on line 23. The header file *"Random.h"* itself is provided in Listing 56 Part 2. The function *Randomize()* is passed a minimum and maximum value. In line 11 we check whether min is really smaller than max (exception handling). Provided that both min and max are within the valid range of numbers for *rand()*, *Randomize()* outputs a pseudorandom number in the interval $R \in [\min, \max]$.

However, one disadvantage of our pseudorandom number generator remains: *rand()* only produces *positive* (unsigned) random numbers. This can be easily changed by modifying the statement x = rand() on line 10 of *Randomize()* as follows:

x = rand() * (1 - 2 * (rand()&1));

What does this line of code do?

Let's first look at rand()&1. In Chapter 2 we learned how to use the bitwise AND operator "&" to isolate single bits. With the instruction rand()&1 we isolate the first bit in the pseudorandom number, i.e. bit 0. Bit 0 represents the binary number $2^0 = 1$. Hence, the resulting value is either 1 if the bit was set, or 0 if the bit was not set. We also learned in Chapter 2 that for any even number the first bit is 0, and for any uneven number the first bit is 1. Hence, as half of the numbers are even and half of them are uneven, the command rand()&1 yields both, 0 and 1 in 50% of all cases. This value

Section 6.2 Simple random numbers

Listing 56. Part 2 – The header file *Random.h*.

```
#ifndef  __RANDOM_H
#define  __RANDOM_H

int Randomize(int min, int max)
{
  int x;

  do
    {
      x = rand();
    } while( (x < min) || ( x > max) );
  return(x);
}
#endif
```

is multiplied by 2, i.e. 2 * rand()&1. This multiplication yields either 0 or 2, which is subtracted from 1. As a result of this subtraction, one obtains 1 or −1, both in 50% of all cases. Thus, as a result one obtains a factor 1 or −1, with which the generated pseudorandom number (the output of *rand()*) is multiplied.

Finally, if you need floating point pseudorandom numbers, you have to add the decimal places to the generated integer by successive additional calls of *rand()* in the function *Random()*. Listing 57 shows the header file *Random.h* which provides this functionality. In Listing 58 we provide the corresponding *main()*-function. This code first generates a signed pseudorandom number. Then, a second pseudorandom number is generated, which is divided by 10 until the result of the division is smaller than 1, e.g. the number 5 467 would become 0.5467. This value is then added to the initially generated number.

The above examples show how to generate pseudorandom numbers, using the options that the language C provides. The quality of such pseudorandom numbers is, however, not good enough by far for scientific computation. Here, one has to use more sophisticated methods for generating random numbers, some of which are covered in the next section.

6.2.1 The linear congruential method

We have seen, that there exist algorithms which generate repeating sequences of m integers which are, to a rough approximation, randomly distributed over the range $R \in [0, \text{RAND\_MAX}]$. Now, we want to look at different algorithms that produce high-quality pseudorandom numbers in the range $R \in [0, m-1]$, where m is a (hopefully) large integer. The best-known algorithm for generating pseudorandom sequences of integers is the so-called *linear congruential method*. The formula linking the nth and

Listing 57. Floating point pseudorandom numbers. Header file *"Random.h"*.

```
1  #ifndef  __RANDOM_H
2  #define  __RANDOM_H
3
4  float Randomize(float min, float max)
5  {
6     float x, y;
7
8     if(max < min)
9       {
10         x   = max;
11        max = min;
12        min = x;
13     }
14
15    do
16       {
17         x=(float)(rand()*(1-2*(rand()&1)));
18         y=(float)(rand());
19         while(y >= 1) y /= 10.0;
20         x += y;
21     } while(( x<min ) || ( x>max ));
22    return(x);
23 }
24 #endif
```

$(n + 1)^{\text{th}}$ integers in the sequence is:

$$I_{n+1} = (aI_n + c) \mod m, \tag{6.3}$$

where a, c, and m are positive integer constants. The first number in the sequence, the so called "seed" value, is selected by the user. Consider an example in which $a = 8$, $c = 0$, and $m = 10$. A typical sequence of numbers generated by equation (6.3) is $I = \{4, 8, 6, 2, 4, 8, 6, 2, 4, 8, \ldots\}$. Obviously, the above choice of values for a, c, and m is a very bad one, since the sequence repeats after only four iterations. However, if a, c and m are chosen properly, the sequence is of maximal length (i.e., of length m), and approximately randomly distributed over $R \in [0, m-1]$. The random function in Listing 60 implements this linear congruential method for the choice of $a = 107$, $c = 1\,283$ and $m = 6\,075$ and Figure 6.1 shows a correlation plot of the first $10\,000$ generated pseudorandom numbers.

The keyword `static` in front of a local variable declaration indicates that the program preserves the value of that variable between function calls. In other words, if the static variable `next` has the value $4\,512$ on exit from the function *Random()* then the next time this function is called, `next` will have exactly the same value. Note that the values of nonstatic local variables are *not* preserved between function calls. The function *Random()* returns a pseudorandom integer in the range 0 to RAND_MAX, where

Section 6.2 Simple random numbers 367

Listing 58. Floating point pseudorandom numbers. *main()*-function.

```
/** This program generates floating point random numbers.
 */

#include <stdio.h>
#include <stdlib.h>
#include <time.h>
#include ''Random.h''

int main(void)
{
   int x;
   float min, max;
   time_t tim;

   srand(time(&tim));

   printf("Lower bound: ");
   scanf("%f",&min);
   printf("Upper bound: ");
   scanf("%f",&max);

   for(x = 1; x <= 20; x++)
      printf("Random Number %2i: %f\n",x,Randomize(min,max));

   return (0);
}
```

RAND_MAX takes the value $m - 1$. In order to obtain a random variable x, uniformly distributed over the range 0 to 1, one can write

x = double (Random ()) / double (RANDMAX).

If x was truly random, there would be no correlation between successive values of x. Thus, a good way of testing our random number generator is to plot x_{i+1} versus x_i, where x_i corresponds to the ith number in the pseudorandom sequence, for many different values of i. For a good random number generator, the plotted points should densely fill the unit square. Moreover, there should be no discernible pattern in the distribution of points.

It can be easily seen in Figure 6.1 that this is a poor choice of values for a, c, and m, since the pseudorandom sequence repeats after a few iterations. In contrast, Figure 6.2 shows a correlation plot for the first 10 000 $x_{i+1} - x_i$ pairs, generated using a linear congruential pseudorandom number generator characterized by $a = 106$, $c = 1283$ and $m = 6075$. It can be seen that this is a much better choice of values, since the pseudorandom sequence is of maximal length, yielding x_i values fairly evenly distributed over the range 0 to 1. However, if we look carefully at Figure 6.2 we can see that there is a slight tendency for the dots to line up in the horizontal and vertical

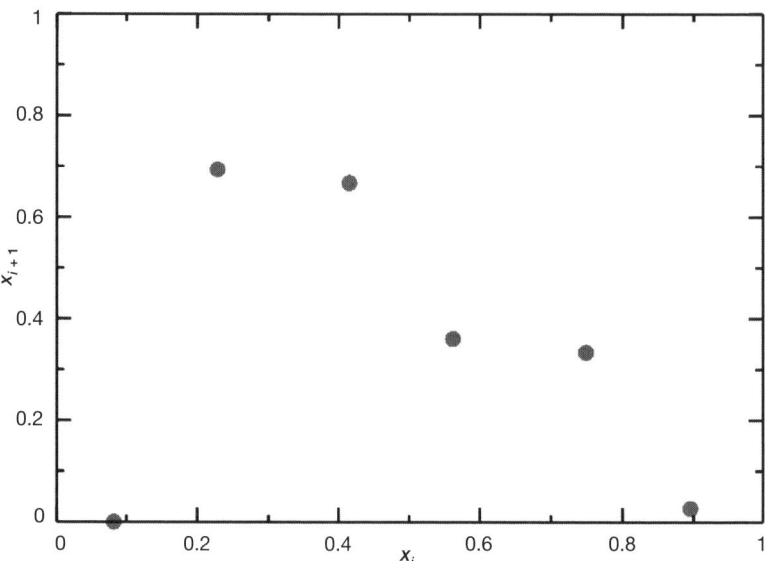

Figure 6.1. Plot of x_{i+1} versus x_i for $i \in [1, \ldots, 10\,000]$. The x_i are pseudorandom variables, uniformly distributed over the range 0 to 1 and generated using a linear congruential pseudorandom number generator, according to equation (6.3), characterized by $a = 107$, $c = 1\,283$, $m = 6\,075$.

directions. This indicates that the x_i are not quite randomly distributed, i.e., there is some correlation between successive x_i values. The problem here is that m is too low, i.e. there is not a sufficiently wide selection of different x_i values in the interval 0 to 1.

Figure 6.3 shows a correlation plot for the first 10 000 $x_{i+1} - x_i$ pairs generated using a linear congruential pseudorandom number generator characterized by $a = 1\,103\,515\,245$, $c = 12\,345$ and $m = 32\,768$. The clumping of points in this figure indicates that the x_i are again not quite randomly distributed. This time the problem is integer overflow, i.e., the values of a and m are sufficiently large that $AI_n > 10^{32}$?1 for many integers in the pseudorandom sequence. Thus, the algorithm in Listing 60 is not executed correctly. Integer overflow can be overcome using Schrange's algorithm. If $y = (Az) \bmod , m$, then

$$y = \begin{cases} A(z \bmod q) - r(z/q) & \text{if } y > 0, \\ A(z \bmod q) - r(z/q) + m & \text{otherwise,} \end{cases} \quad (6.4)$$

where $q = m/a$ and $r = m\%a$. The so called *Park and Miller method* [325] generates a pseudorandom sequence that corresponds to a linear congruential method, characterized by the values $a = 16\,807$, $c = 0$, and $m = 2\,147\,483\,647$. The function

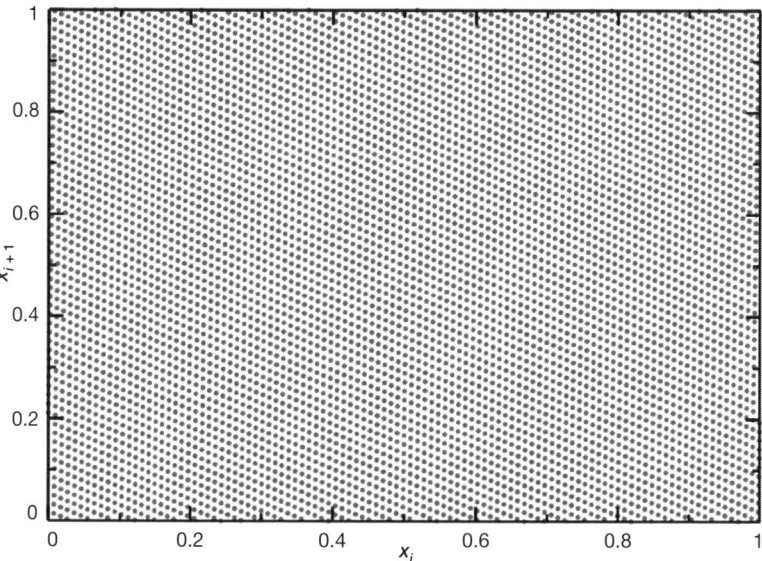

Figure 6.2. Plot of x_{i+1} versus x_i for $i \in [1, \ldots, 10\,000]$. The x_i are pseudorandom variables, uniformly distributed over the range 0 to 1, generated using a linear congruential pseudorandom number generator, according to equation (6.3), characterized by $a = 106$, $c = 1\,283$, $m = 6\,075$.

in Listing 59 implements this method, using Schrange's algorithm to avoid integer overflow.

Figure 6.4 shows a correlation plot for the first 10 000 $x_{i+1} - x_i$ pairs generated using Park and Miller's method. In this figure, one cannot see any pattern whatsoever in the plotted points. This is a visual indication that the x_i are indeed randomly distributed over $R \in [0, 1]$.

6.2.2 Monte Carlo integration – simple sampling

A well-known application of MC simulations is the computation of integrals, particularly *higher dimensional integrals* (we shall see that Monte Carlo is in fact *the* most efficient way to compute higher dimensional integrals). Let us consider the integral of a function $g(x)$ in an interval given by $I = [a, b]$. We may approximate the integral by choosing N points x_i on the x-axis with their corresponding values $g(x_i)$, summing and averaging over these sampled results and multiplying the resulting ex-

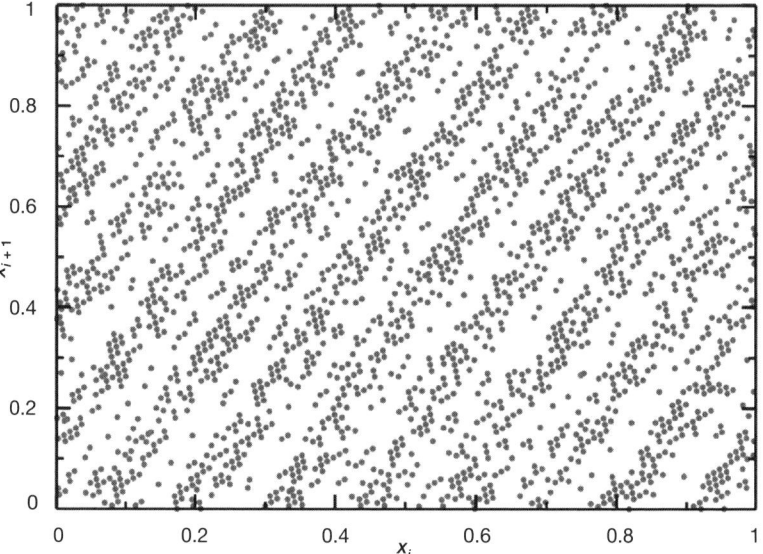

Figure 6.3. Plot of x_{i+1} versus x_i for $i \in [1, \ldots, 10\,000]$. The x_i are pseudorandom variables, uniformly distributed over the range 0 to 1, generated using a linear congruential pseudorandom number generator, according to equation (6.3), characterized by $a = 106$, $c = 1\,283$, $m = 6\,075$.

pression by the length of the interval:

$$\int_a^b g(x)\,dx \approx (b-a)\left[\frac{1}{N}\sum_{i=1}^{N} g(x_i)\right]. \tag{6.5}$$

We now need to say a word about the nature of these randomly chosen points x_i on the x-axis. If we choose them completely at random, the process is called *simple sampling* which works very well if $g(x)$ is smooth, see Figure 6.5 (a). But what if we cannot make the assumption that $g(x)$ is smooth? Let us consider a less advantageous function, for instance one featuring a singularity at a certain value. Due to the limited number of points that we would usually choose around the singularity in simple sampling, we would not be able to really appreciate and understand its behavior. Furthermore, the integral would be a *very* rough approximation. We thus need more precision, which is beyond the possibilities of simple sampling. This additional precision is provided by a second function, a distribution function $\rho(x)$ which makes our condition for successful sampling less strict: only $\frac{g(x)}{\rho(x)}$ needs to be smooth. The

Section 6.2 Simple random numbers 371

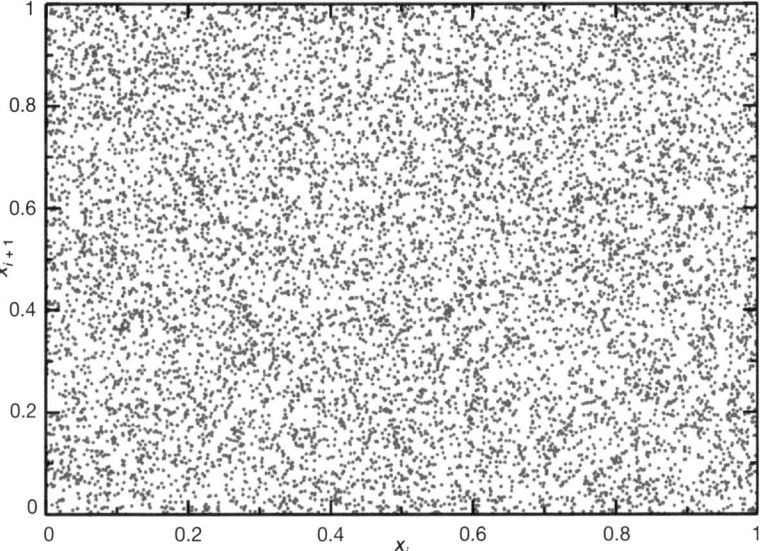

Figure 6.4. Pseudorandom number sequence according to Park and Miller [325]. The first 10 000 random values x_{i+1} versus x_i are plotted, distributed uniformly over the interval $R \in [0, 1]$.

sampling points are now distributed according to $\rho(x)$ and we have.

$$\int_a^b g(x)\,dx = \int_a^b \frac{g(x)}{\rho(x)} \rho(x)\,dx \approx (b-a)\left[\frac{1}{N}\sum_{i=1}^{N}\frac{g(x_i)}{\rho(x)}\right]. \tag{6.6}$$

We have changed our way of sampling by using the distribution function $\rho(x)$, reducing the requirement on $g(x)$. This manifests itself in the summand above. One could state that $\rho(x)$ helps us to pick our sampling points according to their importance. Hence, we select more points close to those regions where the function differs from zero. This kind of sampling is called *importance sampling*.

We end this section with a discussion of a brute force MC program in Listing 61 which integrates the function

$$\int_0^1 dx\,\frac{1}{1+x^2} = \pi, \tag{6.7}$$

where the input is the desired number of MC samples. Note, that we transfer the variable idum in order to initialize the RNG from the function ran0() which is a well known RNG from the book *Numerical Recipes* [342]. The variable idum, which is the seed of the RNG, gets changed for every sampling. What we are doing is to employ

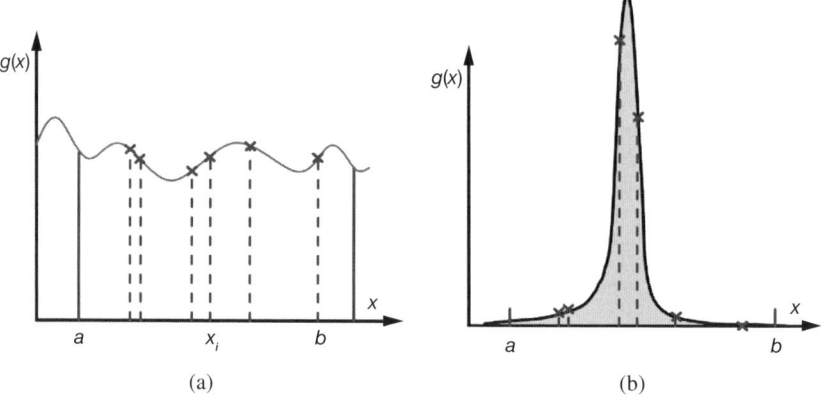

Figure 6.5. MC simple sampling. (a) Simple sampling of a smooth function $g(x)$. (b) Simple sampling of a highly localized function $g(x)$.

an RNG to obtain numbers x_i in the interval $I = [0, 1]$ through a call to one of the library functions *ran0()*, *ran1()*, *ran2()* or *ran3()* which generate random numbers in the interval $x \in [0, 1]$. These functions will be discussed in the next section. Here, we employ these functions simply in order to generate a random variable. In Listing 61, these functions are stored in the included library file *"RandomLib.h"*. In Listing 62 we provide the function *ran0()*.

All *ran()* RNGs produce pseudorandom numbers in the interval $I = [0, 1]$, using the so-called *uniform probability distribution* $P(x)$, defined as

$$P(x) = \frac{1}{b-a} \Theta(x-a) \Theta(b-x), \qquad (6.8)$$

with $a = 0$ and $b = 1$. If one calculates an integral over a general interval $I = [a, b]$, one can still use these random number generators through a change of variables

$$z = a + (b-a)x, \qquad (6.9)$$

with $x \in [0, 1]$. The algorithm in Listing 61 is called "crude" or "brute-force" MC, because the RNG used generates points that are nonhomogeneously distributed over the interval $[0, 1]$, cf. Figure 6.5 (a). If the function is peaked around certain values of x – see Figure 6.5 (b) – one may end up sampling function values where $f(x)$ is small or close to zero. Thus, better schemes, which reflect the properties of the function to be integrated, are needed. The algorithm performs the following steps:

1. choose the number of MC samples n

2. perform a loop over n and for each step use a RNG to generate a random number x_i in the interval $[0, 1]$

Section 6.2 Simple random numbers

Table 6.2. Brute force MC sampling of the integral $I = \int_0^1 dx\, 1/(1+x^2)$. The exact answer is 3.14159 for the integral I and 0.413581 for the variance with six leading digits.

n	I	σ
10^1	3.10263	0.398802
10^2	3.02933	0.404822
10^3	3.13395	0.422881
10^4	3.14195	0.411195
10^5	3.14003	0.414114
10^6	3.14213	0.413838
10^9	3.14162	0.413581

3. use this number to evaluate the function $f(x_i)$

4. evaluate the contributions to the mean value and the standard deviation for each loop

5. after n samples calculate the final mean value and the standard deviation

We note that as n increases, the integral itself never reaches more agreement than to the fourth or fifth digit, see Table 6.2. Note well, that the variance can – with appropriate redefinition of the integral – be made smaller. A smaller variance also yields a smaller standard deviation. As an alternative, we could have used the RNG provided by the C/C++ compiler, through the functions *srand()* and *rand()*. In this case, we initialize it, as in the previous section, through the function *srand()*. The RNG is called through the function *rand()*, which returns an integer from 0 to its maximum value on the system, defined by the variable RAND_MAX, as demonstrated in the next few lines of code:

```
inverseValue = 1. / RAND_MAX;

//Initialize RNG
srand(time(NULL));

//Obtain a floating point number x in [0,1]
x = double (rand()) * inverseValue;
```

Listing 59. Park and Miller pseudorandom number generator. (*)

```c
#include <stdio.h>

/* Park and Miller pseudorandom number generator.
*/

/* RAND_MAX = m - 1 */
#define RAND_MAX 2147483646

int Random (int seed)
{
  static int next = 1;
  static int a    = 16807;
  static int m    = 2147483647; /* 2^31 - 1 */
  static int q    = 127773;     /* m / a    */
  static int r    = 2836;       /* m % a    */

  if (seed) next = seed;
  next = a * (next % q) - r * (next / q);
  if (next < 0) next += m;
  return next;
}

int main(void)
{
  int x;
  double y;

  for (x = 1; x < 10000; x++){
    y = (double) (Random(0)) / (double) (RAND_MAX);
    printf("%.5f ",y);
    y = (double) (Random(0)) / (double) (RAND_MAX);
    printf("%.5f\n",y);
  }
  return (0);
}
```

Section 6.2 Simple random numbers

Listing 60. Linear congruential pseudorandom number generator. (*)

```
1  /* Linear congruential pseudorandom number generator,
2     generating pseudorandom sequences of integers in
3     the range [0...RAND_MAX]
4  */
5  #include <stdio.h>
6
7  /* RAND_MAX = M - 1 */
8  #define RAND_MAX 6074
9
10 int Random (int seed)
11 {
12    static int next = 1;
13    static int a = 107;
14    static int c = 1283;
15    static int m = 6075;
16
17    if (seed) next = seed;
18    next = (next * a + c) % m;
19    return (next);
20 }
21
22 int main(void)
23 {
24    int x;
25    double y;
26
27    for (x = 1; x < 10000; x++){
28       y = (double) (Random(0)) / (double) (RAND_MAX);
29       printf("%.5f ",y);
30       y = (double) (Random(0)) / (double) (RAND_MAX);
31       printf("%.5f\n",y);
32    }
33    return (0);
34 }
```

Listing 61. Brute force MC integration of π. (*)

```
// This is a simple sampling program to calculate the integral
// of Equation ().
#include <iostream>
#include <cmath>

#include "RandomLib.h"

using namespace std;

double FunctionCalculate(double x);

int main()
{
   int i, n;
   long idum;
   double crude_mc, x, sum_sigma, fx, variance;
   cout << "Read the number of Monte-Carlo samples..." << endl;
   cin >> n;
   crude_mc = sum_sigma = 0.; idum = -1;

   for(i = 1; i <= n; i++){
      x          = ran0(idum);
      fx         = FunctionCalculate(x);
      crude_mc   += fx;
      sum_sigma  += fx * fx;
   }
   crude_mc  = crude_mc  / ((double)n);
   sum_sigma = sum_sigma / ((double)n);
   variance  = sum_sigma - crude_mc * crude_mc;
   //This is the final output
   cout << "Variance = " << variance << " Integral = " <<
       crude_mc << " Exact Value = " << M_PI << endl;
} // End of main() program

// This function defines the function to integrate
double FunctionCalculate(double x)
{
   double value;

   value = 4 / (1. + x * x);
   return value;
}//End of FunctionCalculate()
```

Section 6.2 Simple random numbers

Listing 62. The library function ran0(). (*)

```
#ifndef __RANDOM_H
#define __RANDOM_H

#define IA 16807
#define IM 2147483647
#define AM (1.0/IM)
#define IQ 127773
#define IR 2836
#define MASK 123459876

double ran0(long &idum)
// "Minimal" random number generator of Park and Miller with
// Bays-Durham shuffle and added safeguards. Returns a uniform random
// deviate between 0.0 and 1.0. Set or reset idum to any integer value
// (except the unlikely value MASK) to initialize the sequence. idum
// must not be altered between calls for successive deviates in a sequence.
{
    long k;
    double ans;

    idum ^= MASK;         // XORing with MASK allows use of 0 and
    k = idum/IQ;          // other simple bit patterns for idum.
    idum = IA * (idum-k*IQ) - IR*k; // Compute idum = (IA*idum) % IM without
    if (idum < 0) idum += IM;   // overflows by Schrage's method.
    ans = AM * idum;      // Convert idum to a floating result.
    idum ^= MASK;         // Unmask before return.
    return ans;
}
#endif
```

6.3 Case study: MC simulation of harddisks

The source code *Chap6_MCHardDisks.c*, which is provided on the book's website www.degruyter.com, simulates two-dimensional disks confined to a circle. Alternatively, you can write such a program from scratch yourself. All the information about the interaction and the MC moves to be considered is provided in this section. The Hamiltonian for this system may be expressed as

$$H = \sum_{i=1}^{N} H_1(\vec{r}_i) + \sum_{i=1}^{N} \sum_{j=i+1}^{N} H_2(\vec{r}_i, \vec{r}_j) \qquad (6.10)$$

where

$$H_1(\mathrm{r}_i) = \begin{cases} 0 & r_i < R \\ \infty & r_i > R \end{cases} \qquad (6.11)$$

and

$$H_2(\vec{r}_i, \vec{r}_j) = \begin{cases} 0 & \sqrt{(\vec{r}_i - \vec{r}_j)^2} < \sigma \\ \infty & \sqrt{(\vec{r}_i - \vec{r}_j)^2} > \sigma \end{cases}. \qquad (6.12)$$

H_1 acts to keep the particles confined, and H_2 prevents them from overlapping. One nice thing about using harddisk Hamiltonians is that there is never reason to evaluate a Boltzmann factor. Any trial move that results in an overlap or a particle crossing the boundary results in an "infinite" ΔU, so that $e^{-\beta \Delta U}$ is identically 0 and the trial is unconditionally rejected. This makes harddisk simulation rather simple. *Chap6_MCHardDisks.c* accepts as user input any two of the following three parameters: the radius of the circle R, the areal number density of particles (# per square σ) ρ, and the number of particles N. The user may also specify δr, the scalar displacement, and nCycle, the number of MC cycles, where one cycle corresponds to N attempted particle displacements. The output of the code is simply the acceptance ratio:

$$[\text{acceptance ratio}] = \left[\frac{\text{number of successful trials}}{\text{number of trials}} \right] \qquad (6.13)$$

6.3.1 Trial moves

The most common trial move in continuous-space MC, i.e. MC not confined to lattice sites, is a particle displacement. First, a small number ΔR, representing maximum displacement, is set. A trial move consists of

1. randomly selecting a particle, i.

2. displacing x-position coordinate of particle i by a random amount, δx, which is given by

$$\delta x = \Delta R \xi_x, \tag{6.14}$$

where ξ_x is a uniform random variate in the interval $[-0.5, 0.5]$.

3. repeating the procedure for the y and z coordinates, if applicable.

This move guarantees detailed balance, provided the random particle selection is uniform, meaning that, for any given move, selection of all possible particles is equally likely. This means that the probability of suggesting a move that displaces a particle going from a state n to a new state m, has the same probability as selecting the same particle whilst in state m and giving it a displacement that will return the configuration to state n. For a system of simple particles, random displacements are the only necessary trial moves. For more complicated systems, there are zoos of trial moves all through the literature. The question is, how does one choose an appropriate value for ΔR? If ΔR is too small, the system will not explore phase space, given a reasonable amount of computational effort. If it is too large, displacements will rarely result in new configurations, which will be accepted in a Metropolis MC scheme. So it takes a bit of trial and error to find a good value for ΔR, and the rule of thumb is to set ΔR such, that the average probability of accepting a new configuration during a run is about 30%, i.e. one tunes ΔR to achieve a 30% acceptance ratio. We will go through the exercise of determining such an appropriate value for ΔR for a simple continuous-space system, the 2D harddisks confined to a circle, in the case study in Section 6.6.

Rigid rotation is a second common type of trial move is used in systems of more structured molecules than simple, single-center spheres. Consider a diatomic, with a rigid bond length r_0. Clearly, attempting to move one of the two members of the diatomic by random displacement is likely to result in a new bond length, with may be significantly different from r_0. So, for a system of dumbbell molecules, a reasonable set of trial moves would include

1. a small displacement of the molecule's center of mass

2. a small rotation around molecule's center of mass

With more than one kind of move, an attempt to generate a new state must be preceded by a random selection of the trial move. We can weight each kind of move and then use a random number to decide which move to attempt. For example, let's say that we choose 80% of all trial moves to be displacements, and the balance to be rotations. Prior to an attempted move, we select a uniform random variable, ξ, on the interval $[0, 1]$. If $\xi < 0.8$, which will be 80% of the time, we execute a displacement of a randomly chosen molecule. Otherwise, we execute a rotation of a randomly chosen molecule.

6.3.2 Case study: MC simulation of harddisks – suggested exercises

One important aspect of any MC simulation program is how the particle positions are *initialized*. Here, it is best to assign initial positions to the particles such that the initial energy is 0 (i.e., there are no overlaps nor particles out of bounds.)

1. Try to write such a function on your own. If you cannot figure out how to do this, examine the function *Init()* in the program *Chap6_MCHardDisks.c* and study how it accomplishes this.

2. Use the program to determine a reasonable displacement to achieve a 30% acceptance ratio at a density of 0.5. Compare your results across differently sized systems and between runs with different numbers of cycles. For fewer than 10^6 cycles, you will have large acceptance ratios, because the initial condition is not yet fully destroyed.

   Figure 6.6 shows a plot of the acceptance ratio vs. ΔR for densities ρ of 0.2, 0.4, 0.6, from a simulation of 2 000 particles. 10 000 cycles were performed for each run. Check whether your results are consistent with this data.

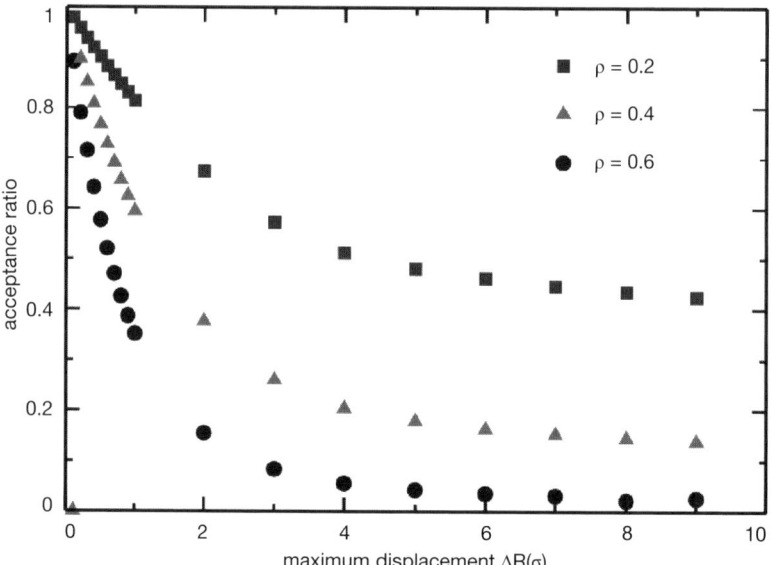

Figure 6.6. Acceptance ratio vs. maximum displacement $\Delta R(\sigma)$, for an MC simulation of harddisks. Various densities are displayed for a system of $N = 200$ particles and 100 000 MC cycles.

3. Consider generating a trial move in the following way:

    (a) Randomly select a particle i

    (b) Randomly choose a direction. In 2D, this is an angle θ chosen uniformly from the interval $[0, 2\pi]$. In 3D, this involves two angles θ and ϕ, where $\cos\theta$ is uniformly chosen from the interval $[-1, 1]$, and ϕ is uniformly chosen from the interval $[0, 2\pi]$.

    (c) Expand the program to 3D.

    (d) Compute the displacement vector components. In 2D: $dx = \Delta R \cos\theta$ and $dy = \Delta R \sin\theta$. In 3D: $dx = \Delta R \sin\theta \cos\phi$, $dy = \Delta R \sin\theta \sin\phi$, $dz = \Delta R \cos\theta$.

    (e) Execute the move by adding the displacement vector components to the position of particle i

The main difference between this scheme and the first one is that, in 2D, only one random number must be chosen per displacement attempt. Explore the acceptance ratio vs. ΔR relation with this new trial move. Does this scheme generate a 30% acceptance ratio for a higher or lower maximum displacement than the original scheme?

6.4 The Metropolis Monte Carlo method

We return to the idea of evaluating the integrand at a discrete set of points, randomly selected from a distribution. Here we call upon the idea of *importance sampling*. Let us try to use whatever we know ahead of time about the integrand, when picking our random distribution, ρ, such that we minimize the number of points (i.e., the expense) necessary to give an estimate of $\langle Q \rangle$ to a given level of accuracy, where Q is some quantity.

Now, clearly the states that contribute most to the integrals that we wish to evaluate by configurational averaging are those states with large Boltzmann factors. That is, those states for which ρ_{NVT} is large. It stands to reason that if we randomly select points from ρ_{NVT}, we will do a pretty good job in approximating the integral. So what we end up computing is the "average of $Q\rho_{NVT}$ over ρ_{NVT}":

$$\langle Q\rho_{NVT}/\rho_{NVT}\rangle \approx \langle Q \rangle, \tag{6.15}$$

which should provide us with an excellent approximation for $\langle Q \rangle$. The idea of using the ρ_{NVT} as the sampling distribution is due to Metropolis [298]. This makes the real work in computing $\langle Q \rangle$ be to generate states that randomly sample ρ_{NVT}.

Metropolis et al. [298] showed that an efficient way to do this involves generating a Markov chain of states constructed such that its limiting distribution is ρ_{NVT}. A Markov chain is just a sequence of trials, where

1. each trial outcome is a member of a finite set called "state space," and

2. every trial outcome depends only on the outcome that immediately precedes it.

By "limiting distribution," we mean that the trial acceptance probabilities are tuned such that the probability of observing the Markov chain atop a particular state is defined by some equilibrium probability distribution, ρ.

A trial is some perturbation (usually small) of the coordinates specifying a state. For example, in an Ising system, this might mean flipping a randomly selected spin. In a system of particles in continuous space, it might mean displacing a randomly selected particle by a small amount δr in a randomly chosen direction. There is a large variety of such "trial moves" for any particular system. In this course, we will only deal with a few simple ones.

The probability that a trial move results in a successful transition from state n to m is denoted by π_{nm} and π is called the "transition matrix", which must be specified ahead of time in order to execute a traditional Markov chain. Since the probability that a trial results in a successful transition to any state, the rows of π add up to unity:

$$\sum_i \pi_{ni} = 1 \tag{6.16}$$

With this specification, we term π a "stochastic" matrix. Furthermore, for an equilibrium ensemble of states in state space, we require that transitions from state to state do not alter state weights as determined by the limiting distribution. So, the weight of state n must be the result of transitions from all other states to state n:

$$\rho_n = \sum_m \rho_m \pi_{mn}. \tag{6.17}$$

For all states n, we can write equation (6.17) as a matrix equation of the form

$$\rho \pi = \rho \tag{6.18}$$

where ρ is the row vector of all state weights. Equation (6.18) constrains our choice of **pi**. This means there is still more than one way to specify ρ. Metropolis et al. [298] suggested to use

$$\rho_m \pi_{mn} = \rho_n \pi_{nm}. \tag{6.19}$$

That is, the probability of transitioning from state m to n is exactly equal to the probability of transitioning from state n to m. This is called the "detailed balance" condition, and it guarantees that the state weights remain static. Observe:

$$\sum_m \rho_m \pi_{mn} = \sum_m (\rho_n \pi_{nm}) = \rho_n \left(\sum_m \pi_{nm} \right) = \rho_n. \tag{6.20}$$

Metropolis et al. [298] chose to construct the matrix π as

$$\pi_{nm} = \alpha_{nm} \mathrm{acc}(n \to m) \qquad (6.21)$$

where α is the probability that a trial move is attempted and acc is the probability that a move is accepted. If the probability of proposing a move from n to m is equal to that of proposing a move from m to n, then $\alpha_{nm} = \alpha_{mn}$, and the detailed balance condition is written as

$$\rho_n \mathrm{acc}(n \to m) = \rho_m \mathrm{acc}(m \to n), \qquad (6.22)$$

from which follows

$$\frac{\mathrm{acc}(n \to m)}{\mathrm{acc}(m \to n)} \frac{\rho_n}{\rho_m} \frac{e^{-\beta U(\Gamma_n)}}{e^{-\beta U(\Gamma_n)}} \qquad (6.23)$$

with Γ_i denoting the various configurations, thus giving

$$\frac{\mathrm{acc}(n \to m)}{\mathrm{acc}(m \to n)} \exp(-\beta[U(\Gamma_m) - U(\Gamma_n)]) \equiv \exp(-\beta \Delta U_{nm}) \qquad (6.24)$$

where we have defined the change in potential energy as

$$\Delta U_{nm} = U(\Gamma_m) - U(\Gamma_n). \qquad (6.25)$$

There are many choices for acc $(n \to m)$ that satisfy equation (6.24). The original choice of Metropolis is used most frequently,

$$\mathrm{acc}(n \to m) = \begin{cases} \exp(-\beta \Delta U_{nm}) & \text{for } \Delta U_{nm} > 0 \\ 1 & \text{for } \Delta U_{nm} < 0. \end{cases} \qquad (6.26)$$

So, suppose we have some initial configuration n with potential energy U_n. We make a trial move, temporarily generating a new configuration m. Now, we calculate a new energy, U_m. If this energy is lower than the original, $(U_m < U_n)$, we unconditionally accept the move and configuration m becomes the current configuration. If it is larger than the original, $(U_m > U_n)$, we accept it with a probability consistent with the fact that the states both belong to a canonical ensemble. How does one decide in practice whether to accept the move? One first picks a uniform random variate x on the interval $I = [0, 1]$. If $x \leq \mathrm{acc}(n \to m)$, the move is accepted.

The next section is devoted to an implementation of the *Metropolis Monte Carlo method* for a 2D Ising magnet.

6.5 The Ising model

The Ising model is a model that originally aimed at explaining ferromagnetism, but is today used in many other areas, such as opinion models and binary mixtures. It is a highly simplified approach to the difficulties of understanding and describing magnetism (e.g. commutation relations of spins). We consider a discrete collection of N binary variables – called spins – which may take on the values ± 1 (representing the up and down spin configurations). The spins σ_i interact pairwise with their nearest neighbors and the energy has one value for aligned spins ($\sigma_i = \sigma_j$), and another for anti-aligned spins ($\sigma_i \neq \sigma_j$). The Hamiltonian and the partition function for this model are given by

$$H = E = -\sum_{\langle ij \rangle}^{N} J_{ij}\sigma_i\sigma_j - \sum_{i}^{N} H_i \sigma_i, \qquad (6.27a)$$

$$Z = \sum_{\{\sigma_i\}}^{N} \exp(-\beta H). \qquad (6.27b)$$

where H_i is a (usually homogeneous) external Field and J_{ij} are the (translationally invariant) coupling constants. Because the coupling constants are translationally invariant, we may drop the indices and simply write $J = J_{ij} \; \forall i, j$. The coupling constant J is half the difference in energy between the two possible states (alignment and anti-alignment). The simplest example is given by the antiferromagnetic one-dimensional Ising model, which has the energy function

$$H = E = \sum_{\langle ij \rangle}^{N} \sigma_i \sigma_j, \qquad (6.28)$$

which can be generalized to two dimensions, see Figure 6.7. One can also add an external field, as in equation (6.27). The notation $\langle ij \rangle$ in the sums of the above equations denotes summing over unique pairs of nearest neighbors. The spins are placed on a cubic lattice with edge length L in D dimensions with the total volume $V = L^D$. For $D = 2$, the Ising model has been solved exactly for both finite and infinite values of L. For $D = 3$ only approximate numerical solutions are currently available.

> How many unique pairs of nearest neighbors are there on a lattice of N spins (assuming periodic boundary conditions)?

To answer the question above, one must know the coordination number z of the lattice. That is, one must know how many nearest neighbors one lattice position has. For a square lattice, $z = 4$. Each spin therefore contributes *two* unique spin pairs to the system, so there are $Nz/2$ unique nearest neighbor pairs.

According to the Hamiltonian, the energy of the system is minimal when all N spins have the same alignment, either all up or all down. Imagine a microscopic

Section 6.5 The Ising model

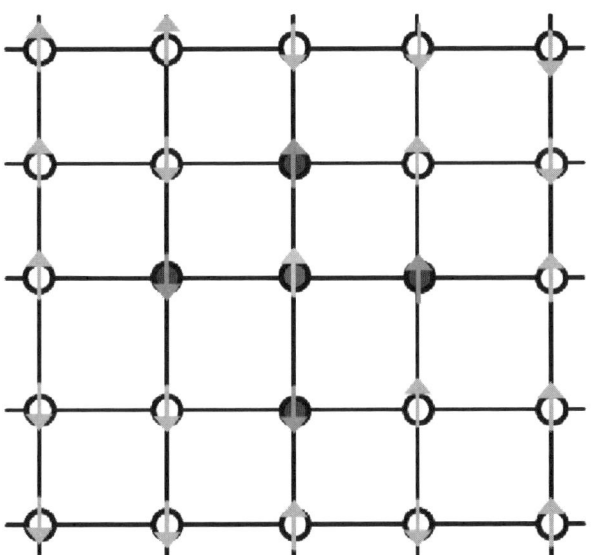

Figure 6.7. Ising spin model on a 2D lattice.

observable called the magnetization, or average spin orientation, M:

$$M \equiv \frac{1}{N} \sum_{i}^{N} \sigma_i \qquad (6.29)$$

Then, $M = \pm 1$ are two energetically equivalent microstates. We can then expect that if there is any thermal energy in the system which randomly flips spins, the amount of this thermal energy (that is, the temperature) will somehow control the observable magnetization. Thus, one could say, that one "observes" magnetization, using an ensemble average given by:

$$\langle M \rangle = \frac{\sum_{s_1=-1}^{+1} \cdots \sum_{s_N=-1}^{+1} \left[\sum_{i}^{N} \sigma_i \right] \exp\left[-\beta H(\sigma_1, \ldots, \sigma_N)\right]}{\sum_{s_1=-1}^{+1} \cdots \sum_{s_N=-1}^{+1} \exp\left[-\beta H(\sigma_1, \ldots, \sigma_N)\right]}. \qquad (6.30)$$

The magnetization of a material vanishes as the temperature goes to zero, i.e. $T \to 0$, $\langle M \rangle \to 1$, and as $T \to \infty$, $\langle M \rangle \to 0$. The interesting thing about an Ising magnet is that there is a *finite temperature*, called the *critical temperature* T_c. In the vicinity of T_c, large regions of mostly up spins compete with relatively large regions of mostly down spins. The *correlation length* ζ is an important observable that characterizes the size of these domains and is a useful measure of criticality. Among the observables that one can compute for the Ising model are the internal energy per site $e = E/V$ and the magnetization $m = M/V = \langle \mu \rangle$, with $\mu = \sum_i^N \sigma_i/V$. If one starts out with a "hot" system and cools it to just below T_c, the absolute value of the magnetization spontaneously jumps from 0 to some finite positive value. In other words, the system undergoes a first-order *phase transition* from a disordered phase to a partially ordered phase at temperature T_c. At this temperature, the correlation length ζ diverges. This means that fluctuations close to the critical point exist on all length scales and thermodynamic observables diverge following a power law dependence of the type

$$\zeta \propto t^{-\nu}, \tag{6.31}$$

$$m \propto t^{-\beta_m}, \tag{6.32}$$

$$t = \left|1 - \frac{T}{T_c}\right|. \tag{6.33}$$

Exactly the same behavior can be seen with macromolecules near the so called Θ-temperature, where attractive and repulsive interactions nearly cancel each other out. It was the French physicist deGennes who realized that the behavior of macromolecules close to the Θ-temperature can be described with exactly the same formalism that describes criticality[3]. For 2D, the values of T_c and the critical exponents β_m and ν, as well as the critical temperature, are known analytically.

A Metropolis MC simulation allows for probing the behavior of an Ising system and learning how the system behaves near criticality. In the remainder of this section, I will discuss the inner workings of a simple Metropolis MC program which simulates the Ising lattice using Metropolis MC.

6.5.1 Case study: Monte Carlo simulation of the 2D Ising magnet

The Monte Carlo simulation of the Ising model is based on a single flip Metropolis algorithm:

1. Choose one site i (having spin σ_i).

2. Calculate $\Delta E = E(Y)E(X) = 2J\sigma_i h_i$.

3. If $\Delta E < 0$ then flip spin, i.e.: $\sigma_i \to -\sigma_i$.

4. If $\Delta E > 0$ then flip spin with probability $exp(-\beta E)$.

Here, I have introduced h_i which is the approximate local field at site i, given by

$$h_i = \sum_{\langle i \rangle^N} \sigma_j. \tag{6.34}$$

---

[3] deGennes did this work in the early 1970s and received the Nobel prize in 1991.

Section 6.5 The Ising model

The sum is characteristic of this approximation. We are summing only over the next neighbors j of i, i.e. when determining h_i, we are only interested in the local fields of the next neighbors. Note that the acceptance probability is evaluated at each step. Given that the acceptance probability is very low, one needs a lot of steps for something to change. A group of N steps is often called a "sweep".

Listing 63 shows the *main()*-function of a C program for simulating the 2D Ising magnet.

As usual, the functions which are called from within *main()* have to be included in header files at the beginning of the module *main.c*, i.e. in the module where *main()* is located. Here, I have named these header files *Energy.h*, *Init.h* and *SampleSystem.h*, according to the purpose of the functions declared within these header files. The definition of the three functions will then be put in corresponding *.c files, for example, "Energy.c", which also include the respective header file with the corresponding function declaration.

In Listing 64 the function *EnergyChange()* is shown, which demonstrates the concise way of writing C code using the logical ? operator. This function computes and returns the change in system energy when spin (i, j) is flipped. The modulo arithmetic (the % operator) ensures periodic boundaries. The syntax i?(i-1):(L-1) performs the following check: if i is nonzero, return i-1, otherwise return L-1. This also ensures periodic boundaries. *EnergyChange()* takes as arguments the 2D array of spins F, the length of the side L, and a position (i,j), and it returns the change in energy upon flipping spin (i,j), without actually flipping it. All variables are of type *int*. The syntax int **F means that F is a pointer to a pointer to an *int*, i.e. a pointer to a field of int-pointers. It is a way of signifying that F is a 2D array. The syntax for accessing the i,j element of F is F[i][j].

The function *SampleSystem()*, which samples the system and computes the average magnetization and the average energy per spin, is provided in Listing 65. This function actually samples the current 2D array of spins and computes the average spin $\langle s_1 \rangle$, and the average energy per spin $\langle \epsilon \rangle$. The syntax x += y is short for x = x + y. The same is true for other "incremental operators", such as -=, *=, and /=. Because s and e are passed by reference (so that their values can be changed by the function), one has to dereference them with the * operator in order to access their contents; that is, *s means "the contents of s".

Finally, Listing 66 introduces the function *Init()* which randomly assigns all spins using the GSL library (see Appendix I for details on how to install it). The function randomly initializes the 2D array of spins. The function gsl_rng_uniform_int() randomly returns either 0 or 1, and we want each spin to be either 1 or -1. Note that the random number generator created in *main()* must be passed as an argument as well.

Lines 1–7 in the main program in Listing 63 include the standard header files (system libraries and the header files of the three functions called in *main()*, including the header for the GSL random number generator. In lines 11–13 F is the 2D array of

Listing 63. Part 1 – *main()*-program for calculating the 2D Ising model, using NVT MC.

```c
#include <stdio.h>
#include <stdlib.h>
#include <math.h>
#include <gsl/gsl_rng.h>
#include ''Init.h''
#include ''SampleSystem.h''
#include ''Energy.h''

int main (int argc, char * argv[]) {
  /* System parameters */
  int ** SpinField;    /* The 2D array of spins */
  int L = 20;          /* The side length of the array */
  int Nspins;          /* The total number of spins = L*L */
  double T = 1.0;      /* Dimensionless temperature = (T*k)/J */

  /* Run parameters */
  int nCycles = 1000000;  /* number of MC cycles to run. One
      cycle is Nspins consecutive attempted spin flips */
  int frequencySample = 1000; /* Frequency with which samples
      are taken */

  /* Computational variables */
  int nSample;         /* Number of samples taken */
  int diffEnergy;      /* energy change due to flipping a spin */
  double kBoltzmann;   /* Boltzman factor */
  double x;            /* random number */
  int i,j,a,c;         /* loop counters */

  /* Observables */
  double spin = 0.0, spinSum = 0.0; /* average magnetization */
  double energyPerSpin = 0.0, energySum = 0.0; /* average
      energy per spin */

  /* This line creates a random number generator
     of the "Mersenne Twister" type, which is much
     better than the default random number generator. */
  gsl_rng * r = gsl_rng_alloc(gsl_rng_mt19937);
  unsigned long int Seed = 23410981;

  /* Here, we parse the command line arguments */
  for (i=1;i<argc;i++) {
    if (!strcmp(argv[i],"-L")) L=atoi(argv[++i]);
    else if (!strcmp(argv[i],"-T")) T=atof(argv[++i]);
    else if (!strcmp(argv[i],"-nc")) nCycles = atoi(argv[++i]);
    else if (!strcmp(argv[i],"-fs")) frequencySample = atoi(
        argv[++i]);
    else if (!strcmp(argv[i],"-s")) Seed = (unsigned long)atoi(
        argv[++i]);
  }
```

Section 6.5 The Ising model

Listing 63. Part 2 – *main()*-program for calculating the 2D Ising model, using NVT MC, continued ....

```
45
46      /* Output some initial information */
47      fprintf(stdout,"# command: ");
48      for (i=0;i<argc;i++) fprintf(stdout,"%s ",argv[i]);
49      fprintf(stdout,"\n");
50      fprintf(stdout,"# ISING simulation , NVT Metropolis Monte Carlo\n");
51      fprintf(stdout,"# L = %i , T = %.3lf , nCycles %i ,
            frequencySample %i , Seed %u\n",
52      L,T,nCycles,frequencySample,Seed);
53
54      /* Seed the random number generator */
55      gsl_rng_set(r,Seed);
56      /* Compute the number of spins */
57      Nspins = L * L;
58
59      /* Allocate memory for the system */
60      SpinField = (int**)malloc(L * sizeof(int*));
61      for(i=0; i<L; i++) SpinField[i]=(int*)malloc(L * sizeof(int));
62
63      /* Generate an initial state */
64      Init(SpinField,L,r);
65
66      /* For computational efficiency , convert T to reciprocal T */
67      T=1.0/T;
68
69      spin = 0.0;
70      energyPerSpin = 0.0;
71      nSample = 0;
72      for (c = 0; c < nCycles; c++) {
73        /* Make Nspins flip attempts */
74        for (a = 0; a < Nspins; a++) {
75          /* randomly select a spin */
76          i=(int)gsl_rng_uniform_int(r,L);
77          j=(int)gsl_rng_uniform_int(r,L);
78          /* get the "new" energy as the incremental change due
79              to flipping spin (i,j) */
80          diffEnergy = EnergyChange(SpinField,L,i,j);
81          /* compute the Boltzmann factor; recall T is now
82              reciprocal temperature */
83          kBoltzmann = exp(diffEnergy * T);
84          /* pick a random number between 0 and 1 */
85          x = gsl_rng_uniform(r);
86          /* accept or reject this flip */
87          if (x < kBoltzmann) { /* accept */
88      /* flip it */
89      SpinField[i][j]*=-1;
90        }
91      }
```

Listing 63. Part 3 – *main()*-program for calculating the 2D Ising model, using NVT MC, continued . . . .

```
92  /* Sample and accumulate averages */
93      if (!(c%frequencySample)) {
94          SampleSystem(SpinField,L,&spin,&energyPerSpin);
95          fprintf(stdout,"%i %.5le %.5le\n",c,spin,energyPerSpin);
96          fflush(stdout);
97          spinSum+=spin;
98          energySum+=energyPerSpin;
99          nSample++;
100     }
101    }
102    fprintf(stdout,"# The average magnetization is %.5lf\n",
           spinSum/nSample);
103    fprintf(stdout,"# The average energy per spin is %.5lf\n",
           energySum/nSample);
104    fprintf(stdout,"# Finished successfully.\n");
105 }
```

Listing 64. Function *EnergyChange()* for the Ising model.

```
1 int EnergyChange ( int ** F, int L, int i, int j ) {
2   return  -2*(F[i][j])*(F[i?(i-1):(L-1)][j]+F[(i+1)\%L][j]+
3           F[i][j?(j-1):(L-1)]+F[i][(j+1)\%L]);
4 }
```

Listing 65. Function *SampleSystem()* for the Ising model.

```
1 double SampleSystem ( int ** F, int L, double * s, double * e )
       {
2   int i,j;
3
4   *s=0.0;
5   *e=0.0;
6   /* Visit each position (i,j) in the lattice */
7   for (i=0;i<L;i++) {
8     for (j=0;j<L;j++) {
9        *s+=(double)F[i][j];
10       *e-=(double)(F[i][j])*(F[i][(j+1)%L]+F[(i+1)%L][j]);
11    }
12  }
13  *s/=(L*L);
14  *e/=(L*L);
15 }
```

Section 6.5 The Ising model

Listing 66. Function *Init()* for the Ising model.

```
void Init ( int ** F, int L, gsl_rng * r ) {
  int i,j;

  /* Visit each position (i,j) in the lattice */
  for (i=0;i<L;i++) {
    for (j=0;j<L;j++) {
      /* 2*x-1, where x is randomly 0,1 */
      F[i][j]=2*(int)gsl_rng_uniform_int(r,2)-1;
    }
  }
}
```

spins, L is the side length of the array (default value is 20), N is the total number of spins = L², T is the dimensionless temperature = $k_B T/J$ (default is 1.0) with J the unit energy of the Hamiltonian. In lines 17–18 the number of cycles to run and the sample interval are defined. A cycle is a set of N attempted spin flips. Lines 21–25 define the computed variables, i.e. the number of samples taken, the change in energy upon spin flip, the Boltzmann factor, the random number, and the loop counters. The observables, average spin, σ_1, and average energy per spin, ϵ, and their respective accumulators for their ensemble averages are all initialized to 0 in lines 28–29. In lines 34–35 a random number generator of the "Mersenne Twister" type is generated, which is much better than the default random number generator. This is why we need the GNU Scientific Library (GSL) in this program. One can see how the parsing of command line arguments is done in lines 36–43. In line 55, the random number generator is assigned a seed. The number of spins N is calculated in line 57 and memory for the 2D array of spins is allocated in lines 60–61. In line 64 the 2D array of spins is initialized and in line 67 the temperature T is converted to its reciprocal value for computational convenience. Lines 72–91 start and end a loop over the number of spins N and the number of cycles, where the energy difference is calculated according to the Boltzmann factor. In lines 93–100 the accumulators are updated, depending on whether the cycle number corresponds to the requested sample frequency. Finally, in lines 102–104 (as well as in 47–52) some output information for the user is provided.

I suggest the following exercises:

1. Run the code for the following values of the temperature: $T = 1.04, 2.0, 3.0, 4.0, 5.0$. Run it several times at each T, with different values for the random number generator seed. Report the average spin and average energy per spin. What happens near $T = 2.0$?

2. Modify the code so that when samples are taken when accumulating statistics for $\langle \sigma_1 \rangle$ and $\langle \epsilon \rangle$, the current sample values are output to the terminal.

3. The current version of the code initializes the Ising lattice with random spins. What temperature does this correspond to? Modify the code so that the initial lattice has two well-defined domains, all spin-up for $i < L/2$ and all spin-down for $i > L/2$. Rerun at the various temperatures. Do you observe any differences?

4. Modify the code to compute the quantity $\langle \sigma_i \sigma_j \rangle - \langle \sigma_i \rangle \langle \sigma_j \rangle$ as a function of various distances between spins i and j.

6.6 Case Study: NVT MC of dumbbell molecules in 2D

In this case study, we consider a system slightly different from the harddisk system in the case study of Section 6.3. Here, we imagine that pairs of disks are tethered together to form dumbbell molecules. The "bond-length" of a dumbbell is a constant parameter, r_0. We confine the dumbbells within a circle, see Figure 6.8.

In this MC problem, one has to reconsider the trial moves for this system. One cannot simply select a random particle and try to displace it, because this is likely to violate the constant bond length of the dumbbell that this particle belongs to. How then does one generate new configurations? A simple idea is to use two kinds of trial moves, translation of entire dumbbells and rotation of dumbbells around their centers of mass. In order to implement an MC code with more than one trial move, one must include a "trial move selection rule", which randomly selects a trial move based on their user-defined "weights".

The code *DumbellMolecules.c* is available online from the book's website[4] and shows how to implement an MC simulation of harddisk dumbbell molecules in 2D, confined to a circle. Let's consider the following question: Does the acceptance ratio of rotational moves depend on the weight given to displacement moves? Below is a plot of this acceptance ratio vs. the maximum displacement for a system of 25 dumbbells, at a density of 0.5, for various maximal displacements 0.1 and 9.

I suggest the following exercises:

1. Run the code and try to reproduce the data in Figure 6.9.

2. Use this code to explore the liquid crystalline nature of a fluid made of dumbbell molecules. In a dense liquid, because the dumbbell molecules are slightly elongated along one direction, they may tend to line up. For dumbbell i, let the quantity θ_i represent the angle made by the inter-particle segment and some global coordinate frame axis (say the x axis). Use this code to compute the average orientation, $\langle \theta \rangle$ as a function of density and temperature. The problem here is that, if the system is truly a liquid, even if at any one time large numbers of dumbbell molecules are lined up, the system will slowly evolve, so that all molecule orientations are eventually realized. Hence, $\langle \theta \rangle$ is probably not the best observable to characterize the orientation of the molecules.

---

[4] www.degruyter.com

Section 6.6 Case Study: NVT MC of dumbbell molecules in 2D   393

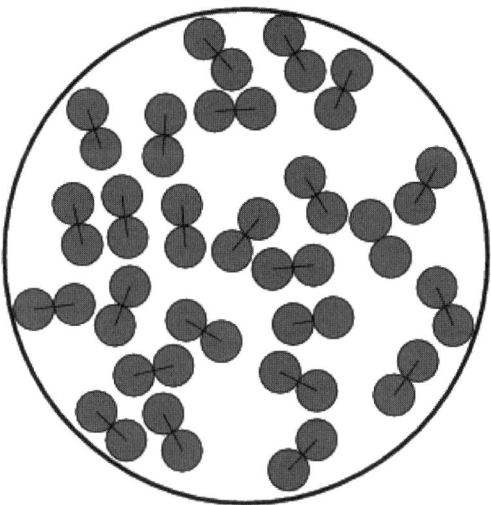

Figure 6.8. MC simulation snapshot of 25 dumbbell molecules in 2D, confined within a circle with $\rho = 0.5$ and $R = 11.3$.

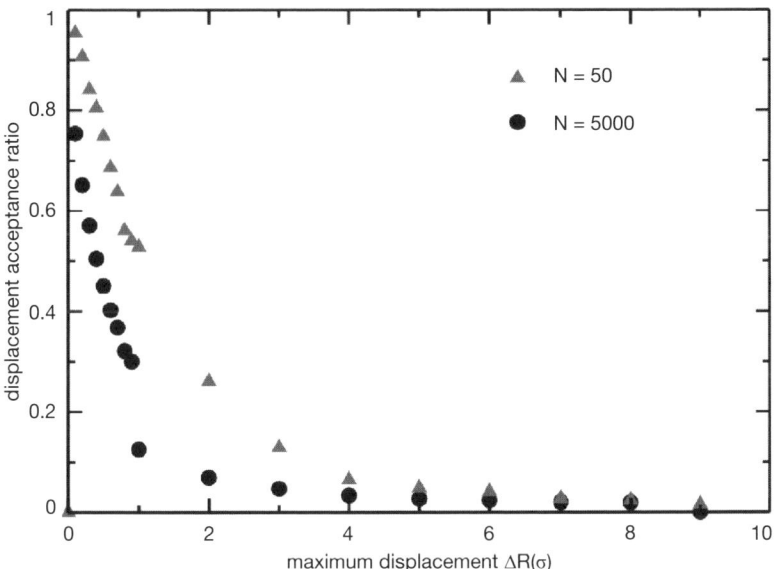

Figure 6.9. Displacement acceptance ratio vs. maximum dumbbell molecule displacement for various maximum displacements between 0.1 and 9. $N = 50, 5\,000$ (25 and 2 500 dumbbell molecules), $\rho = 0.5$, $R = 11.3$.

3. Assess the effect of the confining circle on the spatially resolved orientation field, $\langle\theta\rangle(x, y)$. You could construct a 2D histogram of the orientations and perform MC to populate it. Because the system is nominally *cylindrically symmetric*, it in principle suffices to consider a *one-dimensional* field, $\langle\theta\rangle(r)$. Plot $\langle\theta\rangle(r)$ vs. r for various densities and temperatures.

4. Measure an orientational correlation function. Here, we let the quantity θ_{ij} be the relative angle between the segments of dumbbell molecules i and j. The code *DumbellMolecules.c* can be used to accumulate statistics on θ_{ij} as a function of r_{ij}, where r_{ij} is the center-of-mass to center-of-mass distance of the two dumbbell molecules. It is actually much better to accumulate statistics of the *second Legendre polynomial* in the cosine of θ_{ij}:

$$P_2\left(\cos\theta_{ij}\right) = \frac{1}{2}\left(3\cos^2\theta_{ij} - 1\right). \tag{6.35}$$

$\langle P_2\rangle$ as a function of r reveals the character of the liquid-crystalline-like ordering of the dumbbell molecules. This is the *orientational correlation function*. When two dumbbell molecules are aligned, $\cos\theta_{ij}$ is unity, and therefore P_2 approaches unity. When two dumbbells are perpendicular, P_2 is -0.5. The average, $\langle P_2\rangle(r)$ at a particular distance r ranges between -0.5 and 1.0 for perfectly perpendicular to perfectly aligned, and 0 implies no preferred orientation. Plot $\langle P_2\rangle(r)$ for various densities and temperatures.

6.7 Exercises

6.7.1 The GSL library

The GNU Scientific Library (GSL) (see http://www.gnu.org/software/gsl) and Appendix I is a collection of routines for numerical computing, most notably random number generation. The routines have been written from scratch in C, and present a modern Applications Programming Interface (API) for C programmers, allowing wrappers to be written for many high level languages.

The GSL is open source, i.e. the source code is distributed under the GNU public license. As such, use of this library is free of charge. The library covers a wide range of topics in numerical computing. Routines are available for numerous areas of numerical methods, such as complex numbers, sorting, random numbers, differential equations, minimization, and more. The use of these routines is described in the manual of the GSL. Each chapter provides detailed definitions of the functions, followed by example programs and references to the articles on which the algorithms are based. In addition to the broad range of functions, the most important advantage of the using the GSL is that the functions are already implemented (which saves you work) and checked by many people (you can rely on the correctness of the implementation).

Section 6.7 Exercises

Suggested Exercises:
1. If not yet done, download and install the latest version of the GSL[5].

2. Write a makefile for compiling the program gslDemo.c in Listing 67. Then, modify the makefile so that the compiler can find the location of the GSL libraries (-L switch) and headers (-I switch), see chapter 2.1, depending on where you installed them on your system.

3. Look at the simple program gslDemo.c and try to understand how to generate uniformly distributed random numbers with the aid of the GSL.

4. A nice feature of the GSL random number generator implementation is, that the method to generate random numbers can be chosen at runtime. See the documentation for details.

6.7.2 Calculating π

Consider a circle with diameter d, surrounded by a square with length l ($l > d$), see Figure 6.10. Random coordinates are generated within the square. The value of π can be calculated from the fraction of points that fall within the circle. Estimating this fraction is quite easy: we just look at a set of points in the square and determine what fraction of these lies within the circle. However, this method only works if the trial points are more or less uniformly distributed over the square. Uniformly distributed simply means that the probability for a point to be picked should be the same for all points of the square. Conveniently, typical random number generators, such as those included in the GSL, generate uniformly distributed random numbers.

Suggested Exercises:
1. Complete the small MC program *MCpi.c* provided in Listing 68 to calculate π using the method outlined above.

2. How does the accuracy of the result depend on the ratio l/d and the number of generated trial points? Use the variance of the values of π obtained in the different cycles to estimate the accuracy.

3. Is it a good idea to calculate many decimals of π using this method? Why?

4. The advantage of MC methods over regular numerical integration schemes is that the error in the estimate scales like $n^{-1/2}$, independently of the dimension of the problem. Convince yourself of this by modifying the *MCpi.c* code to compute the volume of the 6-dimensional sphere.

---

[5] Hint: after unpacking the source code archive, run the ./configure command (check the -prefix option!) and then run make, make check, and make install.

5. An improved version of the simple sampling MC is the importance sampling method which we discussed and applied above. Think about how such a method could be applied to the calculation of the integral of a function such as the normal distribution, integrated over the interval $[a, b]$? Why do we expect importance sampling to be better than simple sampling?

Listing 67. File *gslDemo.c* which uses the GSL library. (*)

```c
#include <stdio.h>
#include <gsl/gsl_rng.h>

int main (void)
{
  /* variable for the random number generator */
  const gsl_rng_type * T;
  gsl_rng * r;

  int i, n = 10;
  double u;

  /* initialize random number generator */
  /* If the environment variables GSL_RNG_TYPE
     and GSL_RNG_SEED are set, they will be used
     to initialize the random number generator.
     This happens at RUN TIME. */
  gsl_rng_env_setup ();
  T = gsl_rng_default;
  r = gsl_rng_alloc (T);

  /* output information about the random number generator */
  printf ("generator type: %s\n", gsl_rng_name (r));
  printf ("seed = %lu\n", gsl_rng_default_seed);
  printf ("first value = %lu\n", gsl_rng_get (r));

  /* draw some random numbers */
  for (i = 0; i < n; i++)
  {
    u = gsl_rng_uniform (r);
    printf ("%.5f\n", u);
  }

  /* clean up */
  gsl_rng_free (r);
  return 0;
}
```

Listing 68. Part 1 – Calculating π with the circle method. (*)

```
#include <stdio.h>
#include <gsl/gsl_rng.h>
#include <math.h>

int main (void)
{
  const gsl_rng_type * T;
  gsl_rng * r;

  int i,j,k,NumberOfCycles,NumberOfShots;
  double l,x,y,pi;
  double NumberOfHits,NumberOfTrials,error;

  gsl_rng_env_setup();
  T = gsl_rng_default;
  r = gsl_rng_alloc (T);

  printf("Number of cycles ? (Example: 100) ");
  fscanf(stdin,"%d",&NumberOfCycles);

  printf("Number of random numbers drawn per cycle (in
      thousands) ? (Example: 1000) ");
  fscanf(stdin,"%d",&NumberOfShots);

  printf("Ratio 1/d       ? (always >= 1 !) ");
  fscanf(stdin,"%lf",&l);

  if(l<1.0)
  {
    printf("Ratio must be At least 1 !!!");
    exit(0);
  }

  error=0.0;
  for(k=0;k<NumberOfCycles;k++)
  {
  pi=0.0;
  NumberOfHits=0.0;
  NumberOfTrials=0.0;
```

Listing 68. Part 2 – Calculating π with the circle method. (*)

```
39    // draw random numbers...
40    for(i=0;i<NumberOfShots;i++)
41    {
42      for(j=0;j<1000;j++)// ...in units of thousands.
43      {
44        /* Modify the code here... */
45      }
46    }
47    // calculate PI
48    pi=4*l*l*NumberOfHits/NumberOfTrials;
49
50    printf("- Cycle %i -\n",k);
51    printf("Estimate of Pi : %f\n",pi);
52    printf("Real Pi        : %f\n",M_PI);
53    printf("Relative error : %g\n",fabs(pi-M_PI)/M_PI);
54    }
55
56    gsl_rng_free (r);
57    return 0;
58  }
```

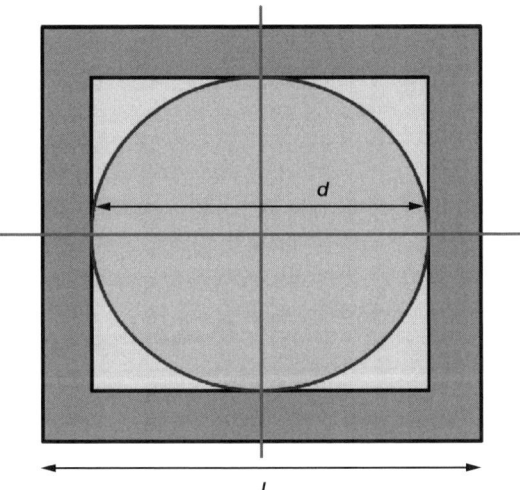

Figure 6.10. Calculating π using a circle of diameter d, surrounded by a square of length l ($l > d$).

6.7.3 Simple and importance sampling with random walks in 1D

The random walk problem is also known as the "drunk man's problem", because a random walk consists of a series of random steps, like the path of a completely drunk man (compare our discussion in Chapter 3). In theoretical physics, random walks are often employed to model systems with no memory. A good example is the classical Brownian motion. Here, we start with a random walk in one dimension. The random walk starts at the origin and can make a fixed number of steps N. Each step is randomly chosen, either to the left or to the right, with equal probability ($p = 0.5$).

Suggested Exercises:
1. Read the code *Randomwalk1d.c* in Listing 69 and understand how it works.

2. Compile and run the sample program for different lengths of the random walk. Confirm that, on average, the final position is close to the origin, i.e. the random walk is not biased.

3. Complete a table with values for the averaged squared end-to-end distance $\langle R_e^2 \rangle$, plotted vs. varying lengths $N = 10, 20, 50, 100, 200, 500, 1\,000, 10\,000$ The end-to-end distance $\langle R_e^2 \rangle$ is the distance between the starting point and the final point, where the random walk is found after N steps. If the starting point is at the origin, then $\langle R_e^2 \rangle = [\text{pos}(N)]^2$, where pos($N$) refers to the position of the random walk in the Nth step. *Hint:* the number of samples required for good statistics increases with N. Try to estimate how many samples you need as a function of N. Plot the data of your table (in log-log scale) and compare them to the theoretical prediction $\langle R_e^2 \rangle \propto N$. Your results should look similar to Figure 6.11.

4. Increase the probability for the random walk to go right to 0.8 (consequently, the probability to go left now is only 0.2). What happens now?

6.7.4 Simple sampling and importance sampling with random walks in 1D

In this section, we study the two dimensional random walk. In two (or more dimensions), it is possible to distinguish two different types of random walks, normal random walks (RW) and self-avoiding random walks (SAW). SAWs have the property that they never return to a place they have been to earlier, i.e. a SAW does not touch or cross itself. This restriction is not imposed on general random walks and leads to significantly different properties. Just as general random walks, which represent for example Brownian motion, SAWs are used in important physical models, most notably to model flexible polymers. Such a polymer can be viewed as a chain of monomers connected by (flexible) bonds. The relative positions of two adjacent beads are essentially uncorrelated, with the exception that monomers of course cannot overlap. This leads naturally to an SAW. We generate two-dimensional random walks by

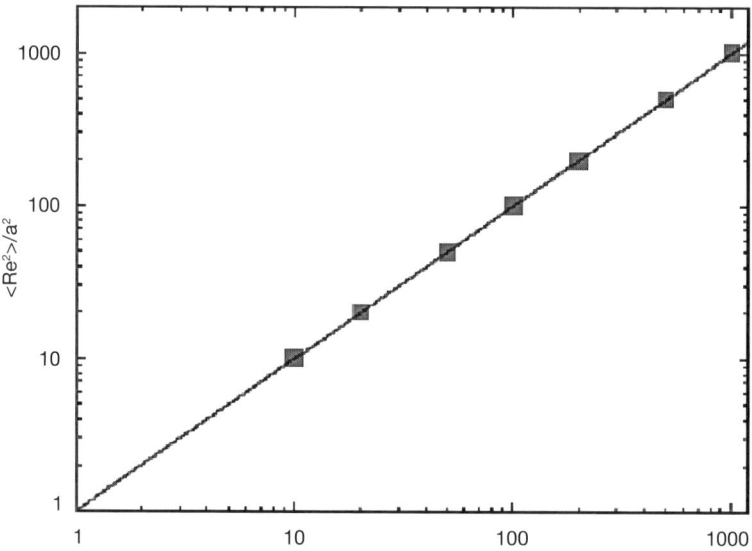

Figure 6.11. Averaged squared end-to-end distance of a one-dimensional random walk.

randomly choosing one of the four directions (up, down, left, right) at each step. The random walk is stored as a sequence of numbers from 0 to 3, indicating the chosen directions. To analyze the properties, the sequence of directions has to be converted to a sequence of coordinates.

Suggested Exercises:

1. Read the code *Randomwalk2d.c* in Listing 70 and understand how it works.

2. Make sure that the constant SAW is set to 1. Compile and run it for different parameters, i.e. length of the walk and number of samples.

3. Look at some of the generated random walks by plotting the file rw.dat which contains the last random walk generated.

4. How does the mean number of self-contacts per random walk and the fraction of self-avoiding walks depend on the length N of the random walk?

5. Plot in log-log scale the mean squared end-to-end distance versus the length N and compare the result to the one-dimensional random walk. What do you observe?

6. Set SAW to 1 and recompile the code *Randomwalk2d.c* in Listing 70. Now, only self-avoiding walks are accepted, whilst non-self-avoiding walks are discarded. How does the average end-to-end distance change with N now? (Only use

Section 6.7 Exercises   401

$N < 30$) Plot your results in log-log scale and be sure to include at least 6–7 points.

7. Can you make this code a bit more efficient by changing the way of storing the RW? How and why?

Listing 69. Part 1 – Random walk in one dimension. (*)

```
1  int main (void)
2  {
3    // variables for the random number generator
4    const gsl_rng_type * T;
5    gsl_rng * r;
6
7    // other variables
8    int N,k;
9    int i, NumberOfSamples;
10   double pos, mean, sd_mean, re2;
11   double* end;
12   double* end2;
13   double p;
14
15   // initialize the random number generator
16   gsl_rng_env_setup();
17   T = gsl_rng_default;
18   r = gsl_rng_alloc(T);
19
20   // get input parameter
21   printf("Number of samples to average? (Example: 1000) ");
22   fscanf(stdin,"%d",&NumberOfSamples);
23
24   printf("Length of random walk? (Example: 50) ");
25   fscanf(stdin,"%d",&N);
26
27   // allocate memory for statistics data (needs to be cleaned up at the end)
28   end = malloc(sizeof(double)*NumberOfSamples);
29   end2 = malloc(sizeof(double)*NumberOfSamples);
30
31   // probability to go right
32   p = 0.5;
33
34   // output
35   printf("Averaging %d one dimensional random walks of length %d.(p = %f)\n", NumberOfSamples, N,p);
```

Listing 69. Part 2 – Random walk in one dimension. (*)

```
37    // loop over all samples
38    for(k=0;k<NumberOfSamples;k++)
39    {
40      // start random walk at origin
41      pos = 0.0;
42        // draw random numbers
43        for(i=0;i<N;i++)
44        {
45
46          // one-dimensional random walk
47          // randomly choose direction
48          if (gsl_rng_uniform(r) < p)
49    pos = pos + 1;
50          else
51    pos = pos - 1;
52        }
53      // save end position
54      end[k] = pos;
55      end2[k] = pos*pos;
56
57    }
58    // use gsl functions to analyze statistics of random walk
59    // check gsl documentation for further information
60    mean = gsl_stats_mean(end,1,NumberOfSamples);
61    sd_mean = gsl_stats_sd(end,1,NumberOfSamples);
62    re2 = gsl_stats_mean(end2,1,NumberOfSamples);
63
64    // output
65    printf("Final position of random walker: %f +- %f\n",mean,
          sd_mean);
66    printf("Average square end-to-end distance: %f\n",re2);
67
68    // clean up
69    free(end);
70    free(end2);
71    gsl_rng_free (r);
72
73    return 0;
74  }
```

Listing 70. Part 1 – Random walk in two dimensions. (*)

```c
#include <stdio.h>
#include <gsl/gsl_rng.h>
#include <gsl/gsl_statistics_double.h>

// two-dimensional random walk
// set to 1 to enforce only self-avoiding random walk to
be accepted
const int SAW = 0;

// saves the current random walk to a file
void save_walk(int* thiswalk, int N) {
  int i;
  int X,Y;
  // open output file
  FILE *fp;
  fp = fopen( "rw.dat", "w");

  X = Y = 0;

  // print start point
  fprintf(fp, "0 0\n");
  // print rest of random walk
  for (i=0;i<N;i++) {
    switch (thiswalk[i]) {
    case 0:
      X++;break;
    case 1:
      X--;break;
    case 2:
      Y++;break;
    case 3:
      Y--;break;
    }
    fprintf(fp, "%d %d\n", X, Y);
  }

  // close output file
  fclose(fp);
}

// calculates the end-to-end distance of the current random
    walk
double end2end_walk(int* thiswalk, int N) {
  int i, X, Y;

  X = Y = 0;
```

Listing 70. Part 2 – Random walk in two dimensions. (*)

```
47  for (i=0;i<N;i++) {
48    switch (thiswalk[i]) {
49    case 0:
50      X++;break;
51    case 1:
52      X--;break;
53    case 2:
54      Y++;break;
55    case 3:
56      Y--;break;
57    }
58  }
59  return X*X+Y*Y;
60 }
61 // counts the number of self-contacts of this random walk
62 int selfcontacts_walk(int* thiswalk, int N) {
63   int i,j;
64   int X, Y;
65   int *XL, *YL;
66   int sc;
67   XL = malloc(sizeof(int)*(N+1));
68   YL = malloc(sizeof(int)*(N+1));
69   X = Y = 0;
70   // generate positions for complete walk
71   for (i=0;i<N;i++) {
72     switch (thiswalk[i]) {
73     case 0:
74       XL[i]=X; YL[i]=Y; X++;break;
75     case 1:
76       XL[i]=X; YL[i]=Y; X--;break;
77     case 2:
78       XL[i]=X; YL[i]=Y; Y++;break;
79     case 3:
80       XL[i]=X; YL[i]=Y; Y--;break;
81     }
82   }
83   // end position of RW
84   XL[i]=X; YL[i]=Y;
85   // count number of self-contacts
86   sc = 0;
87   for (i=0; i<N-1; i++)
88     for (j=i+2;j<N+1; j++)
89       if (XL[i]==XL[j] && YL[i]==YL[j]) sc++;
90   free(XL); free(YL);
91   return sc;
92 }
```

Section 6.7 Exercises   405

Listing 70. Part 3 – Random walk in two dimensions. (*)

```
93  // main program
94  int main (void)
95  {
96    // variables for the random number generator
97    const gsl_rng_type * T;
98    gsl_rng * r;
99
100   // other variables
101   int NumberOfSamples, N;
102   int k,i;
103   int *thiswalk;
104   double *sc;
105   int scfree=0;
106   double mean_sc;
107
108   double *end2;
109   double re2, sc_fraction;
110
111   // initialize the random number generator
112   gsl_rng_env_setup();
113   T = gsl_rng_default;
114   r = gsl_rng_alloc (T);
115
116   // get input parameter
117   printf("Number of samples to average? (Example: 1000) ");
118   fscanf(stdin,"%d",&NumberOfSamples);
119
120   printf("Length of random walk? (Example: 50) ");
121   fscanf(stdin,"%d",&N);
122
123   // allocate memory for statistics data (needs to be cleaned
          up at the end)
124   end2 = malloc(sizeof(double)*NumberOfSamples);
125   sc = malloc(sizeof(double)*NumberOfSamples);
126
127   // allocate memory this random walk
128   thiswalk = malloc(sizeof(int)*N);
129
130   // output
131   printf("Averaging %d two dimensional random walks of length %
          d.(saw = %d)\n", NumberOfSamples, N, SAW);
132
133   // loop over all samples
134   for(k=0;k<NumberOfSamples;k++)
135   {
136
137     printf("%d\r", k);
```

Listing 70. Part 4 – Random walk in two dimensions. (*)

```
138      // generate random walk
139      for(i=0;i<N;i++)
140      {
141        // for each step choose one of the four random directions
142        thiswalk[i] = gsl_rng_uniform_int(r,4);
143      }
144      // check for self-avoiding walks
145      if (isSAW(thiswalk, N) ) scfree++;
146      else if (SAW) {
147        // discard this random walk and generate a new one
148        k--;
149        continue;
150      } else {
151        // count self-contacts for non avoiding random walks
152        sc[k] = selfcontacts_walk(thiswalk, N);
153      }
154
155      // save end position
156      end2[k] = end2end_walk(thiswalk, N);
157
158    }
159    // save last random walk to a file
160    save_walk(thiswalk, N);
161    // use gsl functions to analyze statistics of random walk
162    // check gsl documentation for further information
163    re2 = gsl_stats_mean(end2,1,NumberOfSamples);
164    if (!SAW) {
165      mean_sc = gsl_stats_mean(sc,1,NumberOfSamples);
166      // fraction of self-avoiding random walks
167      sc_fraction = (double)scfree/NumberOfSamples;
168    }
169    // output
170    printf("Average square end-to-end distance: %f\n",re2);
171    if (!SAW) {
172      printf("Mean number of self-contacts per random walk: %f\n"
               ,mean_sc);
173      printf("Self-avoiding walks: %d of %d\n",scfree,
               NumberOfSamples);
174      printf("Fraction of self-avoiding walks: %f\n", sc_fraction
               );
175    }
176    // clean up
177    free(thiswalk);
178    free(sc);
179    free(end2);
180    gsl_rng_free (r);
181
182    return 0;
183  }
```

6.8 Chapter literature

Many good Random Number Generator (RNG) implementations have been proposed in the literature [325, 77, 341, 279]. Schrage contributed an innovation in the method of evaluating RNGs, which guarantees that the intermediate values of the generator remain in range [374]. Wichmann and Hill [438] proposed an RNG with a cycle length exceeding 6.95×10^{12}. An RNG with a period of 2.30584×10^{118} was proposed by L'Ecuyer [266] and Marsaglia and Zaman [284] proposed RNGs based on "run2" of *Numerical Recipes* [342] with periods between 10^{27} and 10^{101}.

Good texts on MC methods are the monographs of Robert and Casella [356], Johnson [226] and Fishman [162]. A recent book on MC is by Binder and Heermann [63], which is yet another monograph that covers some of the basics of MC for applications in many-particle solid state physics.

Chapter 7

Advanced topics, and applications in soft matter

Summary

This chapter briefly discusses the differences between the three major classes of partial differential equations and then touches on a few advanced topics for which corresponding programs are provided online. After a short discussion of MD thermostats, we finally present a discussion of applications of computer simulations in soft matter systems and show some prototypical results.

Learning targets

✓ Learning about different partial differential equations.

✓ Performing several case studies on advanced topics.

✓ Learning about computer simulation of biological macromolecules.

7.1 Partial differential equations

The basic idea in any numerical method for solving a differential equation is to *discretize* the given continuous problem with infinitely many degrees of freedom, in order to obtain a *discrete problem* or system of equations with only finitely many unknowns, that may be solved using a computer. The classical numerical method for partial differential equations (PDEs) is the *difference method* where the discrete problem is obtained by simply replacing the derivatives with difference quotients involving the values of the unknown at certain (finitely many) points. PDEs can generally be solved using finite difference schemes of the type introduced in Chapter 5.

The discretization process using a finite element method (FEM) is different. In this case, one starts from a reformulation of the given differential equation as an equivalent *variational* problem. In the case of elliptic equations, in basic cases, this variational problem is a minimization problem of the form: "Find a function $u \in S$, such that the function $F(u) \leq F(v)$ for all $v \in S$", where S is a given set of admissible functions and $F : V \to \mathbb{R}$ is a *functional*. The functions v in S often represent a continuously varying quantity, such as displacement in an elastic body, temperature, or density. In general, the dimension of S is infinite which means that the functions in S cannot be

Section 7.1 Partial differential equations

described by a finite number of parameters and thus the above minimization problem cannot be solved exactly. To obtain a solvable problem formulation suitable for coding in a computer program, the idea of the FEM is to replace S with a set S_h consisting of simple functions that only depend on finitely many parameters. This idea leads to a finite-dimensional minimization problem of the form: "Find a function $u_h \in S_h$ such that the function $F(u_h) \leq F(v)$ for all $v \in S_h$".

This problem is equivalent to a linear or nonlinear system of equations. With the FEM, we hope that the solution u_h of this discretized problem is a sufficiently good approximation of the solution u of the original minimization problem stated above. Usually, one chooses $V_h \subseteq V$ and in this case the discretized minimization problem corresponds to the so called *Ritz-Galerkin* method, introduced at the beginning 20th century. The special feature of an FEM, as a particular Ritz-Galerkin method, is the fact that the functions in V_h are chosen to be *piecewise polynomial*. PDEs can generally be solved using finite difference schemes of the type introduced in Chapter 5.

In order to approximately solve a given PDE, using the *finite element method* (FEM), one basically has to go through the following four steps:

1. reformulate the problem as a variational problem

2. discretize the problem domain using finite elements, i.e. construct the finite-dimensional space V_h

3. solve the discrete problem

4. implement the method in a high-level language

In Section 1.7.4 on Page 74 we have seen an example of the limiting process, which transforms a theory based on particles to a theory based on continuous fields. As a result of this limiting process we ended up with only *one*, albeit partial, equation instead of $3N$ coupled ODEs. Partial differential equations (PDEs) are differential equations that contain derivatives of several (and not of only one) variables. Many fundamental physical theories lead naturally to the formulation of PDEs, for example, Maxwell's equations of electrodynamics, the Schrödinger equation of elementary quantum mechanics, or the Dirac equation of relativistic quantum mechanics. The most common PDEs in physics contain derivatives up to 2nd order and can be assigned to one of three categories:

- elliptic

- parabolic

- hyperbolic

These names arise due to the algebraic sign of the derivatives, which are evocative of the implicit equations of conic section curves [111]. While there are standard solution procedures for PDEs of first order [153, 154], PDEs of higher order have to be treated individually [305]. In fact, the solution of PDEs by use of analytical methods is only possible in a limited number of cases. Thus, one usually has to resort to numerical methods [103, 14, 72, 188, 342].

7.1.1 Elliptic PDEs

The prototype of an elliptic PDE is the *Laplace equation*

$$\Delta u = 0. \tag{7.1}$$

In three dimensions, and in Cartesian coordinates $\vec{r} = (x, y, z)$, this equation becomes

$$\left(\frac{\partial^2}{\partial x^2} + \frac{\partial^2}{\partial y^2} + \frac{\partial^2}{\partial z^2} \right) u(x, y, z) = 0. \tag{7.2}$$

In electrodynamics, equation (7.1) describes the electric potential $u = \Phi(x, y, z)$ in a region without charges[1]. The charge distribution in the region of interest is determined by the electric potential at the boundaries of the region. In the theory of heat the solution of equation (7.1) provides the temperature distribution $u = T(x, y, z)$ at equilibrium within a region without heat sinks or heat sources, i.e. the static solution of the diffusion equation. Point-like charges $\rho(\vec{x} = \delta(\vec{x}))$ are represented using the delta function.

> **Elliptic PDEs** describe *boundary value problems*, i.e. the solution function $u(t)$ has to fulfill certain conditions at the boundary of its region of definition A. One distinguishes two major forms of boundary conditions:
>
> - **Dirichlet boundary conditions:** The values of the function u are provided at the boundary of A: $u(\partial A) = \ldots$.
>
> - **von Neumann boundary conditions:** The values of the derivative of u in the direction normal to the boundary area or curve are provided: $\nabla u \vec{n}_{|\partial A} = \ldots$.
>
> Mixed forms of boundary conditions are also possible.

---

[1] When you want to introduce charges inside of the region of interest, you obtain the Poisson equation $\Delta u(\vec{r}) = \rho(\vec{R})$

7.1.2 Parabolic PDEs

With this type of PDEs, besides the Laplace operator, there is an additional derivation of first order, in most applications a derivative with respect to time, i.e.,

$$\Delta u(\vec{r},t) = \frac{1}{\kappa} \frac{\partial u(\vec{r},t)}{\partial t}. \tag{7.3}$$

Equation (7.3) has the form of a diffusion or heat conduction equation. In the limit $t \to \infty$, a stationary state is reached. Thus, we have $\lim_{t \to \infty} \partial u/\partial t = 0$, i.e., the Laplace equation. In most physical situations, one has to provide an initial value of the function u at an initial time t_0, i.e.

$$u(x, y, \ldots, t_0) = \ldots. \tag{7.4}$$

Numerical methods to solve differential equations which are essentially defined through *initial* rather than *boundary* values, i.e. which are concerned with time derivatives, are often referred to as *finite difference methods*. Finite difference methods approximate the derivatives that appear in differential equations by a transition to their discrete, finite difference counterparts. Finite difference methods do not use polynomial expressions to approximate functions like it is done with finite element methods (FEM). These methods are designed to numerically solve both complex boundary value and initial-value problems. They have in common the spatial discretization of the region under consideration into a number of finite elements, the time discretization and the approximation of the true spatial solutions in the elements by polynomial trial functions. These features explain why they are referred to as FEM.

Although both the finite difference and the finite element techniques can handle space and time derivatives, the latter approach is more sophisticated in that it uses trial functions and a minimization routine. Thus, finite difference methods can be regarded as a subset of the various more general finite element approximations [453].

Many finite difference methods and particularly most finite element methods are sometimes associated with the solution of macroscopic large-scale problems. Although this association is often true for finite element methods, which prevail when solving meso and macroscale boundary value problems in computational materials science, we emphasize here that these general associations are inadequate. Finite difference and finite element methods represent *mathematical* approximation techniques. They are generally *not* intrinsically restricted to any physical length or time scale. Scaling parameters are introduced by the particular physics of the considered problem, but not by the numerical scheme employed to solve a differential equation.

7.1.3 Hyperbolic PDEs

Hyperbolic PDEs have – besides the Laplace operator – a second derivative with respect to time with an opposite sign compared to the spatial derivatives, i.e. in one dimension x we have

$$\Delta u(x,t) = \frac{1}{v^2}\frac{\partial^2 u(x,t)}{\partial t^2}. \tag{7.5}$$

Equation (7.5) is a wave equation, which describes the propagation of waves through space and time. The parameter v is the propagation velocity. The function $u(x,t)$ could be the elongation of a point of a vibrating thread, which we discussed in the context of field theories in Figure 1.27 on Page 73. In a model of a vibrating thread, all different points are connected with elastic springs – the elongation $u(x,t)$ of the point at position x on the thread is proportional to the force driving the mass point back to its equilibrium position. Thus, the force for three infinitesimal neighboring mass points on an elongated thread is given by.

$$F \propto -[(u(x,t) - u(x-\epsilon,t)) + (u(x,t) - u(x+\epsilon,t))] = \epsilon^2 \frac{\partial^2 u(x,t)}{\partial x^2} + O(\epsilon^3). \tag{7.6}$$

Note the big-O notation and our use of the Taylor expansion of $u(x\pm\epsilon,t)$. According to Newton's second law $F \propto \frac{\partial^2 u}{\partial x^2}$ with the proportionality constant depending on the spring constant, denoting the propagation velocity (e.g. the velocity of sound).

7.2 The finite element method (FEM)

The finite element method is a general technique for the numerical solution of differential and integral equations in science and engineering. The method was introduced by engineers in the late 1950s and early 1960s for the numerical solution of PDEs in structural engineering (e.g. elasticity or plate equations, etc.). At this point the method was thought of as a generalization of earlier methods in structural engineering for beams, frames, and plates, where the structure was subdivided into smaller parts, so called finite elements, with known simple behavior. When the mathematical study of the FEM started in mid 1960s it soon became clear that, in fact, the method is a general technique for the numerical solution of PDEs, with roots in variational calculus, introduced at the beginning of the 20th century. During the 1960s and 1970s the method was further developed by engineers and mathematicians and turned into a general method for numerical solution of PDEs with applications in many areas of science and engineering. Today, the FEM is used extensively and often integrated into CAD systems for problems in structural engineering, materials strength, fluid mechanics, nuclear engineering, wave propagation, scattering, heat conduction and many other areas.

Whilst the finite difference method is simple and straightforward for regular mesh discretizations, it becomes very hard to apply to more complex problems such as

Section 7.2 The finite element method (FEM)

- spatially varying constants, e.g., spatially varying dielectric constants in the Poisson equation.
- irregular geometries, e.g., airplanes or turbines.
- dynamically adapting geometries, e.g., moving pistons.

In such cases the finite element method has big advantages over finite differences, since it does not rely on a regular mesh discretization. We will discuss the finite element method using the one-dimensional Poisson equation

$$\phi''(x) = -4\pi\rho(x) \tag{7.7}$$

with boundary conditions

$$\phi(0) = \phi(1) = 0. \tag{7.8}$$

The first step is to expand the solution $\phi(x)$ in terms of basis functions $\{v_i\}$, $i = 1, \ldots, \infty$ of the function space:

$$\phi(x) = \sum_{i=1}^{\infty} a_i v_i(x). \tag{7.9}$$

For our numerical calculation, the infinite basis set needs to be truncated, choosing a finite subset $\{u_i\}$, $i = 1, \ldots, N$ of N linearly independent, but not necessarily orthogonal, functions

$$\phi_N(x) = \sum_{i=1}^{\infty} a_i u_i(x). \tag{7.10}$$

The usual choice are functions localized around some mesh points x_i which, in contrast to the finite difference method, do not need to form a regular mesh. The coefficients $\vec{a} = (a_1, \ldots, a_N)$ are chosen so as to minimize the residual

$$\phi_N''(x) + 4\pi\rho(x) \tag{7.11}$$

over the whole interval. Since we can choose N coefficients, we can impose N conditions

$$0 = g_i \int_0^1 \left[\phi_N''(x) + 4\pi\rho(x)\right] w_i(x) \, dx, \tag{7.12}$$

where the weight functions $w_i(x)$ are often chosen the same as the basis functions $w_i(x) = u_i(x)$. This is called the *Galerkin method*. In the current case of a linear PDE, this results in a linear system of equations

$$A\vec{a} = \vec{b} \tag{7.13}$$

that can be solved with standard methods.

7.3 Coarse-grained MD for mesoscopic polymer and biomolecular simulations

What is called "soft matter" covers a large variety of different systems, from colloids to polymers, from surfactants to liquid crystals, and from soap bubbles to solutions of macromolecules. Most of these materials are of industrial importance. Generally speaking, all systems that fall under the category of soft matter are, with very few exceptions, made of organic molecules with often complicated (complex) architectures and flexible forms, bound by *weak* intermolecular forces. The range of stability of such structures is, therefore, close to room temperature, and small changes in temperature are often enough to induce changes in morphology or phase transitions, usually accompanied by a small latent heat. Because of the often very flexible forms of the involved macromolecules, entropy rather than enthalpy plays a leading role in the description of soft matter.

To study phase changes, one first has to define an order parameter, and its characteristic phase and amplitude, that varies smoothly in time and space. This approach, which was popularized by David Landau, effectively neglects local atomic or thermal inhomogeneities, but it provides a general framework, applicable to similar mesoscopic problems in magnetism, superfluidity or phase changes in strong solids. *Biological matter*, such as membranes, proteins, DNA, viruses, actin filaments, or microtubules enter the category of "soft materials" when studied by physicists. Soft matter physics is *condensed* matter physics, and so it goes hand in hand with solid state physics. Therefore, in the earlier chapters, a number of examples have been included that pertain to the atomic and molecular arrangement of solids. The material covered in those chapters and the lessons learned can also be applied to soft matter systems.

7.3.1 Why coarse-grained simulations?

With polymer systems, many properties can be successfully simulated by only taking into account the basic and essential features of the chains, thus neglecting the detailed chemical structure of the molecules. Such *coarse grained models*, cf. Figure 7.1, are used because atomistic models of long polymer chains are usually intractable for time scales beyond nanoseconds, although they are important for many physical phenomena and for the comparison with real experiments. Also, the fractal properties of polymers render this type of model very useful for the investigation of scaling properties. Fractality means that some property X of a system (with polymers, e.g. the radius of gyration R_g) obeys a relation $X \propto N^k$, where $N \in \mathbb{N}$ is the size of the system, and $k \in \mathbb{Q}$ is the fractal dimension, which is of the form $k = p/q$ with $p \neq q$,

$q \in \mathbb{N}$ and $p \in \mathbb{N}$. The basic properties sufficient to extract many structural static and dynamic features of polymers are:

- the connectivity of monomers in a chain
- the topological constraints, e.g. the impenetrability of chain bonds
- the Flexibility or stiffness of monomer segments

Using coarse-grained models in the context of lipids and proteins, where each amino acid of the protein is represented by two coarse-grained beads, it has become possible to simulate lipoprotein assemblies and protein-lipid complexes for several microseconds [380].

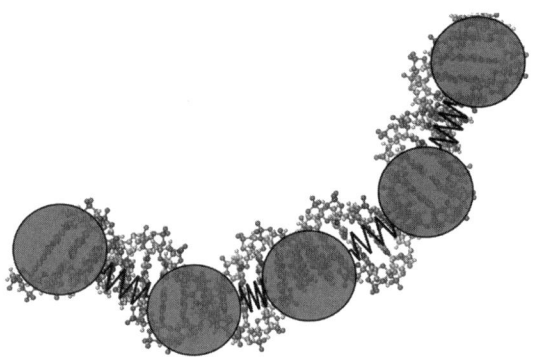

Figure 7.1. A coarse-grained model of a polymer chain where some groups of the detailed atomic structure are lumped into one coarse-grained particle. The individual particles are connected by springs (*bead-spring model*).

7.4 Scaling properties of polymers

Polymers are encountered in a large variety of chemical achievements, from biopolymers like Desoxyribonucleic Acid (DNA), xanthan, cellulose, proteins, and actin filaments, to synthesized polymers like Polyethylene (PE), Polybutadiene (PB), Polystyrene (PS), Polymethylmetacrylate (PMMA)[2], and so on, all of great industrial interest. They are employed in numerous forms, in solutions, as gels, rubbers, synthetic fabrics, molded pieces, and so on.

The first part of this chapter is inspired by the classic presentation of the subject elaborated by de Gennes [124]. The author of this book indeed believes that the simplest way to introduce the fundamentals of the behavior of a polymer in solution or a molten polymer, is through the presentation of the scaling properties, which are the

---

[2] PMMA is used in the production of compact discs.

main characteristics of chains with a large degree of polymerization N. Although the mathematics is easy, the physics is deep. It is not possible to get realistic scaling laws without a thorough analysis of the physical properties. One does not even have the usual help that one can get from the free energy density[3], often used as a black box from which the results pop up as if by magic. On the other hand, the nonlinear effects or the renormalization phenomena that are calculated at such painstaking efforts in the usual theories, appear there in their true light. This is not to underrate the usual presentations, which are absolutely necessary in order to take the material constants into account. In this first part, we limit our interest to the remarkable physical properties of *flexible polymers*, which are akin to critical phenomena arising at phase transitions. The physical properties are described in terms of the conformation of flexible chains as a function of the polymer molecular mass (often expressed as a number of monomers, or monomeric units N per chain), of the volume concentration $c = n/V$ of monomers (n monomers in a volume V), possibly of temperature T, and so on. In the phase transition picture, the analog of the temperature is $1/N$. We shall also find analogs of critical exponents, some of which can be interpreted as fractal dimensions.

Chains in melts are close to the ideal chains defined in Section 7.5, and their physics can be approached within the framework of *mean field theory*. We therefore present the classic Flory–Huggins picture of melts and blends and, in particular, the phenomena of blends phase separation, which are so typical of polymers. Finally, this chapter discusses, in the same vein, rigid and semiflexible polymers, which include all biopolymers, certain anisotropic viruses, and synthetic polymers used as textiles or building elements in high-strength materials used at room temperature, whose mechanical properties result from their trend to form liquid crystalline phases at high temperature, or to align under shear.

We refer the reader to the rich available literature for a detailed description of the *microscopic* dynamical properties (relaxation times, diffusivities, etc.) of polymers (e.g. M. Doi and S. F. Edwards [137]), crystallization (e.g. Strobl [137] and Keller [238]), or macroscopic rheological properties (e.g. Larson [264]). Recommended general textbooks on polymer physics are e.g. Rubinstein and Colby [363], des Cloizeaux and Janninck [129], Doi [136] and the classics by Flory [163], Grosberg and Khoklov [193, 194].

7.5 Ideal polymer chains

We limit ourselves to the description of *main chain flexible polymers*, whose prototype is polyethylene $[-CH_2-]n$. Other standard examples are provided in Figure 7.2.

Each of these polymers can be symbolized by a sequence of bonds linking successive elementary monomers. In the simple case of polyethylene (PE), to which we will

---

[3] Compare the discussion on the free energy function F in Chapter 3.

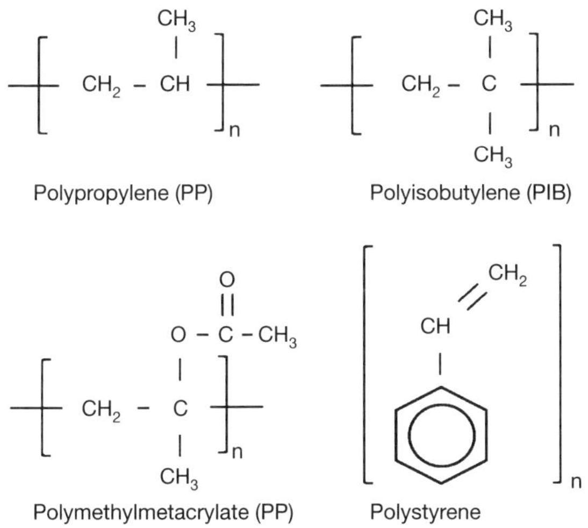

Section 7.5 Ideal polymer chains

Figure 7.2. Chemical structure of some common polymers.

refer as the model case, the angular variation between neighboring carbons is defined as $\Theta_{n-1} = 0$, $\Delta\varphi_n = [0, \pm 120°]$, see Figure 7.3.

$\Delta\varphi_n = 0$ is called a *trans* configuration, and $\Delta\varphi_n = \pm 120°$ is a *gauche* configuration. They do not have the same energy. Such a geometrical picture can be extended to any homopolymer.

Taking the successive values of $\Delta\varphi_n$ at random, the chain appears as a statistical coil, when observed at some scale much larger than the typical average bond length a of the monomers. In that case, the *persistence length* L_p is a measure of the degree of stiffness in such a polymer coil. L_p is the length on which the chain has the appearance of a rigid segment, because the segments are colinear for a succession of trans configurations.

Let $\Delta\epsilon = \epsilon_g - \epsilon_t > 0$ be the difference in energy between the gauche configuration and the trans configuration. Then, the persistence length is given by

$$L_p = a \exp\left(\frac{\Delta\epsilon}{k_B T}\right). \tag{7.14}$$

The concept of L_p is based on the idea that the correlation between the orientations of the local tangent $t(s)$ decays with the distance along the filament contour, according to $\langle t(s) t(s') \rangle = \exp\left(-|s - s'|/L_p\right)$. The *total chemical length* of the chain is given by $L = Na$, where N is the *degree of polymerization*, i.e. the number of repeat units (or monomers) in the polymer. L has to be large compared to L_p for the chain to be considered a statistical coil, i.e. $L_p/L \ll 1$.

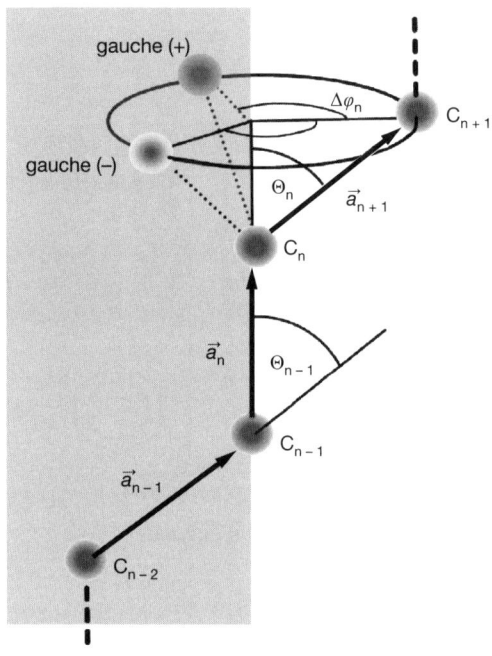

Figure 7.3. Configurational model parameters of an ideal polyethylene chain.

> The degree of polymerization N, i.e. the number of repeating monomer units in a polymer is sometimes also called the *number of segments*, which is not to be confused with the bond vectors $\vec{a}_i$ with the convention that index i refers to the position vector that connects segment number $i-1$ to segment number i. Thus, the counting of bond vectors starts with $i=1$ and ends with $N-1$. In a polymer model, one often just refers to the number N as the *number of monomers*, or – in computer simulation – the number of particles.

An example of polymers with vastly different persistence lengths is displayed in Figure 7.4, which shows a classification of polymer stiffness of the three major polymer components of the cytoskeleton of eukariotic cells, namely *microtubules* (stiff rods), *actin filaments* (semiflexible polymers) and intermediate filaments (flexible polymers).

Phenomenologically, the persistence length can also be thought of as a characteristic length of the thermal fluctuations carried by an elastic rigid rod. Let us assume, for the sake of simplicity that the cross-section of the rod is isotropic. Then, the free

Section 7.5 Ideal polymer chains 419

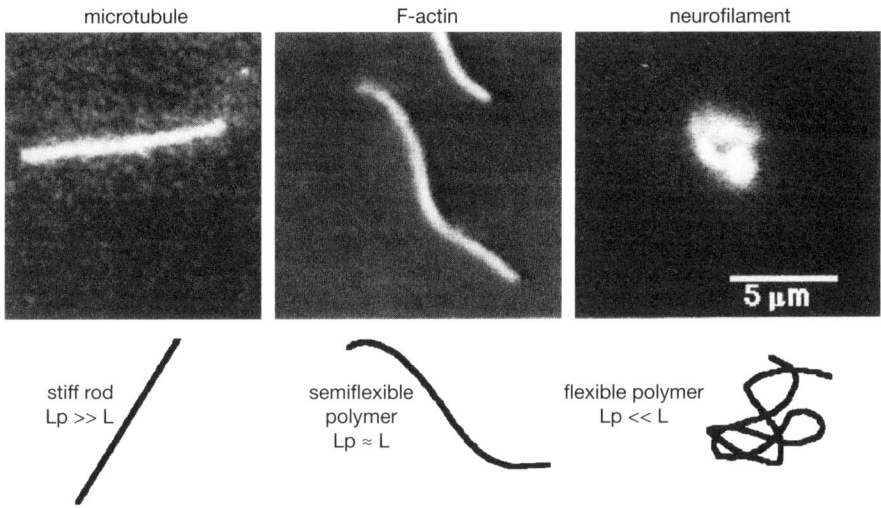

Figure 7.4. Classification of cytoskeleton components by the persistence length L_p. The rhodamine-labeled microtubule, which has a persistence length L_p exceeding the filament length L, is a typical example of a stiff rod. The semiflexible character of the rhodamine-phalloidin-labeled actin filament can be expressed by the fact that L_p and L are comparable. In the case of flexible polymers, like the rhodamine-labeled neurophilament, a random coiled shape, dominated by entropy, can be observed. The high flexibility implies that the persistence length L_p is smaller than the filament length L. Reprinted from Biophysical Journal (1996) [231] with permission from Elsevier.

energy per unit length of the rod reads as [263].

$$F = \frac{1}{2}\kappa r_c^{-2}, \qquad (7.15)$$

where r_c is the radius of curvature. We have $r_c^{-1} = \pm |\partial^2 \vec{u}/\partial s^2| = \partial \omega/\partial s$, where $\vec{u}$ is the small displacement of the rod with respect to the unstrained straight rod, s is the length measured along the rod, and ω is the angle of deflection of the rod. It is possible to show, by Fourier transforming the free energy $F = \int_0^L \kappa r_c^2 \, ds$ of a rod of length L, applying the theorem of equipartition of energy, and finally summing over all modes, that

$$\langle \omega^2 \rangle = 2\frac{k_B T}{\kappa} L. \qquad (7.16)$$

Taking $\langle \cos \omega \rangle \approx 1 - L k_B T/\kappa$ equal to zero, which is the condition at which the elements of the rod at $s = 0$ and $s = L$ are decorrelated [263], one gets

$$L_p = \frac{\kappa}{k_B T}. \qquad (7.17)$$

Eventually, the free energy density can be written as

$$F = \frac{1}{2} k_B T L_p r_c^{-2}. \tag{7.18}$$

Of course, this expression of the free energy is valid for $L < L_p$. For lengths $L > L_p$, the chain is flexible and its free energy is described in a completely different way (see below).

We also mention the concept of a *persistence time* τ_p, i.e., a time characteristic of a change of conformation. Let ΔE be the activation energy necessary for the trans-to-gauche isomerization. In this case, we have

$$\tau_p = \tau_0 \exp\left(\frac{\Delta E}{k_B T}\right). \tag{7.19}$$

The polymer chain appears flexible in any observation on a time duration $\tau > \tau_p$. The characteristic time τ_0 is related to the thermal vibrations of the chain, and it is of the order of 10^{-12} s.

7.6 Single-chain conformations

There are three *rotational isomeric states* per bond: t, g^+, and g^-. There are, therefore, $3N$ different configurations for a polyethylene chain. Among those, some are special, as we will see in the following.

- **Helix Configurations**. These are configurations of the chain in which all monomeric units repeat along a direction of elongation of the chain[4]. Helical configurations have minimal internal energy and are encountered in polymer crystals. For PE, this is the *all-trans* state: The chain shows a zigzag structure along its elongation, rather than a true 3D helical structure. For polytetrafluoroethylene (PTFE), it is a true helix, not far from the all-trans state, but differing from it by a superimposed twist along the elongation, caused by the repulsive interactions between neighboring fluorines. For polyoxymethylene (POM), there are several minimal helical energy states, obtained by twisting the all-gauche configurations in slightly different ways.

- **Coil Configurations**. We will occupy only ourselves with molecules that are *dynamically flexible*, i.e., whose chemical length L is large compared to L_p, and who are observed for durations much larger than τ_p. These molecules run through the $3N$ configurational states. In such a case, a number of properties are not dependent on the local scales, and the chemical features can be ignored. On the other hand, all "mesoscopic" properties depend on N, according to scaling laws of the type

$$R_g = a N^\nu, \tag{7.20}$$

---

[4] most often, this repeat pattern has a helical shape. Hence the denomination.

where R_g is the so-called *radius of gyration*, which will be defined later. R_g is the typical radius of the statistical coil in which the polymeric chain is folded when isolated in some solvent. The next paragraphs are devoted to the calculation of the exponent ν in various theoretical cases.

7.7 The ideal (Gaussian) chain model

Assume that the N elementary segments do not interact and are independent of each other. Such a chain is called ideal. The vector that joins the origin of the chain to its end is called the *end-to-end vector* $\vec{R}_e$ and can be written as

$$\vec{R}_e = \sum_{i=1}^{N} \vec{a}_i. \tag{7.21}$$

Squaring $\vec{R}_e$, and taking averages over all possible configurations, one gets the *mean squared end-to-end vector*:

$$\langle \vec{R}_e \rangle = \sum_{i=1}^{N} \vec{a}_i{}^2 + 2 \sum_{i \neq j}^{N} \langle \vec{a}_i \vec{a}_j \rangle = \sum_{i=1}^{N} \vec{a}_i{}^2 + 2|\vec{a}_i||\vec{a}_j| \underbrace{\sum_{i \neq j}^{N} \langle \cos(\angle(\vec{a}_i, \vec{a}_j)) \rangle}_{=0} = Na^2, \tag{7.22}$$

where we have used the (ideal) condition that segments i and j and thus, $\vec{a}_i$ and $\vec{a}_j$ are statistically independent. Hence, the average separation of the ends, which can be viewed as the *average size* of the chain, is

$$R_e = R_0 = \langle \vec{R}_e{}^2 \rangle^{1/2} = aN^{1/2}, \tag{7.23}$$

and exponent $\nu = 1/2$. Subscript 0 in equation (7.23) indicates ideal chain behavior. For non-ideal chains, the expression R_e is used for the average size[5].

The scaling law in equation (7.23) is unchanged when introducing interactions between the different monomers along the same chain (generally called *bounded interactions*, in contrast to *unbounded interactions*, e.g. between monomers belonging to different chains).

---

[5] Note, that here $R_e \neq |\vec{R}_e|$, as one might expect from the usual use of the same variable for denoting a vector $\vec{v}$ and its absolute length $|\vec{v}|$.

> **The central limit theorem.** Consider N independent random variables $X_1, X_2, \ldots, X_N$ which obey the same probability distribution function with mean value $\bar{X}$ and variance σ_X^2. In this case, the random variable Y, defined as
>
> $$Y = X_1 + X_2 + \cdots + X_N$$
>
> obeys a probability distribution with mean value $\bar{Y} = N\bar{X}$ and variance $\sigma_Y^2 = N\sigma_X^2$. If $N \gg 1$ (in fact, strictly speaking only in the true limit $N \to \infty$, the probability distribution function for Y approaches the following Gaussian distribution
>
> $$P_Y(Y) = \frac{1}{\sqrt{2\pi\sigma_Y^2}} \exp\left[-\frac{1}{2\sigma_Y^2}(Y - \bar{Y})^2\right].$$
>
> This theorem is called central limit theorem, which indicates that, when N is large, i.e. $N \ggg 1$, the width of the distribution of Y is proportional to $\sqrt{N}$, while its mean value is proportional to N. Thus, the fluctuations in Y around its mean value decrease as $1/\sqrt{N}$, and eventually the fluctuations can be neglected in the very large limit of ($N \to \infty$). Such a limit corresponds to the so called *thermodynamic limit*.

One can show that, as a consequence of the *central limit theorem* (see Appendix K for a more detailed derivation), the probability $P_N(\vec{r})$ for the N-link chain to terminate at $\vec{r} = \vec{R}_e$ follows a Gaussian distribution[6]:

$$P_N(\vec{r}) = \left(\frac{3}{2\pi R_0^2}\right)^{3/2} \exp\left(-\frac{3}{2}\frac{r^2}{R_0^2}\right). \tag{7.24}$$

7.8 Scaling of flexible and semiflexible polymer chains

Polymers usually do not exist in vacuum but normally exist in solution. A solvent is referred to as *good* when the prevailing form of the effective interaction between polymer segments in this solvent is the repulsive part of the potential energy at shorter distances. In this case, the chains tend to swell and the size R of the polymer (e.g. the end-to-end distance R_e in the case of linear chains or the radius of gyration R_g) scale with exponent $\nu = 3/5$. In the opposite case of a *poor* solvent, polymers tend to shrink, and R scales with $\nu = 1/3$. The point were the repulsive and attractive interactions just cancel each other out defines the θ-*point* and θ-*temperature*, respectively. Here, the chain configuration is that of a Gaussian random coil with an exponent

---

[6] Compare our discussion of probabilities in Chapter 3.

Section 7.8 Scaling of flexible and semiflexible polymer chains

$v = 1/2$. There are still three-body and higher order interactions in a θ-solvent, but their contribution to the free energy is negligibly small [124]. For the description of the distance of temperature T from the θ-temperature, a dimensionless parameter is used, the *reduced temperature* ζ which is defined as

$$\zeta = \frac{|T - T_\theta|}{T_\theta}. \tag{7.25}$$

A crossover scaling function f serves to describe the scaling behavior of polymer chains in different solvent conditions [124]. The argument of f is given by $\zeta\sqrt{N}$. At θ-temperature,

$$f(\zeta\sqrt{N}) \simeq 1, \quad \zeta\sqrt{N} \ll 1, \quad R = R_0 \propto N^{1/2}. \tag{7.26}$$

At $T < T_\theta$,

$$f(\zeta\sqrt{N}) \simeq (\zeta\sqrt{N})^{1/3}, \quad \zeta\sqrt{N} \gg 1, \quad R \propto N^{1/3}\zeta^{1/3}. \tag{7.27}$$

At $T > T_\theta$,

$$f(\zeta\sqrt{N}) \simeq (\zeta\sqrt{N})^{3/5}, \quad \zeta\sqrt{N} > 1, \quad R \propto N^{3/5}. \tag{7.28}$$

In experiments, it is rather difficult to obtain complete and conclusive results in the study of the collapse transition of chains, because one is often restricted to the three distinct limiting cases of polymer solutions, the extreme dilute regime, the θ-point, and the regime of very high polymer concentrations [96].

At the θ-temperature, the chains behave as $\langle R_g^2 \rangle \propto \langle R_e^2 \rangle \propto (N-1)^{2v_\theta}$ with $v_\theta = 0.5$, besides logarithmic corrections in 3D. Therefore, one expects that a plot of $\langle R^2 \rangle/(N-1)$ vs. T for different values of N shows a common intersection point at $T = T_\theta$, where the curvature changes. For $T > T_\theta$, the larger N is, the larger the ratio $\langle R^2 \rangle/(N-1)$, whilst for $T < T_\theta$, the larger N is, the *smaller* the ratio $\langle R^2 \rangle/(N-1)$. Using the model potential of equation (4.58) – instead of varying temperature (which involves rescaling of the particle velocities), different solvent qualities are obtained by tuning the interaction parameter λ. The corresponding transition curves are displayed in Figure 7.5 which show a clear intersection point at roughly $\lambda = \lambda_\theta \approx 0.65$. Moreover, it can be seen that the transition becomes sharper with increasing chain length N. The different curves do not exactly intersect in one single point, but there is an extended region in which the chains behave in a Gaussian manner. The size of this region is $\propto N^{-1/2}$ [124]. There is a very slight drift of the intersection point towards a smaller value of λ, with increasing chain length N.

Therefore, to obtain a more precise estimate of the θ-temperature in the limit of ($N \to \infty$), one has to chose a different graph, which allows for an appropriate extrapolation. If one draws straight horizontal lines in Figure 7.5, the intersection points of

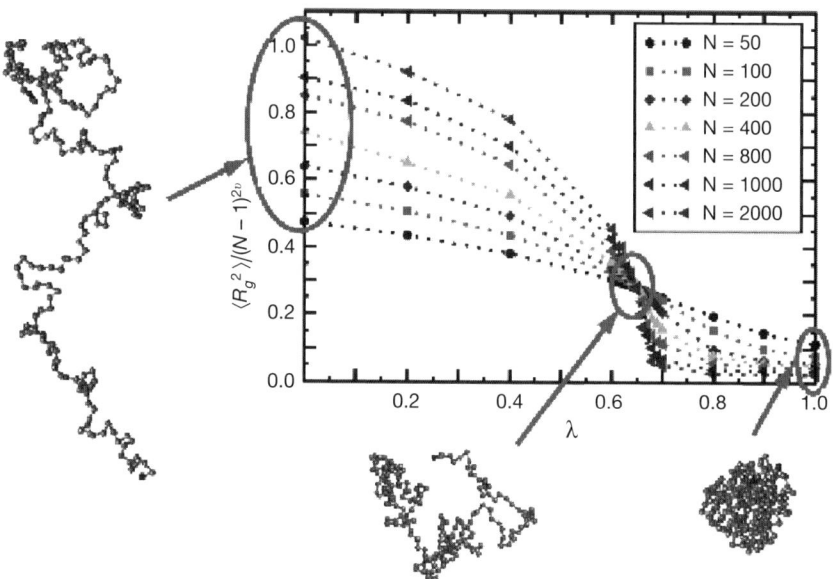

Figure 7.5. Coil-to-globule transition, from good to bad solvent behavior of a polymer chain. Plot of $\langle R_g^2 \rangle/(N-1)^{2\nu}$ vs. interaction parameter λ for linear chains. The points represent the simulated data and the dotted lines are guides for the eye. $\nu = \nu_\theta = 0.5$. Also displayed are simulation snapshots of linear chains for the three cases of a good, a θ-, and a bad solvent.

these lines with the curves are points at which the scaling function $f(\sqrt{N}\zeta)$ of equation (7.27) is constant. Plotting different intersection points over $N^{-1/2}$ therefore yields different straight lines that intersect each other at exactly $T = T_\theta$ and $\lambda = \lambda_\theta$, respectively. This extrapolation ($N \to \infty$) is displayed in Figure 7.6. The different lines do not intersect exactly at $N^{-1/2} = 0$, which is due to the finite length of the chains. As a result of these plots, one obtains the value of λ for which the repulsive and attractive interactions in the used model just cancel each other out,

$$\lambda_\theta = 0.65 \pm 0.02. \tag{7.29}$$

An important property of individual chains is the *structure factor* $S(q)$, the spherical average of which is defined as [395]

$$S(q) = \left\langle \frac{1}{N^2} \left| \sum_{i=1}^{N} e^{-i\vec{q}\vec{r}_i} \right|^2 \right\rangle_{|\vec{q}|}, \tag{7.30}$$

with subscript $|\vec{q}|$ denoting the average over all scattering vectors $\vec{q}$ of the same magnitude $|k| = q$, with $\vec{r}_i$ being the position vector to the ith monomer and N denoting

Section 7.8 Scaling of flexible and semiflexible polymer chains

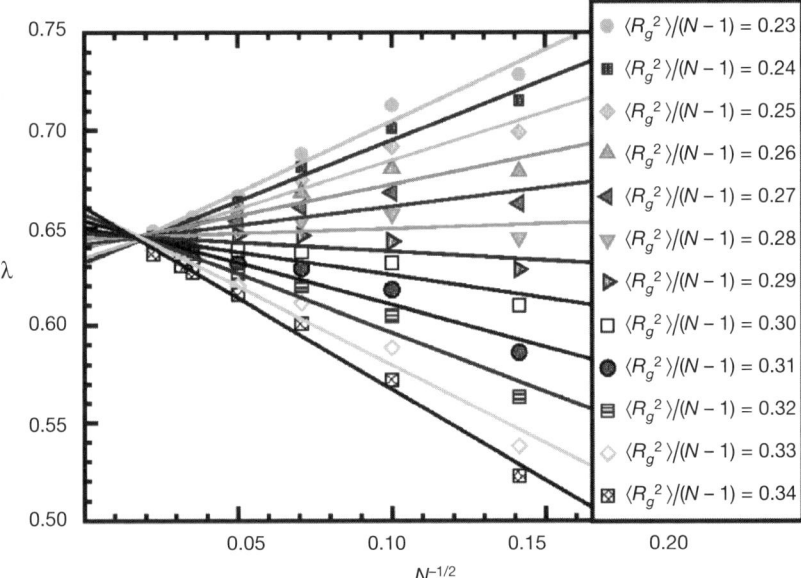

Figure 7.6. Interaction parameter λ in equation (4.58) vs. $N^{-1/2}$ for different values of the scaling function. Data points are based on the radius of gyration of linear chains [396].

the total number of monomers (scattering centers). For different ranges of the scattering vectors the following scaling relations hold [124]:

$$S(q) = \begin{cases} \left(1 - 1/3\, q^2 \langle R_g^2 \rangle\right) N & (2\pi)^2/\langle R_g^2 \rangle \gg q^2, \\ q^{-1/\nu} & (2\pi)^2/\langle R_g^2 \rangle \ll q^2 \ll (2\pi)^2/l_b^2, \\ 1/N & (2\pi)^2/l_b^2 \ll q^2, \end{cases} \quad (7.31)$$

where l_b is the (constant) bond length of the monomers (often also called segment length). The importance of $S(q)$ lies in the fact that it is *directly* measurable in scattering experiments. For *ideal* linear chains, the function $S(q)$ can be explicitly calculated and is given by the monotonously decreasing *Debye function*.

$$S(x) = \frac{2}{x^2}\left(x - 1 + e^{-x}\right), \quad (7.32)$$

where the quantity x is given by $x = q^2 \langle R_g^2 \rangle_0$ with index 0 denoting θ-conditions. For small values of x, corresponding to large distances between scattering units, the Debye function $S(x)$ also provides a good description of a linear chain in a *good* solvent, with the scaling variable x describing the expansion of the chain. For very small scattering vectors q, one obtains the *Guinier approximation* [395] via an expansion

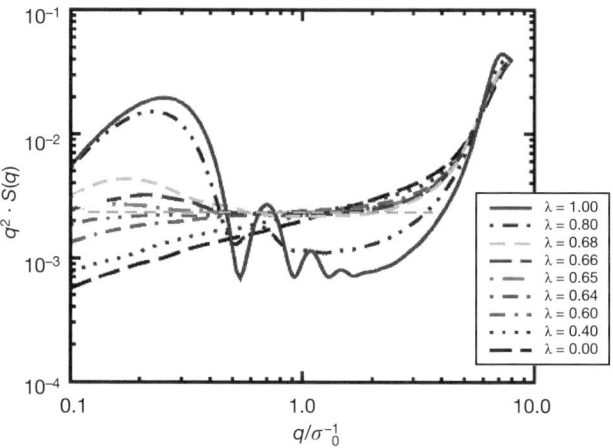

Figure 7.7. Kratky plot of $S(q)$ for linear chains, ($N = 2\,000$) for different values of the interaction parameter λ.

of $S(q)$, which is used in experiments to calculate the radius of gyration $\langle R_g{}^2 \rangle$. In the intermediate regime of scattering vectors, $S(q)$ obeys a scaling law which, in a double-logarithmic plot, should yield a slope of $-1/\nu$. Finally, for large q-values, $S(q)$ is expected to behave as $1/N$. The overall expected behavior of $S(q)$ is summarized in equation (7.31).

In the vicinity of the θ-region, the scaling exponent equals $\nu = \nu_\theta = 0.5$. Therefore, $q^2 S(q)$, plotted against wave vector q, which is called a *Kratky plot*, should approach a constant value. Figure 7.7 shows this behavior with high resolution in terms of λ, for different chain lengths. The respective dotted horizontal line is a guide for the eye. The larger the chains, the smaller the λ-range at which the chains display ideal (Gaussian) random walk behavior. For large values of λ, the chains are collapsed and form compact globules whose local structure is also reflected in the structure function by several distinct peaks for larger q-values. These peaks become more pronounced the longer the chains are, reflecting the fact that the transition curves become ever sharper with increasing chain length. Hence, longer chains are already in the collapsed regime for values of λ at which the smaller chains still exhibit Gaussian behavior. The structure function of the largest system in Figure 7.7 for $\lambda = 1.0$ already very much resembles the scattering pattern of a sphere.

In Figure 7.8, the scaling of $\langle R_g{}^2 \rangle$ for different star polymers as a function of N, and for different functionalities f is displayed. Functionality $f = 2$ corresponds to linear chains, $f = 3$ corresponds three-arm star polymers and so on. The star polymers were generated with the MD simulation package "MD-Cube" developed by Steinhauser [395, 394], which is capable of handling a very large array of branched polymer topologies, from star polymers to dendrimers, H-polymers, comb-polymers

Section 7.8 Scaling of flexible and semiflexible polymer chains

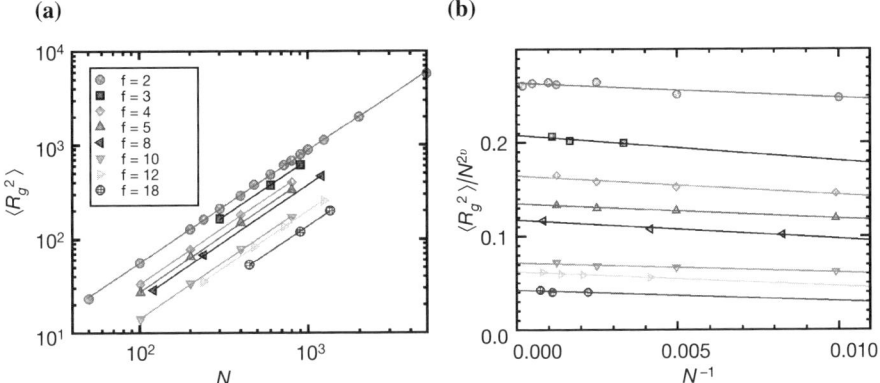

Figure 7.8. (a) Log-Log plot of $\langle R_g^2 \rangle$ vs. N for star polymers with different arm numbers f. For comparison, data for linear chains ($f = 2$) are displayed as well. (b) Scaling plot of the corrections to scaling of $\langle R_g^2 \rangle(f)$, plotted vs. N^{-1} in a good solvent. For clarity, the smallest data point of the linear chains ($f = 2, N = 50$) is not displayed.

Table 7.1. Obtained scaling exponents ν for star polymers in simulations with different arm numbers f.

	2	3	4	5	6	10	12	18
ν	0.5989	0.601	0.603	0.614	0.617	0.603	0.599	0.601

or randomly hyperbranched polymers. Details of the setup of chains, which works the same way for linear and branched polymer topologies, can be found in [395]. Figure 7.8 (a) shows a double-logarithmic plot from which one obtains the scaling exponents of R_g for stars with different numbers of arms. The results for linear chains are displayed as well, where chain lengths of up to $N = 5000$ were simulated. Within the errors of the simulation, the exponents do not depend on the number of arms, as expected from theory. The obtained scaling exponents are summarized in Table 7.1 and exhibit a reasonable agreement with theory. Figure 7.8 (b) exhibits, that the corrections to scaling due to the finite size of the chains are $\propto N^{-1}$. A plot with exponents -2 or $-1/2$ leads to worse correlation coefficients. This result is consistent with lattice-MC simulations on an fcc-lattice [61]. More details on finite-size scaling can be found in [112, 396].

The fundamentals of the dynamics of fully flexible polymers in solution or in the melt were worked out in the pioneering works of Rouse [362] and Zimm [454], as well as of Doi and Edwards [137] and de Gennes [124]. In contrast to fully flexible polymers, the modeling of *semiflexible* and *stiff* macromolecules has recently received recent, because such models can be successfully applied to biopolymers such as proteins, DNA, actin filaments or rodlike viruses [84, 320]. Biopolymers are wormlike

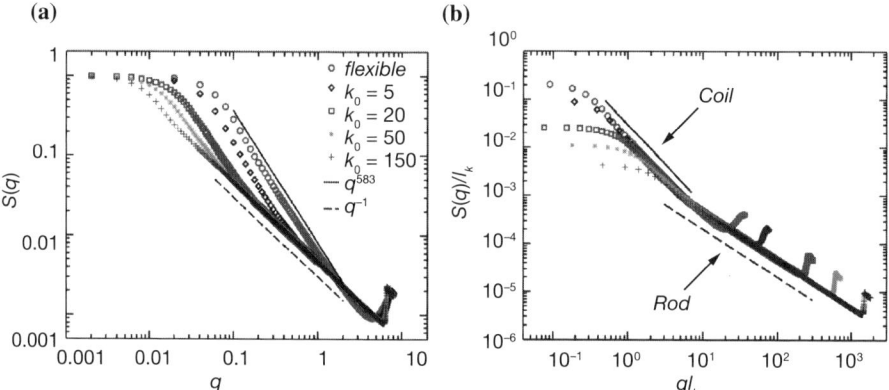

Figure 7.9. (a) $S(q)$ of single linear chains with $N = 700$ and varying stiffness k_θ. The scaling regimes (fully flexible and stiff rod) are indicated by a straight and dashed line, respectively. (b) Scaling plot of $S(q)/l_K$ versus $q \cdot l_K$, using the statistical segment length l_K adapted from [395].

chains with persistence lengths l_p (or Kuhn segment lengths l_K), comparable to or larger than their contour length L, and their rigidity and relaxation behavior is essential to their biological functions.

Using the second term of the bonded potential of equation (4.59), a bending rigidity $\Phi_\text{bend}(\theta)$ can be introduced. Rewriting this term by introducing the unit vector $\vec{u}_j = (\vec{r}_{j+1} - \vec{r}_j)/|\vec{r}_{j+1} - \vec{r}_j|$ along the macromolecule, cf. Figure 4.10, one obtains

$$\Phi_\text{bend}(\theta) = \frac{k_\theta}{2} \sum_{j=1}^{N-1} (\vec{u}_{j+1} - \vec{u}_j)^2 = k_\theta \sum_{j=1}^{N-1} (1 - \cos \theta_j), \qquad (7.33)$$

where θ_i is the angle between $\vec{u}_j$ and $\vec{u}_{j+1}$. The crossover scaling from coil-like, flexible structures on large length scales, to stretched conformations at smaller scales can be seen in the scaling of $S(q)$ when performing simulations with different values of k_θ [395]. Results for linear chains of length $N = 700$ are displayed in Figure 7.9 (a). The chains show scaling according to q^ν. The stiffest chains exhibit a q^{-1}-scaling which is characteristic of stiff rods. Thus, by varying the parameter k_θ, the whole range of bending stiffness of chains, from fully flexible chains to stiff rods, can be covered. The range of q-vectors for which the crossover from flexible to semiflexible and stiff occurs shifts to smaller scattering vectors with increasing stiffness k_θ of the chains. The scaling plot in Figure 7.9 (b) shows that the transition occurs for $q \approx 1/l_K$, i.e. on a length scale of the order of the statistical Kuhn length. In essence, only the fully flexible chains (red data points) exhibit a deviation from the master curve on large length scales (i.e. small q-values), which corresponds to their differ-

Section 7.9 Constant temperature MD   429

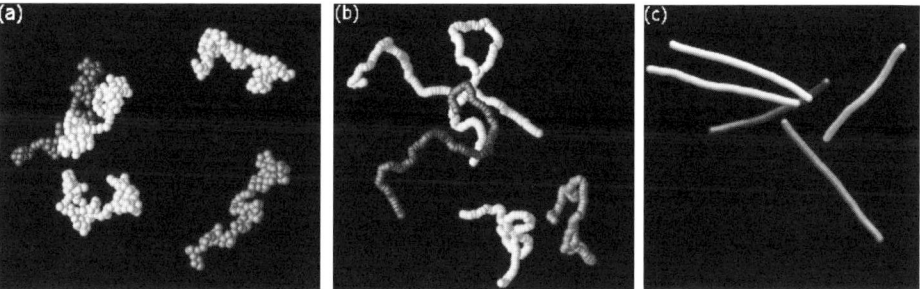

Figure 7.10. Simulation snapshots of (a) flexible chains ($k_\theta = 0$), (b) semiflexible chains ($k_\theta = 20$), (c) stiff, rod-like chains ($k_\theta = 50$).

ent global structure compared to semi-flexible macromolecules. Finally, examples of snapshots of stiff and semiflexible chains are displayed in Figure 7.10.

7.9 Constant temperature MD

In its simplest form, the standard MD simulation conserves total energy. Hence, if the run is long enough and the system is ergodic, the time averages computed from MD simulation are equivalent to the ensemble averages computed from the *microcanonical ensemble*. In contrast to – e.g. FEM simulation techniques, where certain thermodynamic properties of the simulated system or materials are put into the Equation of State (EOS) – in MD, such properties are put into the EOM. The flexibility of MD is greatly enhanced by noting that it is not restricted to NVE ensembles. There exist techniques by which MD can simulate NVT or NPT ensembles as well. In the following section, I briefly introduce such temperature control schemes.

Generally speaking, there are in essence three ways to control the temperature in an MD simulation:

1. scaling velocities (e.g. simple velocity scaling and the Berendsen thermostat)

2. adding stochastic forces and/or velocities (e.g., the Andersen, Langevin, and Dissipative Particle Dynamics thermostats)

3. using "extended Lagrangian" formalisms (e.g., the Nosé-Hoover thermostat)

Each of these classes of schemes has advantages and disadvantages, depending on the specific application. In the following section, we will consider examples of the first two kinds of temperature thermostats.

7.10 Velocity scaling using the Behrendsen thermostat

Velocity scaling schemes do not strictly follow the canonical ensemble, although in practice the amount they deviate from canonical is quite small (this can be measured by comparing the velocity distribution function with a Gaussian). It is relatively easy to implement velocity scaling schemes, because they can be "dropped" into existing programs, using almost any integrator. However, they suffer the drawback of not being time-reversible or deterministic, properties that become important in some advanced MD techniques. We have, in effect, already encountered simple velocity scaling in Section 5.10 in the program *NVE_LJFluid.c*, namely in its initialization function. Here, particle velocities are randomly chosen from $[-0.5, 0.5]$ and rescaled to result in a desired temperature given by the relation

$$\frac{3}{2} N k_\text{B} T = \frac{1}{2} \sum_i m_i v_i^2. \tag{7.34}$$

If we wanted to, we could turn this into a dynamic scheme for continually keeping the velocities scaled such that the total kinetic energy is constant. We can measure the instantaneous temperature immediately after a velocity update, and call it T_i. Equation (7.34) indicates that, if we scale velocities by a constant λ, where

$$\lambda = \sqrt{(T/T_i)} \tag{7.35}$$

we will be left with a system at temperature T. Velocity scaling in order to maintain constant kinetic energy is called an *isokinetic thermostat*. Such a thermostat cannot be used to conduct a simulation in the canonical ensemble, but is perfectly fine to use in a warming-up or initialization phase. We could perform velocity rescaling at every step, or only every few steps.

Suggested exercises:
1. Modify the program *NVE_LJFluid.c* to perform velocity scaling to a user-specified set-point temperature every m time steps, where m is a user-defined interval between velocity scaling events. Begin with the system at $T = 1.0$ and command it to jump to $T = 2.0$ after 1 000 steps. How does the system behave and is it sensitive to your choice of m?

2. Another popular velocity scaling thermostat is that of Berendsen et al. [57]. Here, the scale factor is given by

$$\lambda = \left[1 + \frac{\Delta t}{\tau_T} \left(\frac{T}{T_0} - 1 \right) \right]^{\frac{1}{2}} \tag{7.36}$$

Here, T_0 is the set-point temperature, Δt is the integration time step, and τ_T is a constant called the "rise time" of the thermostat. It describes the strength of the coupling

Section 7.11   Dissipative particle dynamics thermostat   431

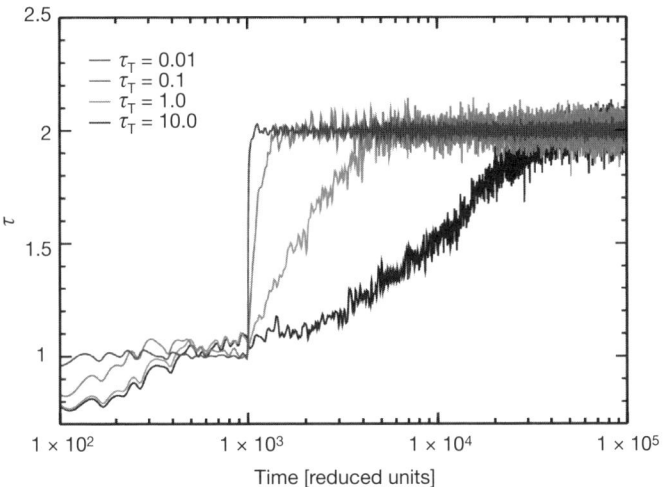

Figure 7.11. Instantaneous temperature, T, vs. time in an MD simulation of $N = 1\,000$ particles at a density of 0.5, with the temperature controlled by a Berendsen thermostat [57], for various values of the thermostat rise time, τ_T.

of the system to a hypothetical heat bath. The larger τ_T, the weaker the coupling. In other words, the larger τ_T, the longer it takes to achieve a given T_0 after an instantaneous change from some previous T_0. The code *Berendsen.c* (available for download on the book's website www.degruyter.com) implements the Berendsen thermostat. As a brief exercise, you can experiment with this program to get a feeling for how various values of the rise time affect the response of the system when the set-point temperature is instantaneously changed from 1.0 to 2.0. Figure 7.11 is a lin-log plot of just such an experiment, with $N = 1\,000$ particles at a density of 0.5. Each curve corresponds to a different value of τ_T, and they increase by factors of 10.

Though relatively simple, velocity scaling thermostats are not recommended for use in production MD runs, because they do not strictly conform to the canonical ensemble.

7.11   Dissipative particle dynamics thermostat

The DPD thermostat [316] adds pairwise random and dissipative forces to all particles, and has been shown to preserve momentum transport. Hence, it is the only stochastic thermostat so far that should even be considered for use, if one wishes to compute transport properties.

The DPD thermostat is implemented by a slight modification of the force routine to add the pairwise random and dissipative forces. For the ij pair, the dissipative force is defined as

$$\mathbf{f}_{ij}^D = -\gamma \omega^D (r_{ij}) \left(\mathbf{v}_{ij} \cdot \hat{\vec{r}}_{ij} \right) \hat{\vec{r}}_{ij}. \tag{7.37}$$

Here, γ is a friction coefficient, ω is a cut-off function for the force as a function of the scalar distance between i and j, which simply limits the interaction range of the dissipative (and random) forces, $\mathbf{v}_{ij} = \mathbf{v}_i - \mathbf{v}_j$ is the relative velocity of i to j, and $\hat{\vec{r}}_{ij} = \vec{r}_{ij}/r_{ij}$ is the unit vector pointing from j to i. The random force is defined as

$$\mathbf{f}_{ij}^R = \sigma \omega^R (r_{ij}) \zeta_{ij} \hat{\vec{r}}_{ij}. \tag{7.38}$$

Here, σ is the strength of the random force, ω^R is a cut-off, and ζ_{ij} is a Gaussian random number with zero mean and unit variance, and $\zeta_{ij} = \zeta_{ji}$.

The update of the velocity uses these new forces:

$$\vec{v}_i (t + \Delta t) = \vec{v}_i (t) + \frac{\Delta t}{m} \nabla_i U + \frac{\Delta t}{m} \vec{F}_i^D + \frac{\sqrt{\Delta t}}{m} \vec{F}_i^R \tag{7.39}$$

where

$$\vec{F}_i^D = \sum_{j \neq i} \vec{F}_{ij}^D, \tag{7.40}$$

$$\vec{F}_i^R = \sum_{j \neq i} \vec{F}_{ij}^R. \tag{7.41}$$

The parameters γ and σ are linked by a fluctuation-dissipation theorem:

$$\sigma^2 = 2\gamma k_B T. \tag{7.42}$$

So, in practice, one must specify either γ or σ, and then a set-point temperature, T, in order to use the DPD thermostat.

The cutoff functions are also related:

$$\omega^D (r_{ij}) = \left[\omega^R (r_{ij}) \right]^2 \tag{7.43}$$

This is the only real constraint on the cutoffs. Other than this, we are allowed to use any cutoff we like. The simplest one uses the cutoff radius of the pair potential, r_c:

$$\omega(r) = \begin{cases} 1 & r < r_c \\ 0 & r > r_c \end{cases}. \tag{7.44}$$

Note that, with this choice, $\left[\omega^R (r_{ij}) \right]^2 = \omega^R (r_{ij}) = \omega^D (r_{ij}) = \omega$.

The code *Chap7_DPD.c* (available for download on the book's website www.degruyter.com) implements the DPD thermostat in an MD simulation of the LJ

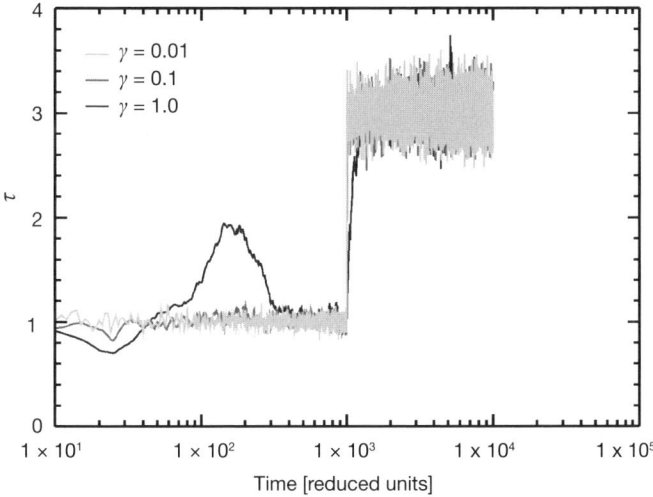

Figure 7.12. Instantaneous temperature, T, vs. time in an MD simulation of 1 000 particles at a density of 0.85, with temperature controlled by the dissipative particle dynamics thermostat, for various values of the parameter γ.

liquid. The major changes (compared to *Berendsen.c*) are to the force routine, which now requires several more arguments, including particle velocities, and parameters for the thermostat. Inside the pair loop, the force on each particle is updated by the conservative, dissipative, and random pairwise force components. The random force is divided by $\sqrt{\Delta t}$, so that the velocity Verlet algorithm need not be altered to implement equation (7.39)

The behavior of the DPD thermostat can be assessed in a similar fashion as the Berendsen thermostat above. Figure 7.12 shows a lin-log plot of the system's temperature after the set-point is changed from 1.0 to 2.0, for an LJ system $N = 1\,000$ and $\rho = 0.85$. Again, each curve corresponds to a different value of γ. You will notice that in all the provided code implementations, the thermostat is applied each and every time step. In principle, one could modulate the strength of a thermostat by controlling the frequency with which it is applied. As an exercise, you can pick one of the provided implementations and modify it so that the thermostat is applied once per M time steps, where M is specified on the command line. You can compare the performance of the thermostat in the same way as we have already seen (time required to go from $T = 1.0$ to $T = 2.0$ upon an instantaneous change in the set-point T) for a given friction or strength and for various values of M.

7.12 Case study: NVT Metropolis MC simulation of a LJ fluid

In this case study I provide source code for the NVT Metropolis MC simulation, an LJ fluid – *the* prototypical system for continuous space computations, i.e. for computer simulations where no underlying lattice is involved. Hence, the particles can occupy any point in three-dimensional space. The primary objective of this MC code is *to predict the pressure* of a sample of an LJ fluid *at a given temperature and density*. That is, one can use MC to map[7] the *phase diagram* of a material. In this program I have also included Periodic Boundary Conditions (PBC), which were introduced and discussed in Chapter 5. Periodic boundaries are used here, because we want to simulate bulk fluid properties and not surface effects. The simplest way to approximate bulk behavior in a system with a finite number of particles is to employ periodic boundaries. This means that the simulation box of length L is imagined as embedded within an infinite space tiled with replicas of the central simulation box. When focusing on the central box, one can watch particles leaving the box reappear in the box at the opposite face. Moreover, particles interact with their "images" and with the images of other particles in all replica boxes. This is the reason why the minimum image convention has to be applied, see Chapter 5. In essence, PBC allow for mimicking an infinite extent of bulk fluid.

One of the caveats that come with the use of PBC is that the total potential, as a sum of pair potentials, in an infinite periodic system, in principle diverges. This can be circumvented by introducing a *finite interaction range* to the pair potential, as discussed in Chapter 5. Usually, one works with systems large enough such that the cutoff r_c of the pair potential is *less than one-half of the box length L* in a cubic box. Hence, the "image" interactions involve only immediately neighboring replicas. The *main()*-function of the program is provided in Listing 71. The major point in introducing a cutoff is that it has to be spherically symmetric, i.e. you cannot simply cut off interactions in each direction beyond the box length L, as this results in a directional bias in the interaction range of the potential. Thus, a hard cutoff radius r_c is required, and it should be less than $L/2$. Once r_c is chosen one has to use correction terms for energy and pressure, if one wants to mimic a potential with infinite range. The considered system in this case study is made of N LJ particles which interact via the potential in equation (4.26). The particles are confined within a cubic box with sides of length L. Length is measured in units of σ and energy in units of ϵ, and we consider particles with $1 - \sigma$ diameters. The code is written in such a way that it calculates the pressure p for a "standard" system, with preset values for temperature T, number of particles N, density ρ, number nc of cycles, and so on, but the user is able to provide these values via *command line options*. For example, if a cutoff radius is chosen by the user, then a truncated and shifted pair potential is used,

---

[7] At least one can do so in principle.

Section 7.12 Case study: NVT Metropolis MC simulation of a LJ fluid   435

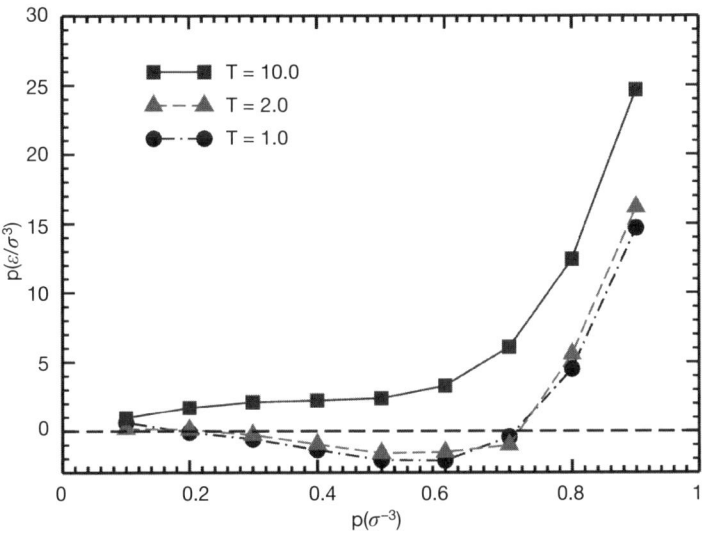

Figure 7.13. Metropolis MC simulation of the phase diagram of an LJ fluid. Pressure vs. density in an LJ-fluid, measured by a Metropolis MC simulation at three different temperatures $T = 1.0, 2.0, 10.0$. As $T = 1.0$ and $T = 2.0$ is *below the critical temperature*, they exhibit *negative* pressures. A small system can sustain negative pressure without phase separating into a vapor/liquid system, because there are not enough particles to nucleate a stable second phase. In these simulations, 2.000 cycles were performed for each run, and the system contained 1 000 particles. Each point is the result of only one single run.

and the following tail corrections are applied:

$$\Phi_{\text{tail}} = \frac{8}{3}\pi\rho\epsilon\sigma^3 \left[\frac{1}{3}\left(\frac{\sigma}{r_c}\right)^9 - \left(\frac{\sigma}{r_c}\right)^3 \right], \tag{7.45a}$$

$$\Delta p_{\text{tail}} = \frac{16}{3}\pi\rho^2\epsilon\sigma^3 \left[\frac{2}{3}\left(\frac{\sigma}{r_c}\right)^9 - \left(\frac{\sigma}{r_c}\right)^3 \right]. \tag{7.45b}$$

The pressure is computed from the virial expression (see Chapter 3):

$$p = \rho T + \frac{1}{V} \times \frac{1}{3} \sum_{i=1}^{N} \sum_{j>i}^{N} \vec{F}(\vec{r}_{ij}) \cdot \vec{r}_{ij}, \tag{7.46}$$

where V is the system volume and $\vec{F}(\vec{r}_{ij})$ is the force exerted on particle i by particle j, defined as

$$\vec{F}(r_{ij}) = -\frac{\partial \Phi_{\text{LJ}}(r_{ij})}{\partial \vec{r}_i} = -\frac{\vec{r}_{ij}}{r_{ij}} \frac{\partial \Phi_{\text{LJ}}(r_{ij})}{\partial r_{ij}}. \tag{7.47}$$

For the LJ potential it is easy to calculate that:

$$\frac{\partial \Phi_{LJ}(r_{ij})}{\partial r_{ij}} = 4\epsilon \left[-12 \frac{\sigma^{12}}{r_{ij}^{12}} + 6 \frac{\sigma^5}{r_{ij}^7} \right] \frac{\vec{r}_{ij}}{r_{ij}^2} \tag{7.48}$$

We note that any particular pair's contribution to the total virial is *positive* if the members of the pair are repelling each other (i.e. if f along the direction of $\vec{r}_{ij}$ is positive, and *negative* if the particles are attracting each other. In the MC program in Listing 71, the initialization of particle positions is accomplished by putting the particles on cubic lattice sites such that an overall density is achieved. It is therefore most convenient to run simulations with numbers of particles that are perfect cubes, such as 128, 216, 512 etc, so that the initial state uniformly fills the box. The user flag -ne allows the user to specify how many equilibration cycles are to be performed before switching into "production" mode, where the actual measurements are done. The most important point with equilibration is that the memory of the system of its artificial initial state is erased. I suggest as a first exercise to try to reproduce the data points in Figure 7.13 that shows the pressure for a system at three different temperatures. One should try different cycles, equilibration cycles and maximum displacements and then further explore different temperatures at different densities, with different system sizes N. The definition of the function *TotalEnergy()* that calculates the energy and the virial, and which is called in *main()* in lines 85 and 114 is provided in Listing 72. Finally, function *Init()* that assigns the particle positions on a cubic grid is displayed in Listing 73.

Section 7.12 Case study: NVT Metropolis MC simulation of a LJ fluid   437

Listing 71. Part 1 – *main( )*-program for calculating the phase diagram of an LJ system. (*)

```
1  #include <stdio.h>
2  #include <stdlib.h>
3  #include <math.h>
4  #include <gsl/gsl_rng.h>
5
6  int main ( int argc, char * argv[] ) {
7
8    double * rx, * ry, * rz;
9    int N = 216, c, a;
10   double L = 0.0;
11   double rho = 0.5, T = 1.0, rc2 = 1.e20, vir, virOld, virSum,
         pcor, V;
12   double Enew, Eold, esum, rr3, ecor, ecut;
13   double dr = 0.1, dx, dy, dz;
14   double rxold, ryold, rzold;
15   int i, j;
16   int nCycles = 10, nSamp, nEq = 1000;
17   int nAcc;
18   int shortOut = 0;
19   int shift = 0;
20   int tailcorr = 1;
21
22   gsl_rng * r = gsl_rng_alloc(gsl_rng_mt19937);
23   unsigned long int Seed = 23410981;
24
25   /* Here, we parse the command line arguments */
26   for (i=1;i<argc;i++) {
27     if (!strcmp(argv[i],"-N")) N=atoi(argv[++i]);
28     else if (!strcmp(argv[i],"-rho")) rho=atof(argv[++i]);
29     else if (!strcmp(argv[i],"-T")) T=atof(argv[++i]);
30     else if (!strcmp(argv[i],"-dr")) dr=atof(argv[++i]);
31     else if (!strcmp(argv[i],"-rc")) rc2=atof(argv[++i]);
32     else if (!strcmp(argv[i],"-nc")) nCycles = atoi(argv[++i]);
33     else if (!strcmp(argv[i],"-ne")) nEq = atoi(argv[++i]);
34     else if (!strcmp(argv[i],"-so")) shortOut=1;
35     else if (!strcmp(argv[i],"+tc")) tailcorr=0;
36     else if (!strcmp(argv[i],"-sh")) shift=1;
37     else if (!strcmp(argv[i],"-seed"))
38       Seed = (unsigned long)atoi(argv[++i]);
39     else {
40       fprintf(stderr,"Error. Argument '%s' is not recognized.\
         n",
41       argv[i]);
42       exit(-1);
43     }
44   }
45
46   /* Compute the side length */
47   L = pow((V=N/rho),0.3333333);
```

Listing 71. Part 2 – *main()*-program for calculating the phase diagram of an LJ system continued....

```
48      /* Compute the tail corrections. It assumes sigma and
           epsilon are both 1 */
49      rr3 = 1.0/(rc2*rc2*rc2);
50      ecor = 8*M_PI*rho*(rr3*rr3*rr3/9.0-rr3/3.0);
51      pcor = 16.0/3.0*M_PI*rho*rho*(2./3.*rr3*rr3*rr3-rr3);
52      ecut = 4*(rr3*rr3*rr3*rr3-rr3*rr3);
53
54      /* Compute the *squared* cutoff, reusing the variable rc2 */
55      rc2*=rc2;
56
57      /* For computational efficiency, use reciprocal T */
58      T = 1.0/T;
59
60      /* compute box volume */
61      V = L*L*L;
62
63      /* Output initial information */
64      fprintf(stdout,"# NVT MC Simulation of a Lennard-Jones fluid\
           n");
65      fprintf(stdout,"# L = %.5lf; rho = %.5lf; N = %i; rc = %.5lf\
           n",
66         L,rho,N,sqrt(rc2));
67      fprintf(stdout,"# nCycles %i, nEq %i, seed %d, dR %.5lf\n",
68         nCycles,nEq,Seed,dr);
69
70      /* Total number of cycles is number of "equilibration" cycles
            plus
71         number of "production" cycles */
72      nCycles+=nEq;
73
74      /* Seed the random number generator */
75      gsl_rng_set(r,Seed);
76
77      /* Allocate the position arrays */
78      rx = (double*)malloc(N * sizeof(double));
79      ry = (double*)malloc(N * sizeof(double));
80      rz = (double*)malloc(N * sizeof(double));
81
82      /* Generate initial positions on a cubic grid,
83         and measure initial energy */
84      Init(rx,ry,rz,N,L,r);
85      Eold = TotalEnergy(rx,ry,rz,N,L,rc2,tailcorr,ecor,shift,ecut
           ,&virOld);
86
87      nAcc    = 0;
88      esum    = 0.0;
89      nSamp   = 0;
90      virSum  = 0.0;
```

Section 7.12 Case study: NVT Metropolis MC simulation of a LJ fluid

Listing 71. Part 3 – *main()*-program for calculating the phase diagram of an LJ system continued . . . .

```
91   for (c = 0; c < nCycles; c++) {
92     /* Randomly select a particle */
93     i=(int)gsl_rng_uniform_int(r,N);
94     /* calculate displacement */
95     dx = dr * (0.5-gsl_rng_uniform(r));
96     dy = dr * (0.5-gsl_rng_uniform(r));
97     dz = dr * (0.5-gsl_rng_uniform(r));
98     /* Save the current position of particle i */
99     rxold = rx[i];
100    ryold = ry[i];
101    rzold = rz[i];
102    /* Displace particle i */
103    rx[i] += dx;
104    ry[i] += dy;
105    rz[i] += dz;
106    /* Apply periodic boundary conditions */
107    if (rx[i]<0.0) rx[i]+=L;
108    if (rx[i]>L)   rx[i]-=L;
109    if (ry[i]<0.0) ry[i]+=L;
110    if (ry[i]>L)   ry[i]-=L;
111    if (rz[i]<0.0) rz[i]+=L;
112    if (rz[i]>L)   rz[i]-=L;
113    /* Get the new energy */
114    Enew = TotalEnergy(rx,ry,rz,N,L,rc2,tailcorr,ecor,shift,
              ecut,&vir);
115    /* Conditionally accept... */
116    if (gsl_rng_uniform(r) < exp(-T*(Enew-Eold))) {
117      Eold=Enew;
118      virOld=vir;
119      nAcc++;
120    }
121    /* ... or reject the move; reassign the old positions */
122    else {
123      rx[i]=rxold;
124      ry[i]=ryold;
125      rz[i]=rzold;
126    }
127    /* Sample: default frequency is once per trial move. We
              must
128        include results of a move regardless of whether the
              move is
129        accepted or rejected. */
130    if (c>nEq) {
131      esum+=Eold;
132      virSum+=virOld;
133      nSamp++;
134    }
135  }
```

Listing 71. Part 4 – *main()*-program for calculating the phase diagram of an LJ system continued . . . .

```
136      /* Output delta-r, the acceptance ratio,
137         and the average energy/particle */
138      if (shortOut)
139        fprintf(stdout,"%.6lf %.5lf %.5lf %.5lf\n",
140          dr,((double)nAcc)/(N*nCycles),
141          esum/nSamp/N,virSum/3.0/nSamp/V+rho*T+pcor);
142      else
143        fprintf(stdout,"NVT Metropolis Monte Carlo Simulation"
144          " of the Lennard-Jones fluid.\n"
145          "_____\n"
146          "Number of particles:              %i\n"
147          "Number of cycles:                 %i\n"
148          "Cutoff radius:                    %.5lf\n"
149          "Maximum displacement:             %.5lf\n"
150          "Density:                          %.5lf\n"
151          "Temperature:                      %.5lf\n"
152          "Tail corrections applied?         %s\n"
153          "Shifted potential?                %s\n"
154          "Results:\n"
155          "Potential energy tail correction: %.5lf\n"
156          "Pressure tail correction:         %.5lf\n"
157          "Potential energy shift at cutoff  %.5lf\n"
158          "Acceptance ratio:                 %.5lf\n"
159          "Energy/particle:                  %.5lf\n"
160          "Ideal gas pressure:               %.5lf\n"
161          "Virial:                           %.5lf\n"
162          "Total Pressure:                   %.5lf\n"
163          "Program ends.\n",
164          N,nCycles,sqrt(rc2),dr,rho,1.0/T,
165          tailcorr?"Yes":"No",shift?"Yes":"No",
166          ecor,pcor,ecut,
167          ((double)nAcc)/(N*nCycles),
168          esum/nSamp/N,
169          rho/T,virSum/3.0/nSamp/V,
170          virSum/3.0/nSamp/V+rho/T+(tailcorr?pcor:0.0));
171    }
```

Section 7.12 Case study: NVT Metropolis MC simulation of a LJ fluid   441

Listing 72. Function *TotalEnergy()* called from *main()* for calculating the phase diagram of an LJ system. (*)

```
1  /* This is an N^2 algorithm for computing the total energy. The
        virial
2     is also computed and returned in *vir. */
3
4  double TotalEnergy ( double * rx, double * ry, double * rz, int
         N, double L, double rc2, int tailCorr, double eCor, int
         shift, double eCut, double * vir ) {
5    int i,j;
6    double dx, dy, dz, r2, r6i;
7    double e = 0.0, hL = L / 2.0;
8
9    *vir = 0.0;
10   for (i = 0; i < (N-1); i++) {
11     for (j = i+1; j < N; j++) {
12       dx  = (rx[i] - rx[j]);
13       dy  = (ry[i] - ry[j]);
14       dz  = (rz[i] - rz[j]);
15  /* Periodic boundary conditions: Apply the minimum image
16     convention. If there is an explicit cutoff this is, of
            course,
17     not used. */
18       if (dx>hL)         dx-=L;
19       else if (dx<-hL) dx+=L;
20       if (dy>hL)         dy-=L;
21       else if (dy<-hL) dy+=L;
22       if (dz>hL)         dz-=L;
23       else if (dz<-hL) dz+=L;
24       r2 = dx*dx + dy*dy + dz*dz;
25       if (r2 < rc2) {
26   r6i  = 1.0/(r2 * r2 * r2);
27   e    += 4*(r6i * r6i - r6i) - (shift?eCut:0.0);
28   *vir += 48 *(r6i * r6i-0.5 * r6i);
29       }
30     }
31   }
32   return e+(tailCorr?(N*eCor):0.0);
33 }
```

Listing 73. Function *Init()* called by *main()* for calculating the phase diagram of an LJ system. (*)

```
1  /* Initialize particle positions by assigning them
2     on a cubic grid, then scale positions
3     to achieve a given box size, and thereby volume
4     and density */
5  void Init ( double * rx, double * ry, double * rz,
6        int n, double L, gsl_rng * r ) {
7    int i, ix, iy, iz;
8
9    int n3 = 2;
10   /* Find the lowest perfect cube, greater than or equal to the
         number of particles */
11   while ((n3 * n3 * n3) < n) n3++;
12
13   ix = iy = iz = 0;
14   /* Assign particle positions */
15   for (i = 0; i < n; i++) {
16     rx[i] = ((double)ix + 0.5) * L/n3;
17     ry[i] = ((double)iy + 0.5) * L/n3;
18     rz[i] = ((double)iz + 0.5) * L/n3;
19     ix++;
20     if (ix == n3) {
21       ix = 0;
22       iy++;
23       if (iy == n3) {
24   iy = 0;
25   iz++;
26       }
27     }
28   }
29  }
```

7.13 Exercise

7.13.1 Dumbbell molecules in 3D

In Section 6.6 an NVT MC study on Dumbbell molecules in 2D was introduced, using more than one kind of trial move. In that case, displacement of randomly selected molecules and rotation of randomly selected molecules. For this exercise, I ask you to perform simulations of dumbbell molecules in 3D in a completely analogous manner. For this, modify the code *DumbellMolecules.c* such that every even-numbered particle is bonded to its odd-numbered neighbor. Each such pair forms a molecule, and trial moves consist of

1. displacements of molecule centers by a distance $\Delta \vec{r}$.

2. rotations of molecules by random angular displacements $(\Delta\theta, \Delta\phi)$.[8]

3. measure the orientational correlation function $P_2(\cos\theta_{ij})$ as a function of density.

---

[8] Note that you have to sample the polar angles uniformly from $\cos\theta$ to maintain a uniform density of direction in 3D space.

Appendix A

The software development life cycle

Developing computer software can be a complicated process, and during many years researchers have identified numerous distinct activities that go into software development. Knowing about these activities can help improve the structure and planning of your own scientific software activities. If you have taught yourself to program or worked mainly on informal projects, you might not have made distinctions among the many activities that go into creating a software product of more than trivial complexity. Mentally, you might have grouped all of these activities together as "programming". If you work on informal projects, the main activity you think of when you think of creating software is probably the activity called "coding and debugging", also commonly referred to as "implementation". As can be seen in Figure A.1, coding and debugging occurs between architectural design and system testing, and is a large part of software development (anything between 30 to 80 percent of the total time spent on the project). Anything that takes up that much project time is bound to affect the success of a software project.

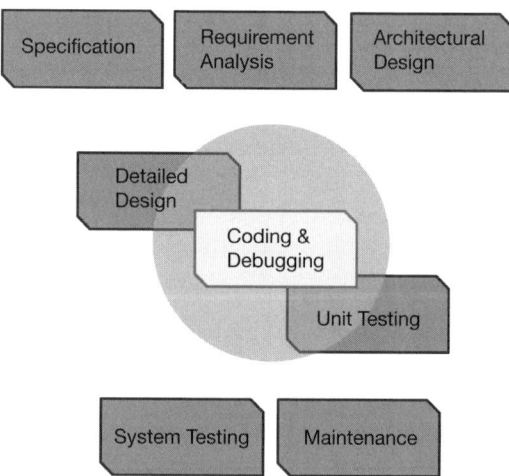

Figure A.1. The software development life cycle denoting the major activities involved in software development. Coding and debugging activities are shown in the circle. These activities also involve some detailed design and some unit testing. They come after architectural designs and before system testing.

Appendix B

Installation guide to Cygwin

Because we want to foster the use of free software, we here provide a very brief description of the procedure to download Cygwin, as well as the steps necessary to install it.

- Download the "setup.exe" installer file from the Cygwin homepage[1]. This program is used for both downloading and setup, as well as for updating the Cygwin environment. Execute this file after download. A wizard appears, which leads you through the rest of the installation process via a series of different screens.
- **Cygwin Net Release Setup Program:** Click the "Next" button.
- **Choose A Download Source:** Select "Install from Internet" (first choice). Click the "Next" button.
- **Select Root Install Directory:** Use the default choice "C:\Cygwin" or pick your own directory. Note, that you should avoid "blanks" within directory install paths, e.g. you should avoid something like
"C:\ Program Files (x86)\" and rather use an install path without blanks. This avoids problems as in UNIX paths, which are used by Cygwin, no blanks are allowed, or rather have to be masked using "\", like in "C:\ Program\ Files\ (x86)\" which makes things rather awkward. Under Cygwin, when typing the command "cd /cygwin", the directory "/cygwin" (please note the slash here vs. backslash in Windows directory structures) allows you to access all Windows directories in a UNIX/Linux-like fashion by e.g. typing "cd /cygwin/c/Windows" which takes you to the directory "C:\Windows\".
- **Select Your Internet Connection:** Should be direct if you are on a university campus but you can also choose a proxy server here.
- **Choose a Download Site:** Select a server close to your location. You can also add servers to this list by entering a new server into the field "User URL" and pressing "Add".
- **Select Local Package Directory:** Use e.g. "C:\temp".
- **Select Packages:** Here is where you get to choose what package to include in the Cygwin installation, see Figure B.1. There are hundreds, if not thousands of packages, so they are grouped into categories. Each category has the word "Default" next to it. That means you will be given the "default" set of packages from each

---

[1] http://www.cygwin.org

category, which corresponds to a minimal installation. You will probably need more than that so you should always select a full installation by clicking directly on the word "Default" next to each category, changing it to the word "Install". The only

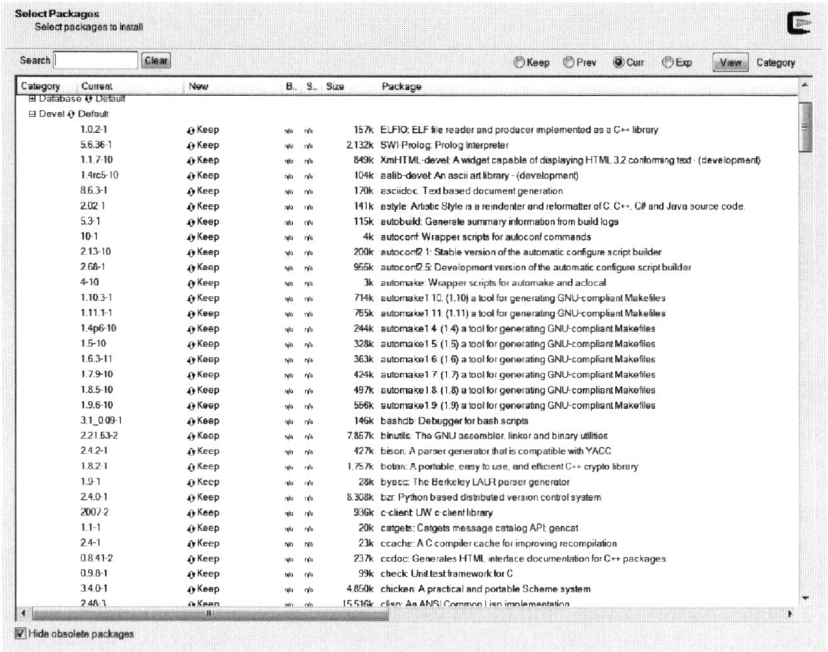

Figure B.1. The package selection on the Cygwin install screen.

crucial category is "Devel", because this contains the compiler gcc. The others will give you some nice functionality. If you don't want to bother with selecting packages, you can simply select everything included in a full installation by clicking on "Default" next to the category "All", right on top of the list. In this case, when you click "Next", the download of packages starts and can take up to an hour depending on the speed of your Internet connection.

- **Finalize:** Here you select whether an icon is put on the desktop and/or the Windows Start menu.

After installing, you can open either a standard Cygwin shell window (Figure B.2), which looks a bit like the DOS command line, or a Cygwin X-terminal (Figure B.3), from which you can start graphical applications such as gnuplot.

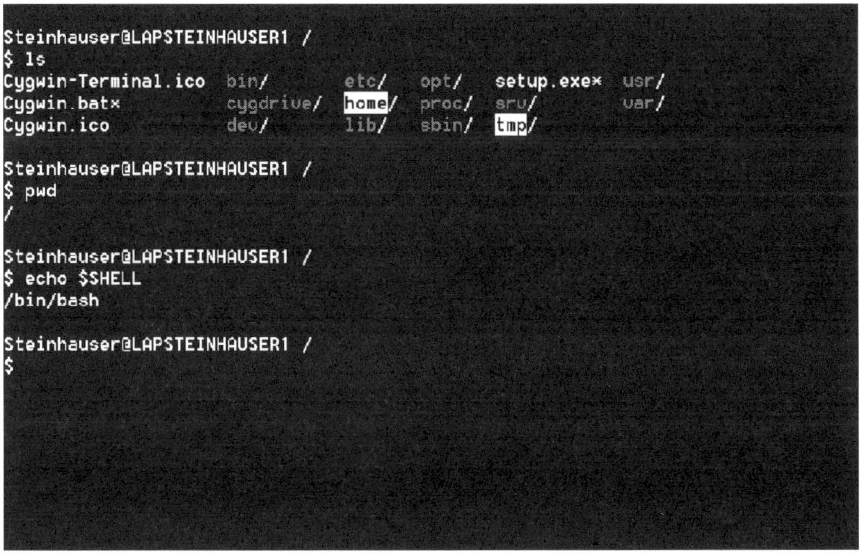

Appendix B Installation guide to Cygwin

Figure B.2. The Cygwin bash terminal.

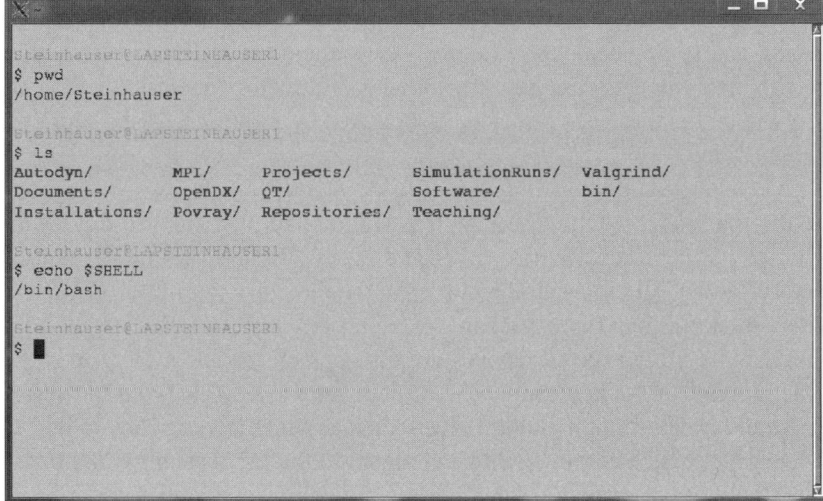

Figure B.3. The Cygwin X-terminal from which graphical applications can be started.

Appendix C

Introduction to the UNIX/Linux programming environment

Most scientific simulations are run under UNIX or UNIX-like systems. UNIX is *the* most common operating system for network and multi-user applications. One example of a UNIX-like system is Linux[1]. The purpose of this tutorial is to provide a short introduction to using the Linux operating system and development environment, so as to enable its very basic use to those who have no prior programming experience and have never used it before. Since Linux is open source software, it can be downloaded for free[2].

C.1 Directory structure

In all Linux systems, the file systems are *mounted* as directories in a tree-structure, see Figure C.1. There are no C, D, ... drives. Therefore, the user does not have to care about where his data are physically stored, he always accesses them in the same way as any other local directory. On Linux systems, the structure of the directory tree follows a standard, and so you can expect (almost) the same on every Linux installation, see Figure C.1. The directory tree starts with the root directory, indicated with a slash /. When going up the tree, each directory is separated by additional slashes, for example "/home/user/bin/".

If one is interested on which devices (disks, memory sticks, RAM, remote computers, etc.) a branch of the directory tree is mounted, one can use the commands df or mount. The command du -H /directory displays the file space usage of "/directory" and of all of its sub-directories in the directory tree. Table C.1 lists some common directories in a Linux system.

Under Linux, all connected devices are also part of the file system, e.g. a hard disk, the monitor or the keyboard. The devices are mounted into the file system (in most systems under "/mnt") and are given a name which corresponds to the path of an ordinary file, e.g. a cdrom could be mounted in the file system as "/mnt/cdrom/". An important command is pwd which prints the current working directory, i.e. the current position in the file system. What this command actually does is printing the value of the system variable PWD which stores the current position of the user in the file system. You can achieve the same result by typing echo $PWD in the shell command line. Linux permits at the most 255 characters for file names and is always case

---

[1] Another example is Mac OS X, the most successful commercial UNIX-like operating system, developed by Apple.
[2] See my remarks on installing a Linux system on a PC in Section 1.3 on Page 13.

Section C.1 Directory structure

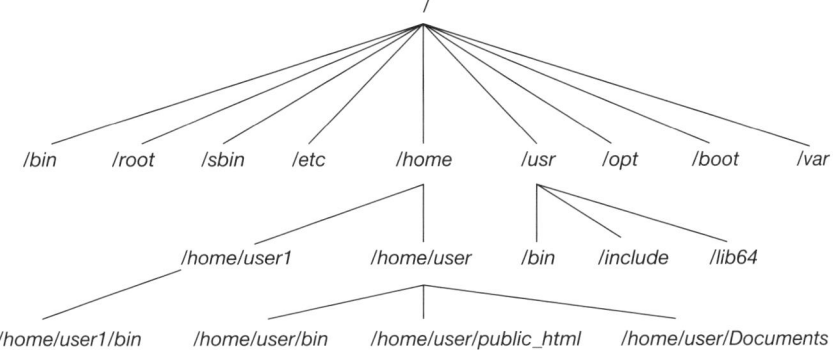

Figure C.1. Detail of a Linux directory tree with some common directories. Note the slash (/) as separator of directory names[3].

sensitive. Special characters such as blank, "$", "\", "&" for file names should be avoided as they are interpreted by the shell and have to be masked with a preceding backslash "\".

Directory paths can be typed as *absolute* or *relative* paths. The absolute path starts with the root directory /, whereas the relative path is given with respect to the current user position in the directory tree. For example, if the user is in his home directory (/home/user), the absolute path to the user's bin directory would be /home/usr/bin, and the relative path is ./bin. The dot-slash (./) represents the current directory where the user is located, whereas dot-dot-slash (../) denotes the directory one leaf higher on the tree (in the direction of /). Hence, for example, the command cd ../ would bring the user to the directory /home. Directories are displayed by using the command ls (list). The syntax of this command is:

ls [options] [path]

If ls is used without any options, it displays the content of the current directory, compare Figure 2.1 on Page 88.

Table C.1. Some common directories in a Linux system.

directory	description
/	system root directory
/sbin	contains programs, only executable by root, the system administrator
/root	the home directory of root
/bin	contains programs and utilities, executable by ordinary users
/lib	contains system libraries
/mnt	usual mount point of other physical drives
/etc	contains many files, e.g. the password file "passwd"
/home/[users]	contains the home directories of all users, except root
/usr	contains documentations, header files, libraries and user programs
/opt	contains add-on packages and software that are not part of the default installation
/boot	contains static files required to boot the Linux kernel
/var	contains files to which the system writes data during the course of its operations

C.2 Users, rights and privileges

Linux, and UNIX-like systems in general, are by construction multi-user/multi-tasking systems. Therefore, each file and directory has a set of permissions which define who is allowed to read, write, and execute them. Each file and each directory has its owner and a group and the permissions for *owner*, *group* and *others* can be set separately. Compare the snapshot of Bash commands in Figure 2.1 on Page 88.

The most privileged user is called "root". The password to login as root is usually known only to one or very few people – the system administrators. The superuser, as root is often called, is allowed to do *any* operation on the system. In contrast, most common users only have read permissions in system directories. Each user is the owner of his /home/[username] directory and has full access permissions to it. By default, if you create a file in your /home/[username] directory, you are given both read and write permissions, while others may read it but are not allowed to change files. As owner, you may change access permissions to all your files. For changing the rights that are associated with a file or directory one uses the command chmod (change mode). Only the owner of a file can change its rights. The syntax of this command is:

chmod [Options] group(s) [+-=] rights filename

An example for the use of chmod is the following command line:

chmod ug+wr file1 file2

Section C.2 Users, rights and privileges

In this example, the owner (user, u) and the group (g) of the files `file1` and `file2` is given write (w) and read (r) access. There are the following `groups`:

- u stands for the user,
- g stands for the group,
- o stands for other users,
- a stands for all (=ugo).

For changing permissions there are the following options:

- + add,
- − remove,
- = exact assignment of rights.

The rights in the `chmod` command are:

- r read access,
- w write access,
- x execute access,

There is also a so called *sticky bit*, which is an access right flag that can be assigned to files and directories on UNIX systems. The sticky bit was introduced in the fifth edition of UNIX in 1974, for use with pure executable files. When set, indicated by the letter t (compare the file /tmp in Figure C.2), it instructs the operating system to retain the text segment of the program in swap space after the process exited. This speeds up subsequent executions by allowing the kernel to make a single operation of moving the program from swap to real memory. The most common use of the sticky bit today is on directories. When the sticky bit is set, only the item's owner, the directory's owner, or the superuser can rename or delete files. Without the sticky bit set, any user with write and execute permissions for the directory can rename or delete contained files, regardless of owner. Typically this is set on the /tmp directory to prevent ordinary users from deleting or moving other users' files.

Files (or directories) the name of which starts with a dot (.) are so called *hidden files*. These are only displayed when the option -a is used with the command `ls`. Often, system and configuration files are hidden files.

Alternatively, rights can also be assigned by using numbers. Read permission (r) is identified by the number 4, write permission is a 2 and execute permission is a 1. The actual rights of a user or group result from the sum of the permissions granted. For example

```
chmod 755 Test
```

```
martin@lxsteinhauser4:~> ls -ls Test
4 -rwxrwxrwx 1 martin users 273 May 15 19:50 Test*
martin@lxsteinhauser4:~> chmod 755 Test
martin@lxsteinhauser4:~> ls -ls
total 72
4 drwxr-xr-t 2 martin users 4096 May 15 21:11 tmp/
4 -rwxr-xr-x 1 martin users  273 May 15 19:50 Test*
0 lrwxrwxrwx 1 martin users   32 May 15 19:47 ThisIsaLink -> /home/martin/Installations/ovito/
0 prw-r--r-- 1 martin users    0 May 15 19:47 pipe|
0 -rw-r--r-- 1 martin users    0 May 15 19:44 ThisIsaFile
4 drwxr-xr-x 3 martin users 4096 Apr 14 16:46 Documents/
4 drwxr-xr-x 2 martin users 4096 Apr  3 14:12 bin/
4 drwxr-xr-x 7 martin users 4096 Apr  3 14:12 Installations/
4 drwxr-xr-x 2 martin users 4096 Apr  3 14:10 Downloads/
4 drwxr-xr-x 6 martin users 4096 Apr  3 09:23 share/
4 drwxr-xr-x 2 martin users 4096 Apr  3 09:15 Desktop/
4 drwxr-xr-x 2 martin users 4096 Apr  1 14:45 mybins/
4 drwxr-xr-x 3 martin users 4096 Apr  1 09:45 lib/
4 drwxr-xr-x 7 martin users 4096 Mar 30 18:22 Repository/
4 drwxr-xr-x 3 martin users 4096 Mar 30 18:20 Projects/
4 drwxr-xr-x 2 martin users 4096 Mar 30 15:05 Music/
4 drwxr-xr-x 2 martin users 4096 Mar 30 15:05 Pictures/
4 drwxr-xr-x 2 martin users 4096 Mar 30 15:05 Public/
4 drwxr-xr-x 2 martin users 4096 Mar 30 15:05 Videos/
4 drwxr-xr-x 2 martin users 4096 Mar 30 15:05 Templates/
4 drwxr-xr-x 2 martin users 4096 Mar 30 13:36 public_html/
martin@lxsteinhauser4:~>
```

attributes no. user group size last change date file name

Figure C.2. Output of the command ls -ls. The number in the first column denotes the number of allocated bytes for this file (in blocks). The first column of the attributes indicate whether the file entry is a "genuine" file (-), a link (l), a directory (d), a named pipe (p), a block-oriented device (b), a character-oriented device (c), or a socket (s). The next group of columns of attributes denotes the access rights of the file. The first three characters denote the user rights, the next three ones denote the rights of the group and the last three denote the rights of other users. If the file is a directory, the number (no.) in the next column denotes the number of sub-directories. If it's a file, the number denotes the number of hardlinks (different names that link to this file). The next column provides the name of the user of the owner of this file (the user), followed by the name of the group to which the file is assigned. This is followed by its size in Kb, the date and time of its last modification, and, finally, the filename.

changes the rights of file Test to rwxr-xr-x (called a *mask*), i.e. the owner of "Test" can read, write and execute the file, whereas the group and others have only read and execute access. The command umask sets a default mask for file access of new files. umask expects the file access rights as a 3-digit number for which different rules apply depending on whether the file is a genuine file or a directory, see Table C.2.

> The command umask expects a 3-digit number as input, which corresponds to the rights that are *not* assigned to the file. For example, "chmod 644 Test" assigns the following access rights to the file Test: -rw-r-r-. To assign a default mask with the same permissions for files using the command umask, one has to subtract the umask value 022 from the value 755, see Table C.2. Hence, for this example, the command umask 022 sets as a default mask the permission -rw-r-r- for all new files.

Table C.2. Different rules for files and directories with the command umask. Files are assigned the default value 666 and directories are assigned 777. The umask value is then subtracted from this default value.

	files	directories
default value	666	777
standard mask	022	022
resulting permissions	644	755

C.3 Some basic commands

A selection of most commonly used Linux commands is listed in Table C.3. Most Linux commands have a number of different possible arguments and switches, which alter their behavior. Nobody can remember all of them. Therefore, in most systems, the man pages, i.e. manuals to all commands, can be accessed by typing

`man commandName`

This gives you the full documentation of the command (in this case, "commandName"), including all its options and peculiarities. To quit a man page, hit q. If you don't know the command related to a given subject, you can try the apropos command, like in

`apropos compiler`

Every command listed in Table C.3 is a small program and is executed in the same way as any other program. The only difference between a system command and other programs is that system commands are found on all Linux systems.

Table C.3. Some common Linux commands.

command	argument type	explanation
cd	directory	change directory
mkdir	directory	create directory
rmdir	directory	delete directory
rm	file	delete file
mv	file1 file2	rename file1 to file2
cp	file	copy file
wc	file	count lines, words and characters in file
ls	file or directory	list directory contents
cat	file	print file contents to standard output
more	file	view file contents interactively
less	file	similar to more
echo	string	print the string
jobs		see the jobs launched from your actual shell
grep	string file	print lines from file which contain string
history	print	commands you have recently executed
cat	file	print file to stdout
echo	string	prints string to stdout
more	file	print file page wise
pstree		show process tree
head	file	print the first lines of file
df		display file system disk space usage
whatis	command	display short description of command
last		display list of last logged in users
who		display list of currently logged in users

C.4 Processes

A *process* in a Linux operating system is a running program, i.e. a program that is being executed. The operating system builds a control block for each process, which is uniquely identified by a process ID (pid – process identification). The shell, which can be seen as the next layer above the operating system, and which allows communication with the user, provides several commands with which the user can interact with the operating system's process management.

C.4.1 Ending processes

Running processes that run in the foreground (which block the shell command line), can be interrupted with the keyboard command [strg+c]. A background process with the process ID *pid* can be stopped with the command kill *pid*. Each user (except root) can only kill processes which he owns. An example of starting a process in the background and subsequently killing it is displayed in Figure C.3.

```
martin@lxsteinhauser4:~> xeyes &
[1] 4928
martin@lxsteinhauser4:~> ps
  PID TTY          TIME CMD
 3575 pts/4    00:00:00 bash
 4928 pts/4    00:00:00 xeyes
 4931 pts/4    00:00:00 ps
martin@lxsteinhauser4:~> kill 4928
[1]+  Terminated              xeyes
martin@lxsteinhauser4:~>
```

Figure C.3. Example of killing a background process. In this example, the program xeyes is started as a background process. The command ps (report process status)shows the currently running user processes, their status and some further information. Here, PID is the process ID, TTY shows from which terminal the process was started, TIME displays the time the process has been running, and CMD is the name of the process.

The command kill can be called with different signals, which indicate to the shell how the process should be finished. The default value is -15. With this signal, it is left to the program to decide how it is ended. With signal -9 a process is ended immediately by the kernel.

C.4.2 Processes priorities and resources

The command top displays the currently running processes on a system and the resources taken up by them, see Figure C.4. The displayed values for CPU, Memory, etc. are averaged over the last 3.0 seconds.

In a multitasking system processes do not really all run concurrently. Rather, each process is interrupted and then resumed several 100 times per second. This is called

```
top - 13:27:15 up  2:39,  8 users,  load average: 0.00, 0.01, 0.05
Tasks: 154 total,   1 running, 153 sleeping,   0 stopped,   0 zombie
Cpu(s):  0.5%us,  0.2%sy,  0.0%ni, 99.3%id,  0.0%wa,  0.0%hi,  0.0%si,  0.0%st
Mem:   3440544k total,  1615976k used,  1824568k free,   232004k buffers
Swap:  2103292k total,        0k used,  2103292k free,   548936k cached

  PID USER      PR  NI  VIRT  RES  SHR S %CPU %MEM    TIME+  COMMAND
 1138 root      20   0  444m 222m 5640 S    1  6.6   0:42.82 Xorg
 3516 martin    20   0  561m  37m  21m S    0  1.1   0:18.82 konsole
 3626 martin    39  19  452m  40m 6688 S    0  1.2   0:03.05 virtuoso-t
 5654 martin    20   0  8864 1228  880 R    0  0.0   0:00.06 top
    1 root      20   0 37260 4316 2004 S    0  0.1   0:01.54 systemd
    2 root      20   0     0    0    0 S    0  0.0   0:00.00 kthreadd
    3 root      20   0     0    0    0 S    0  0.0   0:00.00 ksoftirqd/0
    6 root      RT   0     0    0    0 S    0  0.0   0:00.00 migration/0
    7 root      -2  19     0    0    0 S    0  0.0   0:00.34 rcuc0
    8 root      RT   0     0    0    0 S    0  0.0   0:00.00 rcun0
    9 root      -2   0     0    0    0 S    0  0.0   0:00.00 rcub0
   10 root      RT   0     0    0    0 S    0  0.0   0:00.00 rcun1
   11 root      -2   0     0    0    0 S    0  0.0   0:00.00 rcub1
   12 root      RT   0     0    0    0 S    0  0.0   0:00.07 watchdog/0
   13 root      RT   0     0    0    0 S    0  0.0   0:00.00 migration/1
   15 root      -2  19     0    0    0 S    0  0.0   0:00.28 rcuc1
   16 root      20   0     0    0    0 S    0  0.0   0:00.01 ksoftirqd/1
   17 root      20   0     0    0    0 S    0  0.0   0:00.07 kworker/0:1
   18 root      RT   0     0    0    0 S    0  0.0   0:00.06 watchdog/1
   19 root       0 -20     0    0    0 S    0  0.0   0:00.00 cpuset
   20 root       0 -20     0    0    0 S    0  0.0   0:00.00 khelper
   21 root      20   0     0    0    0 S    0  0.0   0:00.00 kdevtmpfs
   22 root       0 -20     0    0    0 S    0  0.0   0:00.00 netns
   23 root      20   0     0    0    0 S    0  0.0   0:00.05 sync_supers
   24 root      20   0     0    0    0 S    0  0.0   0:00.00 bdi-default
   25 root       0 -20     0    0    0 S    0  0.0   0:00.00 kintegrityd
   26 root       0 -20     0    0    0 S    0  0.0   0:00.00 kblockd
   27 root       0 -20     0    0    0 S    0  0.0   0:00.00 ata_sff
```

Figure C.4. Output of the command top. First displayed are several of the computer's resources. Second, different columns display information about the processes running in the system. The different columns are: PID (process ID), USER (owner of this process), PR (priority of this process), NI(nice value), S (status), %CPU (amount of CPU-time in % which this process uses), %MEM (amount of Memory in % allocated by this process), TIME+ (hitherto running time), COMMAND (name of the process).

time slicing or *round-robin-process* and is managed by the operating system. When looking at the output of the top command, one can notice that the different processes have different priorities. The priority of a process is dynamically calculated by the operating system using different criteria, among them the nice value, which indicates how "nice" a process is to other ones. The default nice value is usually restricted to the interval $[-20, 20]$ and set to 0. The smaller the nice value, the higher the calculated priority of this process, which means that it more often gets a time slice, i.e. that it runs faster. If a user wants to reset the nice values of his commands, he can do so using the command renice, or set the value when starting a program using the command nice. Ordinary users are only allowed to *increase* the nice values of their processes, i.e. to give them less priority, while the super user *root* can change the nice value arbitrarily.

C.5 The Bash

A shell is an interface for executing commands. Besides executing programs for you, it provides a handful of useful features. There are many shells around and the user is free to choose his favorite. One of the most widespread is the Bash[4] (which stands for *Bourne again shell*). One important thing a shell can do for you, is interpreting special characters on a command line, which provides the possibility of extended interaction with the system or between the programs you are executing. Some of these special characters you should be aware of are the pipe (|), the two redirection symbols (> and <), the ampersand (&), the exclamation mark (!), the dot (.), the asterisk (*), the percent (%) and the tilde (~). Here are some examples of their usage:

```
ls / > list
```

List the contents of the root directory "/" and put the output in file "list" (which gets overwritten, if it existed before).

```
cp ./* /tmp
```

Copy all the files (*) from the directory where you are now (./) into the "/tmp" directory.

```
ls /usr/bin/*cc*
```

List all files in "/usr/bin/" containing the string "cc". The asterisk is used here in a *regular expression* as a substitute for any arbitrary sequence of characters and numbers, followed by "cc", followed again by an arbitrary sequence of characters and numbers.

```
~/myprogram < input | grep error
```

Feed program "myprogram", located in your home directory (~) with the contents of the file "input" (instead of typing the input by hand) and search for the string "error" in the output using the command `grep` to which the output is piped (|).

```
./program > output &
```

Launch the program "program" in background (&) and redirect the output to file "output". You can see your jobs running in background with the `jobs` command. With `fg` %1 you take job number 1 back to the foreground. Use the combination ˆz (control+z) to stop it to get back to the shell. Now the program is not running anymore. Let it keep running in background by issuing the command bg %1.

```
./program 2> /dev/null 1> log
```

---
[4] Developed by Stephen Bourne at AT&T Bell Laboratories in the 1970s and released in 1977.

Run "program" and redirect the output messages (stdout) to file "log" and the error messages (stderr) to the special file "/dev/null" (i.e., discard them). *stdout* and *stderr* are two special files (in Linux, everything is a file!) that are used to output to the screen. They can be redirected using the special "1>" and "2>" shell commands. The only difference between *stdout* and *stderr* is that the first one is buffered, whilst the second is not.

`!127`

Repeat the 127th command in your command history. Remember that in principle, the complete path of a program has to be given, e.g. /bin/ls, to run it, but if you don't (and type just `ls`, Bash tries to infer the path from a list of common ones, stored in the $PATH environment variable (type `echo $PATH` to see it, and `export PATH=$PATH:/newpath` to add /newpath to the search list). When you compile a new executable that is not located in one of your $PATH directories, call it like this:

`./myprogram`

or add the ./ directory to your path like this: `export PATH=$PATH:./`.

C.6 Tips and tricks

Here, we list several useful features provided by most modern Linux systems, which make the life of a scientific programmer much easier and help prevent typing mistakes.

- **Tab completion:** When one starts typing a command and hit a tab key after one has typed the first few letters, the system automatically completes the rest, provided there is one unique possibility. If there are more possibilities and one hits the tab twice, it displays all of them.

- **Copy-paste using mouse:** Use a mouse to select a piece of text by the cursor. It is automatically stored in a buffer and when one clicks the middle button, it is pasted in the current cursor position.

- **Command-line history:** Suppose one executed a very complicated and long command, and one wants to execute it again, perhaps with a small change. By hitting the ↑ and ↓ keys, one can browse the history of commands executed recently. If the command has been executed a long time ago, call a command history by using the command `history` to get the whole history of commands on the screen.

C.7 Useful programs

In this section, we describe some programs which you will need for everyday work in computer simulations on a Linux system. You are free to use any program you like, but the ones listed below are known to work well and are widely used in the

Section C.7   Useful programs   459

computational science community. They are all freely available for download and run under a number of operating systems. They are also parts of all major Linux distributions.

C.7.1   Remote connection: ssh

There are several ways to access a computer from another (remote) computer, provided that both are connected to the Internet (and they are set up so that remote access is allowed). A common way of doing it is a program called ssh which stands for *secure shell*. It is used as follows:

```
ssh username@computerAddress
```

After executing this command, the user has to type his password for the machine and, if he succeeds, obtains access to his shell command line as if he was sitting directly at the machine. One feature of ssh is that, for security reasons, all communications between the remote and local computer, which are passed through the network, are encrypted. On many servers, ssh or any other remote access is forbidden or limited by system settings, so as to avoid attempts to break the security.

C.7.2   Gnuplot

Data from simulations and scientific data in general are often presented in the form of two-dimensional plots, displaying relations between different data sets. Therefore, one needs an appropriate plotting program. The one that I recommend and which is standard on any Linux system is gnuplot. It is relatively easy to use and offers a broad functionality particularly suited to scientific applications. In principle, powerful mathematical programs such as MAPLE or MATLAB can also be used for plotting and processing data,but they are primarily designed for symbolic math computations and are not as easy to learn as gnuplot. In principle, one could also use MS Excel or OpenOffice.org Calc to plot the data, but this is not recommended. These programs are *not* primarily designed for plotting scientific data and offer very limited possibilities in this respect (besides being utterly slow).

C.7.3   Text editors: vi, EMACS and others

Since most files one works with when writing computer simulation programs are simple text files, a text editor is a necessary tool to write, view and modify them. The most widespread and powerful text editors under Linux are vi and EMACS. I strongly recommend that you learn to use at least one of them. They are not intuitive to use at first glance, but they offer a multitude of very helpful features, especially when writing programs. Other common editors include for example nano, a simple text editor with context-sensitive help on available commands, and kate, an editor with a graphical frontend. Just type the editor name in a command line and try it out. Note, that for example MS Word or OpenOffice Writer are *not* text editors but *word processors*, which is something completely different!

The vi editor

The vi (visual editor) is a powerful editor just like EMACS and is usually available by default on *all* UNIX/Linux systems (just like the editors ex or ed), and has many helpful utilities, such as indentation or automatic line numbering. There are many versions of vi, for example vim (vi-improved), elvis, xvi, stevie, etc. Most often, vi is used in a console or terminal window of the shell, but there are also graphical interfaces available such as gvim and kvim. To start editing using vi just type

vi <filename>

on the command line. If a file filename exists in the current directory or in the provided path, it is loaded into the buffer and displayed in the terminal window of the shell. If you just type vi without providing a filename, vi starts with an unnamed buffer. Empty lines are displayed with a ∼ at the beginning of line. The last line of the buffer serves as command and status line. After starting vi, here, the cursor position, the number of lines and the number of characters in the file are displayed. But it is also used to display errors or status messages, for example when changing from *command mode* to *insert mode*.

After starting vi it is in *command mode* by default, which means that one cannot type any text for now. In command mode, one can quickly move about within the text file, copy and insert text, open new files or save files, see Table C.4 for the most important commands in this mode. The insert mode or text mode is used to actually type text. Switching between modes is done by pressing the [esc] key. The most important vi commands for text editing are displayed in Table C.5.

The EMACS editor

This section is only meant as a quick reference, in order to get you started with using EMACS. To start editing a new or existing file using EMACS, simply type this at the Linux prompt:

emacs <filename>

where filename is the file to be edited. Once EMACS has started, text entry works much the same way as in any other text editor, i.e. the default mode is INSERT mode where text being typed is inserted at the current cursor position. All the fancy editing commands, such as find-and-replace, are invoked by typing special key sequences (there really isn't a great simple mouse-driven Linux text editor for writing code).

Two important key sequences to remember are: ˆx (holding down the "ctrl" key while typing 'x') and [esc]-x (simply pressing the "esc" key followed by

Table C.4. Important commands available in `command` mode of vi.

command	action
h	one character backwards
l	one character forward
j	one line down
k	one line up
w	move forward in text by one word
b	move backwards in text by one word
e	go to end of this word
0	go to beginning of line
+/-	go to beginning of next/previous line
G	go to the last line in buffer
nG	go to line n in buffer
n	go n lines further from current position
H	move cursor to first line on screen
L	move cursor to last line on screen
M	move cursor to the middle of screen
strg+F	move one screen page forward in file
strg+B	move one screen page backwards in file
/	search forward in file
?	search backwards in file

Table C.5. Important commands available in `insert` mode of vi.

command	action
c	change
C	substitute until end of line
r	substitute character below cursor
x	delete character below cursor
	exchange upper and lower case
$n$dd	delete line, or n lines, respectively
$n$yy	copy line, or n lines, respectively
D	delete from current position until end of line
p	insert in next line
P	insert at current position
u	undo last change
U	undo all changes in current line
.	repeat last issued command
strg+G	display filename

typing 'x'), both of which are used to start command sequences. Note that, in most user manuals for EMACS, the "esc" key is actually referred to as the "Meta" key. Therefore, you will often see commands prefaced with the key sequence M-x, as opposed to the [esc]-x that we will use here. Since the two are pretty much synonymous, we'll stick with explicitly using the "esc" key. Now let's see some examples of these "command sequences".

For instance, to save the file being edited the sequence is ^x^s. To exit (and to be prompted to save) EMACS, the sequence is ^x^c. To open another file within EMACS, the sequence is ^x^f ('f' for "find file"). This sequence can be used to open an existing file as well as a new file. If you have multiple files open, EMACS stores them in different "buffers." To switch from one buffer to another[5], you use the key sequence ^x followed by typing 'b' (without holding down "ctrl" when typing 'b'). You can then enter the name of the file to switch to the corresponding buffer (a default name is provided for fast switching). The arrow keys usually work as the cursor movement keys, but in the event that they don't, ^f is for moving forward (right), ^b is for moving backward (left), ^p is for moving to the previous line (up), and ^n is for moving to the next line (down). [esc]-< ([esc] followed by '<') moves to the beginning of the file, and [esc]-> moves to the end of the file. Finally, ^v does a "page down" and [esc]-v does a "page up." Note that the [esc] key is simply pressed and not held, whereas the [ctrl] key is held in all cases. Quite frequently, you'll make mistakes in typing. The backspace key usually deletes the character before the text cursor[6]. In case this fails, try the delete key (the one on the mini-keypad that also contains page up, page down, etc). You can also delete a whole line at a time by using ^k. Another important editing feature is copy/cut and paste. To copy or cut, you first have to select a region of text. To tell EMACS where this region begins, use [esc]-@ or ^[space] (control and space simultaneously). Then move the text cursor to the end of the region. Alternatively, if you're running EMACS under X-Windows, you can use your mouse by pressing down on the left button at the start of the region and "drag" your mouse cursor to the end. To copy the region, use [esc]-w, while to cut the region, use ^w. To paste, use ^y. EMACS does command completion for you. Typing M-x <spc> will give you a list of EMACS commands. There is also a man page on EMACS. Type man EMACS in a shell. A variant of EMACS is XEMACS, which is essentially the same but with a more modern-looking user interface (buttons, menus, dialogs for opening files, etc). Nearly all commands that EMACS accepts are valid in XEMACS. Table C.6 displays a list of the most important EMACS commands.

---

[5] This is very handy when you are editing a .c source file and need to refer to the prototypes and definitions in the .h header file.

[6] Sometimes, the backspace key is strangely bound to invoke help.

Table C.6. List of useful EMACS commands. In this table, "^z" means hit the "z" key while holding down the "ctrl" key. "M-z" means hit the "z" key while hitting the "META" or after hitting the "ESC" key.

command	explanation
	Running EMACS
^z	Suspend EMACS.
^x ^c	Quit EMACS.
^x ^f	Load a new file into EMACS.
^x ^v	Load a new file into EMACS and unload previous file.
^x ^s	Save the file.
^x ^k	Kill a buffer.
Moving about[7]	
^f	Move forward one character.
^b	Move backward one character.
^n	Move to next line.
^p	Move to previous line.
^a	Move to beginning of line.
^e	Move to end of line.
^v	Scroll down a page.
M-v	Scroll up a page.
M-<	Move to beginning of document.
^x-[	Move to beginning of page.
M->	Move to end of document.
^x-]	Move to end of page.
^l	Redraw screen centered at line under the cursor.
^x-o	Move to other screen.
^x-b	Switch to another buffer.
^s	Search for a string.
^r	Search for a string backwards from the cursor.
M-%	Search-and-replace
Deletion	
^d	Deletes letter under the cursor.
^k	Kill from the cursor all the way to the right.
^y	Yanks back all the last kills[8].

Continued on next page ...

---

[7] Note: the standard arrow keys also usually work
[8] Using the ^k ^y combination you can get a cut-paste effect to move text around.

Table C.6. List of useful EMACS commands – continued from previous page.

command	explanation
Screen splitting	
ˆx-2	Split screen horizontally.
ˆx-3	Split screen vertically.
ˆx-1	Make active window the only screen.
ˆx-0	Make other window the only screen.
Compiling	
M-x compile	Compile code in active window using "make".
ˆcˆc	Scroll to the next compiler error.
Miscellaneous	
ˆg	Cancel and go back to normal command.
ˆx-u	Undo.
M-x shell	Start a shell within EMACS.
Getting help	
ˆh	EMACS help.
ˆh t	Run the EMACS tutorial.

The GNU debugger (gdb)

To start the debugger from the shell just type:

gdb <TargetName>

where <TargetName> is the name of the executable that you want to debug. If you do not specify a target then gdb will start without a target and you will later need to specify one before you can do anything useful. As an alternative, from within XEMACS you can use the command M-x gdb which will then prompt you for the name of the target file. You cannot start an inferior gdb session from within XEMACS without specifying a target. The XEMACS window will then split between the gdb "window" and a buffer containing the current source line. Once started, the debugger will load your application and its symbol table (which contains useful information about variable names, source code files, etc.). This symbol table is the map that the debugger reads as it is running your program. The debugger is an interactive program. Once started, it will prompt you for commands. The most common commands in the debugger are: *setting breakpoints*, *single stepping*, *continuing after a breakpoint*, and *examining the values of variables*.

> If you forget to specify the "-g" flag (debugging info) when compiling your source files, the symbol table will be missing from your program and gdb (and you) will be "in the dark" as your program runs, i.e. debugging then is not possible, so you have to compile your sources with this option.

Running the program When a target application is first selected (usually on startup), the current source file is set to the file with the main function in it, and the current source line is the first executable line of this function. As you run your program, it will always be executing some line of code in some source file. When you pause the program (using a "breakpoint" by hitting Control-C to interrupt), the "current target file" is the source code file in which the program was executing when you paused it. Likewise, the "current source line" is the line of code in which the program was executing when you paused it. The following is a list of important commands when running gdb.

- run resets the program, i.e. it runs (or reruns) from the beginning. You can supply command-line arguments to run the same way you can supply command-line arguments to your executable from the shell.

- step runs next line of source and returns to the debugger. If a subroutine call is encountered, step into that subroutine.

- step *count* runs *count* lines of source.

- next is similar to step, but doesn't step into subroutines.

- finish runs until the current function/method returns.

- return makes the selected stack frame return to its caller.

- jump *address* continues the program at the specified line or *address*.

Breakpoints You can use breakpoints to pause your program at a certain point. On creation, each breakpoint is assigned an identifying number so that you can later refer to that breakpoint, should you need to manipulate it. A breakpoint is set by using the command break, specifying the location of the code where you want the program to be stopped. This location can be specified in a variety of different ways, such as with the filename and either a line number or a function name within that file[9]. If the filename argument is not specified, the file is assumed to be the current target file,

---

[9] It is a good idea to specify lines that are really code, since comments and whitespace will not do the right thing.

and if no arguments are passed to break, the current line of source code will be the breakpoint. gdb provides the following commands to manipulate breakpoints:

- `info break` prints a list of all breakpoints with numbers and status.

- `break` *function* places a breakpoint at the start of the specified *function*.

- `break` *line number* prints a breakpoint at *line number*, relative to current source file.

- `break` *filename:line number* places a breakpoint at the specified *line* within the specified source file. You can also specify an if clause to create a conditional breakpoint:

- `break` *fn* `if` *expression* stops at the breakpoint only if *expression* evaluates to true. *Expression* is any valid C expression, evaluated within the current stack frame when hitting the breakpoint.

- `disable` *breaknum* / `enable` *breaknum* disables/enables breakpoint identified by *breaknum*.

- `delete` *breaknum* deletes the breakpoint identified by *breaknum*.

- `commands` *breaknum* specifies commands to be executed when *breaknum* is reached. The commands can be any list of C statements or gdb commands. This can be useful to fix code on the fly in the debugger without recompiling.

- `cont` continues a program that has been stopped.

For example, the commands

```
break program.c:120
break DoGoofyStuff
```

sets a breakpoint on line 120 of the file `program.c` and another on the first line of the function `DoGoofyStuff`. When control reaches these locations, the program will stop and give you the chance to look around in the debugger. If you're running gdb from XEMACS, it will load the source file and put an arrow (->) at the beginning of the line that is to be executed next. Also, from inside XEMACS, you can set a breakpoint simply by going to the line of the file where you want to set the breakpoint, and hit ^x-[space]. gdb (and most other debuggers) provides mechanisms to determine the current state of the program and how it got there. The things that we are usually interested in are "where are we in the program?" and "what are the values of the variables around us?".

Examining the stack To answer the question of "where are we in the program?", we use the `backtrace` command to examine the runtime stack[10]. The runtime stack

---

[10] The concepts of memory stack and heap are explained in Section 2.4.6.

Section C.7 Useful programs

is like a "trail of breadcrumbs" in a program. Each time a function call is made, a "crumb is dropped" (a stack frame is pushed). When a return from a function occurs, the corresponding runtime stack frame is popped and discarded. These stack frames contain valuable information about where in the source code the function was called (line # and filename), what the parameters for the call were, etc. gdb assigns numbers to stack frames counting from zero for the innermost (currently executing) frame. At any time gdb identifies one frame as the "selected" frame. Variable lookups are done with respect to the selected frame. When the program being debugged stops (at a breakpoint), gdb selects the innermost frame. The commands below can be used to select other frames by number or address.

- `backtrace` shows stack frames, useful to find the calling sequence that produced a crash

- `frame` *frame number* starts examining the frame with *frame number*. This does not change the execution context, but allows to examine variables for a different frame.

- `down` selects and prints the stack frame called by this sequence

- `up` selects and prints the stack frame called by this sequence

- `info` *args* shows the argument variables of the current stack frame

- `info` *locals* shows the local variables of the current stack frame

Examining source files Another way to find our current location in the program and other useful information is to examine the relevant source files. gdb provides the following commands:

- `list` *linenum* prints ten lines centered around *linenum* in current source file.

- `list` *function* prints ten lines centered around beginning of function (or method).

- `list` prints ten more lines.

The `list` command will show the source lines with the current source line centered in the range. (Using gdb from within XEMACS makes these commands obsolete since it automatically loads the sources into XEMACS for you).

Examining data It is also useful to answer the question, "what are the values of the variables around us?" In order to do so, we use the following commands to examine variables:

- `print` *expression* prints value of *expression*. Expression is any valid C expression and can include function calls and arithmetic expressions, all evaluated within current stack frame.

- `set` *variable* = *expression* assigns the value of *variable* to *expression*. You can set any variable that is within the current scope. Variables which begin with $ can be used as convenience variables in gdb.

- `display` *expression* prints the value of *expression* each time the program stops. This can be useful in order to watch the change in a variable as you step through the code.

- `undisplay` cancels previous display requests.

In gdb, there are two different ways of displaying the value of a variable: a snapshot of the variable's current value and a persistent display for the entire life of the variable. The print command will print the current value of a variable and the display command will make the debugger print the variable's value on every step, for as long as the variable is "alive". The desired variable is specified by using C syntax, for example,

`print x.y[3]`

will print the value of the fourth element of the array field named y of a structure variable named x. The variables that are accessible are those of the currently selected function's activation frame, plus all those whose scope is global or static to the current target file. Both the `print` and `display` functions can be used to evaluate arbitrarily complicated expressions, even those containing function calls, but be warned that if a function has side-effects, a variety of unpleasant and unexpected situations can arise.

Debugging strategy and shortcuts If your program has been crashing spectacularly, you can just run the program by using the `run` command right after you start the debugger. The debugger will catch the signal and allow you to examine the program (and hopefully find the cause and remedy it). More often, the bug will be something more subtle. In these cases the "best" strategy is often to try to isolate the source of the bug, using breakpoints and checking the values of the program's variables before setting the program in motion by using `run`, `step`, or `continue`. A common technique for isolating bugs is to set a breakpoint at some point before the offending code and to slowly continue toward the crash site, examining the state of the program along the way.

Finally, there are some things that make using gdb a bit simpler. All of the commands have shortcuts so that you don't have to type the whole command name every time you want to do something simple. A command shortcut is specified by typing just enough of the command name so that it unambiguously refers to a command, or for the special commands `break`, `delete`, `run`, `continue`, `step`, `next` and `print` you need only use the first letter. Additionally, the last command you entered can be repeated by just hitting the <return key> again. This is really useful for single stepping through a range whilst watching variables change. If you're not running gdb from XEMACS, the up and down arrow keys will jump to the previous or next commands that you've issued, the left and right arrows will allow you to move through the command line for editing.

Section C.7 Useful programs   469

The Data Display Debugger (ddd)

The ddd debugger is based on the GNU gdb debugger and its syntax, but has a fancy GUI attached to it that makes it more comfortable to use. The snapshot in Figure C.5 shows a typical window of ddd with the following window elements.

- **Data window** Displays the current content of the local (and global) variables in the program stack.

- **Source window** Displays the actual source code and which line of code is executed. Here, you can define halting points (called "break points") and have single variable contents displayed by pointing on top of it with the mouse.

- **Machine code window** Displays the assembly code corresponding to the high level language code. You can set breakpoints here too, and have the content of individual memory registers displayed.

- **GDB console** Here you type in commands just like you do when using gdb without GUI.

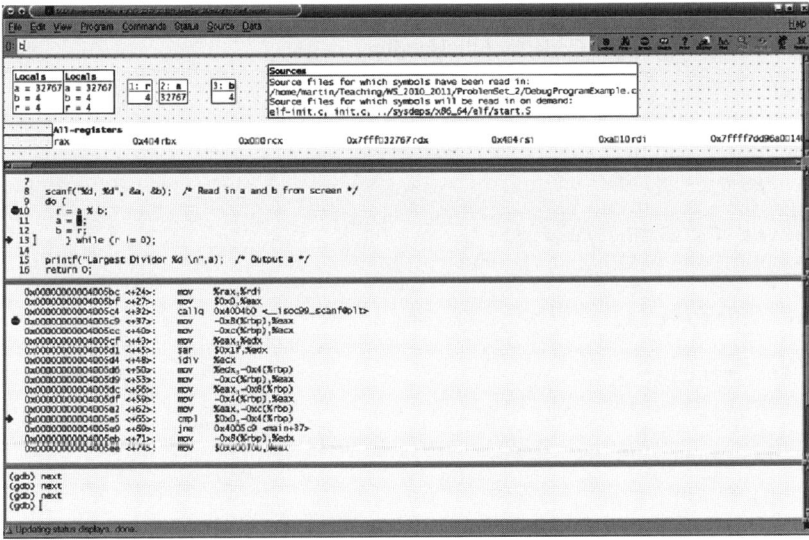

Figure C.5. The ddd debugger GUI.

Appendix D

Sample program listings

D.1 Sample code for file handling

This code refers to the initial discussion of I/O functionality in C, in Chapter 2 on Page 84, and demonstrates the use of the *fopen()* function to open and *fclose()* to close a binary or ASCII text file. The program uses the *puts()* (put string) and *gets()* (get string) functions, which are declared in <*string.h*>. The program asks you whether you want to read or write a file named "test.txt" First chose option (2) and write several lines of text as input. When you type a single period as the only input of a line, this is recognized as the last line. Then, the program writes the input lines into a file "test.txt', which you can now read, using an ordinary text editor or by restarting the program and choosing option (1).

Listing 74. Part 1 – Sample Code for File Handling.

```
1
2  /* Sample code for file handling */
3  #include <stdio.h>
4  #include <string.h>
5
6  #define DATNAME "test.txt"
7  #define MAX_LINE_LENGTH 160
8
9  void write(void)
10 {
11     FILE *fhd;
12     char s[MAX_LINE_LENGTH];
13
14     fhd=fopen(DATNAME,"w");
15     if(!fhd)
16        {
17           printf("File could not be generated!\n\n");
18        }
19     else
20        {
21           printf("Please write several lines and input not more than \
22  %i characters per line.\n",MAX_LINE_LENGTH);
```

Listing continued on next page ...

Section D.1 Sample code for file handling

Listing 74. Part 2 – Sample Code for File Handling continued ....

```
23         printf("End your input by typing only a \".\" on a line and \
24     pressing <return >\n\n");
25         do
26         {
27             printf(">");
28             gets(s);
29             if(strcmp(s,"."))
30             {
31                 fputs(s,fhd);
32                 fputs("\n",fhd);
33             }
34         } while(strcmp(s,"."));
35
36         fclose(fhd);
37         printf("\nEnd of input!\n");
38     }
39 }
40
41 void read(void)
42 {
43     FILE *fhd;
44     char s[MAX_LINE_LENGTH];
45     int x=1;
46
47     fhd=fopen(DATNAME,"r");
48     if(!fhd)
49     {
50         printf("File could not be opened!\n\n");
51     }
52     else
53     {
54         printf("The file has the following content\n");
55         fgets(s,MAX_LINE_LENGTH,fhd);
56         do
57         {
58             printf("%i:%s",x++,s);
59             fgets(s,MAX_LINE_LENGTH,fhd);
60         } while(!feof(fhd));
61
62         fclose(fhd);
63         printf("\nEnd of file!\n");
64     }
65 }
```

Listing continued on next page ...

Listing 74. Part 3 – Sample Code for File Handling continued . . . .

```
66  void main(void)
67  {
68     int input;
69
70     printf("Do you want to (1) read or (2) write a file ?\n");
71     scanf("%i",&input);
72     if(input==1)
73        {
74           read();
75        }
76     else
77        {
78           if(input==2)
79              {
80                 write();
81              }
82           else
83              {
84                 printf("\nWrong input !\n\n");
85              }
86        }
87  }
```

Appendix E

Reserved keywords in C

The following Table E.1 lists the reserved keywords in C.

Table E.1. Reserved keywords in C. Keywords printed in bold face were only added with the *ANSI-C99* standard.

auto	break	case	char	const
continue	default	do	double	else
enum	extern	float	for	goto
if	**inline**	int	long	register
restrict	return	short	signed	sizeof
static	struct	switch	typedef	union
unsigned	void	volatile	while	**_Bool**
_Complex	**_Imaginary**			

Appendix F

Functions of the standard library *<string.h>*

In Table F.1, we provide an overview of the most important functions of the header file *<string.h>*.

Table F.1. The most important functions of *<string.h>*.

function name	parameter list
char *stpcpy[a]	(const char *dest,const char *src)
int strcmp[b]	(const char *string1,const char *string2)
char *strcpy[c]	(const char *string1,const char *string2)
char *strerror[d]	(int errnum)
int strlen[e]	(const char *string)
char *strncat[f]	(const char *string1, char *string2, size_t n)
int strncmp[g]	(const char *string1, char *string2, size_t n)
char *strncpy[h]	(const char *string1,const char *string2, size_t n)
int strcasecmp[i]	(const char *s1, const char *s2)
int strncasecmp[j]	(const char *s1, const char *s2, int n)
void *memchr[k]	(void *s, int c, size_t n)
int memcmp[l]	(void *s1, void *s2, size_t n)
void *memcpy[m]	(void *dest, void *src, size_t n)
void *memmove[n]	(void *dest, void *src, size_t n)
void *memset[o]	(void *s, int c, size_t n)
char *strchr[p]	(const char *string, int c)

[a] copy one string into another
[b] compare string1 and string2, to determine alphabetic order
[c] copy string2 to string1
[d] get error message corresponding to specified error number
[e] determine the length of a string
[f] append n characters from string2 to string1
[g] compare first n characters of two strings
[h] copy first n characters of string2 to string1
[i] case insensitive version of strcmp()
[j] case insensitive version of strncmp()
[k] search for a character in a buffer
[l] compare two buffers
[m] copy one buffer into another
[n] move a number of bytes from one buffer to another
[o] set all bytes of a buffer to a given character
[p] find first occurrence of character c in string

Appendix G

Elementary combinatorial problems

For counting elementary states of physical systems, the results of the following basic counting problems are very useful. Consider N distinguishable objects. In order to have something concrete in mind, suppose you have N numbered beads.

G.1 How many differently ordered sequences of N objects are possible?

The answer is $N!$.

Proof. Seeing this is not hard and one should test the result with a few easy examples. Generally speaking, one has to convince oneself that in the beginning (when the very first choice is made) there are N different choices of objects. So, having picked the first object leaves $(N-1)$ objects as second choice, and so forth until, for the Nth choice, there is only one object left. Hence, the total number of choices is $N \times (N-1) \times (N-2) \times \cdots 1 = N!$. □

G.2 In how many ways can N objects be divided into two piles, with n and m objects, respectively?

We assume that the ordering within the piles is unimportant. The answer obviously is zero, unless $n + m = N$. If the numbers do add up, then the answer is $N!/(n! \times m!)$. The zero answer may seem trivial, but it is analogous to the requirement of getting right the total number of particles in a statistical distribution. There are several ways of seeing the correct answer, one of which is as follows.

Proof. Call the required number t. We can recognize that the $N!$ arrangements (see Section G.1) can be partitioned such that the first n placed objects are put into the first pile and the remaining m placed objects into the second pile. However, these $N!$ arrangements will not all give distinct answers to the present question since the order within each pile is unimportant. Hence, we can see that

$$N! = t \times n! \times m!, \tag{G.1}$$

where the $n!$ and $m!$ factors allow for this disordering (again using the result of (G.1)). □

We note, that this problem is identical to the binomial theorem problem of determining the coefficient of $x^n y^n$ in the expansion of $(x + y)^N$. The answer here is the same: zero unless $n + m = N$, but if $n + m = N$, the required number is $N!/(n! \times m!)$.

G.3 In how many ways can N objects be arranged in $r + 1$ piles with n_j objects in pile number j with $j \in [0, 1, \ldots, r]$?

The answer is zero, unless $N = \sum_{j=1}^{N} n_j$, i.e. if the n_js do sum up to N objects then the required number of ways is $N!/(n_0! n_1! \cdots n_r!)$. This result follows as a straightforward extension of the proof of problem (refappendix:OrderedPiles) from equation (G.1), setting $N! = t \times \prod n_j!$.

We note that this problem may again be thought of as the question of the multinomial theorem, with the identical answer, i.e., what is the coefficient of $y_0{}^{n_0} y_1{}^{n_1} y_2{}^{n_2}$ in the expansion of $(y_0 + y_1 + y_2 + \cdots)^N$?

G.4 Stirling's approximation of large numbers

Stirling's approximation gives a useful method for dealing with factorials of large numbers. The form in which it should be known and used in statistical physics is:

$$\ln x! = x \ln x - x \tag{G.2}$$

for any large integer x. Another useful result may be obtained by differentiating equation (G.2) to give:

$$d(\ln x!)/dy = (dx/dy)\{\ln x + 1 - 1\} = (dx/dy) \ln x, \tag{G.3}$$

where one assumes the x may be considered a continuous variable.

Equation (G.2) is remarkably accurate for the sort of numbers that one uses in statistical physics. Even for a modest $x = 1\,000$, the error in using Stirling's approximation is less than 0.1%, and the fractional error roughly scales as $1/x$.

Appendix H
Some useful constants

Table H.1. Some useful constants. Source: The NIST Reference on constants, units, and uncertainty [3].

constant	symbol	value	units
atomic mass unit	u	$1.66605402 \times 10^{-27}$	kg
Avogadro's number	N_A, L	6.021367×10^{23}	mol^{-1}
Bohr magneton / (magnetic constant)	μ_0	$4\pi \times 10^{-7}$	NA^{-2}
Bohr radius	a_0	$0.52917721092 \times 10^{-10}$	m
Boltzmann constant	k	$1.3806488 \times 10^{-23}$	JK^{-1}
Compton wavelength	λ_C	$2.4263102389 \times 10^{-12}$	m
dielectric constant	ϵ_0	$8.854187817 \times 10^{-12}$	Fm^{-1}
electron mass	m_e	$9.10938291 \times 10^{-31}$	kg
elementary charge	e	$1.602176565 \times 10^{-19}$	C
molecular gas constant	R	8.3144621	Jmol^{-1}K^{-1}
Newton's constant of gravitation	G	6.67384×10^{-11}	m^{3}kg^{-1}s^{-2}
Planck's constant	h	6.63×10^{-34}	Js^{-1}
Planck length	l_p	1.616199×10^{-35}	m
Planck mass	m_p	2.17651×10^{-8}	kg
Planck time	t_p	5.39106×10^{-44}	s
proton mass	m_p	$1.672621777 \times 10^{-27}$	kg
speed of light in vacuum	c	299 792 458	ms^{-1}
standard atmosphere	atm	101 325	Pa
von Klitzing constant	R_K	25 812.8074434	Ω

Appendix I
Installing the GNU Scientific Library, GSL

In this section, we provide concise instructions for downloading and installing the GNU Scientific Library GSL on a UNIX/Linux system. This library is very useful for many scientific calculations. It provides predefined functions for many purposes, among them several high-quality random number generators. Some of the programs presented in this book rely on this library to be installed on the system. If you want to learn more about GSL, visit `http://www.gnu.org/software/gsl`.

1. Use one of the "Downloading GSL" links from the website `http://www.gnu.org/software/gsl` to download the GSL source code package `gsl-<Version>.tar.gz`. You have to substitute "<Version>" by the actual version number of the file you are downloading. Let's say the latest version number is 1.15. You now download the file "gsl-1.15.tar.gz" and store it, for example, in your home directory.

2. Go to your home directory with the command "`cd $HOME`" and type "`tar zxvf gsl-1.15.tar.gz`". This extracts a large number of files into the directory called "gsl-1.15".

3. Change into this new directory by typing "`cd gsl-1.15`".

4. If you like, you can follow the instructions in the file "INSTALL". You can browse this file with the command "`more INSTALL`".

5. If you don't understand the file INSTALL, just do the following: type "`./configure; make; make install`".

GSL has extensive documentation available, both online and as part of the distribution. You can display a text-only hyperlinked manual by typing "`info gsl-ref`".

Appendix J

Standard header files of the *ANSI-C* library

The following Table J.1 lists the standard header files in C.

Table J.1. Standard header files in C. Header files printed in bold face were only added with the *ANSI-C99* standard.

assert.h	limits.h	**stdint.h**
complex.h	locale.h	stdio.h
ctype.h	math.h	stdlib.h
errno.h	setjmp.h	string.h
fenv.h	signal.h	tgmath.h
float.h	stdarg.h	time.h
inttypes.h	**stdbool.h**	wchar.h
iso646.h	stddef.h	wctype.h

Appendix K

The central limit theorem

This theorem states that the sum of N random independent variables x_i with ($i = 1, \ldots, N$) is a *Gaussian* random variable, when N goes to infinity, if the moments $\langle x_i{}^n \rangle$ exist. A Gaussian random variable ξ is a variable that follows a probability law of the exponential type, as displayed in equation (K.7). The variance σ^2 is defined as $\sigma^2 = \int \xi^2 p(\xi)\,d\xi$.

Let $X = \sum_{i=1}^{N} x_i$. In this case, the mean-square value of X is $\langle X^2 \rangle = \langle \sum_{i,j} x_i x_j \rangle$. Because the variables are independent, one has $\langle x_i x_j \rangle_{i \neq j} = 0$. We now assume that all mean square values $\langle x_i^2 \rangle = \sigma^2$. Defining $\xi = N^{-1/2} X$, one obtains

$$\langle \xi^2 \rangle = \sigma^2, \tag{K.1}$$

which is independent of N.

The Fourier transform of the distribution function $p(\xi)$ of a random variable ξ is called a *characteristic function* $\varphi_\xi(k)$, i.e.

$$\varphi_\xi(k) = \langle \exp(ik\xi) \rangle = \int \exp(ik\xi) p(\xi)\,d\xi. \tag{K.2}$$

The inverse transformation is

$$p(\xi) = \frac{1}{2\pi} \int \exp(-ik\xi) \varphi_\xi(k)\,dk. \tag{K.3}$$

With the above expression of ξ, one obtains

$$\varphi_\xi(k) = \left\langle \left(ikN^{-1/2} \sum x_i \right) \right\rangle = \left[\varphi\left(kN^{-1/2} \right) \right]^N, \tag{K.4}$$

as all x_i have the same probability distribution. The characteristic function φ of equation (K.4) can be expanded in the form

$$\varphi(kN^{-1/2}) = 1 - \frac{k^2}{2N} \langle x_i^2 \rangle + \cdots = 1 - \frac{k^2}{2N} \sigma^2 + O(k^3 N^{-3/2}). \tag{K.5}$$

Taking the limit ($N \to \infty$) finally yields

$$\varphi_\xi(k) = \lim_{N \to \infty} \left[1 - \frac{k^2}{2N} \sigma^2 \right]^N = \exp\left(-\frac{k^2 \sigma^2}{2} \right). \tag{K.6}$$

Finally, the Fourier transform of equation (K.6) is

$$p(\xi) = \frac{1}{(2\pi\sigma^2)}^{1/2} \exp\left(-\frac{\xi^2}{2\sigma^2} \right), \tag{K.7}$$

which is Gaussian.

Bibliography

[1] http://cgns.sourceforge.net/. The CGNS documentation homepage. Last access date: 10 March 2012.

[2] http://dlang.org/. The D language website. Last access date: 16 March 2012.

[3] http://physics.nist.gov/cuu/index.html. The NIST website. Last access data: 16 June 2012.

[4] http://plasmagate.weizmann.ac.il/Grace/. The Xmgrace website. Last access date: 16 March 2012.

[5] https://wci.llnl.gov/codes/visit/. The VisIt website. Last access date: 12 March 2012.

[6] http://www.gnuplot.info/. The Gnuplot webpage. Last access date: 16 March 2012.

[7] http://www.hdfgroup.org/. The HDF5 homepage. Last access date: 11 March 2012.

[8] www.ivec.org/gulp/. The GULP website. Last access date: 12 June 2012.

[9] http://www.ks.uiuc.edu/Research/vmd/. The VMD homepage. Last acces date: 12 March 2012.

[10] http://www.ovito.org/. The Ovito website. Last access date: 16 March 2012.

[11] D. J. W. Aastuen, N. A. Clark, and L. K. Cotter. Nucleation and growth of colloidal crystals. *Phys. Rev. Lett.*, 57:1733–1736, 1986.

[12] Farid F. Abraham and Huajian Gao. How fast can cracks propagate? *Phys. Rev. Lett*, 84:3113, 2000.

[13] Farid F. Abraham, Robert Walkup, Huajian Gao, Mark Duchaineau, Thomas Dias De La Rubia, and Mark Seager. Simulating Materials Failure by Using up to One Billion Atoms and the World's Fastest Computer: Brittle Fracture. *Proc. Natl. Acad. Sci.*, 99(9):5777–5782, April 2002.

[14] M. Abramovitz and I. A. Segun. *Handbook of Mathematical Functions*. Dover Publications, New York, NY, 1964.

[15] D. J. Adams, F. M. Adams, and G. J. Hills. The Computer Simulation of Polar Liquids. *Mol. Phys.*, 38:387–400, 1979.

[16] James L. Adams. *Conceptual Blockbusting: A Guide to Better Ideas*. Norton, New York, NY, USA, 1980.

[17] Alfred V. Aho, John E. Hopcroft, and Jeffrey D. Ullman. *The Design and Analysis of Computer Algorithms*. Adison-Wesley, 1974.

[18] Alfred V. Aho, John E. Hopcroft, and Jeffrey D. Ullman. *Data Structures and Algorithms*. Addison-Wesley, 1983.

[19] W. L. Alba and K. B. Whaley. Monte Carlo studies of grain boundary segregation and ordering. *J. Chem. Phys.*, 97:3674–3688, 1992.

[20] G. Alber, T. Beth, M. Horodecki, P. Horodecki, M. Rötteler, H. Weinfurter, R. F. Werner, and A. Zeilinger. *Quantum Information – an introduction to basic theoretical concepts and experiments*, volume 173 of *Springer Tracts in Modern Physics*. Springer, 2001.

[21] B. J. Alder and T. E. Wainwright. Phase Transition for a Hard Sphere System. *J. Chem. Phys.*, 27:1208–1029, 1957.

[22] B. J. Alder and T. E. Wainwright. Phase Transition in Elastic Disks. *Phys. Rev.*, 127:359–361, 1962.

[23] Andrei Alexandrescu. *The D Programming Language*. Addison-Wesley Professional, 2010.

[24] M. P. Allen and D. J. Tildesly. *Computer Simulation of Liquids*. Oxford University Press, Oxford, London, New York, 1991.

[25] Nicholas Allsoppa, Giancarlo Ruoccoc, and Andrea Fratalocchia. Molecular dynamics beyonds the limits: Massive scaling on 72 racks of a BlueGene/P and supercooled glass dynamics of a 1 billion particles system. *J. Comp. Phys.*, 231:3432–3445, 2008.

[26] A. M. Alsayed, M. F. Islam, J. Zhang, P. J. Collings, and A. G. Yodh. Premelting at defects within bulk colloidal crystals. *Science*, 309:1207–1210, 2005.

[27] H. C. Andersen. Molecular dynamics simulation at constant pressure and/or temperature. *J. Chem. Phys.*, 72(4):2384–2393, February 1980. This paper introduces the Andersen thermostat for NpT ensembles.

[28] H. C. Andersen. Rattle: A "velocity" version of the shake algorithm for molecular dynamics calculations. *J. Comp. Phys.*, 52:24–34, 1983.

[29] H. L. Anderson. Metropolis, Monte Carlo and the MANIAC. Los Alamos Science 14, 96–108, Los Alamos National Laboratory, 1986.

[30] H. L. Anderson. Scientific uses of the MANIAC. *J. Stat. Phys.*, 43(5/6):731–748, 1986.

[31] S. Aoki, T. Hatsuda, and N. Ishii. The nuclear force from Monte Carlo simulations of lattice quantum chromodynamics. *Sci. Disc.*, 1:015009, 2008.

[32] A. Appel. An efficient program for many-body simulation. *SIAMJ. Sci. Stat.Comput.*, 6:85–103, 1985.

[33] David L. Applegate, Robert E. Bixby, Vasck Chvátal, and William J. Cook. *The Traveling Salesman Problem: A Computational Study*. Princeton University Press, Pinceton, NJ, 2006.

[34] V. V. Aristov. *Methods of direct solving the Boltzmann equation and study of nonequilibrium flows*, volume 68 of *Fluid Mechanics and Its Applications*. Springer, 2001.

[35] Markus Arndt, Olaf Nairz, Julian Vos-Andreae, Claudia Keller, Gerbrand van der Zouw, and Anton Zeilinger. Wave-particle duality of C_{60} molecules. *Nature*, 401:680–682, 1999.

[36]   V. Arnold. *Mathematical methods of classical mechanics*. Springer, New York, 1978.

[37]   Alain Aspect, Jean Dalibard, and Gérard Roger. Experimental test of Bell's inequalities using time-varying analyzers. *Phys. Rev. Lett.*, 49:1804–1807, 1982.

[38]   A. Aspect, P. Grangier, and G. Roger. Experimental realization of Einstein-Podolsky-Rosen-Bohm Gedankenexperiment: a new violation of Bell's inequalities. *Phys. Rev. Lett.*, 49:91–94, 1982.

[39]   Sara Baase and Alan Van Gelder. *Computer Algorithms: Introduction to Design and Analysis*. Addison-Wesley, 2000.

[40]   Maurice J. Bach. *The Design of the Unix Operating System*. Prentice-Hall, Englewood-Cliffs, NJ, 1986.

[41]   P. Bachmann. *Zahlentheorie. Versuch einer Gesamtdarstellung dieser Wissenschaft in ihren Hauptteilen*. B. G. Teubner, Leipzig, 1892.

[42]   M. Baines. *Moving Finite Flements*. , Oxford University Press, Oxford, UK, 1994.

[43]   P. Ball. What toys can tewll us. *Nature*, November 2009. published online.

[44]   Pietro Ballone, Wanda Andreoni, Roberto Car, and Michele Parrinello. Equilibrium structures and finite temperature properties of silicon microclusters from ab initio molecular-dynamics Calculations. *Phys. Rev. Lett.*, 60:271, 1988.

[45]   J. Barker, R. Fischer, and R. Watts. Liquid argon: Monte Carlo and molecular dynamics calculations. *Mol. Phys.*, 21:657–673, 1971.

[46]   J. Barnes and P. Hut. A hierarchical $O(N \log(N))$ force-calculation algorithm. *Nature*, 324:446–449, 1986.

[47]   A. Baumgärtner. *Applicaions of the Monte Carlo Method*, pages 137–192. Soringer, Berlin, Heidelberg, 1984.

[48]   D. Beazley, P. Lomdahl, N. Gronbech-Jensen, R. Giles, and P. Tamayo. *Annual Reviews of Computational Physics*, volume 3, chapter Parallel algorithms for short-range molecular dynamics, pages 119–175. World Scientific, 1996.

[49]   R. Becker and S. Panchanadeeswaran. Effects of grain interactions on deformation and local texture in polycrystals. *Acta Metall.*, 39:1211, 1991.

[50]   John S. Bell. On the Einstein Podolsky-Rosen paradox. *Physics*, 1:195–200, 1964.

[51]   John S. Bell. On the problem of hidden variables in quantum mechanics. *Rev. Mod. Phys.*, 38:447–452, 1966.

[52]   John S. Bell. On the impossible pilot wave. *Foundations of Physics*, 12:989–999, 1982.

[53]   John S. Bell. *Speakable and Unspeakable in Quantum Mechanics*. Cambridge Univeristy Press, Cambridge, 1987.

[54]   T. Belytschko, Y. Lu, and L. Gu. Element-free Galerkin methods. *Int. J. Numer. Meth. Eng.*, 27:229–256, 1994.

[55]   Jon L. Bentley. *Writing Efficient Programs*. Prentice-Hall, Englewood-Cliffs, NJ, 1982.

[56] Jon L. Bentley. *Programming Pearls*. Adison-Wesley, 1986.

[57] H. J. C. Berendsen, J. P. M. Postma, W. F. van Gunsteren an A. DiNola, and J. R. Haak. Molecular dynamics with coupling to an external heat bath. *J. Chem. Phys.*, 81:2384–2393, 1984.

[58] J. D. Bernal. The Bakerian Lecture 1962. Structure of liquids. *Proc. Roy. Soc. A*, 280:299–322, 1964.

[59] C. Bichara and G. Inden. Monte Carlo calculation of the phase diagram of bcc FeAl alloys. *Scr. Metall*, 25:2607–2611, 1991.

[60] Kurt Binder. *Applicaions of the Monte Carlo Method in Condensed Matter Physics*. Springer, Berlin, 1984.

[61] Kurt Binder. *Monte Carlo Methods in Condensed Matter Physics*, volume 71 of *Topics in Applied Physics*. Springer, Berlin, 1992.

[62] K. Binder and D. W. Heermann. *Monte Carlo Simulations in Statistical Physics*. Springer Verlag Berlin, Heidelberg, New York, Tokio, 1988.

[63] Kurt Binder and D. W. Heermann. *Monte Carlo Simulation in Statistical Physics*. Springer, Berlin, Heidelberg, 2010.

[64] K. Binder and D. Stauffer. *A Simple Introduction to Monte Carlo Simulation And Some Specialized Topics*. Springer, Berlin, 1987.

[65] J. F. W. Bishop and R. Hill. A theory of the plastic distortion of a polycrystalline aggregate under combined stresses. *Phil. Mag.*, 42:414–427, 1951.

[66] Niels Bohr. *Atomic Theory and the Description of Nature*, volume 1. Ox Bow Press, Woodbridge, 1934. reprinted as "The Philosophical Writings of Niels Bohr", 1987.

[67] Ludwig Boltzmann. *Wissenschaftliche Abhandlungen*. Barth, Leipzig, 1909.

[68] Max Born. Zur Quantenmechanik der Stossvorgänge. *Z. Physik*, 38:803–827, 1926.

[69] M. Born, W. Heisenberg, and P. Jordan. Zur Quantenmechanik II. *Zeitschrift für Physik*, 35:557–615, 1925.

[70] M. Born and P. Jordan. Zur Quantenmechanik. *Zeitschrift für Physik*, 34:858–888, 1925.

[71] Max Born and Robert Oppenheimer. Zur Quantentheorie der Molekeln. *Annalen der Physik*, 20:457–484, 1927.

[72] J. F. Botha and G. F. Pinder. *Fundamental Concepts in the numerical Solution of Differential Equations*. John, 1983.

[73] D. Bouwmester, A. K. Ekert, and A. Zeilinger. *The Physics of Quantum Information*. Springer, 2000.

[74] Paulo S. Branicio, Rajiv K. Kalia, Aiichiro Nakano, and Privy Vashishta. Shock-induced strucutral phase transition, plasticity, and brittle cracks in aluminium nitride ceramic. *Phys. Rev. Lett.*, 96:065502, 2006.

[75] G. Brassard and P. Bratley. *Fundamentals of Algorithms*. Prentice-Hall, Englewood-Cliffs, NJ, 1996.

[76] Donald W. Brenner. Empirical potential for hydrocarbons for use in simulating the chemical vapor deposition of diamond films. *Phys. Rev. B*, 42:9458–9471, 1990.

[77] H. Bright and R. Enison. Quasi-random number sequences from a long-period TLP generator with remarks on application to cryptography. *Computing Surveys. 11(4): 357–370.*, 11:257–370, 1979.

[78] F. L. H. Brown. Simple models for biomembrane structure and dynamics. *Comp. Phys. Comm.*, 177:172–175, 2007.

[79] Bernd Bruegge and Allen H. Dutoit. *Object Oriented Software Engineering Using UML, Patterns, and Java: International Version*. Prentice Hall Internatio n, 2009.

[80] R. A. Buckingham. The Classical Equation of State of Gaseous Helium, Neon and Argon. *Proc. Roy. Soc*, A106:264, 1938.

[81] M. J. Buehler, A. Hartmaier, H. Gao, M. Duchaineau, and F. A. Abraham. Atomic plasticity: Description and analysis of a one-billion atom simulation of ductile materials failure. *Comput. Methods Appl. Engrg.*, 193:5257–5282, 2004.

[82] Vasiliy V. Bulatov, Luke L. Hsiung, Maejie Tang, Athanasios Arsenlis, Maria C. Bartelt, Wei Cai, Jeff N. Florando, Masato Hirtani, Moon Rhee, Gregg Hommes, Tim G. Pierce, and Thomas de la Rubia. Dislocation multi-junctions and strain hardening. *Nature*, 440:1174–1178, 2006.

[83] Timothy S. Bush, Julian D. Gale, C. R. A. Catlow, and Peter D. Battle. Self-consistent interatomic potentials for the simulation of binary and ternary oxides. *J. Mater. Chem.*, 4:831–838, 1994.

[84] C. Bustamante, J. F. Marko, E. D. Siggia, and S Smith. Entropic Elasticity of Lambda-Phage DNA. *Science*, 265:1599–1600, 1994.

[85] G. Cantor. *Gesammelte Abhandlungen mathematischen und philosophischen Inhalts*. E. Zermelo, reprint edition, 1980.

[86] Jones Capers. *Software Engineering Best Practices: Lessons from Successful Projects in the Top Companies*. Mcgraw-Hill, 2009.

[87] R. Car and M. Parinello. Unified Approach for Molecular Dynamics and Density-Functional Theory. *Phys. Rev. Lett.*, 55:2471, 1985.

[88] A. E. Carlsson. Beyond pair potentials in elemental transition metals and semiconductors. *Soild State Physics*, 43:1–91, 1990.

[89] Carlo Cercignani. *The Boltzmann equation and its applications*, volume 68 of *Applied Mathematical Sciences*. Springer, 1988.

[90] J. L. Chabert. *A History of Algorithms – From the Pebble to the Microchip*. Springer, 1999.

[91] J. W. Chan and J. E. Hilliard. Free energy of a non-uniform system I: Interfacial energy. *J. Chem. Phys.*, 28:258–266, 1958.

[92] B. Chen, M. A. Gomez, M. Sehl, J. D. Doll, and David L. Freeman. Theoretical studies of the structure and dynamics of metal/hydrogen systems: Diffusion and path integral Monte Carlo investigations of nickel and palladium clusters. *J. Chem. Phys.*, 105:9686–9694, 1996.

[93] W. F. Chen and D. J. Han. *Plasticity for Structural Engineers*. Springer-Verlag Berlin, Heidelberg, 1994.

[94] Kaare Christian and Susan Richter. *The Unix Operating System*. John Wiley & Sons, Chichester, New York, 1994.

[95] J. Christiansen. Vortex methods for flow simulations. *J. Comput. Phys.*, 13:363–379, 1973.

[96] B. Chu, R. L. Xu, and J. Zuo. Transition of Polystyrene in Cyclohexane From Theta To The Collapsed State. *Macromolecules*, 21:273–274, 1988.

[97] Isaac L. Chuang, Neil Gershenfeld, and Mark Kubinec. Experimental Implementation of Fast Quantum Searching. *Phys. Rev. Lett.*, 80:3408–3411, 1998.

[98] G. Ciccotti, G. Frenkel, and I. R. McDonald. *Simulation of Liquids and Solids*. North-Holland, Amsterdam, 1987.

[99] P. Cifra, F. E. Karasz, and W. J. MacKnight. Expansion of polymer coils in miscible polymer blends of asymmetric composition. *Macromolecules*, 25:192–194, 1992.

[100] Barry Cipra. The Best of the 20th Century: Editors Name Top 10 Algorithms. *SIAM News*, 33(4):1–2, 2000.

[101] J. F. Clauser and A. Shimony. Bell's theorem: experimental tests and implications. *Rep. Prog. Phys.*, 41:1883–1927, 1978.

[102] E. Clementi and C. Roetti. Roothaan-Hartree-Fock Atomic Wavefunctions. In *Data Nuclear Tables*, volume 14, page 177. 1974.

[103] A. Cohen. *Numerical Analysis*. McGraw-Hill, London, 1962.

[104] Douglas Comer. Principles of Program Design Induced from Experience with Small Public Programs. *IEEE Transactions on Software Engineering SE-7*, 2:169–74, 1981.

[105] David B. Cook. *Handbook of Computational Quantum Chemistry*. Dover Publications, New York, NY, 2005.

[106] J. W. Cooley and J. W. Tukey. An algorithm for the machine calculation of complex fourier series. *Math. Comput.*, 19:297–301, 1965.

[107] T. H. Cormen, C. E. Leiserson, R. L. Rivest, and C. Stein. *Introduction to Algorithms*. MIT Press, 2001.

[108] Wendy D. Cornell, Piotr Cieplak, C. I. Baylay, Ian R. Gould, Kenneth M. Merz, D. M. Ferguson, D. C. Spellmeyer, T. Fox, J. W. Caldwell, and P. A. Kollman. A second generation force field for the simulation pf proteins, nucleic acids, and organic molecules. *J. Am. Ceram. Soc.*, 117:5179–5197, 1995.

[109] G. Cottet and P. Koumoutsakos. *Vortex methods: Theory and Practice*. Cambridge University Press, Cambridge, UK, 2000.

[110] R. Courant. Variational Methods for the Solution of Problems of Equilibrium and Vibrations. *Bull. Amer. Math. Soc.*, 49:1–23, 1943.

[111] R. Courant and D. Hilbert. *Methods in Mathematical Physics, Vol. II: Partial Differential Equations*. John Wiley & Sons, Chichester, New York, 1962.

[112] B. Dünweg, M. O. Steinhauser, D. Reith, and K. Kremer. Corrections to scaling in the hydrodynamics of dilute polymer solutions. *J. Chem. Phys.*, 117:914–924, 2002.

[113] Sivarama P. Dandamudi. *Guide to Assembly Language Programming in Linux*. Springer US, 2005.

[114] T. Darden, D. York, and L. Pedersen. Particle Mesh Ewald: An N-log N method for sums in large systems. *J. Chem. Phys.*, 103:8577–8592, 1993.

[115] Peter A. Darnell and Phillip E. Margolis. *C. A Software Engineering Approach*. Springer, 1993.

[116] Sanjoy Dasgupta, Christos Papadimitriou, and Umesh Vazirani. *Algorithms*. McGraw-Hill, 2008.

[117] Michel Daune. *Moleckulare Biophysik*. Vieweg Verlag, Braunschweig, Wiesbaden, 1997.

[118] M. S. Daw. Model of metallic cohesion: The Embedded-Atom Method. *Phys. Rev. B*, 39:7441–7452, 1988.

[119] M. S. Daw. Model for energetics of solids based on the density matrix. *Phys. Rev. B*, 47:10895–10898, 1993.

[120] M. S. Daw and M. I. Baskes. Embedded-Atom Method: Derivation and application to impurities, surfaces, and other defects in metals. *Phys. Rev. B*, 29:6443, 1983.

[121] M. S. Daw, S. M. Foiles, and M. I. Baskes. The Embedded-Atom Method: a review of theory and applications. *Mater. Sci. Rep.*, 9:251–310, 1993.

[122] I. Dawson, P. D. Bristowe, M.-H. Lee, M. C. Payne, M. D. Segall, and J. A. White. First-principles study of a tilt grain boundary in rutile. *Phys. Rev. B*, 54:13727–13733, 1996.

[123] Loius de Broglie. *Recherches sur la théorie des quanta*. PhD thesis, Paris, 1924.

[124] Pierres Gilles de Gennes. *Scaling Concepts in Polymer Physics*. Cornell University Press, Ithaca, London, 1979.

[125] Pierre Simon de Laplace. *Exposition du système du monde*. L'imprimerie du Cercle Social, Paris, 1796.

[126] R. Dedekind. *Stetigkeit und Irrationale Zahlen*. Friedrich Vieweg & Sohn, Braunschweig, unabridged 6th edition, 1960.

[127] R. Dedekind. *Was sind und was sollen die Zahlen / Stetigkeit und Irrationale Zahlen*. Friedrich Vieweg & Sohn, Braunschweig, 1969.

[128] F. M. Dekking, C. Kraaikamp, H. P. Lopuhaä, and L. E. Meester. *A Modern Introduction to Probability and Statistics: Understanding Why and How*. Springer Texts in Statistics. Springer, London, 2005.

[129] J. des Cloizeaux and G. Janninck. *Polymers in Solution. Their Modeling and Structure*. Oxford University Press, Oxford, London, New York, 1990.

[130] J. W. Dettman. *Mahtematical Methods in Pyhsics and Engineering*. John Wiley, 1969.

[131] B. Devincre. Three Dimensional Stress field Expressions for Straight Dislocation Segments. *Solid State Commun.*, 93:875–878, 1993.

[132] M. Diaz Peña, C. Pando, and J. A. R. Renuncio. Combination rules for two-body van der Waals coefficients. *J. Chem. Phys.*, 72:5269–5275, 1980.

[133] Edsger W. Dijkstra. A case against the GOTO statement. *Communications of the ACM*, 3:147–148, 1968.

[134] Edsger W. Dijkstra. *How do we tell truths that might hurt?*, pages 129–131. Selected Writings on Computing: A Personal Perspective. Springer, 1982.

[135] Edsger W. Dijkstra. On the Cruelty of Really Teaching Computer Science. *Communications of the ACM*, 32:1397–1414, 1989.

[136] M. Doi. *Introduction to Polymer Physics*. Oxford Science Publications, 1995.

[137] M. Doi and S. F. Edwards. *The Theory of Polymer Dynamics*. Clarendon Press, Oxford, 1986.

[138] P. Doucet and P. B. Sloep. *Mathematical Modelling in the Life Sciences*. Ellis Horwood, London, 1992.

[139] S. Drake. *Dialogue concerning the two chief world systems. Ptolemaic and Copernican*. Modern Library Science. University of California Press, Berkely, 1967. The original was published in Florence in 1632.

[140] Jeff Duntemann. *Assembly Language Step-by-Step: Programming with Linux*. John Wiley & Sons, Chichester, New York, 2009.

[141] J. Eastwood. Optimal particle-mesh algorithms. *J. Comput. Phys.*, 18:1–20, 1975.

[142] J.-P. Eckmann and D. Ruelle. Ergodic theory of chaos and strange attractors. *Rev. Mod. Phys.*, 57:617–656, 1985.

[143] Albert Einstein. Die Plancksche Theorie der Strahlung und die Theorie der spezifischen Wärme. *Annalen der Physik*, 22:180–190, 1907.

[144] Albert Einstein. Zur allgemeinen Relativitätstheorie. *Sitzungsberichte der Königlich Preussischen Akademie der Wissenschaften*, 98:778–786, November 1915.

[145] A. Einstein, B. Podolsky, and N. Rosen. Can quantum-mechanical description of physical reality be considered complete? *Phys. Rev.*, 47:777, 1935.

[146] H. Eisenschitz and Fritz London. Über das Verhältnis der Waalschen Kräfte in den homöopolaren Bindungskräften. *Z. Physik*, 60:491–526, 1930.

[147] F. Ercolessi and J. Adams. Interatomic potentials from first-principles calculations: The force-matching method. *Europhys. Lett.*, 26:583–588, 1994.

[148] F. Ercolessi, M. Parinello, and E. Tossati. Simulation of gold in the glue model. *Phil. Mag. A*, 58:213–226, 1988.

[149] J. D. Eshelby. *Progress in Solid Mechanics*. North Holland, Amsterdam, 1961.

[150] K. Esselink. A comparison of algorithms for long-range interactions. *Comp.Phys. Comm.*, 87:375–395, 1995.

[151] U. Essmann, L. Perera, M. Berkowitz, T. Darden, H. Lee, and L. Pedersen. A smooth particle mesh Ewald method. *J. Chem. Phys.*, 103:8577–8593, 1995.

[152] P. P. Ewald. Die Berechnung optischer und electrostatischer Gitterpotentiale. *Ann. Phys.*, 64:253–287, 1921.

[153] S. J. Farlow. *Partial Differential Equations for Scientists and Engineers*. John Wiley & Sons, Chichester, New York, 1982.

[154] S. J. Farlow. *An Introduction to Differential Equations and their Applications*. McGraw-Hill, New York, 1994.

[155] M. Farooq and F. A. Khwaja. Monte Carlo calculation of order-disorder phase diagram of Cu-Au. *IInt. J. Mod. Phys. B*, 7:1731–1743, 1993.

[156] S. E. Feller. An Improved Empirical Potential Energy Function for Molecular Simulations of Phospholipids. *J. Phys. Chem. B*, 104:7510–7515, 2000.

[157] R. Ferell and E. Bertschinger. Particle-mesh methods on the Connection Machine. *Int. J. Modern Physics C*, 5:993–956, 1994.

[158] Alexander L. Fettter and John Dirk Walecka. *Theoretical Mechanics of Particles and Continua*. Dover Publications, New York, NY, 2004.

[159] Jay Fineberg, S. P. Gross, M. Marder, and Harry L. Swinney. Instability in the propagation of fast cracks. *Phys. Rev. B*, 45(10):5146–5154, 1992.

[160] M. W. Finnis, A. T. Paxton, D. G. Pettifor, A. P. Sutton, and Y. Ohta. Interatomic forces in transition metals. *Phil. Mag. A*, 58:143–164, 1988.

[161] M. W. Finnis and J. E. Sinclair. A simple empirical N-body potential for transition metals. *Phil. Mag. A*, 50:45, 1984.

[162] G. S. Fishman. *Monte Carlo, Concepts, Algorithms and Applications*. Springer, Berlin, Heidelberg, 1996.

[163] B. J. Flory. *Principles of Polymer Chemistry*. University Press, Ithaca, New York, 1953.

[164] V. Fock. Näherungsmethoden zur Lösung des quantenmechanischen Mehrkörperproblems. *Z. Physik*, 61:126–148, 1932.

[165] S. M. Foiles, M. I. Baskes, and M. S. Daw. Embedded-Atom-Method functions for the fcc metals Cu, Ag, Au, Ni, Pd, Pt, and their alloys. *Phys. Rev. B*, 33:7983–7991, 1986.

[166] G. E. Forsythe and W. R. Wasow. *Finite Difference Methods for Partial Differential Equations*. John Wil, 1960.

[167] W. M. C. Foulkes and R. Haydock. Tight-binding models and density-functional theory. *Phys. Rev. B*, 39:12520, 1989.

[168] Peter L. Freddolino, Feng Liu, Martin Gruebele, and Klaus Schulten. Ten-Microsecond Molecular Dynamics Simulation of a Fast-Folding WW Domain. *Biophys. J.*, 94:L75–L77, 2008.

[169] Stuart J. Freedman and John F. Clauser. Experimental test of local hidden-variable theories. *Phys. Rev. Lett.*, 80:938–941, 1972.

[170] D. Frenkel and B. Smit. *Understanding Molecular Simulation: From Algorithms to Applications*. Academic Press, 2002.

[171] Carlos Frontera, Eduard Vives, and Antoni Planes. Monte Carlo study of the relation between vacancy diffusion and domain growth in two-dimensional binary alloys. *Phys. Rev. B*, 48:9321–9326, 1993.

[172] E. S. Fry and R. C. Thompson. Experimental test of local hidden-variable theories. *Phys. Rev. Lett.*, 37:465–468, 1976.

[173] Kurt Gödel. Über formal unentscheidbare Sätze der Principia Mathematica and verwandter Systeme I. *Monatsheft für Math. und Physik*, 38:173–198, 1931.

[174] Mike Gancarz. *Linux and the Unix Philosophy. Operating Systems*. Elsevier Science, Woburn, Massachusetts, 2003.

[175] U. Gasser, E. R. Weeks, A. Schofield, P. N. Pusey, and D. A. Weitz. Real space imaging of nucleation and growth in colloidal crystallization. *Science*, 292:258–262, 2001.

[176] Murray Gell-Mann. The eightfold way: a theory of strong interaction symmetry. Technical report, Caltech Synchrotron Laboratory Report No. CTSL-20, 1961.

[177] M. Gell-Mann. A schematic model of baryons and mesons. *Phys. Lett.*, 8:214–215, 1964.

[178] E. Giannetto. Quantum entanglements: Selected papers. *J. Phys. A: Math. Gen.*, 38:7599, 2005.

[179] J. B. Gibson, A. N. Goland, M. Milgram, and G. H. Vineyard. Dynamics of radiation damage. *Phys. Rev.*, 120:1229–1253, 1960.

[180] G. Gladyszewski and L. Gladyszewski. Monte Carlo simulation of surface diffusion and monolayer completion. *Phys.Stat. Sol. B*, 166:K11–K14, 1991.

[181] P. Glasserman. *Monte Carlo Methods in Financial Engineering*. Springer, New York, 2004.

[182] W. A. Goddard M. M. Goodgame. Modified generalized valence-bond method: A simple correction for the electron correlation missing in generalized valence-bond wave functions; Prediction of double-well states for Cr_2 and Mo_2. *Phys. Rev. Lett.*, 54:661–664, 1985.

[183] W. A. Goddard, T. H. Dunning, W. J. Hunt, and P. J. Hay. Generalized valence bond description of bonding in low-lying states of molecules. *Accounts of Chemical Research*, 6:368–376, 1973.

[184] Gene H. Golub and James M. Ortega. *Scientific Computing and Differential Equations: An Introduction to Numerical Methods*. Academic Press, 1991.

[185] G. H. Gonnet. *Handbook of Algorithms and Data Structures*. Adison-Wesley, 1984.

[186] Robert J. Good and Christopher J. Hope. Test of combining rules for intermolecular distances. Potential function constants from second virial coefficients. *J. Chem Phys.*, 55:111–116, 1971.

[187] Michael T. Goodrich and Roberto Tamassia. *Algorithm Design: Foundations, Analysis, and Internet Examples*. John Wiley & Sons, Chichester, New York, 2001.

[188] H. Gould and J. Tobochnik. *An Introduction to Computer Simulation Methods, Parts 1, 2*. Addison-Wesley Publ. Co., Reading, MA, 1988.

[189] L. Greengard and W. Gropp. A parallel version of the fast multipole method. *Comp. Math. Applic.*, 20:63–71, 1990.

[190] L. Greengard and V. Rokhlin. A fast algortihm for particle simulations. *J. Comput. Phys.*, 73:325–378, 1987.

[191] David Gries. *The Science of Programming*. Springer, 1981.

[192] Simon Gröblacher, Tomasz Paterek, Rainer Kaltenbaek, Caslav Brukner, Marek Zukowski, Markus Aspelmeyer, and Anton Zeilinger. An experimental test of non-local realism. *Nature*, 446:871–875, 2007.

[193] A. Yu. Grosberg and A. R. Khoklov. *Statistical Physics of Macromolecules*. AIP Press, San Die, 1994.

[194] A. Yu. Grosberg and A. R. Khoklov. *Giant Molecules*. AIP Press, San Diego, CA, 1997.

[195] Jürgen Gulbins. *UNIX System V.4: Begriffe, Konzepte, Kommandos, Schnittstellen*. Springer Berlin Heidelberg, 1995.

[196] Dan Gusfield. *Algorithms on Strings, Trees and Sequences: Computer Science and Computational Biology*. Cambridge University Press, 1997.

[197] F. M. Haas, R. Hilfer, and K. Binder. Layers of semiflexible chain molecules end-grafted at interfaces: An off-lattice Monte Carlo simulation. *J. Chem. Phys.*, 102:2960–2969, 1995.

[198] J. M. Haile. *Molecular Dynamics Simulation: Elementary Methods*. Wiley, New York, 1992.

[199] J. M. Hammersley and D. C. Handscomb. *Monte Carlo Methods*. Methuen's / John Wiley & Sons, New York, 1964.

[200] Jean-Pierre Hansen and Loup Verlet. Phase transitions of the Lennard-Jones system. *Phys. Rev.*, 184:151–161, 1969.

[201] J. Harris. Simplified method for calculating the energy of weakly interacting fragments. *Phys. Rev. B*, 31:1770–1779, 1985.

[202] D. R. Hartree. The Wave Mechanics of an Atom With a Non-Coulomb Central Field. *Proc. Cambridge Phil. Soc.*, 24:89–132, 1928.

[203] D. W. Heermann. *Computer simulation methods in theoretical physics*. Springer, Berlin, 1990.

[204] W. Heisenberg. Über quantentheoretische Umdeutung kinematischer und mechanischer Beziehungen. *Z. Physik*, 33:879–893, 1925.

[205] C. S. Helvig, Gabriel Robins, and Alex Zelikovsky. The moving-target traveling salesman problem. *Journal of Algorithms*, 49:153–174, 2003.

[206] H. Heuser. *Lehrbuch der Analysis*, volume I of *Mathematische Leitfäden*. B. G. Teubner, Stuttgart, 6th edition, 1989.

[207] David Hilbert. Mathematische Probleme. *Nachr. d. König. Gesellsch. d. Wiss. zu Göttingen, Math-Phys. Klasse*, 3:253–297, 1900. Lecture given at the International Congress of Mathematicians, Paris, 1900.

[208] D. Hilbert. *Axiomatisches Denken*, volume D. Hilbert. Gesammelte Werke. Springer, Berlin, 1901.

[209] David Hilbert. Die Grundlagen der Physik (Erste Mitteilung). *Nachr. d. König. Gesellsch. d. Wiss. zu Göttingen, Math-Phys. Klasse*, 3:395–407, 1915.

[210] R. Hill. *Plasticity*. Oxford University Press, Oxford, London, New York, 1950.

[211] J. O. Hirschfelder, C. F. Curtis, and R. B. Bird. *Molecular Theory of Gases and Liquids*. John Wiley & Sons, Chichester, New York, 1954.

[212] R. W. Hockney and J. W. Eastwood. *Computer Simulation Using Particles*. McGraw-Hill, New York, 1981.

[213] Micha Hofri. *Analysis of Algorithms*. Oxford University Press, Oxford, London, New York, 1995.

[214] P. Hohenberg and W. Kohn. Inhomogeneous electron gas. *Phys. Rev.*, 36(3B):864–871, 1964.

[215] B. L. Holian and P. S. Lomdahl. Plasticity Induced by Shock-Waves in Nonequilibrium Molecular-Dynamics Simulations. *Science*, 280:2085, 1998.

[216] B. L. Holian and R. Ravelo. Fracture simulations using large-scale molecular dynamics. *Phys. Rev. B*, 51:11275–11279, 1995.

[217] Dominic Holland and Michael Marder. Cracks and Atoms. *Advanced Materials*, 11:793–806, 1999.

[218] Sun-Min Hong, Ann-Tuam Phan, and Christof Jungemann. *Deterministic Solvers for the Boltzmann Transport Equation*. Computational Microelectronics. Springer, 2011.

[219] W. Hoover. Canonical dynamics: Equilibrium phase-space distributions. *Phys. Rev. A*, 31:1695–1697., 1985.

[220] Ellis Horowitz, Sartaj Sahni, and Sanguthevar Rajasekaran. *Computer Algorithms*. Computer Science Press, 1998.

[221] P. Jackel. *Monte Carlo Methods in Finance*. John Wiley & Sons, 2002.

[222] John David Jackson. *Classical Electrodynamics*. John Wiley & Sons, Inc., New York, 1975.

[223] K. Jacobsen, J. Norkov, and M. Puska. Interatomic interactions in the effective medium theory. *Phys. Rev. B*, 35:7423–7442, 1987.

[224] Max Jammer. *The Philosophy of Quantum Mechnics*. John Wiley & Son, 1974.

[225] Claes Johnson. *Numerical Solution of Partial Differential Equations by the Finite Element Method*. Dover Publications, New York, NY, 2009.

[226] J. L. Johnson. *Probability and Statistics for Computer Science*. Wiley-Interscience, 2003.

[227] W. Johnson and H. Kudo. *O. Hoffman and G. Sachs*. McGraw-Hill, New York, 1953.

[228] Richard Johnsonbaugh and Marcus Schaefer. *Algorithms*. Prentice-Hall, Englewood-Cliffs, NJ, 2004.

[229] D. S. Jones. *Assembly Programming and the 8086 Microprocessor*. Oxford University Press, Oxford, London, New York, 1988.

[230] W. L. Jorgensen and J. Tirado-Rives. The OPLS potential functions for proteins. Energy minimizations for crystals of cyclic peptides and crambin. *J. Am. Chem. Soc.*, 110:1657–1666, 1988.

[231] J. Käs, H. Strey, J. X. Tang, D. Finger, R. Ezzell, E. Sackmann, and P. A. Janmey. F-Actin, a Model Polymer for Semiflexible Chains in Dilute, Semidilute, and Liquid Crystalline Solutions. *Biopyhsical Journal*, 70:609–625, 1996.

[232] Martin Kühn and Martin O. Steinhauser. Modeling and Simulation of Microstructures Using Power Diagrams: Proof of the Concept. *Appl. Phys. Lett.*, 93:1–3, 2008.

[233] K. Kadau, T. C. Germann, and P. S. Lomdahl. Large-Scale Molecular Dynamics Simulation of 19 Billion Particles. *J. Modern. Phys. C*, 15(1):193–201, 2004.

[234] Kai Kadau, Timothy C. German, and Perter S. Lomdahl. Molecular dynamics comes of age: 320 billion atom simulation on BlueGene/L. *Int. J. Modern Physics C*, 17:1755–1761, 2006.

[235] M. H. Kalos and P. H. Whitlock. *Monte Carlo Methods*. John Wiley & Sons, New York, 1986.

[236] E. Kamke. *Differentialgleichungen, Lösungsmethoden und Lösungen, I. Gewöhnliche Differentialgleichungen*. Akad. Verlagsges. Geest& Protig K.-G., Leipzig, 1961.

[237] W. K. Kegel and A. van Blaaderen. Direct observation of dynamical heterogeneities in colloidal hard-sphere suspensions. *Science*, 287:290–293, 2000.

[238] A. Keller. *Chain-Folded Crystallization of Polymers from Discovery to Present Day: A Personalized Journey*. Sir Charles Frank, an Eightieth Birthday Tribute. IOP Publishing, 1991.

[239] Lord Kelvin. Nineteenth Century Clouds over the Dynamical Theory of Heat and Light. *Phil. Mag.*, 2:1–40, 1901.

[240] B. W. Kernighan and D. M. Ritchie. *The C programming language*. Prentice-Hall, Englewood-Cliffs, NJ, 1978.

[241] Michael Kerrisk. *The Linux Programming Interface: A Linux and UNIX System Programming Handbook*. No Starch Press, San Francisco, 2010.

[242] J. Kew, M. R. Wilby, and D. D. Vvedensky. Continuous-space monte-carlo simulations of epitaxial-growth. *J. Cryst. Growth*, 127:508–512, 1993.

[243] D. Kincaid and W. Cheney. *Numerical Analysis*. Brooks/Gole Publishing Company, 1996.

[244] K. N. King. *C Programming: A Modern Approach*. W. W. Norton & Company, 2 edition, 2008.

[245] Jeffrey H. Kingston. *Algorithms and Data Structures: Design, Correctness, Analysis*. Addison-Wesley, 1997.

[246] S. Kirkpatrick, C. D. Gelatt, and M. P. Vecchi. Optimization by Simulated Annealing. *Science*, 220:671–680, 1993.

[247] Charles Kittel. *Introduction to Solid State Physics*. John Wiley & Sons, 1995.

[248] J. Lechuga, D. Drikakis, and S. Pal. Molecular dynamics study of the interaction of a shock wave with a biological membrane. *Int. J. Numer. Meth. Fluids*, 57:677–692, 2008.

[249] Jon Kleinberg and Éva Tardos. *Algorithm Design*. Addison-Wesley, 2006.

[250] Donald E. Knuth. *Fundamental Algorithms*, volume 1 of *The Art of Computer Programming*. Addison-Wesley, 1968.

[251] Donald E. Knuth. *Seminumerical Algorithms*, volume 2 of *The Art of Computer Programming*. Addison-Wesley, 1981.

[252] D. Knuth. An empirical study of FORTRAN programs. *Software-Practice and Experience*, 1:105–133, 1971.

[253] Donald E. Knuth. *Sorting and Searching*, volume 3 of *The Art of Computer Programming*. Addison-Wesley, 1973.

[254] Andrew Koenig and Barbara E. Moo. *Accelerated C++: Practical Programming by Example*. Addi, 2000.

[255] W. Kohn and L. J. Sham. Self-consistent equations including exchange and eorrelation effects. *Phys. Rev.*, 140:1133–1138, 1965.

[256] S. E. Koonin. *Computational Physics*. Benjamin/Cumming Publishing, 1986.

[257] Dexter C. Kozen. *The Design and Analysis of Algorithms*. Springer, 1992.

[258] E. Kröner. On the plastic deformation of polycrystals. *Acta Metall.*, 9:155–161, 1961.

[259] K. W. Kratky and W. Schreiner. Computational techniques for shpherical boundary conditions. *J. Comput. Phys.*, 47:313–323, 1982.

[260] R. Kress. *Numerical Analysis*. Springer, Berlin, 1998.

[261] L. D. Landau and E. M. Lifshitz. *Mechanics*, volume 1 of *Course of Theoretical Physics*. Butterworth-Heinemann, 3rd edition, 1976.

[262] L. D. Landau and E. M. Lifshitz. *Quantum Mechnics (non-relativistic theory)*. Elsevier Science Limited, 3rd edition, 1977.

[263] L. D. Landau and E. M. Lifshitz. *Statistical Physics*. Pergamon Press, Oxford, 1986.

[264] R. G. Larson. *The Structure and Rheology of Complex Fluids*. Oxford University Press, Oxford, London, New York, 1999.

[265] E. L. Lawler, J. K. Lenstra, A. Rinnooy Kan, and D. B. Shmoys. *The Traveling Salesman Problem: A Guided Tour of Combinatorial Optimization*. John Wiley & Sons, Chichester, New York, 1995.

[266] P. L'Ecuyer. Efficient and portable combined random number generators. *Communications of the ACM*, 31:742–749, 1988.

[267] H. Lee, T. Darden, and L. Pedersen. Accurate crystal molecular dynamics simulations using particle-mesh-Ewald: RNA dinucleotides – ApU and GbC. *Chem. Phys. Lett.*, 243:229–235, 1995.

[268] J. A. N. Lee. Computer Pioneers. *IEEE Comp. Sci. Press*, 17(4):24–45, 1995.

[269] A. Leonard. Vortex methods for flow simulations. *J. Comput. Phys.*, 37:289–335, 1980.

[270] I. Levine. *Quntum Chemistry*. Prentica-Hall Inc., Englewood Cliffs, N.J., 2000.

[271] Anany Levitin. *Introduction to the Design & Analysis of Algorithms*. Addison-Wesley, 2007.

[272] G. V. Lewis and C. R. A. Catlow. Potential models for ionic oxides. *J. Phys. C: Solid State Phys.*, 18:1149–1116, 1985.

[273] Y. Limoge and J. L. Boquet. Monte Carlo Simulation in diffusion studies: Time scale problem. *Acta Metall.*, 36:1717–1722, 1988.

[274] Ray Lischner. *C++ in a Nutshell*. O'Reilly, 2003.

[275] R. K. Livesley. *Finite Elements – An Introduction for Engineers*. Cambdrige University Press, Cambridge, 1983.

[276] H.-K. Lo, S. Popescu, and T. Spiller. *Introduction to Quantum Computation and Information*. World Scientific, Singapore, 1998.

[277] A. M. Lyapunov. *Stability of Motion*. Academic Press, New York, London, 1966.

[278] Perter C Müller. Calculation of Lyapunov exponents for dynamic systems with discontinuities. *Chaos, Solitons & Fractals*, 5:1671–1681, 1995.

[279] J. Macomber and C. White. An n-dimensional uniform random number generator suitable for IBM-compatible microcomputers. *Interfaces*, 20:49–59, 1990.

[280] Lawrence E. Malvern. *Introduction to the Mechanics of a Continuous Medium*. Prentica-Hall Inc., Englewood Cliffs, N.J., 1969.

[281] Udi Manber. *Introduction to Algorithms: A Creative Approach*. Addison-Wesley, 1989.

[282] C. Marchioro and M. Pulvirenti. *Vortex Methods in Two-Dimensional Fluid Dynamics*, volume 203 of *Lecture Notes in Physics*. Springer, Berlin, 1984.

[283] M. Marder. New dynamical equation for cracks. *Phys. Rev. Lett.*, 66:2484–2487, 1991.

[284] G. Marsaglia and A. Zaman. Some portable very-long period random number generators. *Computers in Physics*, 8:117–121, 1994.

[285] D. Marx and J. Hutter. Ab Initio Molecular Dynamics – Basic Theory and Advanced Methods. Nic series, John von Neumann Institute for Computing, 2000.

[286] E. A. Mason, M. Islam, and S. Weissmann. Composition dependence of gaseous thermal diffusion factors and mutual diffusion coefficients. *Phys. Fluids*, 7:1011–1022, 1964.

[287] Thomas R. Mattsson, Urban Engberg, and Göran Wahnström. H diffusion on Ni(100): A quantum Monte Carlo simulation. *Phys. Rev. Lett.*, 71:2615–2618, 1993.

[288] P. Mazzanti and R. Odorico. A Monte Carlo program for QCD event simulation in e^+e^- annihilation at LEP energies. *Zeitschrift für Physik C, Particles and Fields*, 7:61–72, 1980.

[289] McCarthy. Recursive Functions of Symbolic Expressions and Their Computation by Machine, Part I. Communications of the acm, 1960. "Part II" was never published.

[290] J. L. McCauley. *Dynamics of Markets, Econophysics and Finance*. Cam, 2004.

[291] Kolby McHale. *Understanding Assembly Programming Languages and Their Early Beginnings*. Webster's Digital Services, 2011.

[292] P. Meakin. Models for Material Failure and Deformation. *Science*, 251:226–234, 1991.

[293] Kurt Mehlhorn. *Sorting and Searching*, volume 1 of *Data Structures and Algorithms*. Springer, 1994.

[294] Kurt Mehlhorn. *Graph Algorithms and NP-Completeness*, volume 2 of *Data Structures and Algorithms*. Springer, 1984.

[295] Kurt Mehlhorn. *Multidimensional Searching and Computational Geometry*, volume 3 of *Data Structures and Algorithms*. Springer, 1984.

[296] A. Messiah. *Quantum Mechanics*. North-Holland, Amsterdam, 1961.

[297] N. Metropolis. The beginning of the Monte Carlo method. Los Alamos Science 125, Los Alamos National Laboratory, 1987. Special issue dedicated to Stanislaw Ulam.

[298] Nicholas Metropolis, Arianna W. Rosenbluth, Marshall N. Rosenbluth, Augusta H. Teller, and Edward Teller. Equation of state calculations by fast computing machines. *Journal of Chemical Physics*, 21:1087, 1953.

[299] N. Metropolis and J. Worlton. A trilogy on errors in the history of computing. *Annals of the History of Computing*, 2:49–55, 1980.

[300] N. Metropolis and S. Ulam. The Monte Carlo method. *Journal of the American Statistical Association*, 44:335–341, 1949.

[301] M. Milik and A. Orszagh. Monte Carlo model of polymer chain attached to an interface in poor solvent conditions: Collapse to dense globular state. *Polymer*, 31:506–512, 1990.

[302] John C. Mitchell. *Foundations for Programming Languages*. The MIT Press, 1996.

[303] Peter Mittelstaed. *The Interpretation of Quantum Mechnics and the Measurement Process*. Cambdrige University Press, Cambridge, 1998.

[304] C. Moore. Braids in classical gravity. *Phys. Rev. Lett.*, 70:3675–3679, 1993.

[305] P. M. Morse and H. Feshbach. *Methods of Theoretical Physics*. McGraw-Hill, New York, 1953.

[306] W. van Megen, T. C. Mortensen, and S. R. Williams. Measurement of the self-intermediate scattering function of suspensions of hard spherical particles near the glass transition. *Phys. Rev. E*, 58:6073–6085, 1998.

[307] K. W. Morton and D. F. Mayers. *Numerical Solution of Partial Differential Equations, An Introduction*. Cambridge University Press, 2005.

[308] T. Moto-Oka, editor. *Fifth Generation Computer Systems: International Conference Proceedings*. Elsevier Science Ltd, 1982.

[309] Richard P. Muller, Jean-Marc Langlois, Murco N. Ringnalda, Richard A. Friesner, and W. A. Goddard. A generalized direct inversion in the iterative subspace approach for generalized valence bond wave functions. *J. Chem. Phys.*, 100:1226–1235, 1994.

[310] Glenford J. Myers. *Software Reliabilty*. John Wiley, New York, 1976.

[311] A. Nakano, R. Kalia, and P. Vashishta. Scalable molecular dynamics, visualization, and data-management algorithms for material simulations. *Comp. Sci. Eng. 1 (1999)*, 1:39–47, 1999.

[312] B. Nayroles, G. Touzot, and P. Villon. Generalizing the finite element method: Diffusive approximation and diffusive elements. *Comput. Mech.*, 10:307–318, 1992.

[313] Isaac Newton. *Principia: Mathematical Principles of Natural Philosophy*. 1. University of California Press, October 1999. The text is based on the final 1726 third edition.

[314] M. A. Nielsen and I. L. Chuang. *Quantum Computation and Quantum Information*. Cambdrige University Press, Cambridge, 2001.

[315] R. M. Nieminen, M. J. Puska, and M. J. Manninen, editors. *Many-Atom Interactions in Solids*, volume 48 of *Proceedings in Physics*. Springer, New York, 1990.

[316] P. Nikunen, M. Karttunen, and I. Vattulainen. How would you integrate the equations of motion in dissipative particle dynamics simulations? *Comput. Phys. Comm*, 153:407–423, 2003.

[317] E. Noether. Invariante Variationsprobleme. *Nachr. d. König. Gesellsch. d. Wiss. zu Göttingen, Math-Phys. Klasse*, pages 235–257, 1918. English translation of the origianl article in German: Invariant variation problems, reprinted in M. A. Travel, Transport theory and statistical physics 1 (3):183–207, 1971.

[318] S. Nosé. A unified formulation of the constant temperature Molecular Dynamics methods. *J. Chem. Phys.*, 81:511, 1984.

[319] S. Nosc. A Molecular Dynamics method for simulations in the canonical ensemble. *Mol. Phys.*, 52:255, 1984.

[320] C. K. Ober. Shape Persisitence of Synthetic Polymers. *Science*, 288(5465):448–449, April 2000.

[321] Carlos E. Otero. *Software Engineering Design: Theory and Practice*. Applied Software Engineering. Taylor & Francis Books Ltd, Raton, FL, USA, 2012.

[322] Steve Oualline. *Practical C Programming*. O'Reill, 3 edition, 1997.

[323] L. Padulo and M. A. Arbib. *System Theory*. Saunders, Philadelphia, 1974.

[324] T. Pang. *Computational Physics*. Cambdrige University Press, Cambridge, 1997.

[325] Stephen K. Park and Keith W. Miller. Good random number generators are hard to find. *Communications of the ACM*, 31:1192–1201, 1988.

[326] David Parnas. Designing Software for Ease of Extension and Contraction. *IEEE Transactions of the ACM 5*, 12:1053–1058, 1979.

[327] Jerry Peek. *Learning the Unix Operating System (In a Nutshell)*. O'Reilly Media, Sebastopol, CA, 2001.

[328] Pavel A. Pevzner. *Computational Molecular Biology*. The MIT Press, 2000.

[329] Roberto Piazza, Tommaso Bellini, and Vittorio Degiorgio. Equilibrium sedimentation profiles of screened charged colloids: A test of the hard-sphere equation of state. *Phys. Rev. Lett.*, 71:4267–4270, 1993.

[330] M. Planck. Über irreversible Strahlungsvorgänge. In *Sitzungsberichte der Preußischen Akademie der Wissenschaften*, volume 5, pages 440–480. May 1899.

[331] S. Plimpton. Fast parallel algorithms for short-range molecular dynamics. *J. Comput. Phys.*, 117:1–19, 1995.

[332] Klaus Pohl, Günter Böckle, and Frank J. van der Linden. *Software Product Line Engineering: Foundations, Principles and Techniques*. Springer, Berlin, Heidelberg, 2005.

[333] Henri Poincaré. Sur les solutions périodiques et le principe de moindre action. *C.R.A.S. Paris*, 123:915–918, 1896.

[334] E. Pollock and J. Glosli. Comments on P^3M, FMM and the Ewald method for large periodic Coulomb systems. *Comp. Phys. Comm.*, 95:93–110, 1996.

[335] D. Porezag, D. Porezag, Th. Frauenheim, Th. Köhler, G. Seifert, and R. Kaschner. Construction of tight-binding-like potentials on the basis of density-functional theory: Application to carbon. *Phys. Rev. B*, 51:12947, 1995.

[336] H. A. Posch and W. G. Hoover. Lyapunov instability of dense Lennard-Jones fluids. *Phys. Rev. A*, 38, 1988.

[337] R. B. Potts. Some Generalized Order-Disorder Transformations. *Proc. Cambidge Phil. Soc.*, 48:106–109, 1952.

[338] Shelley Powers. *UNIX Power Tools*. O'Reilly Media, Sebastopol, CA, 2002.

[339] V. Prasad, D. Semwogerere, and E. R. Weeks. Confocal microscopy of colloids. *J. Phys. Condens. Matter*, 19:113102, 2007.

[340] Stephen Prata. *C Primer Plus*. SAMS, 2005.

[341] W. Press and S. Teukolsky. Quasi- (that is, sub-) random numbers. *Computers in Physics*, pages 76–79, Nov/Dec 1989.

[342] William H. Press, Saul A. Teukolsky, William T. Vetterling, and Brian P. Flannery. *Numerical Receipes*. Cambridge University Press, Cambridge, 3 edition, 2007.

[343] Jr Paul W. Purdom and Cynthia A. Brown. *The Analysis of Algorithms*. Holt, Rinehart and Winston, 1985.

[344] P. N. Pusey and W. van Megen. Phase behaviour of concentrated suspensions of nearly hard colloidal spheres. *Nature*, 13:340–342, 1986.

[345] A. Rahman and F. H. Stillinger. Molecular Dynamics Study of Liquid Water. *Phys. Rev.*, 55:3336, 1971.

[346] A. Rahman. Correlations in the Motion of Atoms in Liquid Argon. *Phys. Rev.*, 136:405–411, 1964.

[347] S. S. Rao. *The Finite element Method in Engineering*. Per, 1989.

[348] D. C. Rappaport. *The Art of Molecular Dynamics Simulation*. Cambridge University Press, UK, 1995.

[349] D. C. Rappaport. Large-scale molecular dynamics simulation using vector and parallel computers. *Comp. Phys. Reports*, 9:1–53, 1988.

[350] Michael Redhead. *Incompleteness, Nonlocality and Realism*. Clarendon, 1987.

[351] D. Reidel. *The Physicist's Conception of Nature*, chapter Subject and Object, pages 687–690. Dordrecht, Holland, 1973.

[352] Edward M. Reingold, Jürg Nievergelt, and Narsingh Deo. *Combinatorial Algorithms: Theory and Practice*. Prentice-Hall, Englewood-Cliffs, NJ, 1977.

[353] Jane S. Richardson. Schematic Drawings of Protein Structures. *Methods in Enzymology*, 115:359–380, 1985.

[354] Jane S. Richardson. Early ribbon drawings of proteins. *Nature Structural Biology*, 7(8):624–625, 2000.

[355] Arnold Robbins. *Unix in a Nutshell*. O'Reilly Media, Sebastopol, CA, 2005.

[356] C. P. Robert and G. Casella. *Monte Carlo Statistical Methods*. Springer, 2004.

[357] V. Rokhlin. Rapid solution of integral equations of classical potential theory. *J. Comput. Phys.*, 60:187–207, 1985.

[358] Christopher Roland and Martin Grant. Monte Carlo renormalization-group study of spinodal decomposition: Scaling and growth. *Phys. Rev. B*, 39:11971–11981, 1989.

[359] B. Roos, editor. *Lecture Notes in Quantum Chemistry I*, volume 58 of *Lecture Notes in Chemistry*. Springer, Berlin, 1992.

[360] B. Roos, editor. *Lecture Notes in Quantum Chemistry II*, volume 64 of *Lecture Notes in Chemistry*. Springer, Berlin, 1992.

[361] Arturo Rosenblueth and Norbert Wiener. The role of models in science. *Phil. Sci.*, 12:316–321, 1945.

[362] Jr. Prince E. Rouse. A Theory of the Linear Viscoelastic Properties of Dilute Solutions of Coiling Polymers. *J. Chem. Phys.*, 21:1281–1286, 1953.

[363] Michael Rubinstein and Ralph H. Colby. *Polymer Physics*. Oxford University Press, Oxford, London, New York, 2003.

[364] J. Ryckaert and A. Bellmans. Molecular dynamics of liquid n-butane near its boiling point. *Chem. Phys. Lett*, 30:123–125, 1975.

[365] D. Saari. A Visit to the Newtonian N-Body Problem via Elementary Complex Variables. *Am. Math. Monthly*, 89:105–119, 1990.

[366] Laure Saint-Raymond. *Hydrodynamic Limits of the Boltzmann Equation*, volume 1971 of *Lecture Notes in Mathematics*. Springer, 2009.

[367] J. J. Sakurai. *Modern quantum mechanics*. Addison Wesley Publishing, revised edition, 1994.

[368] Klaus Schätzel and Bruce J. Ackerson. Observation of density fluctuations. *Phys. Rev. Lett.*, 68:337–340, 1992.

[369] M. Schlenkrich, J. Brinckmann, A. D. MacKerell, and M. Karplus. *Empirical Potential Energy Function for Phospholipids: Criteria for Parameter Optimization and Applications*. Birkhäuser, Boston, 1996.

[370] E. Schrödinger. Quantisierung als Eigenwertproblem (Erste Mitteilung). *Ann. Phys.*, 79(4):361–376, 1926. For an English translation, see [373]

[371] Erwin Schrödinger. Quantisierung als Eigenwertproblem (Zweite Mitteilung). *Ann. Phys.*, 79(4):489–527, 1926.

[372] Erwin Schrödinger. The statistical law in nature. *Nature*, 153:704, 1944.

[373] Erwin Schrödinger. *Collected Papers on Wave Mechanics*. Chelsea Pub Co, 3rd edition, April 1982.

[374] L. Schrage. A more portable fortran random number generator. *ACM Transactions on Mathematical Software*, 5:132–138, 1979.

[375] C. N. Schutz and A. Warshel. What are the Dielectric Constants of Proteins and How to Validate Electrostatic Models? *Proteins*, 44:400–417, 2001.

[376] Robert Sedgewick. *Algorithms*. Addison-Wesley, 1988.

[377] Robert Sedgewick and Philippe Flajolet. *An Introduction to the Analysis of Algorithms*. Addison-Wesley, 1996.

[378] Joao Setubal and Joao Meidanis. *Computational Molecular Biology*. PWS Publishing Company, 1997.

[379] Eran Sharon and Jay Fineberg. Confirming the continuum theory of dynamic brittle fracture for fast cracks. *Nature*, 397:333–335, 1998.

[380] Amy Y. Shih, Anton Arkhipov, Peter L. Freddolino, and Klaus Schulten. Coarse Grained Protein-Lipid Model With Application to Lipoprotein Particles. *J. Phys. Chem. B*, 110:3674–3684, 2006.

[381] Y. A. Shreider. *Y. A. Shreider*. Pergamon Press, New York, 1966.

[382] Oktay Sinanoğlu. Many electron theory of atoms and molecules. I. Shells, electron pairs vs many electron correlations. *J. Chem. Phys.*, 36:706–717, 1962.

[383] A. Silverman, A. Zunger, R. Kalish, and J. Adler. Effects of configurational, positional and vibrational degrees of freedom on an alloy phase diagram: a Monte-Carlo study of Ga(1-x)In(x)P. *J. Phys. C*, 7:1167–1180, 1995.

[384] Herbert Simon. *The Sciences of the Artificial*. MIT Press, Cambridge MA, 1969.

[385] H. Sitter M. Tagwerker, W. M. Plotz. Three-dimensional model calculation of epitaxial growth by Monte Carlo simulation. *J. Cryst. Growth*, 146:220–226, 1995.

[386] Shirley W. I. Siu, Robert Vácha, Pavel Jungwith, and Rainer A. Böckmann. Biomolecular simulations of membranes: Physical properties from different force fields. *J. Chem Phys.*, 128:125103, 2008.

[387] Shirley W. I. Siu, Robert Vácha, Pavel Jungwirth, and Rainer A. Böckmann. Biomolecular Simulations of Membranes: Physical Properties from Different Force Fields. *J. Chem. Phys.*, 125:125103, 2008.

[388] Steven S. Skiena. *The Algorithm Design Manual*. Springer, 1998.

[389] Mark G. Sobell. *A Practical Guide to Linux Commands, Editors, and Shell Programming*. Prentice-Hall, Englewood-Cliffs, NJ, 2009.

[390] D. Solvason, J. Kolafa, H. Peterson, and J. Perram. A rigorous comparison of the Ewald method and the fast multipole method in two dimensions. *Comp. Phys. Comm.*, 87:307–318, 1995.

[391] D. Sornette. *Why Stock Markets Crash*. Princeton University Press, Pinceton, NJ, 2002.

[392] C. Störmer. Sur les trajectoires des corpuscles 'eletris' es dans l'espace sous l'action du magnetisme terrestre avec application aux aurores boréales. *Arch. Sci. Phys. Nat.*, 24:221–247, 1907.

[393] Martin Oliver Steinhauser. A Review of Computational Methods in Materials Science: Examples from Shock-Wave and Polymer Physics. *Int. J. Mol. Sci.*, 10:5135–5216, 2009.

[394] M. O. Steinhauser. Computational Methods in Polymer Physics. *Recent Res. Devel. Physics*, 7:59–97, 2006.

[395] Martin Oliver Steinhauser. *Computational Multiscale Modeling of Solids and Fluids – Theory and Applications*. Springer, Berlin, Heidelberg, New York, 2008.

[396] Martin Oliver Steinhauser. A Molecular Dynamics Study on Universal Properties of Polymer Chains in Different Solvent Qualitites. Part I: A Review of Linear Chain Properties. *J. Chem. Phys.*, 122:094901, 2005.

[397] M. O. Steinhauser and K. Grass. Failure and Plasticity Models of Ceramics – A Numerical Study. In A. Khan, S. Kohei, and R. Amir, editors, *Dislocations, Plasticity, Damage and Metal Forming: Materials Response and Multiscale Modeling (The 11th Int. Symposium on Plasticity and Current Applications, (PLASTICITY 2005), Kauai, Hawaii 03.-08.01.2005)*, pages 370–373. Neat Press, 2005.

[398] Martin Oliver Steinhauser, Kai Grass, Elmar Strassburger, and Alexander Blumen. Impact Failure of Granular Materials – Non-Equilibrium Multiscale Simulations and High-Speed Experiments. *International Journal of Plasticity*, 25:161–182, 2009.

[399] Martin Oliver Steinhauser, Kai Grass, Klaus Thoma, and Alexander Blumen. A Nonequilibrium Molecular Dynamics Study on Shock Waves. *Europhys. Lett.*, 73:62, 2006.

[400] Martin Oliver Steinhauser and Martin Kühn. *Anisotropy, Texture, Dislocations, Multiscale Modeling in Finite Plasticity and Viscoplasticity and Metal Forming*, chapter Numerical Simulation of Fracture and Failure Dynamics in Brittle Solids, pages 634–636. Neat Press, Maryland, 2006.

[401] M. O. Steinhauser and M. Kühn. Modeling of Shock-Wave Failure in Brittle Materials. In Peter Gumbsch, editor, *MMM Multiscale Materials Modeling (3rd Intl. Conference on Multiscale Materials Modeling (MMM), Freiburg, Germany, 18.-22. 09. 2006)*, pages 380–382. Fraunhofer IRB Verlag, 2006.

[402] M. O. Steinhauser and M. Kühn. Numerical Simulation of Fracture and Failure Dynamics in Brittle Solids. In A. Khan, S. Kohei, and R. Amir, editors, *Anisotropy, Texture, Dislocations, Multiscale Modeling in Finite Plasticity and Viscoplasticity and Metal Forming*, pages 634–636. Neat Press, 2006.

[403] F. H. Stillinger and T. A. Weber. Computer simulation of local order in condensed phases of silicon. *Phys. Rev. B*, 31(8):5262–5271, 1985.

[404] J. Stoer and R. Bulirsch. *Introduction to Numerical Analysis*, volume 12 of *Tests in Applied Mathematics*. Springer, New York, 1983.

[405] Bjarno Stroustrup. *The C++ Programming Language*. Addison Wesley, 2000.

[406] A. P. Sutton and R. W. Balluffi. *Interfaces in Crystalline Materials*. Oxford University Press/Clarendon Press, Oxford, 1995.

[407] A. Sutton and J. Chen. Long-range Finnis-Sinclair potentials. *Phil. Mag. Lett.*, 61:139–146, 1990.

[408] K. T. Tang and J. Peter Toennies. New combining rules for well parameters and shapes of the van der Waals potential of mixed rare gas systems. *Z. Phys. D*, 1:91–101, 1986.

[409] G. I. Taylor. Plastic strain in metals. *J. Inst. Met.*, 62:307–324, 1938.

[410] Morris Tenenbaum and Harry Pollard. *Ordinary Differential Equations*. Dover Books on Mathematics. Dov, 1985.

[411] J. Tersoff. Modeling solid-state chemistry: Interatomic potentials for multicomponent systems. *Phys. Rev. B*, 39:5566–5568, 1989.

[412] J. Tersoff. New empirical approach for the structure and energy of covalent systems. *Phys. Rev. B*, 37:6991–7000, 1988.

[413] T. Theuns. Parallel P^3M with exact calculation of short range forces. *Phys. Comm.*, 78:238–246, 1994.

[414] Ian Torrens. *Interatomic Potentials*. Academic Press, 1972.

[415] A. Toukmaji and J. Board. Ewald summation techniques in perspective: A survey. *Comp. Phys. Comm.*, 95:73–92, 1996.

[416] J. Tully. *Modern Methods for Multidimensional Dynamics Computations in Chemistry*, chapter Nonadiabatic Dynamics, pages 34–79. World Scientific, Singapore, 1998.

[417] C. Uebing and R. Gomer. A Monte Carlo study of surface diffusion coefficients in the presence of adsorbate-adsorbate interactions. I. Repulsive interactions. *J. Chem. Phys.*, 95:7626–7635, 1991.

[418] S. Ulam. Random processes and transformations. In *Proceedings of the International Congress of Mathematicians*, volume 2, pages 264–275, 1950.

[419] Rupert Ursin, Thomas Jennewein, Markus Aspelmeyer, Rainer Kaltenbaek, Michael Lindenthal, Philip Walther, and Anton Zeilinger. Communications: Quantum teleportation across the Danube. *Nature*, 430:849, 2004.

[420] Alfons van Blaaderen and Pierre Wiltzius. Real-space structure of colloidal hard-sphere glasses. *Science*, 270:1177–1179, 1995.

[421] W. F. van Gunsteren and H. J. C. Berendsen. Algorithms for Brownian Dynamics. *Molec. Phys.*, 45:637, 1982.

[422] Marc W. van der Kamp, Katherine E. Shaw, Christopher J. Woods, and Adrian J. Mulholland. Biomolecular simulation and modelling: status, progress and prospects. *J. R. Soc. Interface*, 5:173–190, 2008.

[423] L. Verlet. Computer "experiments" on classical fluids. I. thermodynamical properties of Lennard-Jones molecules. *Phys. Rev.*, 159:1098–1003, 1967.

[424] Loup Verlet. Computer "experiments" on classical fluids II. Equilibrium correlation functions. *Phys. Rev.*, 159:201, 1968.

[425] J. Voit. *The Statistical Mechanics of Financial Markets*. Springer Verlag Berlin, Heidelberg, New York, 2005.

[426] Johann von Neumann. *Collected works of J. von neumann*, volume 5. Perg, 1963.

[427] Johann von Neumann. *Mathematical Principles of Quantum Mechanics*. Princeton Landmarks in Mathematics and Physics. Princeton University Press, Princeton, New Jersey, reprint edition, Dez. 1996. The German original "Die mathematischen Methoden der Quantenmechanik" was published by Springer, Berlin 1932.

[428] H. J. von Bardeleben, D. Stievenard, D. Deresmes, A. Huber, and J. C. Bourgoin. Identification of a defect in a semiconductor: EL2 in GaAs. *Phys. Rev. B*, 34:7192–7202., 1986.

[429] A. F. Voter, F. Montelenti, and T. C. Germann. Extending The Time Scale in Atomistic Simulations of Materials. *Annu. Rev. Mater. Res.*, 32:321–346, 2002.

[430] J. Wang and A. P. Young. Monte Carlo study of the six-dimensional Ising spin glass. *J. Phys. A*, 26:1063–1066, 1993.

[431] Y. F. Wang, J. M. Rickman, and Y. T Chou. Monte Carlo and analytical modeling of the effects of grain boundaries on diffusion kinetics. *Acta Mater.*, 44:2505–2513, 1996.

[432] S. Wansleben and D. P. Landau. Monte Carlo investigation of critical dynamics in the three-dimensional Ising model. *Phys. Rev. B*, 43:6006–6014, 1991.

[433] M. Warren and J. Salmon. A portable parallel particle program. *Comp. Phys. Comm*, 87:266–290, 1995.

[434] Michael S. Waterman. *Introduction to Computational Biology, Maps, Sequences and Genomes*. Chapman & Hall, 1995.

[435] Gerald M. Weinberg. *The Psychology of Computer Programming*. Van Nostrand Reinhold, New York, 1971.

[436] S. J. Weiner, P. A. Kollman, D. A. Case, U. C. Singh, C. Ghio, G. Alagona, S. Profeta Jr., and P. Weiner. A new force field for molecular mechanical simulation of nucleic acids and proteins. *J. Am. Chem. Soc.*, 106:765–784, 1984.

[437] M. S. Wertheim. Analytic solution of the Percus-Yevick equation. *Phys. Rev. Lett.*, 10:321–323, 1964.

[438] B. Wichmann and I. Hill. Algorithm AS 183. An efficient and portable pseudo-random number generator. *Applied Statistics*, 31:188–190, 1982.

[439] Herbert S. Wilf. *Algorithms and Complexity*. A. K. Peters, 2002.

[440] A. Wintner. *The Analytical Foundations of Celestial Mechanics*. Princeton University Press, Pinceton, NJ, 1941.

[441] A. Wolf, J. B. Swift, H. L. Swinney, and J. A. Vastano. Determining Lyapunov exponents from a time series. *Physica D*, 16:285–317, 1985.

[442] J. T. Wulf, S. Schmauder, and H. F. Fischmeister. Finite element modelling of crack propagation in ductile fracture. *Comput. Mater. Sci.*, 1:297–301, 1993.

[443] J. T. Wulf, T. Steinkopf, and H. F. Fischmeister. Fe-simulation of crack paths in the real microstructure of an Al(6061)/SiC composite. *Acta. Metall.*, 44:1765–1779, 1996.

[444] W. Yang, J. Cai, Y.-S. Ing, and C.-C. Mach. Transient Dislocation Emission From a Crack Tip. *J. Mech. Phys. Solids*, 49:2431–2453, 2001.

[445] J. M. Yeomans. *Statistical Mechanics of Physe Transitions*. Clarendon Press, Oxford, 1992.

[446] Edward Yourdon and Larry L. Constantine. *Structured design: Fundamentals of a discipline of computer program and systems design*. Yourdon Press, Englewood Cliffs, New Jersey, 1979.

[447] X. Yuan, C. Salisbury, D. Balsara, and R. Melhem. A load balance package on distributed memory systems and its application to particle-particle particle-mesh (P^3M) methods. *Parallel Comp*, 23:1525–1544, 1997.

[448] N. G. Zamkova and V. I. Zinenko 1994 J. Phys.: Condens. Matter 6 9043. Monte Carlo investigation of the phase transition in $CsLiSO_4$ and $CsLiCrO_4$ crystals. *J. Phys*, 6:9043–9052, 1994.

[449] E. Zermelo. Über Grenzzhalen und Mengenbereiche. *Fund. Math.*, 16:29–47, 1930.

[450] Guo-Ming Zhang and Chuan-Zhang Yang. Monte Carlo study of the order of the phase transition in Ising systems with Multispin Interactions. *Phys. Stat. Sol. B*, 175:459–463, 1993.

[451] F. Zhao and S. Johnsson. The parallel multipole method on the Connection Machine. *SIAM J. Sci. Stat. Comput.*, 12:1420–1437, 1991.

[452] S. J. Zhou, D. M. Beazley, P. S. Lomdahl, and B. L. Holian. Large-scale molecular dynamics simulations of three-dimensional ductile fracture. *Phys. Rev. Lett.*, 78(3):479, 1997.

[453] O. C. Zienkiewicz and K. Morgan. *Finite Elements and Approximation*. Jo, 1983.

[454] Bruno H. Zimm. Dynamics of Polymer Molecules in Dilute Solution: Viscoelasticity, Flow Birefringence and Dielectric Loss. *J. Chem. Phys.*, 24:269–278, 1956.

Glossary of Acronyms

BOA	Born-Oppenheimer approximation	**MC**	Monte Carlo
		MD	Molecular Dynamics
CGNS	CFD General Notation System	**MFA**	Mean Field Approximation
		MO	Molecular Orbital
CPU	Central Processing Unit	**NMR**	Nuclear Magnetic Resonance
DFT	Density Functional Theory		
DNA	Desoxyribonucleic Acid	**PB**	Polybutadiene
EAM	Embedded Atom Method	**PBC**	Periodic Boundary Conditions
EOM	Equations of Motion	**PDB**	Protein Data Bank
EOS	Equation of State	**PDF**	Probability Distribution Function
FEM	Finite Element Method	**PE**	Polyethylene
FENE	Finitely Extensible Non-linear Elastic	**PME**	Particle Mesh Ewald
		PMMA	Polymethylmetacrylate
FPU	Floating Point Unit	**PRNG**	Pseudo Random Number Generator
GTO	Gaussian Type Orbitals	**PS**	Polystyrene
HF	Hartree-Fock	**RNG**	Random Number Generator
IDE	Integrated Development Interface	**SCF**	Self Consistent Field
IEEE	*I*nstitute of *E*lectrical and *E*lectronic *E*ngineers	**SPH**	Smooth Particle Hydrodynamics
		STL	(C++) Standard Library
LDA	Local Density Approximations	**STO**	Slater Type Orbitals
LJ	Lennard-Jones	**TB**	Tight Binding

Index

main() 99, 100, 113, 114
main()-function 89, 100, 101, 113, 143, 332
printf() 122
scanf() 122–124, 142, 154, 160
strcpy() 148

ab initio 45, 70, 294, 299, 305, 306, 308, 357
   MD 299
ab initio method 272, 273, 306
abstract datatype 301
actin filaments 415
ALU 19, 28
ANSI-C 85, 144
assembly 5–11, 40, 187

Bash 87
   command line 87
Behrendsen thermostat 429
best coding practices 332
binary 5, 8
   number 27, 32
BOA 265, 266, 298
Boltzmann distribution 249, 255–257
Boltzmann equation 360
bond order potentials 292
Born-Oppenheimer approximation *see* BOA
BSD 6

C 84, 85, 99–102, 116, 118, 119, 139, 142, 183, 210
   keyword 100
   libraries 85
   program 99, 101, 103, 112, 113
   programming 144
   standard 85, 102
   statement 120
   version 111

C/C++ 3, 8, 12, 16, 23, 29, 30, 33, 35, 40, 182, 183
C99 11, 85, 100, 139
Car-Parinella MD 307
coarse-grained methods 302
COBOL 6–8, 11, 22
code 2, 5, 8, 10–13, 16, 29, 31, 38, 40, 181–185, 187–192
coding 444
computing 5, 8, 11, 17, 21, 41, 187
computing history *see* history of computing, 23
correlation length 385
CPU 8, 10, 18, 19, 22, 23, 182
critical phenomena 416
criticality 386

data management 301
debugging 444
Debye temperature 304
density functional theory *see* DFT, 298
DFT 273
differential equation
   partial 44
diffusion equation 410
dissipative particle dynamics 429
DNA 4, 415
domain decomposition 301

EAM 292, 307
eigenvector 64
electrodynamics 301
EMACS 87, 96, 460, 462
executable 86, 89
   code 157
   file 86, 87, 89, 101
   program 87, 99, 100
   statements 103, 145
experiment 423

Index 507

FEM 411
field equations
  relativistic 67
finite difference
  methods 411
finite difference method 412
finite difference methods 411
finite element methods *see* FEM
flexible chains 416
FORTRAN 6–8, 11, 13, 22, 187
  77 356
FORTRAN 77 11, 13
FORTRAN 90 11
fractal dimensions 416

Galerkin method 413
glue model 292
Graphical User Interface *see* GUI
GUI 5

Hamiltonian $\mathcal{H}$ 306
Hartree-Fock approach 298
Hartree-Fock method 273, 306
Helmholtz free energy 260
high level language 186
high-level language 5, 9–11, 13, 186, 189
Hilbert space 64
history of computing 1, 6, 17
hypersurface 273

ideal chains 416
importance sampling 371, 381
inline functions 121

Legendre functions 207
Linux 7
LISP 7
Lyapunov
  exponents 338
  instability 338
  spectrum 338

machine code 5, 6, 8, 10–12, 40
MC method 356
MD 5, 12, 38, 44, 45, 65, 70, 304
memory 85
minimum image convention 338
molecular dynamics *see* asoMDv

molecular mass 416
Monte Carlo 65
multipole expansion 280, 301

ODE 309, 311
operating system 90, 115, 116, 138, 148, 157, 186
ordinary differential equation *see* ODE
orientational correlation function 394

parallelization 301
partial differential equations *see* PDEs
PDEs 409–411
phase transition 386
phases transition 416
Poisson equation 410
polybutadiene, PB 415
polyethylene, PE 415
polymers 416
polymethylmethacrylate, PMMA 415
polystyrene, PS 415
potential 273, 299
  function 301
pseudocode 43, 47–49
pseudorandom 365
pseudorandom number 362, 363
  generator 363

quantum mechanics 249

random numbers 362
random sampling 360
random walk 360
running time 49–52, 55, 56

Schrödinger equation 273, 294, 304
scientific computing 1, 5, 10–13, 15, 41, 184
scientific programming 84, 144
second virial coefficient 278
seed value 366
self-consistent field theory 306
semi-empirical methods 307
simple sampling 369, 370
simulated annealing 303
software 2, 12, 181, 182, 188, 189, 191
  design 13, 84, 182, 187, 189
  package 10
software development 444
statistical mechanics vi

step length 311
step size 311
Stirling's approximation 251, 476
STL 159
supercomputing 85

Taylor expansion 301
thermodynamics vi
third virial coefficient 278
thread 85

tight binding 306
tight binding MD 307

uncertainty principle 64
Unicode 85
UNIX 86, 87

Verlet velocity scheme 306
Vortex method 295

XEMACS 87, 96, 462

Authors

Alder, Bernie 299
Allen, Michael P. 356
Andersen, Hans C. 300

Babbage, Charles 17
Bell, John S. 64
Bernal, James D. 279
Bohr, Niels 64
Brenner, Donald W. 292

Ciccotti, Giovanni 356
Courant, Richard 356

de Broglie, Louis 64
deGennes, Pierre-Gilles 386
Dijkstra, Edsger 183
Drude, Paul K. L. 287

Einstein, Albert 64, 65, 67, 74
Euler, Leonhard 311
Ewald, Paul P. 300

Finnis, Michael W. 292
Forsythe, George Elmer 356
Frenkel, Dan 356

Galilei, Galileo 60
Gulbins, Jürgen 87

Heermann, Dieter W. 356
Heisenberg, Werner 64
Hilbert, David 62, 356
Holerith, Hermann 18
Hoover, William G. 300, 356

Kelvin, Lord 360
Kernighan, Brian 6, 210
Knuth, Donald 83, 183

London, Fritz W. 287

McDonald, Ian R. 356
McIlroy, Douglas 6
Metropolis, Nicholas 21, 381
Miller, Keith W. 368

Newton, Sir Isaac 60

Ossanna, Joe 6

Park, Stephen K. 368
Pascal, Blaise 7
Press, William H. 210

Rahman, Aneesur 300
Rappaport, Dennis C. 356
Riese, Adam 17
Ritchie, Dennis 6, 210

Schickard, Wilhelm 17
Steinhauser, Martin Oliver 67
Stroustroup, Bjarne 210

Tersoff, Jerry 292
Thompson, Ken 6
Tildesley, Dominic J. 356
Torvalds, Linus 6

Verlet, Loup 300, 306

Wainwright, Thomas 299
Wirth, Niklaus 7